Water Infrastructure for Sustainable Communities

Water Infrastructure for Sustainable Communities
China and the World

Editors
*Xiaodi Hao, Vladimir Novotny
and Valerie Nelson*

Publishing
London • New York

Published by   **IWA Publishing**
                Alliance House
                12 Caxton Street
                London SW1H 0QS, UK
                Telephone: +44 (0)20 7654 5500
                Fax: +44 (0)20 654 5555
                Email: publications@iwap.co.uk
                Web: www.iwapublishing.com

First published 2010
© 2010 IWA Publishing

Originated by The Manila Typesetting Company
Cover by designforpublishing.co.uk
Printed by Lightning Source

British Library Cataloguing in Publication Data
A CIP catalogue record for this book is available from the British Library

Library of Congress Cataloging-in-Publication Data
A catalog record for this book is available from the Library of Congress

ISBN: 9781843393283
ISBN 10: 184339328X

Table of Contents

Contents

Preface
Universal Recycling for the Benefit of Society

The Industrial Revolution of the eighteenth and nineteenth centuries accelerated urbanisation, which is generally considered a positive thing; but it did so at the cost of causing problems of hygiene and water pollution within the new cities. To cope with these problems, flushing toilets, sewers and even wastewater treatment plants were invented and applied extensively in urban areas. Such centralised systems, consisting of collection, transportation and treatment, have, ever since, been seen as an integral part of modern civilization.

From the point of view of sustainability, however, these centralized systems are unreasonable, as it is difficult to reclaim and recycle the water, nutrients and even energy contained in wastewater. For this reason, modern civilization has, in some areas, been seriously criticised by global experts.

And so a new model for water management is emerging worldwide in response to water, nutrients and energy shortages, polluted waterways, climate changes and a loss of biodiversity. Cities and towns are not only questioning the ecological and financial sustainability of big-pipe water supplies, stormwater and sewer systems, but are also searching for "lighter footprint" solutions. For these purposes, a decentralized concept is being explored, one that can make water, nutrients and even energy reclaimable for onsite reuse. In recent years, some decentralized systems, such as stormwater harvesting, source separation/ecological sanitation, nutrient/energy recovery, green buildings/infrastructure, and so on, have been widely applied in both cities and towns, and pilot projects are being built that use, treat, store and reuse water, nutrients and energy locally, building distributive designs into restorative hydrologies.

As the largest developing country in the world, China is undergoing extensive urbanization after its economy having been through a process of development for thirty years. Cities are becoming bigger and bigger; villages are developing into towns, and traditional habits of cultivation in villages are gradually disappearing. But in fact, such traditional habits of cultivation have a great deal to teach us about ecological sustainability. Food comes from the land and wastes (urine/manure and wastewater), and goes back to the land – in essence, a simple ecologically recycling model. Such a traditional model is in fact the very thing we have been searching for. The flushing toilets and sewers which have been promoted until now, while hygienic and convenient, also have the undesirable effect of keeping nutrients far away from their sources of origin. Because of this, a new model for water management and/or the traditional habits for cultivation have become especially important for the rapid urbanization of China. Destroying centralized water/wastewater systems already constructed in cities would seem to be an unreliable method, and is not recommended, but it is possible to simply retain the habits of farmers and to develop decentralized systems instead. In other words, there is a vast potential in China for developing its green infrastructure in urbanization, as about sixty percentage of the total population of China (1.34 billion people in 2009) consists of farmers. China is also today by cooperating with several countries up front in the development of sustainable cities/towns and will greatly benefit from collaboration with internationally thinking societies in ideas, techniques, and even policies.

With the 2009 International Conference on Sustainable Water Infrastructure for Cities and Villages of the Future, held in Beijing (November 6th – 9th 2009), we built an academic platform to attract top research and science papers, helping to advance the water sciences regarding sustainable development. Even under the poor global economic conditions in place at the time of the conferences' planning and execution, many top experts from the world still attended the conference and presented their ideas, research and demonstration projects, in an inspiring display of their enthusiasm for this field.

From about ninety speakers, the program committee has selected half of the papers for publication – a difficult choice, given their quality, but necessary. We hope that this book, a record of the conference proceedings, will encourage regulators, politicians, the scientific community and even farmers and citizens who own houses (a less common situation in China than other countries) to reconsider their choices in the development of infrastructures, and to move towards applying decentralized sustainable systems.

This conference has been invaluable in its learning and sharing of ideas, techniques and policies. It is our hope and belief that this will be a step towards an ecologically sustainable future of available water, nutrients and energy, both for China and for the world.

Dr. Wentang ZHENG, Professor,
SWIF2009 Conference Honorary Chair,
President of Beijing University of Civil Engineering and Architecture,
Beijing, P. R. of China.

序言：
万物循环，造福社会

十八至十九世纪间的工业革命加速了城市化进程，而这样一个人们普遍认为的人类进步却使新城市出现了卫生及水污染问题。面对这些问题，水冲厕所、下水道系统、甚至污水处理厂应运而生并且在城市中广泛应用。这样一些包括收集、输送与处理为核心的集中式污水系统，从那时起一直被视为现代文明不可或缺少的组成部分。

然而，从可持续角度看，这些集中式污水系统具有其不合理性，它们难以循环水资源，回收污水中营养物质、甚至能量。有鉴于此，这种现代文明在世界一些地方已遭致专家们的严肃批评。

针对水、营养物质与能源短缺、河流污染、气候变暖与生物多样性丧失等问题，一个全新的水管理模式正在全球范围内出现。人们在质疑城镇中大管径供水、雨水和排水管道系统在生态与经济方面是否具有可持续性的同时，也在积极寻找"低排放"解决方案。以此为目的，人们正在探索一种能够使水、营养物质、甚至能源原位回收/利用的分散式方法。近年来，许多分散式系统，如，雨水收集、源分离/生态卫生、营养物质/能量回收、绿色建筑/基础设施等已广泛应用于城市和乡镇；许多利用、处理、存储、原位再利用水、营养物质与能源的中试项目正在建设之中，以构建恢复各种水文循环的分布设计模式。

中国作为世界上最大的发展中国家，在经历了30年经济发展历程之后，正在进行着大规模城市化进程。城市变得越来越大，乡村正朝着城镇化方向发展，农村传统耕作习俗正在逐渐消失。而事实上，这种传统农业耕作习俗恰好向我们展示了一种生态的可持续性。食物来自于土地和废物（尿液/粪便与污水），利用之后再返回土地——本质上看，这是一种最简单的生态循环模式。这种传统模式实际上正是我们一直寻求的生态方式。水冲厕所与下水道的确是卫生与方便，但与此同时也给我们带来了相当的负面影响，正是这些集中式污水系统使营养物质远离其源头（土地）。正因如此，建立一种新的水管理模式和/或保持传统耕作习俗对快速发展的中国城市化进程突显重要。撤除木已成舟的城市集中式污水系统似乎并不可取，也不宜建议，而简单保留农民传统耕作习俗并鼓励他们发展分散式系统则相对容易。中国总人口（2009年为13.4亿）中近6成是农民，因此，在中国城市化进程中发展绿色基础设施存在着巨大潜力。中国目前正在与一些发达国家合作，以发展可持续的城市/城镇；与国际思考型社会的合作必将使我们在观念、技术甚至政策方面大为受益。

在2009年于北京举行的城镇/乡村未来可持续水基础设施国际大会（2009年11月6日至9日）上，我们建立了一个吸引国际顶尖研究与科学论文的学术平台，这将有助于推动水科学朝着可持续方向发展。在本次会议筹备和召开期间恰遇全球金融危机爆发，但是，世界上许多顶级专家、学者积极参会，并展示了他们的观点、研究以及示范项目，显示出他们对这一领域的积极热情和鼓舞人心的行动。

在近90位会议发言者中，会议组委会筛选出约一半论文进行出版——这是一个非常困难的抉择，但为保证本书质量，这样做也是必要的。我们希望这本记录会议论文集的书能够鼓励立法者、政府官员、科学界、甚至农民和拥有别墅的市民（尽管中国比其他国家还较为少见）重新审视他们在选择基础设施时的考虑，尽可能朝着分散式可持续水系统方向发展。

这次会议必将在获取和分享观点、技术和政策的知识中起到重要作用。我们希望并坚信，这将是中国与世界在水、营养物质与能源朝着生态可持续方向发展中迈出的重要一步。

郑文堂　博士　教授
SWIF2009大会　名誉主席
北京建筑工程学院　校长
中华人民共和国　北京

Committees

LOCAL ORGANIZING COMMITTEE

Wentang Zheng — Conference Honorary Chair, Beijing University of Civil Engineering and Architecture (BUCEA), China
Xiaodi Hao — Conference Executive Chair, BUCEA, China
Guohua Song — Coordinator, BUCEA, China
Deying Li — Director, BUCEA, China
Dayu Zhang — Scientific Program Manager, BUCEA, China

FINAL PROGRAM COMMITTEE

Xiaodi Hao — BUCEA, China
Vladimir Novotny — Northeastern University, USA
Valerie Nelson — Coalition for Alternative Wastewater Treatment, USA
Jianwei Liu — BUCEA, China
Jianlong Wang — BUCEA, China
Fuguo Qiu — BUCEA, China

INTERNATIONAL TECHNICAL SCIENTIFIC COMMITTEE

Xiaodi Hao (Chair) — BUCEA, China
Vladimir Novotny (Co-Chair) — Northeastern University, USA
Valerie Nelson (Co-Chair) — Coalition for Alternative Wastewater Treatment, USA
Glen Daigger — CH2MHill, USA
Mark C. M. van Loosdrecht — TU Delft, the Netherlands
Steve Moddemeyer — CollinsWoerman, USA
Goen Ho — Murdoch University, Australia
Patrick Lucey — Aqua-tex, Canada
Jerry Stonebridge — National Onsite Wastewater Recycling Association, USA
Guanghao Chen — Hong Kong University of Science and Technology, H.K., China
Xiaochang Wang — Xi'an University of Architecture and Technology, China
Min Yang — Research Centre for Eco-Environmental Sciences, Chinese Academy of Sciences, China
Yajun Zhang — BUCEA, China
Wu Che — BUCEA, China

Acknowledgements

Sponsored by

- *Beijing University of Civil Engineering and Architecture, China*
- *Water Environment Research Foundation (WERF), USA*
- *National Onsite Wastewater Recycling Association (NOWRA), USA*

Funded by

- *Beijing Municipality (a Special Conference Fund and PHR20100508)*
- *MOST & MOHURD (2006BAJ01B03-02), China*

Supported by

- *Delft University of Technology (TU Delft), the Netherlands*
- *The Hong Kong University of Science and Technology, China*
- *Research Centre for Eco-Environmental Sciences, Chinese Academy of Sciences, China*
- *Xi'an University of Architecture and Technology, China*
- *Journal of China Water & Wastewater, China*
- *EnvrionSystems, Germany & China*
- *China Water Industry Net, China*

Introduction
Water Infrastructure for Sustainable Communities: China and the World

Xiaodi Hao[a], Vladimir Novotny[b] and Valerie I. Nelson[c]

[a]Beijing University of Civil Engineering and Architecture, Beijing 100044, P. R. China
E-mail: xdhao@hotmail.com
[b]Northeastern University, Boston, MA 02458, USA
E-mail: v.novotny@comcast.net
[c]Coalition for Alternative Wastewater Treatment, USA
E-mail: valerie508@aol.com

CHALLENGES AND VISIONS FOR THE FUTURE SUSTAINABLE CITIES AND VILLAGES

Cities and villages across the globe are confronting growing challenges of supplying safe clean water and sanitation services to residents and industry. Urban populations are expanding, climate change is beginning to create more erratic patterns of droughts and storms, and per capita consumption is on the rise. The water and wastewater profession is beginning to understand traditional linear infrastructure that transports clean water into and wastewater out of urban neighborhoods wastes water, uses too much energy, disrupts waterways and ecosystems, and is very expensive to build and maintain.

The Beijing conference was a follow-up of two international events that took place in the USA: (1) The Wingspread (Racine, WI) workshop, "Cities of the Future - Bringing Blue Water to Green Cities" in July 2006 and (2) the Baltimore (Maryland) conference, "Water for All Life: A Decentralized Infrastructure for a Sustainable Future" in March 2007.

The Wingspread workshop and Baltimore conference forged a coalition of experts in urban water and drainage infrastructure, landscape, economy, environmental law, NGOs, sewerage utilities, academia, practice and foreign partners. At these two events a new (fifth) paradigm of ecocity urbanism was formulated, emerging from the past successes and failures of efforts to control pollution and reduce floods, offering adequate amounts of clean water for all beneficial uses, water and energy reclamation, and reduction of the carbon footprint. Next-generation designs come from a different engineering model: use, treat, store, and reuse water efficiently on a smaller scale, and blend these designs into restorative water hydrology. This paradigm of sustainable urban waters and watersheds is based on the premise that urban waters are the lifeline of cities and focus of the movement towards more sustainable and emerging "smart and green" cities. The concepts include reuse of highly treated effluents and urban stormwater for various purposes including landscape irrigation and aquifer replenishment; nutrients and energy recoveries from wastewater; environmental flow enhancement of effluent-dominated and flow-deprived streams; and ultimately a source for safe water supply. The new paradigm treats reclaimed water, nutrients and energy as a resource. Experts at both events proposed that this new paradigm of water/stormwater/wastewater infrastructure is based on decentralization and de-regionalization of the infrastructure systems and its management.

Participants in a workshop of speakers at the Baltimore Conference drafted and signed a document, the Baltimore Charter for Sustainable Water Systems, that asserted:

> Water is at the heart of all life. In the past, we built water and wastewater infrastructure to protect ourselves from diseases, floods, and droughts. Now we see that fundamental life systems are in danger of collapsing from the disruptions and stresses caused by this infrastructure.
>
> New and evolving water technologies and institutions that mimic and work with nature will restore our human and natural ecology across lots, neighborhoods, cities, and watersheds. We need to work together in our homes, our communities, our workplaces, and our governments to seize the opportunities to put these new designs in place.

The new water management will make a switch from strictly engineered systems (sewers) to ecologic systems (rain gardens, surface wetlands, ponds, restored and daylighted water bodies) and ecosanitation. Municipal stormwater and sewage management is expected to be decentralized into city clusters rather than regionalized. This book is another step forward in the development of a new paradigm for sustainably managing urban water in the future. Water management will shift in interlocking ways: water will not be used once and then thrown away, but rather treated and reused over and over; nutrients in wastewater will be captured and reused in fertilizer; biogas digesters will recover energy from wastewater and other solid waste and, in the near future, convert directly to hydrogen and electricity by fuel cells, rainwater will be captured for local use; green roofs and trees will capture stormwater, cool down buildings, improve air quality, and beautify cities; water will be returned to urban rivers and safe downstream uses; and the private sector will step forward to build and manage water in green buildings and integrated neighborhood infrastructure. The communities built or retrofitted according to the new paradigm will use less fossil fuel-based energy and emit far less green house gases.

OBJECTIVES AND GENERAL OUTCOME OF THE CONFERENCE

Specific objectives of the Beijing Conference were:

- Update and expand upon the insights and lessons of the Wingspread Workshop and Baltimore Conference
- Present case studies and identify lessons from sustainable water demonstration projects in China and other countries
- Consider and make recommendations for how new water systems can be integrated into Chinese and other national policies and initiatives for urban and village development

With urban waters as a focal point, this conference explored the links between urban water quality and hydrology, landscape, and the broader concepts of green cities and smart growth. The conference focused on decentralized concepts of potable water/stormwater/used water management that would provide clean water to multiple uses and result in sustainable water management systems.

The 2009 Beijing conference included leading experts who presented and summarized the ongoing efforts and movement towards the "Cities and Villages of the Future – the Ecocities" that are now emerging from the laboratories and design studios into reality in some countries, including Australia, Belgium, Canada, China, Germany, Singapore, Sweden, United Kingdom and the US. The purpose of the Beijing Conference in 2009 was to further the learning process begun at the Wingspread Workshop and the Baltimore Conference and to advance the commitments made by signatories to the Baltimore Charter.

The subtitle of this book "China and the World" highlights the role of China and emphasizes collaboration between China and the United States and other countries. China is building new cities and new neighborhoods in older cities and has the opportunity to design sustainable infrastructure from the start. The Chinese Ministry of Environmental Protection issued "Guidelines on building ecological provinces, ecological cities and ecological country" in May 2003 (revised in 2007). This has been followed by an upsurge in building sustainable cities. China is looking for urban housing for up to 300 million people in the next 30 years because of intensification of agriculture (loss of jobs of indigenous population) and a large increase of GNP being derived by industries in the cities. Essentially, it is a planned attempt to manage migration from rural to urban areas and to accommodate population growth that has been so devastating in several other fast developing countries, including Brazil, Mexico, India, *etc.*

The design of the area of Dongtan at the Yangtze River's mouth to the sea near Shanghai included for the first time the ecocity concepts and outlined the difference between the now traditional Low Impacts Developments (LID) popular in the US and the concepts of water centric ecocities. In November 2007, governments of China and Singapore signed an agreement under which Singapore will export its ecocity know-how and technologies and will build an ecocity in Tianjin, 100 km southeast of Beijing. The University of California in Berkeley has teamed up with Siemens industry giant to provide know how to the Chinese planners in building the cities. Sweden is providing assistance to China on developing the new ecocity urbanism concepts in constructing a new site of the Capital Iron & Steel Co. at Caofeidian near Tangshan, 150 km east of Beijing. Now more and more Chinese cities from the north to the south are planning construction of ecocities, including Harbin, Shenyang, Beijing, Chengdu, cities cluster in the Pearl River Delta, etc. An ecological agriculture demonstration project has been almost completed near Xiaotangshan, 35 km north of Beijing, which combines the traditionally cultivating habits of Chinese farmers with ecological sanitation (Ecosan), as can be seen in the article by Qiu and Hao of this book.

Other planned ecocities are emerging in Sweden, Canada (Victoria, Vancouver), the Netherlands, Germany, Australia, and Japan. Components of the "Cities of the Future" have been implemented in Singapore, Masdar in Abu Dhabi (UAE) and Israel. Australia, South Africa and Israel have made tremendous advances in the field of water conservation. Singapore is a relatively small island city/state (4.5 million inhabitants, 626 km²) in South Asia that has minimal natural water resources. Until recently it imported most water for its potable needs from Malaysia. It is being converted into an ecocity that will derive much of its drinking water by collecting and reusing stormwater. Today, Singapore is a pioneer of water conservation, reclamation and reuse. These endeavours and many that will follow will create more than a trillion dollar market throughout the world.

In the United States, urban water challenges are different. Water lines, sewer mains, and treatment plants, many built over a hundred years ago, are leaking, collapsing, and overflowing. Furthermore, they are incapable to handle extreme events that are expected to be magnified by global warming. Because much of the water and wastewater infrastructure is well on its way to breaking, there is a golden opportunity to leapfrog into the future—as developing countries like China and India are beginning to do. Calling the US essential infrastructure failure an "opportunity" may seem counter-intuitive, but if these systems had been kept in good shape, there would actually be fewer openings to shift to something new. The worldwide economic potential of building new or retrofitting old cities to switch to the new paradigm of sustainable water/stormwater/wastewater management, to build the ecocities, is enormous and it is expected to reach several trillion dollars over the next 25 years.

European countries, Australia, and Canada have been ahead of the US in exploring radical shifts in water management. Since the Australian efforts in the early 1990s to implement water conservation, reclamation and reuse efforts based on the total hydrologic water cycle, the "green" city initiatives have been sprouting throughout the world. Notable examples already being realized include Hammarby Sjöstad, a waterfront district of Stockholm (Sweden) built on a former brownfield, where integrated water/stormwater/wastewater management and reclamation as well as solid waste recycle, biogas and heat/cooling energy recovery and use of other renewable energy (e.g., solar) are implemented. In addition, Hammarby Sjöstad is a "green" city in many other aspects such as reliance on bicycles and public transportation, solar energy, use of ecotones as parks and buffers and others. Other notable developments are being realized in Malmő and Gőteborg (Sweden), in Holland, Denmark and Germany which are also leaders in the implementation of energy from renewable sources with zero GHG emissions.

The concept of "ecocity" cannot be limited to large urban areas. Many of the same concepts can also be applied to smaller communities. Rural ecovillages also need water conservation and reclamation, energy recovery from waste and manure, and provide protection to regional water resources and the ecology of the area. Family size solar energy heated water tanks or biogas production units are being implemented in China along with small household or community wastewater treatment and reclamation units. Effluent reuse for irrigation in small rural communities has been practiced for decades.

TOPICS COVERED IN THE BOOK

This book compiles the presentations at the *Conference on Sustainable Water Infrastructure for Villages and Cities of the Future* (*SWIF2009*) held on Novemeber 6–9, 2009 in Beijing, China. Over 200 delegates from 15 countries attended the conference. 20 international scientists and experts were invited to present keynote lectures. Speakers from China included chief planners and university and national academy researchers working on the new ecocities, leaders in eco-sanitation, urban eco-development, energy and nutrient recovery, climate change impacts, and other emerging topics. After peer review, 48 peresentations were selected for inclusion into this book.

The book is divided into sections covering the following topics:

Sustainability of Urban Water Systems and Infrastructure; Water: Energy Nexus

This section covers the concepts of sustainable urban development, the driving forces for change, limitation of resources as impacted also by population increase and future outlook. It also focuses on the connection between water use, transport and treatment and its effect on the use of energy and GHG emissions. Some planned or already being built eco-cities and eco-villages are thriving for the net zero GHG emissions and promote water conservation, energy savings and production of energy from renewable sources. Natural landscape and ecologic engineering are emphasized. Finally, in some developing countries that do not have safe infrastructure for providing potable water, a preliminary system of water distribution by public utilities has to be developed as a first step.

Precipitation, Stormwater Drainage and Hydrologic Cycle

Climate changes, rainwater harvesting, stormwater management using low impact development (LID) concepts in several Chinese communities, solids in storm sewers and innovative highway stormwater management are the topics covered in chapters of this section.

Used Water Source Separation and Decentralized Management

Considering used urban water, currently still called wastewater, as a resource may require decentralization and source water separation into brown water (water containing mostly fecal matter), yellow water (urine) and gray water (cleaner but still polluted used water without brown water and urine). Black water contains most of the organic matter while a small volume of urine flow (about 1 % of the total used water flow) contains most of nitrogen and about 50% of phosphorus. The chapters discuss the options for source separation, treatment technologies and reuse.

Ecological/Small Community Sanitation

Hundreds of millions of people in China, Japan and other countries are still living in small communities. Therefore, it is natural that the book includes chapters on "ecosanitation" that is pertinent to smaller communities or urban clusters. Ecosanitation minimizes underground pipe infrastructure and eliminates the need for centralized western style treatment plants. It includes also a chapter on innovative small systems that treat and biodegrade organics based on natural principles and also on used water source separation. Examples and case studies of ecosanitation systems in China and Japan are presented in this section.

Nutrient Management and Recovery

Nutrients (nitrogen and phosphorus) are pollutants causing excessive algal developments in receiving waters, especially in impounded bodies, including lakes and reservoirs providing potable water. Noxious algal blooms by green and blue green algae diminish beneficial uses of inland and coastal water bodies such as recreation and fishing and render impoundments unfit for water supply. The nutrient section of the book contains chapters on the concepts of nutrient recovery from used water, a case study on recovering phosphorus in the form of struvite, nitrogen removal by nitrification, and converting biomass to a mineral fertilizer struvite.

Treatment of Separated and Combined Used Water and Solids

The section includes a variety of chapters describing design of composting toilets, separate treatment of brown water, and nitrification processes in urine and processes of particle separation. Topics also include measurements of endocrine disruptors, wetland construction with recycled materials and sludge treatment.

The book concludes with the future outlook and the current and planned programs by the International Water Association which is now in the forefront of the worldwide "Cities of the Future" efforts.

In overall, this book provides a wealth of vision and implementation both for the international communities interested in Cities of the Future and specialists interested in the new development in China and other countries.

The conference organizers and the editors of this book appreciate the generosity of the host of the Beijing conference - Beijing University of Civil Engineering and Architecture, and especially thank the Municipal Government of Beijing for a special financial support. The conference was also supported by a project from both the Ministry of Housing and Urban-Rural Development and the Ministry of Science and Technology. Finally, we also give our appreciation to international sponsors including the Water Environment Research Foundation (WERF) and the National Onsite Wastewater Recycling Association (NOWRA) from the US.

PART ONE

Sustainability of Urban Water
Systems and Infrastructure; Water
Energy Nexus

Integrating Water and Resource Management for Improved Sustainability

G. Daigger

CH2M HILL, 9191 South Jamaica Street, Englewood, CO 80112, USA
E-mail: gdaigger@ch2m.com

Abstract Integrated urban water and resource management systems offer significant opportunities to deliver more sustainable urban water service with much greater resource efficiency than the current, independently operating, systems. These systems rely largely on local water resources, thereby minimizing or eliminating the need to import water into the urban area, while becoming net producers of energy and recovering significant quantities of nutrients. They are built upon a diverse toolkit of technologies and practices which are configured into unique combinations to meet the specific needs of each urban area. These systems combine centralized and decentralized components based on guiding principles which meet water, salt, nutrient, and biosolids management constraints. Implementation of these systems requires a change in mindset by practicing professionals which needs to be integrated into both the educational and professional practice systems. Institutional changes can enable, but are not necessary, to accomplish the transition to these higher performing systems. They do require proper economic evaluations, along with institutional arrangements which properly allocate revenues to support the costs of these systems. Legacy centralized systems can be transitioned to these higher performing ones by aggressively implementing the decentralized elements as continued development and redevelopment occurs and by repurposing existing centralized assets.

Keywords Integrated, water, stormwater, wastewater, energy, nutrients, sustainability

INTRODUCTION AND OBJECTIVES

Urban water management has traditionally focused on independently supplying water, managing stormwater, and collecting and managing water-borne wastes, resulting in superior provision of these services to the great benefit of modern society (British Medical Journal, 2007; Constable and Sommerville, 2003). Thus, one may question why this practice should not continue. The answer is that our current approach to urban water management is not sustainable given changed circumstances (Daigger, 2009; 2007). Table 1 contrasts the historical context within which current urban water management systems evolved with the future context for such systems. Said simply, population growth, increased affluence, and resource limitations are increasingly creating water stress around the planet (Daigger, 2009).

Table 1 Comparison of Historical and Future Context for Urban Water Management Systems

Item	Historical Context	Future Context
Global Population	< 1 Billion	+/- 10 Billion
Urban Population	< 15 percent	Two-thirds or More
Life Style	Simple	Affluent
Water Resources	Abundant	Severely Limited
Natural Resources	Abundant	Severely Limited
Technology	Basic	Advanced

Fortunately, new technologies and practices are available which allow more water and resource efficient approaches to urban water management (Brown and Novotny, 2007; Daigger et al., 2005; DiGiano et al., 2004; Daigger, 2003; Novotny and Brown, 2007). A key element of these new approaches is integrated management of stormwater, drinking water, and urban wastes (Daigger and Crawford, 2007). Integrated management also offers the potential to significantly reduce the net energy and resource requirements for urban water management, and to facilitate nutrient recovery rather than removal (Daigger, 2009). We refer to these as integrated urban water and resource management systems. This paper reviews the objectives of such systems in comparison to specified sustainability criteria. The evolving toolkit of technologies and practices upon which successful systems are built is then summarized. Each urban area is different and requires a site-specific combination of these toolkit elements. Approaches for integrating the appropriate elements into an integrated system are presented, along with approaches to successfully implement these systems.

OBJECTIVES OF URBAN WATER AND RESOURCE MANAGEMENT SYSTEMS

The objectives of urban water and resource management systems include water supply, public health, and environmental protection, as summarized in Table 2. Urban water management has historically focused on meeting these objectives at the least cost and, as a result, have contributed significantly to the economic development of the communities they serve while providing significant social contributions by enhancing public health. These contributions will not be sufficient in the future, as resources become more limited. It is also recognized that, while urban water and waste management has been effective in developed countries, effective systems have not evolved in developing countries where approximately 1 billion people lack access to safe drinking water and 2.5 billion lack access to appropriate sanitation.

Table 2 Urban Water System Objectives

Water Supply	Public Health Protection	Environmental Protection
• Domestic • Commercial • Industrial	• Flooding • Pathogens • Toxics	• Oxygen Demand • Nutrients • Toxics • Micro-Pollutants

Daigger (2009) has suggested the triple bottom line goals to achieve sustainable urban water and resource management listed in Table 3. The objectives listed in Table 2 must be met, but future urban water and resource management systems must also address the broader goals listed in Table 3. Utilities providing these services must be viewed as providing sufficient value so that system users are willing to adequately fund them to achieve long-term fiscal sustainability. Broader environmental goals related to water quantity and resource utilization must be met, and approaches to provide service to all must be developed and implemented.

Table 3 Triple Bottom Line Urban Water and Resource Management Sustainability Goals (From Daigger, 2009)

Sustainability Area	Goal
Economic	• Financially stable utilities with the ability to maintain their infrastructure.
Environmental	• Locally sustainable water supply (recharge exceeds net withdrawal). • Energy neutral (or positive if possible) with minimal chemical consumption. • Responsible nutrient management that minimizes dispersal to the aquatic environment.
Social	• Provide access to clean water and appropriate sanitation for all

Accomplishing these goals will require transformations in approaches to urban water and resource management. Monsma (2009) have outlined key principals, as follows:

• The traditional definition of water infrastructure must evolve to embrace a broader, more holistic definition of sustainable water infrastructure that includes both traditional man-made water and wastewater infrastructure and natural watershed systems.

- This definition of sustainable water infrastructure should be embraced by all public and private entities involved in water management, and these same entities have a shared role in ensuring their decisions consider and integrate a set of criteria that include environmental, economic and social considerations (the Sustainable Path).
- A watershed-based management approach is required for drinking water, wastewater and stormwater services to ensure integrated, sustainable management of water resources.

They further identified the elements of sustainable urban water management systems, as listed in Table 4. These objectives, goals, key principals, and elements form the basis for developing sustainable urban water and resource management systems.

Table 4 Elements of Sustainable Urban Water Management Systems (From Monsma, 2009)

• Transparency • Public outreach & stakeholder involvement • Good governance • Full cost pricing • Allocation of cost of development • Asset management	• Security & emergency preparedness • Conservation & water efficiency • Environmental stewardship • Energy management • Climate change mitigation & adaptation	• Advanced procurement & project delivery methods • Modernized plant operations • Management of environmental impacts • Watershed & regional optimization	• Network optimization • Regulatory optimization • Workforce management • Affordability • Research and technological, managerial innovation

URBAN WATER AND RESOURCE MANAGEMENT TOOLKIT

While urban water and resource managers face significant challenges, a broad and diverse toolkit of technologies and practices has evolved, as summarized in Table 5. A diverse set of stormwater management and rainwater harvesting technologies captures rainwater and either store it for direct use or infiltrate it into the groundwater for later use and/or to restore urban terrestrial and aquatic habitats. As illustrated in Figure 1, capture of a modest amount of the rain falling on many urban areas can provide much of the water supply for that urban area. Coupled with water conservation and water reclamation and reuse, local water supplies provided by rainwater harvesting can be sufficient to meet the needs of many urban areas. Water reclamation and reuse technologies upgrade various water sources to meet the quality requirements for a variety of uses, ranging from non-potable to potable to ultra-pure water for industrial purposes. These technologies enable the concept known as "fit for purpose" water supply where water is treated to meet the quality requirements of specific purposes. The production of high-quality water has been further enabled in recent years by the development of cost-effective membrane and advanced oxidation technologies (Daigger et al., 2004, DiGiano et al., 2004, Daigger, 2003).

Energy management technologies directly produce energy from the organic matter present in the waste stream, such as anaerobic treatment to produce biogas or microbial fuel cells. It also includes the extraction of heat from (or rejection to) the waste stream. System efficiencies can be achieved through combining centralized and decentralized system components (Daigger, 2009; Daigger and Crawford, 2007). Analysis demonstrates that these technologies can be combined appropriately to produce urban water systems that are net producers of energy. Likewise, the nutrients nitrogen and phosphorus can be recovered from the waste stream by technologies ranging from ammonia stripping and recovery, calcium phosphate precipitation, and struvite precipitation. The historical practice of agricultural, residential, and commercial reuse of biosolids also reuses the nutrients present in the waste stream.

Dual distribution and source separation practices compliment water reclamation and reuse by delivering "fit for purpose" water for various uses and separating the components of the typical waste stream to facilitate energy and nutrient recovery. Table 6 illustrates segregation of the typical residential waste stream into greywater (laundry, bath), blackwater (feces, kitchen), and yellowater (urine). Greywater is the largest volume and is relatively uncontaminated, facilitating low-energy treatment for reuse. Blackwater is relatively low in volume and high in concentration when greywater is removed and can be treated directly by anaerobic processes for energy production. Yellowater is very low in volume (about 1 L/capita/day) and contains most of the nutrients. In short, source separation segregates the water, organic matter, and nutrient components for efficient recovery and reuse.

Table 5 Urban Water and Resource Management Toolkit (From Daigger, 2009)

Toolkit Element	Contribution
Stormwater Management and Rainwater Harvesting	Provide local water supplies.
Water Conservation	Reduces the volume of water needed and reduces the associated energy and chemical requirements to convey and treat water and wastewater.
Water Reclamation and Reuse	Provides local water supplies and can reduce energy requirements in many instances as pumping requirements are reduced to a greater extent than treatment energy requirements are increased.
Energy Management	A number of technologies that can directly extract energy from the wastewater stream, including anaerobic treatment, microbial fuel cells, and the direct extraction of heat.
Nutrient Recovery	A growing number of technologies which allow nitrogen and especially phosphorus to be recovered from the wastewater stream.
Dual Distribution and Source Separation	Numerous efficiencies are created by allowing water of appropriate qualities to be distributed for individual purposes and the separation of relatively uncontaminated water, organic matter, and nutrients at the source.

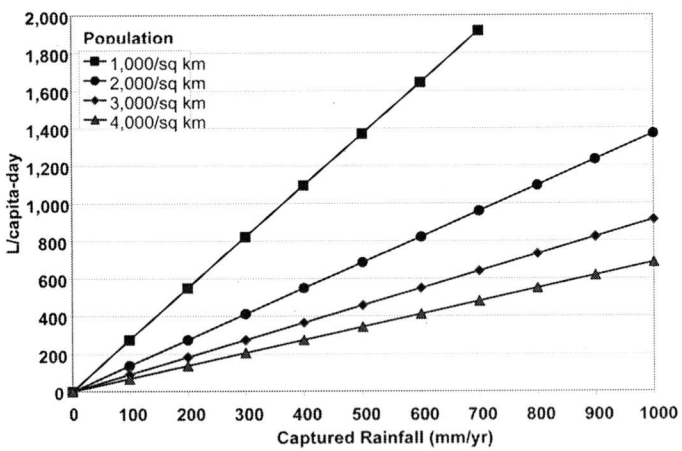

Figure 1 Water Supply Provided by Capture of Local Rainfall as a Function of Population Density

Table 6 Distribution of Organic Matter and Nutrients in Typical European Wastewater (Henze and Ledin, 2001)

Source	BOD$_5$ (g/(person·day)	Total Nitrogen (g-N/ (person·day)	Total Phosphorus (g-P/(person·day)	Potassium (g-P/ (person·day)
Toilet Waste				
Feces	20	1.1	0.6	1.1
Urine	5	11.0	1.4	2.5
Kitchen	30	0.8	0.3	0.4
Bath/ Laundry	5	1.1	0.3	0.4
Total	60	14.0	2.6	4.4

INTEGRATED URBAN WATER AND RESOURCE MANAGEMENT SYSTEMS

Daigger (2009) and Daigger and Crawford (2007) have demonstrated how the toolkit elements presented in Table 5 can be combined into higher performing integrated urban water and resource management systems. These systems will often incorporate both distributed and centralized system components. Their analysis is that, given the technologies currently available, the sustainability goals listed in Table 3 can best be accomplished through management of water on a more distributed basis and organic matter and nutrients for energy production and nutrient recovery on a more centralized basis. Figures 2 and 3 represent such systems.

The example presented in Figure 2 was developed to maximize the use of local water supplies. Both potable and non-potable water is provided for residential and commercial purposes, and industrial supply is from the non-potable supply. This example assumes that a local potable water aquifer of sufficient capacity is available. If this supply is insufficient, its capacity can be augmented by treatment of the captured rainwater, with the aquifer providing needed system storage capacity and water distribution. Non-potable water supply is provided by rainwater harvesting and water reclamation and reuse, and non-potable water storage is provided in another aquifer. Salt build-up can occur in a system such as this if a high fraction of net rainfall is collected, resulting in minimal export from the area. In such instances, salt must be removed from the waste stream (for example by reverse osmosis) and either reclaimed or exported from the system (as illustrated in Figure 2).

Figure 3 presents another example, this one focused on maximizing energy and nutrient recovery. Dual water supplies are again provided, but in this case along with separation of greywater, blackwater, and yellow water to facilitate low-energy reclamation and reuse along with energy and nutrient capture from the waste stream. Distributed water management facilitates the recovery of heat from the waste stream. The use of local water supplies is maximized to avoid the energy associated with the importation of water supplies.

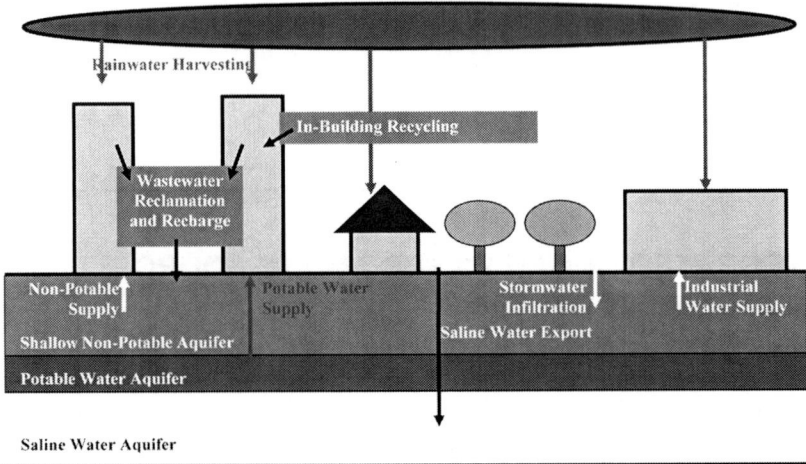

Figure 2 Example Integrated Urban Water and Resource Management System – Maximize Use of Local Water Resources

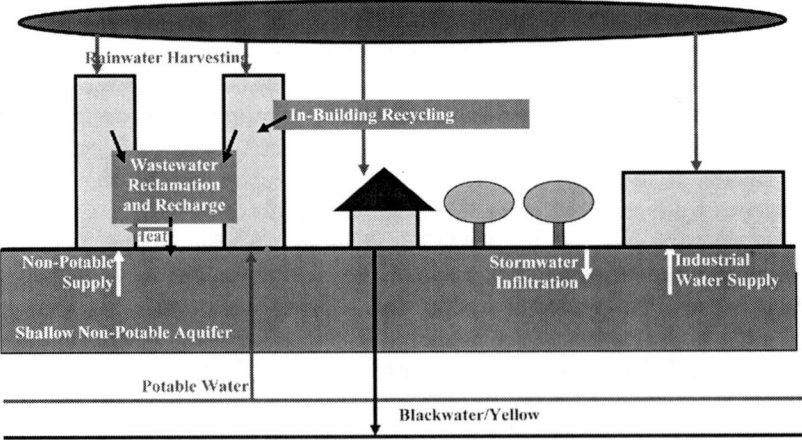

Figure 3 Example Integrated Urban Water and Resource Management System – Maximize Energy and Nutrient Recovery

The most appropriate system for a particular urban area will incorporate many of the same elements, but the specific combination will differ. Thus, the question arises of how to develop the most appropriate system for a particular urban area. As illustrated in Table 7, guiding principles can lead to the formulation of viable system alternatives which meet necessary constraints, as also listed in Table 7. These options can then be subject to triple bottom line analysis using multi-criteria approaches to select the option which best meets the needs of the particular urban area (Daigger and Crawford, 2005). In addition to the sustainability goals listed in Table 3, such systems can also result in improved public health protection, easier system upgrade and expansion, greater system resiliency, and an enhanced urban environment. Greater public health protection is provided because the water quality appropriate for each use is provided, rather than attempting to meet all water needs from the potable water supply. These systems are easier to upgrade and expand because they consist of a larger number of elements which can be added as needed. System resiliency is improved because the failure of any one element will have much smaller overall effects. The urban environment is enhanced because of the terrestrial and aquatic habitat produced.

Table 7 Guiding Principles and Constraints to Formulate Decentralized Systems (From Daigger, 2009)

Guiding Principles	Constraints
1. Protect and subsequently use locally available water resources whenever possible.	1. Maintain water balance for both typical and extreme (wet and dry) conditions.
2. Mimic local hydrologic patterns prior to development. Generally means infiltration of high frequency, low-intensity storms and allowing high-intensity storms to create run-off.	2. Maintain long-term salt balance.
3. Public health is protected by incorporating multiple barriers, especially considering pathogens and trace contaminants.	3 Maintain nutrient balance.
4. Consider total resource consumption and potential for resource recovery when formulating systems.	4. Manage residuals, both over short and long time scales.

IMPLEMENTATION

While the approaches discussed above may seem to be a radical departure from past practices, they are being implemented in various locations thereby demonstrating their feasibility (Daigger, 2009; 2007). These examples also provide the initial learning needed to accomplish broader application. The question arises, however, concerning changes needed to allow further application of these approaches so that their benefits may be more generally realized. First of all, the mindset of the water profession must evolve from dealing separately with elements of the urban water cycle to integrated management, and from a focus on treatment to resource recovery and reuse. This transition in mindset is beginning, but it needs to be accelerated at both the educational and professional levels.

Economic evaluations must consider the integrated nature of these systems so that the full range of benefits and costs are assessed. For example, reclamation and reuse options must consider not only their cost but also the water supply, treatment, and distribution option which they off-set. In practical terms, this means that such options should not be evaluated relative to the average cost of water in the urban area but to the next water supply which must be developed (which by definition will be more expensive than the current average). At the same time, such evaluations can be hampered by institutional arrangements if different utilities provide different elements of the urban water management system. Integrating these institutions may bring system efficiencies, but only if integrated thinking and actions can be fostered in the combined utility. There are numerous examples of how inter-agency agreements suffice.

Some view the existing centralized "legacy" systems present in developing countries to be a barrier to implementing the more distributed systems discussed above. This is not necessarily the case as on-going development and re-development is occurring which provides the opportunity to retrofit the ditributed elements into the existing centralized system, thereby achieving the hybrid system described above. Two actions are required to achieve this. One is aggressive implementation of the decentralized elements. The second is selective repurposing of existing facilities, for example of the water distribution and waste collection systems and the centralized treatment facilities.

SUMMARY AND CONCLUSIONS

In summary, integrated urban water and resource management systems offer significant opportunities to deliver more sustainable urban water service with much greater resource efficiency. These systems rely largely on local water resources, thereby minimizing or eliminating the need to import water into the urban area, while becoming net producers of energy and recovering significant quantities of nutrients. These systems are built upon a diverse toolkit of technologies and practices which are configured into unique combinations to meet the specific needs of each urban area. These systems combine centralized and decentralized components based on guiding principles which meet water, salt, nutrient, and biosolids management constraints. Implementation of these systems requires a change in mindset by practicing professionals which needs to be integrated into both the educational and professional practice systems. Institutional changes can enable, but are not necessary, to accomplish the transition to these higher performing systems. They do require proper economic evaluations, along with institutional arrangements to properly allocate revenues to support the costs of these systems. Legacy centralized systems can be transitioned to these higher performing ones by aggressively implementing the decentralized elements as continued development and redevelopment occurs and by repurposing existing centralized assets.

REFERENCES

British Medical Journal (2007). Medical milestones. *British Medical Journal*, **334**, s1–s20.
Constable G. and Sommerville B. (2003). *A Century of Innovation: Twenty Engineering Achievements that Transformed Our Lives*. Joseph Henry Press, Washington, DC.
Daigger G. T. (2007). Wastewater management in the 21st century. *Journal of Environmental Engineering*, ASCE, **133**(7), 671–680.
Daigger G. T. (2009). Evolving urban water and residuals management paradigms: Water reclamation and reuse, decentralization, resource recovery *Water Environment Research*, **81**(8), 809–823.
Daigger G. T. and Crawford G. V. (2005). Wastewater treatment plant of the future – Decision analysis approach for increased sustainability. *2nd IWA Leading-Edge Conference on Water and Wastewater Treatment Technology, Water and Environment Management Series*. IWA Publishing, London, pp. 361–369.
Daigger G. T. and Crawford G. V. (2007). Enhanced water system security and sustainability by incorporating centralized and decentralized water reclamation and reuse into urban water management systems. *Journal of Environmental Engineering Management*, **17**(1), 1–10.
Daigger G. T., Rittmann B. E., Adham S. and Andreottola G. (2005). Are membrane bioreactors ready for widespread application? *Environmental Science & Technology*, **39**(19), 399A–406A.
Daigger G. T. (2003). Tools for future success. *Water Environmental & Technology*, **15**(12), 38–45.
DiGiano F. A., Andreottola G., Adham S., Buckley C., Cornel P. M., Daigger G. T., Fane A. G., Galil N., Jacangelo J. G., Pollice A., Rittmann B. E., Rozzi A., Stephenson T. and Ujani Z. (2004). Safe water for everyone. *Water Environmental & Technology*, **16**(6), 31–35.
Guest J. S., Skerlos S. J., Barnard J. L., Beck M. B., Daigger G. T., Hilger H., Jackson S. J., Karvazy K., Kelly L., Macpherson L., Mihelcic J. R., Pramanik A., Raskin L., van Loosdrecht M. C. M., Yeh D. and Love N. G. (2009). A new planning and design paradigm to achieve sustainable resource recovery from wastewater. *Environmental Science & Technology*, **43**(16), DOI 10.1021/es9010515.
Monsma D., Nelson R. and Bolger R. (2009). *Sustainable Water Systems: Step One – Redefining the Nation's Infrastructure Challenge*, A Report of the Aspen Institute's Dialogue on Sustainable Water Infrastructure in the U.S., Aspen Institute, Washington, DC.
Novotny, V. and Brown P. (2007). *Cities of the Future: Towards Integrated Sustainable Water and Landscape Management*. IWA Publishing, London.

Sustainable Water Infrastructure of the Future – The Contest of Ideas and Ideals in Sustainability

Goen Ho

Environmental Technology Centre, Murdoch University, Perth, Australia
E-mail: g.ho@murdoch.edu.au

Abstract Sustainability can be considered as a process of balancing the need for economic development, environmental protection and social justice. Viewed in this way sustainability is a journey because the needs and wishes of society evolve over time. This sustainability journey for water infrastructure is briefly reviewed beginning with provision for piped drinking water, sewerage for removing human wastes and stormwater management infrastructure to protect against flooding. Development of infrastructure for these, while creating societal and economic benefits, has impacted considerably on environmental flow of rivers and water quality of rivers receiving wastewater and stormwater discharges. While efforts have been made to mitigate these environmental impacts, there have been many ideas, often conflicting, about what sustainability is and how it can be achieved. There has been even less on the end or ideals of the sustainability journey. Three ideals are put forward in the paper, covering the ideal of self-sufficiency, mimicking nature and source separation of wastewater. The first aims to have development relying solely on rainfall within land development/village or urban boundaries. The second raises the ideal further to bring the water cycle within the boundaries towards the natural undeveloped land, while the third follows the natural discharge of urine and faeces separately and onto land for land based development.

Keywords Sustainability, water infrastructure, water self-sufficiency, source separated wastewater, imitating nature

INTRODUCTION

The subject of sustainable water infrastructure of the future is indeed a vast one encompassing four topics of sustainability, water, infrastructure and the future, each of which deserves a separate discourse even in the context of the integration between the four. I wish to share my reflection on the topic of sustainability in the context of the integration.

It is perhaps easier to first discuss about what is not sustainable, because we tend to identify and feel unhappy about what is not sustainable. If we do not have, or do not have adequate, water infrastructure, we have contaminated drinking water and polluted streams and rivers. This causes endemic water borne diseases in the community, which occasionally become epidemic. The economic loss, due to loss productivity and costs of health treatment, and the associated human suffering and loss of environmental amenity is clearly undesirable. This is the situation faced by many in the developing part of the world, which is inhabited by two thirds of the world population.

In the developed part of the world, we usually have safe drinking water piped to the community. Drinking water for a city may be supplied from surface water by building a water reservoir (dam across a river) upstream of the city. A consequence of building the dam is that the river downstream of the dam does not receive water in the same quantity and seasonal variation compared to before the dam was built, thus altering the ecology of the river to its disadvantage. Collected wastewater is usually treated and discharged to a river downstream of the city or the sea. Because treatment does not usually remove all pollutants, the river environment becomes degraded. This together with the need to pump water and wastewater over large distances is unsustainable environmentally.

A city is usually built by clearing vegetation and replacing it with impermeable surfaces (roads, driveways, car parks and roofs), thus rainfall runoff increases more rapidly following rainfall and with a higher peak compared to the original surface. The runoff is more likely to cause flooding at a lower rainfall intensity and the quality of runoff is poorer compared to runoff from naturally vegetated surfaces. Flooding is controlled by building underground or surface storages at high costs. This scenario can also be considered unsustainable. Building a city may also mean draining of landscape, removing wetlands and covering of streams. A consequence is loss of environmental amenity.

Unsustainable situations illustrated above are caused by lack of infrastructure or infrastructure that results in negative environmental or social consequences. Hence there has been the development of the concept of sustainable development, which must take into account and balance economic, environmental and social needs (commonly referred to as the triple bottom line).

In this paper I trace the development of the concept of sustainability as it applies to the water sector. I will first outline some historical background, because it provides the backdrop to current advances in our thinking, then explore further the triple bottom line concept before discussing in particular three topics that may guide us towards the future. These are the concepts of self-sufficiency, mimicking nature and source separation. Following this I will consider application of the concepts to different scales of development and the need to weave water sustainability to sustainability in the other infrastructure sectors (energy, food, materials, transport). Finally I will consider the policy framework that could make sustainability happen, and whether the developing part of the world can leap-frog in their development and avoiding the mistakes that have been experienced in the developed world.

SUSTAINABILITY

Historical Background

The beginnings of the conversation in sustainability in the water sector are over the same period as the beginnings of the environmental movement about four decades ago when concerns were raised about the degradation of the environment by developments. Developments make economic sense and are desired by society, but may be at the cost of environmental degradation. Developments require new sources of water to be developed and for wastewater to be collected, treated and disposed. They also require more land surfaces to be covered by buildings and roads resulting in a changed stormwater run-off pattern. These three are the undesirable consequences mentioned in the Introduction section.

If we step back a further six decades to the beginning of the 20th century we see that it made good sense to sewer cities, because people could be separated from their wastes. It breaks the faecal oral cycle for the transmission of water borne diseases. Sewering a city incurs investment and operation costs, but these costs are justifiable because of the savings from avoided health costs. Recent costings by the World Health Organization show that an investment of one dollar results in an economic benefit of at least five dollars (Hutton and Haller, 2004).

The collected wastes were not generally treated until a few decades later. Investment in treatment was the result of the deterioration of water quality of the receiving water environment. The loss of amenity from the deterioration of the environment is more difficult to quantify in dollar terms. Reduced dissolved oxygen in a river results in odour, the absence of fish and unpleasant appearance of the water. A more prosperous community would want to overcome this loss of environmental amenity. Wastewater is then treated in stages initially to remove its oxygen demand and suspended solids. This is followed by removal of nutrients, then there are heavy metals and trace concentrations of organic compounds (pharmaceuticals, endocrine disruptors) that have been raised as possible concerns. If this historical trend is projected to its ultimate, we may have to purify the water further. Again this may reach its economic unsustainability.

For stormwater runoff from urban impermeable surfaces sewering makes sense, because it removes the water from the developed site and reduces the possibility of flooding at the site, but increases the likelihood of flooding downstream. Investment for sewer and flood control is therefore required and needs to be balanced against the benefits from avoiding the costs due to flooding impairment. These historical trends force us to focus on the concept of the triple bottom line.

Triple Bottom Line

The concept of the triple bottom line is an extension of the concept of the economic bottom line. The latter takes into account the economic benefits from development solely from the consideration of income from a development and the costs incurred in producing the income. The latter do not include social costs (e.g. costs of displaced people due to development) or environmental costs (e.g. degradation of environmental quality due to development). The latter are borne by the community instead of by the enterprise. The process of considering these three is dependent on the system of governance of the community and therefore varies from community to community and from jurisdiction to jurisdiction. What is clear is that the process involves balancing between the need for economic development and the social and environmental needs of the community. Compromises are usually made irrespective of system of governance.

The concept of sustainability in this context can be considered to be a journey, because the community may change by becoming more prosperous through development but with deteriorating environmental amenities. The community may then want both prosperity and restored environmental amenities and improved social conditions. The ideal sustainability goal is intergenerational equity so that future generations can still determine their development without being lumbered by social and environmental costs unpaid from previous generations (Brundtland, 1987).

The sustainability as a journey for the water sector has been fairly well described as one from the water supply city, to a sewered city, drained city, waterways city, water cycle city and finally water sensitive city (Wong and Brown, 2009). The progression reflects the priority that communities have given in first of all supplying good quality water to the community, then installing sewerage to improve public health, followed by drainage to protect against flooding. Environmental amenities are then addressed by controlling pollution, and this appears to be where cities are currently in the developed world. The water cycle city is proposed to ensure that the city diversifies its sources of water (including from urban stormwater) and uses water of only the appropriate quality for the intended purpose coupled with water use efficiency. The ideal water sensitive city is suggested as a city with urban planning and an infrastructure designed to facilitate and reinforce all of the above. Case studies that demonstrate aspects of the ideal city are given, but there does not appear to be a definitive final destination. This must be the case if sustainability is indeed a journey, so that as knowledge increases and community priority changes we will continue to modify our definition of what we consider to be sustainable.

There are however ideals that we can aspire to. These can be put forward to communities to consider and to inspire and influence our efforts towards sustainability. I will discuss three of them, and they are not mutually exclusive and may overlap or inter-relate.

Self-Sufficiency

The idea of self-sufficiency in water reflects the ideal or desire for security of water supply for all our needs from within the boundary of our properties. People who live in isolated or remote areas already operate on this basis and perhaps assisted by the large area of their properties. The question is whether, for example, a householder in a city can be fully self-sufficient within the small area of their land plot. The only source of water is rainfall falling within the boundary of the property. Some of this rainfall can be harvested from the roof and the remainder from runoff from other surfaces within the boundary. Groundwater, recharged from rainfall from within the property (e.g. from an unconfined aquifer beneath the property), can also be harvested provided the amount does not exceed its natural recharge.

Two major factors that determine the feasibility of this concept are the climate of the locality and availability of technology. Where rainfall is adequate, then the answer to whether the householder can be self-sufficient is affirmative. Storage may be required if the climate is, for example, Mediterranean with long periods without rainfall. Where rainfall is more than adequate, then there will be water export from the property. How this export can be handled is discussed below under Mimicking Nature.

Technology can assist with achieving the ideal of water self-sufficiency. Water efficient appliances (e.g. tap/faucet water aerator, low flow shower head, dual flush toilet cistern, water efficient washing machine and dishwasher) are becoming readily available. Technology is also available that can convert wastewater to high purity water. Reliability of water purification products is improving rapidly and costs are decreasing. Costs of water purification technology are, however, still relatively high and coupled with energy requirement will limit the use of this technology to those who see this as a societal objective. The need for technology may also be reduced through water reduction and conservation measures, for example, shorter shower and having a garden that uses plants adapted to the local climate.

Application of the concept has therefore progressed along the line of using water of the appropriate quality for the intended use. 'Fit for purpose' is the term that has been coined, and examples are roof rainwater used for drinking and indoor water uses; water from bathroom and laundry (greywater) used for garden irrigation. Similarly wastewater, treated to only secondary effluent standard, can be used for garden irrigation.

Water balance around the property boundaries generally indicates that for single or cluster of houses water self-sufficiency is feasible. With high-rise buildings the strict application of the concept becomes difficult. High-rise buildings are, however, usually located within a precinct that includes parks and gardens and under the jurisdiction of a local authority. If we extend the self-sufficiency concept to the whole local authority boundaries, then it could become entirely feasible and would even provide alternatives that would not be feasible for single house lots, as well as economy of scale advantage.

Mimicking Nature

Mimicking nature and its natural processes provides us with an ideal to guide us towards environmental sustainability. Nature in its natural state is the setting for ecological systems that are complex, dynamic and sustaining and relying solely on water from precipitation and its natural variability. It is also solely reliant on the energy derived from the sun and the plant nutrients available to it.

Construction of infrastructure considerably alters the routes and flow variation of rainfall precipitation. The proportion evaporated or evapotranspired, stored in the soil profile, infiltrating to groundwater, flowing as surface run-off is biased towards reduced evapotranspiration and more rapid and increased surface run-off.

The implementation of the self-sufficiency concept returns a property back towards the water flow pattern of a natural system. The water flow dynamic balance of the natural state can then be used as a guide on how much water can be harvested from natural precipitation. Ideally the water import and export from the boundaries of a development will mimic the natural state so that the water environment outside the boundaries, especially when still in its natural state, is not affected by the development (Ho *et al.*, 2008).

The water flow balance of the natural state will guide whether there is sufficient water available for the implementation of the self-sufficiency concept, if we do not wish to rely on sophisticated technology for water purification. It will also guide us on how much water should be exported from the property boundaries if there is more than sufficient rainfall precipitation available.

The concept of mimicking nature is also a useful guide on how to use nutrients contained in wastewater. Rather than removing the nutrients as is required if the wastewater is to be disposed, using the wastewater after treatment for irrigation of the garden is logical, because plants require nutrients. We may also consider growing food in the garden, so that we can reduce the import of 'virtual water' through the consumption of food. Virtual water is water that is required in growing food elsewhere, especially if the crop is irrigated.

Vegetated surfaces also return the water balance of a property closer to the pre-clearing natural state through increased evapotranspiration. The roof of a building can be used to grow plants as well as to collect roof run-off. An extension of this is to grow plants on the south facing side (or north facing side in the southern hemisphere) of a building to grow plants. In this way the surface area of a developed property may resemble more the vegetated surface of the natural surface before its clearing.

Mimicking nature also provides us with a way to return the amenities provided by nature such as streams, wetlands, lakes which are up to now engineered away through channelisation, drainage, filling and covereing. Surface and groundwater flows, which are closer to their pre-development state, are conducive to restoring these amenities.

Source Separation

Source separation relates to wastewater and how it can be separated into streams with differing characteristics. These are wastewater streams from the toilet, kitchen, bathroom and laundry. Toilet wastewater can be further separated to faecal materials and urine. Colours have been assigned with blackwater denoting wastewater from the toilet (can be separated into brown (faecal materials) and yellow (urine)) and greywater from bathroom and laundry, and kitchen wastewater either to grey or blackwater depending on treatment method and intended use.

Separation at source into different streams makes logical sense, because these streams differ in quantity and quality, and are suited to different treatment and reuse. Urine can be used as a liquid fertiliser simply by diluting it (Matsui et al., 2001). It contains a significant portion of nitrogen and phosphorus in wastewater. Brownwater contains most of the solids and pathogens. Treatment by composting produces humus materials that can be applied into the garden. Greywater can be used for garden irrigation with only filtration if used immediately.

Source separation sits comfortably within the concept self-sufficiency and mimicking nature. All components of wastewater can be reused within the property boundaries. And just as in nature wastes are deposited into the land environment rather than disposed into the water environment. Both the water cycle and the nutrient cycle can be then considered to have achieved a close loop within the property boundaries.

The concept of source separation extends to the separation of wastewater from industry from wastewater from homes. Industry is required to close its water cycle, and the cycle of materials contained in its wastewater. This separation will avoid contamination of domestic type wastewater with the more contaminated industrial wastewater.

DISCUSSION

Consequence of Longevity of Infrastructure

Water infrastructure has a relatively long life, of the order of decades, and the consequences of making a wrong choice will be difficult to remedy in the short term. This is particularly so for large centralised infrastructure with a correspondingly large investment by government. With the infrastructure comes the institutional arrangement for cost recovery for the investment and ongoing operation. The institutional arrangement consists of government agencies or corporations and the legislation that has set them up. The momentum is in continuing with what has been well set up. This is the current situation in the developed world with existing large centralised water infrastructure.

The choice of sustainable water infrastructure for the future is therefore crucial, because we may make a choice that may be proven to be incorrect in the future. We may impose on future generations the consequences of our choice, which will again be difficult to remedy, because of the longevity of infrastructure.

Water Quality and Public Health

In discussing the application of the above concepts the protection of public health should not be forgotten. It is a primary reason why we have water infrastructure. In implementing the ideals of sustainability we should ensure that water quality for drinking, cooking and human contact uses should meet the standards or guidelines for drinking water. The risks of human contact with water containing contaminants, particularly pathogens, and of cross-connections between streams of different water quality should be closely managed. Management for this is clearly better carried by a body that is independent of owners or occupiers of buildings, which may not be familiar or interested with the operation or maintenance of the water infrastructure.

Concerns about public health from reusing or recycling water are justified. Management of water reuse or recycling to achieve sustainability should be given a high priority to ensure that risks are not higher than from the current situation in the developed world with centralised management of centralised water infrastructure.

Scale of Application

The ideals of self-sufficiency, mimicking nature and source separation can be applied at the scale of a single dwelling lot, cluster of dwellings, high rise building, village, municipality or indeed a whole city. There are now examples of water self-sufficient dwellings in urban areas. Key factors are availability of land, rainfall precipitation and water use per person. Technology and energy can, however, compensate for the other two factors. In high density urban setting the ideals can only be met by including land adjoining the city. The latter needs to be recognised and the land accordingly protected for the specified purpose.

There appears to be an optimum size for the achieving the ideals covered in the above concepts. Single dwellings do not offer the advantages of a cluster of dwellings. The latter may include features such as wetlands and public open space. It has the advantage of economy of scale in using a common treatment plant for, say greywater, which is then used for irrigation of the common open space. A larger scale of operation can also cater for emergency purposes, for example water for fire fighting. A large highly populated urban area, on the other hand, has the disadvantage of requiring land area that may not be available nearby, thus requiring piping and pumping of water, likely through several pipes carrying different quality water. There is a need for research into the optimum scale based on local conditions of rainfall, topography, size of water catchments and other special local water features.

There is also a need to research the area of land that is required to grow food for the city, unless this is carried out within the city as suggested in the concept of self-sufficiency above. The latter strengthens the sustainability case, because the produce does not need to be transported large distances, and the nutrients from the wastewater can be recycled for food production within the city.

Not all the food required in a city can be grown within the city, unless the city boundaries are enlarged to include not only water for domestic or commercial purposes, but also for agricultural purposes. Water used in irrigated agriculture can be greater than water used in a city.

Linkage to Sustainability in Other Infrastructure Sectors

Sustainability of the water sector cannot be considered in isolation from the sustainability in other infrastructure sectors. These are infrastructure for energy, agriculture, transport, buildings and material resources. The interconnection between water and agriculture has been mentioned above. Energy is required for pumping water and for operating the technology to treat and purify water. Water is also required for power generation from fossil fuels, and this has not been taken into account in our discussion thus far, but must clearly be included.

Water is also required for industry and this has also not been discussed in this paper. Water requirement varies considerably between industry type, and each deserves a separate consideration. Again the sustainability concepts discussed above are applicable.

What we need to realise is that sustainability in the water sector is complex, not only in the areas that are traditionally covered by water authorities (water supply, sewerage and drainage) and municipalities (open spaces, wetlands), but that it is intrinsically linked to other infrastructure sectors. Further compromises will need to be made in balancing social, economic and environmental factors when sustainability in water infrastructure is linked to sustainability in other infrastructure sectors.

Policy Framework

If we agree that the ideals or concepts for sustainability in the water infrastructure sector are sound, and we wish to implement them, the question then is the policy framework that is required to implement them. Water authorities in the developed world are generally responsible for water supply, wastewater and drainage. They are in a position to implement, for example, much of the concept of self-sufficiency. There is growing evidence that a number of water authorities are beginning to implement large scale water reuse. Because of the need for treatment, piping and pumping for the water to be reused, there may not be great economic advantage for doing so. They also need to negotiate with large water users, such as industry, municipal authorities.

It is much easier to implement the concept of water sustainability at the single household level, cluster of houses or single multi-storey buildings. There are an increasing number of examples of water self-sufficient households and buildings. They have, however, met barriers in implementation, because of the planning and approval processes.

There is an imperative for all of us therefore to debate concepts and ideals in water infrastructure sustainability, and promote the concept and implementation of sustainability to all those involved in the water sector. These include water professionals, land developers, builders, regulators, researchers and educators. As with any process of change it is likely to be led by progressive individuals (champions) and land developers in new developments or re-developments.

While we debate the concept of sustainability of water infrastructure in the developed world, we may also ask whether in the developing world there is an opportunity for their leap frogging to sustainability. This is development from a situation where there is inadequate water infrastructure to infrastructure that is based on the ideals of sustainability and thus avoiding the unsustainable situations faced by the developed world. In theory it is entirely feasible, but it appears to depend on whether we can demonstrate in the developed world the implementation of the sustainability concepts. This is because of the apparent desire of the developing world to use the developed world as a model for its own development.

Contest of Ideas and Ideals

The need to consider equally economic, social and environmental factors to achieve sustainable development is now generally accepted. Arriving at the right balance between these factors will always be a matter for the local community. Environmental considerations will favour the ideal of self-sufficiency and closing the loop for water and nutrients onsite. Implementation of this ideal will benefit from mimicking nature and considering separation of water streams.

The contest of ideas and ideals for sustainable water infrastructure is therefore in their implementation. It can appear as a contest between implementation at the centralised versus decentralised scale. Further examination into the issues reveals, however, that it is not necessarily the case. The ideas and ideals can be applied at both scales. Application at the large centralised scale appears to be more feasible, because the institutional system that has been established to manage water infrastructure and therefore manage public health risks is associated with centralised systems.

It is clear that management for water infrastructure, whether for centralised or decentralised infrastructure, should be by a common (centralised) institutional arrangement to ensure proper operation of the infrastructure and protection of public health. The present institutional arrangement, which has been set up for centralised large scale systems, does not facilitate implementation at the decentralised scale. There is a need to change this arrangement so that sustainability can be facilitated at all scales.

Decentralised water systems appear to be more sustainable with some reasons already discussed above (Ho and Anda, 2006). Transport of water over long distances is avoided. Decentralised system is also less prone to large scale disruption (e.g. act of terrorism). Research is, however, required into quantitative comparison between centralised and decentralised systems for costs, energy requirement, carbon footprint and other intangible environmental impacts (such as environmental flows).

CONCLUSION

Sustainable water infrastructure for the future involves a journey if we consider the need for balancing economic, social and environmental factors. Ideals for the end of the journey are suggested, and include water self-sufficiency and mimicking nature. The latter includes separation of wastewater streams at source for reuse or recycling.

Water professionals have a crucial role in implementation of sustainable water infrastructure by debating the issues surrounding sustainability, promoting the concepts and ideas to those that can implement them.

There is a need for research into water balance in single and cluster of dwellings, village and city scale to provide the data for water availability from rainfall, water use, and possible water reuse either for self-sufficiency or for mimicking nature. Research into comparison between centralised and decentralised water infrastructure systems for sustainability is also required.

REFERENCES

Brundtland H. (1987). *Our Common Future*. Oxford University Press, for the World Commission on Environment and Development, Oxford.

Ho G. and Anda M. (2006). Centralised versus decentralised wastewater systems in an urban context: the sustainability dimension. In: *2nd IWA Leading Edge on Sustainability in Water-Limited Environments*, M. B. Beck and A. Speers (eds.), IWA Publishing, London, pp. 81–89.

Ho G., Anda M. and Hunt J. (2008). Rainwater harvesting at urban land development scale: Mimicking nature to achieve sustainability, *IWA World Water Congress and Exhibition*, 7–12 September, Austria Center, Vienna.

Hutton G. and Haller L. (2004) *Evaluation of the Costs and Benefits of Water and Sanitation Improvements at the Global Level*. World Health Organization, Geneva.

Matsui M., Henze M., Ho G. E. and Otterpohl R. (2001). Chapter 5. Emerging paradigms in water supply and sanitation. In: *Frontiers of Urban Water Management: Deadlock or Hope?*, C. Maksimovic and J. A. Tejada-Guibert (eds.), International Water Association Publishing, London.

Wong T. H. and Brown, R. R. (2009). The water sensitive city: principles for practice. *Water Science and Technology*, **60**(3), 673–682.

Water Energy Nexus- towards Zero Pollution and GHG Emission Effect of Future (Eco) Cities

V. Novotny

Northeastern University, Boston, Ma 02458, USA
E-mail: v.novotny@comcast.net

Abstract The presentation and the ensuing paper will describe the energy water connection in the ecocities, outline quantitatively the steps towards the reduction of energy use and their limits. This analysis will synthesise the investigative analysis of seven ecocities in China, United Arab Emirates, Sweden, and US (Novotny and Novotny, 2009). These cities are already being built or are in advanced stages of planning. Three other existing urban areas are being retrofitted to achieve a high degree of water reclamation and reuse (Singapore, El Paso (TX) and Orange County (CA)) and will also be included in the analysis. The analysis of the water/used water/stormwater systems in these cities revealed that they range from linear system to closed systems, with a relatively narrow range of population density for the ecocities but different degrees of energy saving. Two cities are expected to meet the net zero GHG emissions and most will achieve a high level of pollution reduction by implementing water conservation and reuse. By the overall energy balance it was estimated that the achieving the net zero GHG emission footprints in 2030 is achievable.

Keywords One Planet Living Criteria, Greenhouse gas emissions, Water conservation, Water reclamation, LEED criteria, Green development, Water demand

CITIES OF THE FUTURE (ECOCITY) VISION AND GOALS

The Cities of the Future or Ecocities represent a major paradigm shift in the way new cities will be built or older ones retrofitted to achieve a change from the current unsustainable status to sustainability. A working definition of an ecocity and the goal of future new urban developments as well as retrofitting the old ones are as follows (Register, 1987; Novotny *et al.*, 2010):

> *An ecocity is a city or a part thereof that balances social, economic and environmental factors (triple bottom line) to achieve sustainable development. A sustainable city or ecocity is a city designed with consideration of environmental impact, inhabited by people dedicated to minimization of required inputs of energy, water and food, and*

waste output of heat, air pollution - CO₂, methane, and water
pollution. Ideally, a sustainable city powers itself with renewable
sources of energy, creates the smallest possible ecological footprint,
and produces the lowest quantity of pollution possible. It also uses
land efficiently; composts used materials, recycle or convert waste-
to-energy. If such practices are adopted, overall contribution of the
city to climate change will be none or minimal below the resiliency
threshold. Urban (green) infrastructure, resilient and hydrologically
and ecologically functioning landscape, and water resources will
constitute one system

Green House Emissions Related to Urban Areas

The CO_2 emissions vary widely among nations. Until recently the US was the largest emitter of GHG gases but recently was overtaken by China. If statistics are presented in emissions per person, the Middle East states are the largest emitters (Table 1) but the US, Australia and Canada are in the top ten. It should be pointed out that various statistics differ and the emissions vary year from year, but generally in the US and European Community, they seem to levelling off in this century. In the latest data for the US, the GHG emissions have begun to decrease after 2007 (Brown, 2009). Dodman (2009) found that large cities emit per capita less GHG than the national average. For example London emissions (6.2 tons/capita-year) are 50% less than the national average (9.4 tons/capita-year). Same is true for US data. The average of 100 largest cities analyzed by Gleaser and Kahn (2008) is 8.5 tons/capita-year, without considering industries, while the national average is 19 tons/capita-year (which includes industries). The highest per capita emissions are in low density suburbs.

In 2007, 55 billions m³ of water was used by the population of the 301.3 million in the US. Using the US EPA estimate of 3% energy use for water would result in the unit energy use of 2.26 kW-hr/m³ attributed to water. Corresponding carbon emission is of 1.37 kg CO_2/m³.

Achieving the Goal of Net Zero Carbon Footprint in New Ecocities and Retrofits

The current criteria and guidelines used for ecocity certification are the LEED (Leadership in Energy and Environment Design) of the US Green Building Council and One Planet Living (OPL) by the World Wildlife Fund. However, the requirement for ecocities is not just to reduce GHG emissions such as expressed in

Table 1 Per capita CO_2 emissions statistics

Top ten countries in the CO_2 emissions in tons/person-year in 2006[1]									
Qatar	UAE	Kuwait	Bahrain	Aruba	Luxembourg	USA	Australia	Canada	Saudi Arabia
56.2	32.8	31.8	28.8	23.3	24.5	19	18.1	16.7	15.8

Selected world cities total emissions of CO_2 equivalent in tons/person-year[2]									
Washington, DC	Glasgow UK	Toronto CA	Shanghai, China	New York City	Beijing China	London UK	Tokyo Japan	Seoul Korea	Barcelona Spain
19.7	8.4	8.2	8.1	7.1	6.9	6.2	4.8	3.8	3.4

Selected US cities domestic emissions of CO_2 equivalent in tons/person-year[3]									
San Diego CA	San Francisco	Boston MA	Portland OR	Chicago IL	Tampa FL	Atlanta GA	Tulsa OK	Austin TX	Memphis TN
7.2	4.5	8.7	8.9	9.3	9.3	10.4	9.9	12.6	11.06

[1]Wikipedia (2009); [2]Dodman (2009); [3]Gleaser and Kahn (2008).
[2,3]Values include transportation, heating, and electricity.

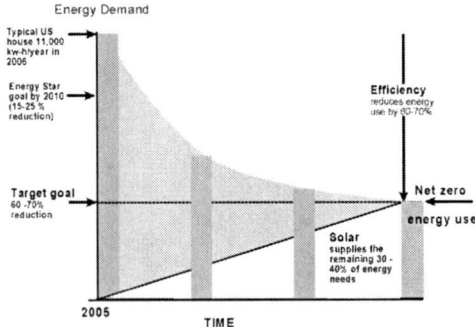

Figure 1 A path to achieving the net zero energy goals (NSTC, 2008)

the LEED energy efficiency standards. The demand elucidated in the One Planed Living (OPL) criteria by the World WildLife Fund calls for ecocities to become carbon neutral. The same call has also been issued by the National Science and Technology Council (NSTC) (2008) of The US President and by the British government which called for development and implementation of both net zero carbon footprint and high performance building technologies.

Figure 1 shows the possible paths towards the net zero GHG emissions goal. Current scientific research indicates that 60 to 70% of energy reductions can be achieved with more efficient appliances such as better water and space heaters, heat pumps, significant reduction of water demand by water conservation and other improvements. 30 to 40% energy can be produced by renewable sources, including heat recovery from used water or extracted from the ground and groundwater.

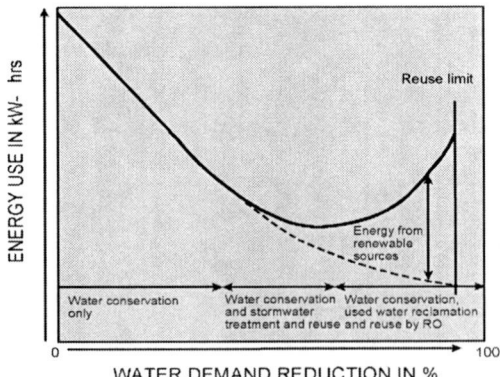

Figure 2 Relationship between water conservation, reuse and energy (Novotny, 2010)

Water and energy nexus. In general, people and professionals believe that implementing water conservation, reuse, and recycle also reduces GHG emissions and a fully closed water cycle is the most optimal solution to both water and carbon footprint problems. This may be true up to a certain limit. Figure 2 presents the possible relation of water demand reduction leading to a closed urban water cycle and energy. The plot suggests that there is a minimum beyond which further reduction of water use will increase energy demand. In the water conservation phase, energy use and GHG emission reduction of water demand by using more efficient appliances, xeriscape and plugging the leaks and losses does not require large amount of extra energy, hence, the energy use reduction is directly proportional to the reduction of the water demand. Several current ecocities are located or being planned in areas with meagre water resources which necessitates using desalinated, brackish and reused water. To further close the water cycle, energy demanding water reclamation processes are needed such as nanofiltration and reverse osmosis. Consequently, larger dependence on renewable zero carbon energy sources (wind, solar, geothermal, energy recovery from used water organic solids) will ensue. The recycle systems cannot be fully closed to prevent accumulation of nondegradable potentially harmful compounds that may pass the reverse osmosis and other high degree treatment processes.

ALTENAṬIVE PATHS TOWARDS NET ZERO GHG EMISSIONS

To achieve the green net-zero GHG and pollution emissions goals the following alternative options and measures in the ecocity developments and retrofits are Novotny *et al.*, (2010):

- Passive architectural features for heating and cooling
  ◊ Southern exposure with large windows equipped regulated by shutters
  ◊ Cross ventilation
  ◊ Green roofs
  ◊ A lot of insulation
  ◊ Energy efficient lighting
- Renewable energy sources (solar, wind, extracted from used water and stormwater)
- Water conservation and reuse, addressing the entire water (hydrologic) cycle within the development, including rainwater harvesting and storage
- Distributed stormwater and used (waste) water management to enable efficient used water and reuse and renewable energy production

- Xeriscape of the surroundings that reduces or eliminates irrigation and collects and stores runoff from precipitation
- Energy efficient appliances (e.g., water heaters), treatment (e.g., reverse osmosis) and machinery (e.g., pumps, aerators)
- Connecting to off-site renewable energy sources such as solar power plants and wind farms
- Organic solids management for energy recovery
- Connection to low or no GHG net emissions heat/cooling sources such as heat recovered from used water or from ground
- Smart metering of energy and water use and providing flexibility between the sources of water and energy
- Sensors and cyber infrastructure for smart real time control
- Restoration and maintaining integrity of urban water resources

SEVEN ECOCITIES CASE STUDY

The report by Novotny and Novotny (2009) analyzed seven ecocities and looked at common features of the key parameters such as population density, energy use and carbon imprint, water use and reuse, and cost We looked whether the water system is linear and centralized or closed and decentralized. The results are summarized in Table 2. A full list of sources is in the report.

Hammarby Sjőstad (Sweden)

Hammarby Sjostad, located on Lake Hammarby Sjö in Stockholm, has attracted international acclaim for the quality of habitat it created and convinced many that significcqnt reductuiin of GHG emssions does not require lifestyle changes. The development concept successfully connects the historic landscape with aquatic areas which act as storm water drainage, encourages biodiversity, creation of new habitats, informal amenity areas and formal areas of public open space. Sustainability is also enhanced through the use of green roofs, solar panels, and eco-friendly construction products. The use of glass as a core material maximizes sunlight and views of the water and green spaces. The city has a fully integrated underground sanitary (separated) waste collection system conveying wastewater to the local district treatment and heat recovery plant. The development has its own ecosystem, known as the Hammarby Model. In spite of the fact that the ecocity is still based on the linear model and extensive water reclamation and water reuse is not included, Hammarby Sjöstad is the first city built on ecological principles that broke the barrier toward sustainable urban development.

Dongtan (China)

Dongtan was planned to be at the eastern tip of Chongming Island at the mouth of the Yangtze River in the middle of a designated nature reserve with outstanding biodiversity about 40 km north of downtown Shanghai. The approach to the design of the city by the architectural/engineering firm, Arup, was different from other designs submitted by various other firms in the mode of "low impact" spread out subdivisions. Arup envisioned Dongtan to be a vibrant city with green 'corridors' of public space ensuring a high quality of life for residents. The city was designed to attract employment locally across all social and economic demographics in the hope that people will choose to live and work there. In this way, Arup's approach is pioneering and initiated the new paradigm of ecocity building that was then adopted by other ecocity developers. Although the city design is exceedingly water centric, the canals and lagoons within the development have mostly architectural functions of aesthetics, recreation and transportation.

The development of Dongtan that was supposed to have first 5000 inhabitants moved in during the 2010 World Exhibition in Shanghai but the project realization was indefinitely postponed.

Figure 3 Plan and view of the Ecoblock module (Source H. Fraker, UC Berkeley and ARUP)

Qingdao (China) Ecoblock and Ecocity

The ecoblock concepts were developed by the team of Dean Harrison Fraker of the University of California College of Environmental Design specifically for urban developments in China but could be used anywhere in the world, especially in the countries with a rapid population increase. *Ecoblock* is a city block much smaller than the super block. It is self sustained and semi - independent with its water and energy needs. It generates its own energy from renewable sources, harvests rainwater, produ-ces its own water and processes and reclaims its used water (wastewater).

A typical standardized ecoblock (Figure 3) has 600 units on 3.5 hectares and will house 1500 – 1800 residents. The Qingdao ecocity layout consist of 16 ecoblocks. Similarly to Dongtan, the Qindao ecocity is a concept with uncertain realization; however, the ecoblock concept has been also incorporated into the Tianjin new development (H. Fraker, personal communication).

Tianjin (China)

The site of the new city development is about 150 kilometres southeast of Beijing and 40 km from the historic Tianjin City which is the regional center and the largest port city in northeast China. The city will be a part of a huge regional development of the Tianjin – Binhai New Area. The city is a joint project of China and Singapore.

The city layout is divided into (eco) blocks. The smallest block unit has an area of 400 x 400 metres (16 ha). Dean Harrison Fraker testified that Tianjin would also include the Quindao ecoblocks. The city is water centric and will features an "Eco-valley" which is the main north–south green connector in the city which will retain a large ecological wetland set aside as a habitat for birds' migration, and preserve the former watercourses.

The primary sources of water for Tianjin are desalinated water and rainwater which constitute more than 50% of water used in the city. An extensive system of rainfall collection and sewage reuse will be established relying heavily on the landscape. The city will have a centralized treatment of sewage and wastewater treatment and recycling and will develop and utilize non-conventional water resources such as recycled water and desalted seawater in a water supply infrastructure that will reduce the need for conventional water resources. The system is heavily water centric.

Table 2 Summary table of seven ecocity evaluation

City	Population Total	Population Density #/ha	Water use L/cap-day	% water recycle	Water System	% Energy savings renewable	Green area m²/cap	Cost US$/ unit*
Hammarby Sjöstad	30,000	133	100	0	Linear	50	40	200,000
Dongtan	500,000 (80,000)++	160	200	43	Linear	100	100	~40,000
Qingdao	1500+	430 – 515	160	85	Closed loop	100	~15	?
Tianjin	350,000 (50,000)++	117	160	60	Partially closed	15	15	60,000 – 70,000
Masdar	50,000	135	160	80	Closed loop	100	<10	1 million
Treasure Island	13,500	170	264	25	Mostly Linear	60	75	550,000
Sonoma Valley	5,000	62	185	22	Linear	100	20	525,000

*Based on average 2.5 members per household.
+Qingdao ecoblock.
++Phase I.

Masdar (UAE)

Masdar is designed in a tradition of a typical historic Arabic city called "Medina" with a square layout, separated from the desert surroundings by a wall. The city will be built and was designed to follow the "One Planet Living" (OPL) ten principles. When the city is completed it will be home to 50,000 people and 40,000 people are expected to commute to work in the city. Water used inside the city will be provided by a desalinization plant run by solar power. The plant will produce two types of high quality water: one fit for drinking and the other fit for personal uses such as showering and washing dishes. The desalinization is expected to be 80% more efficient than other plants built elsewhere.

Masdar city will use a plethora of water management principles in order to treat all parts of the water cycle and use them as a water source. As many as nine water conveyance systems will be used in 12 different ways and treat used water at three treatment levels. The variety of water sources to be used include groundwater, seawater, surface runoff, rainwater harvesting, dew/fog capture, grey and black water reuse and resource recovery from urine streams.

Treasure Island (California, USA)

Treasure Island is a manmade island built by dredging sediments from the San Francisco Bay. The water/wastewater system is linear with some reuse after treatment in a central WWTP for irrigation. 75% of the effluent will be discharged into the bay. Potable water will be imported from San Francisco municipal grid. Storm water management will center on xeriscape, permeable surfaces and pavements, green roofs and routing excess runoff to be treated in a wetland. Once the excess runoff is collected it will be routed to a constructed treatment wetland and water reused for irrigation and other nonpotable uses.

Sonoma Mountain Village

Sonoma Mountain Village with its 5000 inhabitants is the smallest of the seven ecocities. It has applied for OPL certification. The goal for water used within the village is a reduction in water consumption by 60% from a general norm for single family homes in the region. This will be accomplished through water reduction devises, education, rainwater harvesting and reuse of water. The municipal drinking water supply will be used inside of all building and irrigation in private backyards. Reclaimed water will be used for irrigation of all public parks, medians, and street trees along with high efficiency (sub-drip) irrigation of all common areas, private front yards and for use in fire. Storm water reuse will be used for habitat maintenance, groundwater recharge and as a supplemental irrigation supply for all landscape areas. There will be habitat protected bioswales acting as wetlands connected to an underground reservoir from which water will be recycled for irrigation purposes.

Synthesis of the seven cities study

The summary of the key synthesis parameters is in Table 2.

Population Density. With the exceptions of Sonoma Valley Village, the smallest development, and Qingdao ecoblock, the development with the highest population density, the density of the remaining five developments varied between 117 to 170 people/ha. From the presentations and literature findings it was evident all design teams used some kind of a proprietary model which balances the population and its energy use based on probability of walking and biking instead driving, energy insulation of buildings and exposure to sun, renewable energy sources and other determinants for GHG emissions from urban areas. Three sites, Dongtan, Tianjin, and Treasure Island were designed by Arup teams. Literature indicates low density "American style" suburban areas with one oversized house on 0.4 ha (1 acre) land are the most wasteful regarding energy use and efficiency (Newman, 2006). The fact of medium design density development being the most optimal refutes, to some degree, the utility of the "low impact" subdivisions which in most cases have an objective of minimizing stormwater impacts and discharges and generally results in low density developments.

Green House Gas Emissions (carbon footprint). Dongtan, Qingdao, Masdar and Sonoma Valley designs are proving ecocities can fulfil the OPL criterion of zero GHG emissions from infrastructure heating and cooling, electricity consumption and traffic.

Water Reclamation and Reuse. All cities use the latest technology for in-house water savings such as low flush toilets, showers, etc. Hammarby Sjöstad is almost a 100 % linear system with recovery of phosphorus. Stockholm is water rich and there is apparently no need for recycle yet they expect to reduce the per capita water use to the "magic" limit of 100 L/capita-day. All other cities use various degrees of water reclamation and reuse but start with a higher per capita water use reduced by reclamation of used water and stormwater.

A high density Qingdao ecoblock with 430–515 people/ha appears to be an anomaly which should be further researched as to the feasibility and sustainability of the concept regarding the used water reclamation. For one, locating free surface wetlands that are supposed to provide treatment to partially treated black wastewater next to the high-rise buildings may not be acceptable in many countries because of health concerns. Qingdao's treatment of black water consists of "sequential batch reactors" described in a promotional video (Green dragon, 2008) as septic tanks, followed by wetland treatment. Because of health reasons, the acceptable wetland type would have to have a fully submerged flow. Based on the WEF (2001) manual the minimum area of the wetland serving 1500–1800 people will have to be about ½ hectare or one football field and could not be accessible as a park. Also the wetlands will have a relatively large evapotranspiration during dry summer days which might not have been included in the water balance analyses, especially in Qingdao. Constructed wetlands also emit large quantities of GHGs methane, nitrous oxide and carbon dioxide (Sovak *et al.*, 2006).

Surface drainage for runoff and clean water. All ecocities use surface drainage for collecting urban runoff and clean water inputs and will use extensively best management practices for urban runoff such as pervious pavements for infiltration, capture and storage in underground basins, and reuse for various purposes such as irrigation, fire protection, and some plan to tap into the groundwater resources for reclaimed water. All cities are planning reuse of the captured stormwater for irrigation and in some cities for reuse in the community for nonpotable water supply. Use of green roofs has not been planned on a large scale with exception of Hammarby Sjöstad.

Water Centric Development Opportunities. Hammarby Sjöstad, Dongtan, and Tianjin are clearly water centric whereby water and canals are the architectural centerpieces of the development and will have an aesthetic role, provide recreation and local transportation. By locating their advanced wastewater treatment plant at the fringe of the city and directly discharging their treated used water into the Hammarby Lake connected to the Stockholm Bay without water reclamation, the city has missed its opportunity for water reuse. The other two cities in China considered using the water bodies inside of the city for discharge and treatment of reclaimed water and reuse in their mostly recreated canals. The desert city Masdar will apparently create small artificial streams transecting the city. It is not clear whether or not these streams in Masdar will be used for conveyance of reclaimed used water. Qingdao, Sonoma Mountain Village, and Treasure Island will not have permanent streams, natural or artificial, planned within the ecocity boundary. Sonoma Valley Village is planning to create habitat bioswales with wetlands for stormwater conveyance transecting the village and connected to a storage basin from which water will be reused. Qingdao created two conveyance systems for reuse: one for the reclaimed black water via a chain of wetlands, the other for stormwater both ending in an underground storage facility, followed by reuse. The architectural rendering of the Qingdao ecoblock does not show any surface stormwater conveyance to the central storage basin.

CURRENT ECOCITY TECHNOLOGIES OF REDUCING CARBON FOOTPRINT

Planners of water frugal ecocities in Qingdao (China) and Masdar (UAE) consider a fully closed water loop similar to that shown on Figure 4. The original Qingdao double loop (Fraker, 2008) was modified to avoid direct potable reuse. The numbers on the plot represent daily water use in Liters/person-day living in the cluster of the ecocity. The Qingdao ecocity cluster has about 1500 to 2000 inhabitants living in several high rise and medium height buildings. The figure shows the total water use in the cluster as 130 Liters/capita-day but the municipal grid supplies only 50 Liters/capita-day. It is assumed maximum water saving practices are implemented in the cluster ecoblock. The water reclamation and reuse is carried in a double loop consisting of black and grey water reclamation and reuse. Black water flow containing water from toilets, kitchen sinks and dishwashers will be treated by a solids separation unit and an anaerobic treatment and energy recovery reactor (digester) from which methane can be recovered, followed by a subsurface flow wetland. While the constructed submerged wetland treatment does not use energy (it produces vegetation residues for digestions) it emits methane and nitrous oxide as well as smaller amounts of carbon dioxide. However, these emissions are less than those emitted by conventional treatment plants (Sovik et al., 2006).

The concentrated flow from the solids recovery reactors (primary settling and membrane filter) and a portion of the black water stream could be diverted to a central integrated resource recovery facility which will produce energy in a form of biogas, hydrogen and electricity, fertilizer in a form of struvite, heat energy, and reusable organic solids (see the next section).

In addition to providing water to inhabitants, the double loop system also could provide some ecological flow to the surface water bodies within the ecocity and garden irrigation. It can be seen that 50 L/capita-day water input from the municipal grid is not sufficient to sustain the total demand of 140 L/cap–day within the ecoblock during dry weather. Rainwater harvesting and stormwater capture and infiltration (via pervious pavements and infiltration raingardens) are needed to supplement the dry weather flow (Novotny and Novotny, 2009). Hence, one can consider the 50 L/cap-day as the minimum inflow from the grid and 140 L/cap-day as the optimal water demand after implementing a suite of water conservation measures.

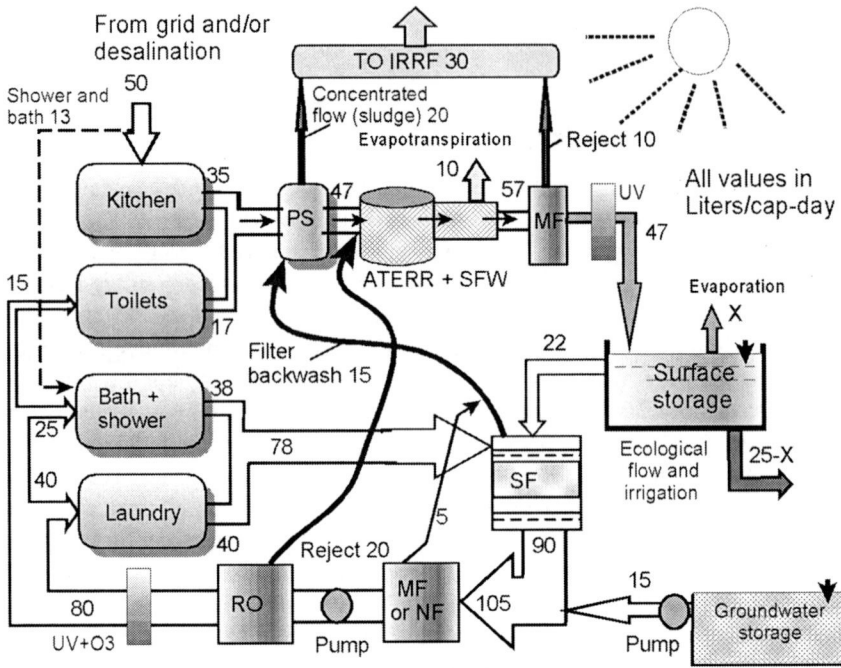

Figure 4 Mass balance of flows in L/cap-day for the system in a closed water cycle ecocity on a dry day. Legend: ATERR – anaerobic treatment and energy recovery reactor; SFW-H horizontal submerged flow wetland; IRRF – integrated resource recovery facility; SFW-V – Submerged flow wetland - vertical flow; HWT – reclaimed hot water tank; ST- storage tank; PS – primary settler with solid removal; MF-membrane filter; SF-sand- filter; NF-nanofilter; RO-reverse osmosis; UV-ultraviolet disinfection; O_3 – ozone addition

The Qingdao ecoblock also saves energy by passive heating and cooling, producing energy by solar panels, voltaics, and wind turbines. It will could also produce biogas from digested sludge and organic solids harvested from the wetland, fallen leaves and gardens. In the overall scheme, the planners claim the ecoblock to have a net zero carbon emission footprint.

NEW AND OLD TECHNOLOGIES OF ENERGY RECOVERY FROM WATER

Used water and solids treated in the water reclamation process are a resource. Energy in the form of heat and electricity can be produced and nutrients and soil conditioning solids can be recovered. The technologies are known and have been practiced for decades. However, since the beginning of this century a new look at resource recovery is emerging, driven by the goal of reducing GHG emissions, water shortages and discoveries (or rediscoveries) of new technologies that could revolutionize the reuse, recovery and management of used water. These technologies will produce energy without adding green house gasses to the atmosphere, recover ammonium and phosphate, and heat (and cool) communities and industries, including those providing water and reclaiming used water for reuse. This is another component of urban sustainability and net-zero carbon footprint efforts.

In general, energy recovery from urban water/stormwater, used water systems can be from

- Heat contained in used water (e.g., Hammarby Sjöstad) by heat pumps
  ◊ Water to water (hot water heating)
  ◊ Water to air (space heating)
- Biogas produced by anaerobic treatment processes. Biogas can be either used as a fuel in a combustion process or converted to hydrogen and electric energy by a hydrogen fuel cell. The process unites are
  ◊ Upflow Anaerobic Sludge Blanket (UASB) reactor (Lettinga *et al.*, 1980)
  ◊ Anaerobic sludge and organic solids digester.
- Converting biogas into hydrogen and subsequently into electricity by hydrogen fuel cell.
- Hydraulic energy capture of water and used water flow and pressure
  ◊ Energy in drop shaft manhole in steep terrains
  ◊ Energy released in pressure valve of pressurized water supply distribution systems
  ◊ Energy in pressurized reject water of reverse osmosis systems
- Wind and solar energy capture by the water utilities and conversion of excess energy into hydrogen by electrolysis of water. Hydrogen can be then converted back to electricity by hydrogen fuel cell or provided as fuel for vehicles.
- Electricity and/or hydrogen harvesting by microbial cells from decomposing organic matter in microbial fuel cells.

Anaerobic digestion (fermentation) of biodegradable materials such as sewage sludge, manure, strong wastewater (leachate from landfills, wastewater from high density operations), green waste from gardens, energy crops and agricultural residues, grass, algae, etc.) is the oldest energy from waste conversion process (besides burning). The biogas product of anaerobic fermentation is methane and carbon dioxide. This type of process and biogas production occurs in digesters or landfills and in nature in wetlands (swamp gas) and anoxic sediments. The final product is relatively inert organic humus type residue and a liquid concentrate that contains very high concentrations of dissolved organic solids (BOD/COD) and recoverable high concentrations of ammonium and phosphates that can be precipitated into struvite. Figure 5 shows digesters in the Changi treatment plant in Singapore that produces biogas for heating the digesters and plant buildings.

Figure 5 Sludge digesters in Changi water reclamation plant in Singapore designed by CH2M-Hill (photo V. Novotny)

The fundamentals of the three step anaerobic digestion/fermentation are well covered in many professional textbooks. Gas produced by gasification or biogas or natural gas can be converted into mechanical energy and then electricity using an internal combustion engine which runs a generator producing electricity. The efficiency of the internal combustion engine is relatively low, about 42 to 45%. Hence most of the energy is converted into heat that is normally wasted but could be recovered. All organic carbon in the combustion energy producing processes is converted into carbon dioxide which is green house gas (GHG) pollutant.

In regional or cluster (ecoblock) integrated resource reclamation systems biogas production units accepting both sludge and other organic residues can produce significant amount of energy for the community in a form of biogas and/ or electricity. The biogas after treatment and in-situ purification can be added to the natural gas grid or provided as gas fuel to public transportation or converted into electricity and sold to the electric utility grid. The methane content of biogas is 65% as compared to 97% methane content of natural gas (fossil fuel). Other gases present in appreciable quantities in the biogas are carbon dioxide, hydrogen sulphide and nitrogen. The energy content of biogas is about 6 kW-hrs/m^3.

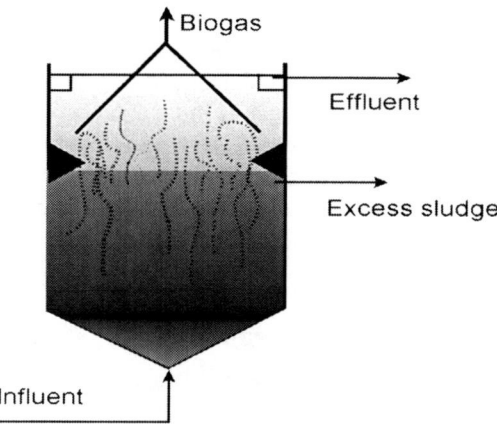

Figure 6 A variant of an upflow anaerobic sludge blanket (UASB) reactor

Upflow anaerobic sludge blanket (UASB) reactors (Figure 6) treat a variety of domestic and industrial wastes of different strength, from typical sewage to highly concentrated industrial organic wastewater (brewery, sugar refineries) with high input COD values. While anaerobic digesters work best in the mezophilic (35 – 45°C) range, the UASB reactors work well at temperatures of 20° C and up. The UASB process was introduced more than thirty years ago (Lettinga *et al.*, 1980) and by the end of the last millennium, UASB technology has become the most popular anaerobic treatment process. Methane production is 0.3 – 0.4 m³ of methane per one kilo-gram of COD removed in the reactor (Tchobanoglous *et al.*, 2003).

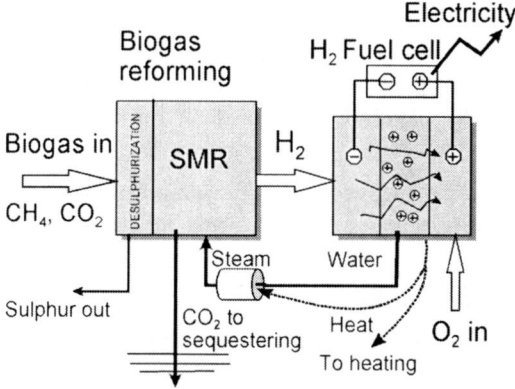

Figure 7 Concept of the hydrogen fuel cell (US Department of Energy)

Hydrogen fuel cells enable conversion of biogas intro electricity without combustion and, potentially, without carbon emission. Hydrogen is generated from water (Trafton, 2008; Kanan and Nocera, 2008) or from hydrocarbons (natural gas, methyl alcohol, or biogas (Figure 7). Conversion of biogas into electricity is in two steps. (1) Biogas is converted by fuel reforming process into hydrogen and carbon monoxide, which is then converted by steam methane reforming (SNR) gas shift process into more hydrogen and carbon dioxide; and (2) hydrogen the in the fuel cell is combined with oxygen to produce electricity, heat and water.

Integrated resource recovery facility

A variant of the Natural Biological Mineralization Concept proposed by Lettinga (2009) for biogas production can be the fundamental block of the integrated resource recovery facility (IRRF) which could include (Novotny *et al.*, 2010)

- an upflow anaerobic sludge blanket (UASB) reactor unit;
- a pre – digester decomposing biomass other than sludge (decaying vegetation, food waste, woodchips) to acetates with suppressed methane fermentation. Hydrogen from the pre-acidification reactor from biomass could be conveyed directly to the fuel cell to produce electricity;
- Residual solids dewatering such as belt press filter. Solids can be used as soil conditioner. Low levels of nutrients incorporated in microbial biomass remain in the dry solids. Most of the nutrients are dissolved in the supernatant.
- Fluidized bed reactor for struvite production. Magnesium is added to the influent to precipitate struvite and increase pH.
- Polishing treatment units such as a trickling and membrane filter.
- Disinfection by UV
- Hydrogen fuel cell unit to produce electricity and heat. Carbon dioxide can be used to adjust pH of the effluent from the fluidized bed reactor.
- Concentrated solar panels to provide heat for digestion and SMR biogas reforming.

The integrate resource recovery facility would produce biosolids, biogas, hydrogen, electricity, heat and, above all, water for reuse or ecological flow.

OVERALL ENERGY OUTLOOK

The average carbon footprint of the larger urban areas in the US is about 8.5 tons CO_2 equivalent/persons – year. Gleaser and Kahn (2008) statistical data averages of GHG emissions for 100 larger US cities shown in Table 3 separate household energy use into transportation by vehicles, public transportation, household heating and use of electricity. Heating is by natural gas or oil. Because data are from before 2005, hybrid or electric cars were not available and were not considered in these estomates.

Table 3 Average statistics of energy use in 100 large US cities (recalculated and modified from Gleaser and Kahn, (2008)

Energy use for	CO_2 emissions in tons/cap-year	% of total
Transportation by cars	4.091	47.0
Public transportation	0.388	4.4
Home heating by gas or oil	1.470	17.0
House electricity including that for cooling	2.751	31.6
Total	8.71	100

A look into future 20 or more years ahead

In 2007 the carbon emissions trend in the US, after 50 years of increasing, reversed and started to decrease 9% in 2007 alone, continuing to decrease in 2008 and 2009 (Brown, 2009). The article by Earth Policy Institute noted a number of near future trends that raise optimism:

- automobile fuel-economy standards are becoming progressively better;
- higher appliance efficiency standards (Energy Star program);
- financial incentives supporting large-scale development of wind, solar, and geothermal energy;
- public awareness and movement to green lifestyle of climate-conscious, cost-cutting Americans who are altering their lifestyles to reduce energy use;
- federal government—the largest US energy consumer, is reducing vehicle fleet fuel use 30 % by 2020, recycling at least 50% of waste by 2015, and buying environmentally responsible products;
- virtual de facto moratorium on new coal plants; many plants are expected to close and be replaced by wind farms, natural gas plants, wood chip plants, or efficiency gains. Shifting to wind, solar, and geothermal energy drops carbon footprint to zero.
- US solar cell installations are growing at 40% a year;
- oil use and imports are both declining as the new fuel economy standards raise the fuel efficiency of new cars 42 % and light trucks 25 % by 2016;
- 42 % of diesel fuel burned in the rail freight sector is used to haul coal; hence, falling coal use means falling diesel fuel use; and

- federal government provides incentives to development train networks run by electricity instead of diesel fuel.

Assumptions for the Future Cities and Retrofits

Electric energy production

Electric energy production from sources that do not emit GHGs (nuclear, hydropower, and renewable wind, solar) will be increased from 30 to 60%. This will reduce GHG emissions attributable to power production by 43% from 0.62 kg CO_2/kW-hr in early 2000s to 0.35 kg CO_2/kW-hr which is anticipated to be attainable in 2030. Most of the reduction would be realized by phasing out coal power plants and replacing them by renewable energy. This estimate is on a conservative side because US DOE claims just wind energy potential will reach at least 20 % of the total energy production in the US. Jacobson and Delucchi (2009) in *Scientific American* presented a realistic plan by which all electric power would be produced by renewable sources that would include hydro, solar and wind.

Vehicular traffic assumptions

The majority of cars will be electric plug-ins or hybrids which will double mileage and cut the GHG emissions from automobiles by half. It is anticipated that with better public transportation, less miles (kilometers) will be driven by drivers. Anticipated GHG emissions will be cut by 60% or to about 1.5 GHG tons/cap-year.

Public transportation

Public transportation will be mostly by electric trains, light rail; and buses powered by biogas, electricity or green hydrogen. Because most buses will be powered by green fuels, GHG emissions from public transportation should be dramatically reduced to less than 0.1 GHG tons/cap-year.

Heating

GHG emissions from heating buildings can be reduced by 50 % to 60% by switching to heat pumps deriving heat from air or, better, from used water or the ground. Passive energy house measures (green roofs; good insulation, southern exposure) can also dramatically reduce heating needs. By comparing heating needs in Florida (Tampa, Orlando, New Orleans, Phoenix) and northern cities (New York, Boston, Chicago, Portland) the difference in GHG emissions from heating represents about 2 tons of GHGs. On average, heating GHG emission can be less than 0.3 ton GHG tons/cap-year, in northern climatic conditions less than 1 GHG to/cap-year.

Electricity demand

NSTC (2008) estimated 60–70% household electric energy use reduction can be achieved by more efficient appliances, heated water savings, water savings by conservation, reuse of rainwater and reduction of cooling energy demand by cross-ventilation, green roofs, insulation and shading. Furthermore, the electric energy production from renewable sources based on the assumption given above would increase from 30% to 60% (still less than France and several European countries today) which will reduce the carbon footprint or power production to 0.35 kg CO_2/kW-hr. These measures would bring the GHG equivalent urban emissions by the use of electricity from 4.45 tons/cap-year to about 1.0 ton/cap-day.

In overall, these reductions of the total GHG emission in US cities could be reduced to about 3 tons GHG equivalent/cap-year, which is slightly more than today in Barcelona (Spain).

Water related energy savings (from Novotny, Ahern, and Brown, 2010)

The energy equivalent of one cubic meter of *water service* (providing, transporting, treating and disposal) of 2.26 kW-hr/m³ with corresponding carbon emission was estimated as 1.37 kg CO_2/m³. Reducing water demand by water conservation from 0.5m³/cap-day to 0.2m³/cap-day and making the same assumption of the reduction of GHG emissions by the power industry could bring the CO_2 reduction of 0.54 kg CO_2/cap-day or 0.2 ton CO_2/cap-year.

The contribution of **the integrated resources facility** to reduction of energy use and improving carbon footprint was estimated as 0.79 kW-hrs/cap-day (Novotny *et al.*, 2010). Assuming the future carbon/energy conversion as 0.35 kg CO_2/kW-hr this would represent a reduction of the carbon footprint by 0.27 kg CO_2/cap-day or 0.1 ton CO_2/cap-year.

The effect of *extracting heat from used water* was estimated as 10.47 kW-hr/m³. The future per capita water use is conservatively estimated as 0.2 m³/cap-day; hence, the heat extracted from used water will have energy value of 2.1 kW-hrs/capita-day and the reduction of the carbon footprint by 0.73 kg CO_2/cap-day or 0.268 ton CO_2/cap-year.

Additional energy savings can be achieved by bringing surface runoff drainage to the surface which would eliminate pumping from deep interceptions and by lift stations. Conservatively, such pumping and miscellaneous measures can save another 0.3 ton CO_2/cap-year.

In total, one can expect minimum reduction of carbon footprint by urban water sector as almost 1 ton CO_2/cap-year and the total urban carbon footprint to be reduced to 2 ton CO_2/cap-year.

Finding savings of 2 ton CO_2/cap-year

The 2 ton CO_2/cap-year energy gap requires, with the future energy/carbon conversion of 0.35 kg CO_2/kW-hr, finding 15.6 kW-hr/capita-day additional savings or new renewable sources of energy. The above look into the future does not consider cluster wide and other local energy sources in the future ecocities. The additional energy savings and new sources of energy will come from:

1. More far reaching reduction of the carbon emissions by private automobiles. The ecocities of the today and future drastically reduce driving by a switch to walking, biking and public transportation. Public transportation stops should be within 15 minutes walking distance from every home. Reducing driving of an average driver from 20,000 km/year (12,437 mi/year) to 10,000 km/year (6,220 mi/year) and increasing mileage from 8.25 km/Liter (20 mi/gal) of gasoline to 21 km/liter (50 mi/gal) would reduce GHG emissions to 0.8 tons CO_2/cap-year or additional savings of 0.7 tons CO_2/cap-year.

2. Including photovoltaics and solar panels on every house which, even under the current technology, would eliminate most of the house electric energy needs. In warmer sunny climatic conditions with the solar radiation rate or 4 kW-hr/m^2 and panel efficiency of 40% (double the current efficiency) a panel array on the roof of a family building with a size of 5 m^2 would provide 8 kW-hrs of electricity per day on average. This would be equivalent to 1.0 tons CO_2/cap-year savings, assuming the 0.35 kg of CO_2/kW-hr conversion. The same panel would provide all the electricity needed for one person.

3. Installing small wind turbines through out the city. A small 1 kW turbine operating 50% of time could provide 12 kW-hrs/day or 4380 kW-hrs/year, which would provide enough renewable energy to reduce GHG emissions by 1.5 tons CO_2/year in 2030.

4. Energy recovery by burning flammable refuse and dried sludge solids in incinerators without carbon sequestering is carbon neutral.

CONCLUSION

Reaching the net zero carbon footprint is realistic even in the high energy consumption countries like the US. This will require a change in the paradigm of how the cities are built and their energy, transportation, and building infrastructures are developed, restored and retrofitted. Most of the technologies have been developed, are available today or in the near future and, using the triple bottom line assessment, are more beneficial and efficient than the current unsustainable paradigm.

REFERENCES

Brown L. R. (2009). U.S. Headed for a Massive Decline in Carbon Emissions, Plan B Update, Earth Policy Institute. http://www.earthpolicy.org/index.php?/plan_b_updates/2009/ (accessed October 2009).

Dodman D. (2009). Blaming cities for climate change? An analysis of urban greenhouse gas emissions inventories. *Environment and Urbanization*, **21**(1), 185–2001.

Fraker H., Jr. (2008). The Ecoblock-China Sustainable Neighbourhood Project, Power point presentation – Connected Urban Development Conference, September 24, 2008, Amsterdam. http://bie.berkeley.edu/ecoblocks.

Glaeser E. L. and M. E. Kahn (2008). *The Greenness of Cities: Carbon Dioxide Emissions and Urban Development*. Working Paper 14238, National Bureau of Economic Research, Cambridge, MA. http://www.nber.org/papers/w14238 (accessed March 2009).

Green Dragon (2008). http://www.greendragonfilm.com/qingdao_ecoblock_project.html (accessed March 2009).

Jacobson M. Z. and M. A. Delucchi (2009). A path to sustainable energy by 2030. *Scientific American*, **301**(5), 58–65.

Kanan M. W. and D. G. Nocera (2008). In situ formation of an oxygen – evolving catalyst in neutral water containing phosphate and Co2+. *Science*, **321**(5892), 1072–1076.

Lettinga G., A. F. M. Van Velsen, S. W. Hobma and W. de Zeeuw (1980). Use of upflow sludge blanket (USB reactor for biological wastewater treatment, especially for anaerobic treatment. *Biotechnology and Bioengineering*, **22**, 699–734.

National Science & Technology Council (2008). *Federal Research and Development Agenda for Net-Zero Energy, High Performance Green Buildings*, Committee on Technology, Office of the President of the United States, Washington, DC.

Newman P. (2006). The environmental impact of cities. *Environment and Urbanization*, **18**(2), 275–295.

Novotny V. (2010). *Urban Water and Energy Use: From Current US Use to Cities of the Future*, Proc. International Conference Cities of the Future, Water Environment Federation, Cambridge, MA, March 7–10, 2010.

Novotny V. and E. V. Novotny (2009). Ecocities – Evaluation and Synthesis. http://www.coe.neu.edu/environment.

Novotny V., J. F. Ahern, and P. R. Brown (2010). *Water Centric Sustainable Communities: Planning, Retrofitting and Constructing the Next Urban Environments*. J. Wiley & Sons, Hoboken, NJ.

Register R. (1987). *Ecocity Berkeley: Building Cities for a Healthy Future*. North Atlantic Books. ISBN 1556430094.

Sovik A. K., J., Augustin, K. Heikkinen, J. T. Huttunen, J. M. Necki, S. M. Karjalainen, B. Kløve, A. Liikanen, U. Mander, M. Puustinen, S. Teiter and P. Wachniew (2006). Emissions of greenhouse gases nitrous oxide and methane from constructed wetlands in Europe. *Journal of Environmental Quality*, 35(6), 2360–2375.

Tchobanoglous G., F. Burton and D. Steusel (2003). *Metcalf and Eddy, Wastewater Engineering: Treatment and Reuse,* 4th edn, McGraw – Hill, New York.
Trafton, A. (2008) 'Major discovery' from MIT primed to unleash solar revolution. *MIT News,* July 2008. http://web.mit.edu/newsoffice/2008/oxygen-0731.html?tmpl= component&print=1 (accessed September 2009).
Water Environment Federation (2001) *Natural Systems for Wastewater Treatment.* Manual of Practice FD-16, Alexandria, VA.
Wikipedia (2009) List of Countries by Carbon Dioxide Emissions Per Capita. http:// wikipedia.org/wiki/List_of_countries-by_carbon_dioxide_emissions_per_capita (accessed October 2009).

Closed-Loop Water and Energy Systems: Implementing Nature's Design in Cities of the Future

Wm. Patrick Lucey, Cori L. Barraclough and Sarah E. Buchanan

390 7th Ave., Kimberley, B.C., Canada, V1A 2Z7
E-mail: aqua-tex@islandnet.com

Abstract During the latter half of the 20th century western-influenced countries began to develop and manage municipal water and energy infrastructure in response to public health concerns and sanitation issues. This effected broad increases in living standards, coupled with reductions in infant mortality and enhanced longevity. The 21st Century has seen an insatiable demand to expand the West's social vision, for an urban-based living standard, to hundreds of millions in Brazil, Russia, India and China. This accelerated movement of humanity into new towns and cities will require substantial new exploitation of ecosystem services and Natural Capital, including petro-chemical resources, whose extant globally disruptive characteristics – ecological, social and economic – are not sustainable. The achievement of sustainable and healthy cities will require designing water and waste systems in accordance with the ecological principle of integration and the fundamental notion that nature has no wastes (liabilities), only resources (values). An understanding of the fundamental links between water, energy, climate change and the feedback loops that drive them, is essential to accelerate the adoption of policies that support smart, clean and green cities. A new urban design model, using engineered ecology principles and valuation economics, based upon a profitable business case, has been shown to be regenerative and adaptive, whilst creating healthy, livable and functional landscapes.

Keywords Closed-loop, cities of the future, water-energy nexus, engineered ecology, integrated resource management, decentralized, resilience, freshwater, wastewater, Dockside Green, South East False Creek.

INTEGRATING AND CLOSING LOOPS

Nature is not random, but highly organized and purposeful. Each component of an ecosystem fulfills a mosaic of roles in the community. Making each component multi-functional maximizes efficiency of the system while ensuring that there are a number of overlapping niches to create resiliency. This same principle of multi-functionality applies to infrastructure and urban design.

Riparian vegetation, for example, provides a number of structural and functional services. The roots of riparian vegetation hold soil in place, preventing erosion, and purify water by taking up nutrients and trapping suspended sediments. Reduced nutrients and turbidity have positive effects on water quality and trapped sediments contribute to the building of stream banks and soils. The plants themselves, both above and below ground, provide habitat for other organisms and facilitate biodiversity. Riparian vegetation slows the velocity of stream flow, thus storing water and maximizing the contact time for biofiltration. Finally, riparian vegetation is vital to temperature moderation, shading waterways from the sun and cooling the air through evapotranspiration (Postel 2008; Parkes and Horwitz 2009; Akbari *et al.* 2001).

Nature does not have any wastes, but instead utilizes a completely closed-loop system. Nature recognizes wastes as valuable resources and the recovery of energy and nutrients is essential to sustaining successive generations. In a forest ecosystem, the death of an old giant signifies new life for a multitude of organisms. A fallen tree is not a waste site, but a bustling community of resource recovery specialists. There are armies of opportunists who have been waiting for such a chance and immediately they begin to stake their claim on the newly available resources.

It is the biosphere that maintains homeostasis for the planet. "Nature depends on connections through different levels of biological organization . . . no organism is an island unto itself" (Todd and Todd 1993). Optimal ecosystem function is achieved through synergy, where the combined effect resulting from the interaction of components is greater than the sum of the individual effects and cannot be achieved if the components act independently. Ultimately, sustainable design can only be achieved through the emulation and integration of living systems, working with natural processes and restoring ecosystem function within our built environments by adopting an 'engineering ecology' design process.

Water

"Water is a finite resource which nature recycles constantly, yet modern cities are designed to use water only once before it is considered wastewater" (Lucey 2009). Water supply issues do not necessarily stem from a lack of water, but a failure to properly manage available water resources. Nature does not differentiate between wastewater, groundwater, stormwater, rainwater, greywater or drinking water, but conventional urban water management treats them as completely separate entities.

Urban water supply systems require an extensive amount of infrastructure to harvest and deliver treated potable water to users where it fulfills all urban water needs. The water is used once, sent to a wastewater facility and then treated for release back into the environment. Conventional urban developments have a large proportion of impervious surfaces that prevents infiltration and generates large volumes of stormwater. Conventional management of stormwater has been to remove the excess water from urban areas as quickly as possible by funnelling the water into underground pipes and releasing it, untreated, directly into nearby waterways. This significantly compromises the health of receiving aquatic habitats by introducing pollutants and eroding stream channels with excessive flows.

Closing the loop in our urban water cycle keeps more water on the land and allows water to infiltrate the soil and recharge groundwater. Capturing rainwater, and using reclaimed water for non-potable applications, significantly reduces consumption of treated drinking water and reduces the volume of water entering sewage treatment facilities, thus decreasing water treatment, transport and energy costs. Better water management practices contribute to strengthening water security by reducing overall consumption and increasing community self-sufficiency.

Implementation of sustainable water infrastructure creates an opportunity to greatly affect the health and well-being of people living within a community. Maintaining, creating, and promoting functional landscapes can result in healthy urban environments, provide an opportunity to celebrate natural systems, and promote education.

HEALTHY URBAN ENVIRONMENTS

As the character of the world changes and more and more people become concentrated in cities and urban areas, it is increasingly important that these cities support and promote the health of the population living within them. Clean, plentiful water is a key element of a healthy city. By recycling water of a lower quality for non-potable uses, the demand on potable water sources, for activities other than drinking or cooking, can be significantly reduced. Two examples of this approach are seen at the Dockside Green development in Victoria, B.C. and at the Vancouver 2010 Olympic Village at South East False Creek.

Dockside Green has rejected conventional water management practices by adopting a 'fit-for-purpose' approach to water use. Water is first used for potable purposes, and then reused as many times as possible for non-potable purposes. Wastewater for the entire development is collected and treated onsite using a membrane bioreactor to a level of treatment fit for unrestricted public access. This reclaimed water is reused directly onsite for suitable, non-potable uses including toilet flushing, irrigation of green roofs and balcony planters, and augmentation of an ecologically functional constructed watercourse. The Dockside Green site was designed to mimic the natural hydrologic cycle and rainwater is intercepted, stored and used onsite. Rainwater is captured and filtered by green roofs, bioswales and cisterns, and stored in the waterway for final polishing. Annually, the volume of reclaimed water that is collected and reused is equivalent to one day's consumption for the entire Greater Victoria region (pop. 300,000) on the driest day of the year. Consequently, recycling water in one development has created a benefit to the entire region. Similarly, at the Vancouver Olympic Village at Southeast False Creek, when rainwater is plentiful during the winter months, it is used to flush toilets thus allowing potable water to remain in the reservoirs for use in the dry summer months by both the Village and the rest of the region. We term this 'water banking' (The Challenge Series 2009a).

Public green spaces used for on-site water management, can also provide contemplative and recreational areas that further urban health benefits. For example, green space provides improved neighbourhood walkability associated with lower body mass index, a reduction in the severity of Attention Deficit Disorder (ADD) symptoms, and a reduction in violence and stress levels (Smith *et al.* 2008; Taylor *et al.* 2001; Sullivan and Kuo 1996).

Sustainable approaches to water infrastructure protect water sources for human populations as well as ecosystem functions and organism requirements, which also ultimately support healthy urban communities.

CELEBRATING NATURAL SYSTEMS THROUGH DESIGN

Celebrating natural systems and the functions they provide the human community creates understanding, appreciation, and ownership of these natural systems. One of the best ways to promote appreciation is to make natural infrastructure, such as creeks and wetlands, visible to the public eye.

Dockside Green

Making water infrastructure visible has proven to be successful in many locations. For example, at Dockside Green, a man-made creek which uses reclaimed water and rainwater inputs is now a centre-piece of the development available to the public as well as private home owners. This creek acts as a barrier between public and private spaces, provides an educational opportunity, a play place for children, and habitat for local wildlife including ducks, crayfish, three-spine stickleback, Great Blue herons and river otters (Figure 1).

Celebrating natural systems by making them visible in the community also promotes and facilitates education. This education can be informal such as taking a walk and observing the natural places, or the education can be made more formal through the provision of tours, information plaques and kiosks. This information allows the public to take notice of features and functions that they may have previously overlooked. At Dockside Green, information plaques line the greenway and interactive models of the reclaimed water system can be found inside the buildings. Most importantly, the wastewater treatment plant is not at the margins of the development or the city– it is located centrally within the development and adjacent to an outdoor public patio that is a popular year-round space. This provides an excellent educational opportunity against the typical "NIMBYism" which plagues many projects wishing to treat waste and reuse resources on site.

Figure 1 Artist's rendering of Dockside Green (top) (Dockside Green, 2005) and a 2009 photo showing the actual development (bottom).

South East False Creek

The Vancouver 2010 Olympic Village at South East False Creek (SEFC) is an innovative development built on a brownfield site historically contaminated by heavy industry. Founded on the four pillars of sustainable community building, SEFC was designed to "acknowledge social, economic, and cultural values alongside a deep respect for the environment" (The Challenge Series 2009b). Managing water, both freshwater and marine, as a valuable resource and reduction of potable water use were fundamental principles that guided the design process. The privately-developed portion of the project is named "Millennium Water", which serves as a reminder of the principles of rainwater management, domestic water management, ecosystem function and ecosystem regeneration (particularly on the marine foreshore). The following key design principles were highlighted in the story of the project called The Challenge Series (2009a):

- Infrastructure is designed as an amenity. Wherever possible water is made visible to the public rather than buried in pipes and water is celebrated through design. For example, the Hinge Park Wetland treats runoff before it enters the marine environment (False Creek) and serves as a public greenspace.
- Roof structures are designed to collect water for rooftop gardening, while overflow is stored in cisterns. 50% of roofs are vegetated.
- Intensive green roofs include eight inches of soil and are designed for larger species. These roofs mitigate heat island effects, are energy efficient, create habitat and social benefits, and provide rainwater management.
- Rainwater is captured and stored in cisterns to be used for toilet flushing in the winter months, allowing drinking water to remain in the public drinking water reservoir ("water banking").
- Smart irrigation systems use soil sensors and weather stations to supply water based on plant needs.
- Water for domestic use is pre-heated through heat recovery from sewage.
- One kilometre of formerly contaminated marine shoreline was regenerated, including a newly constructed habitat island. Herring have returned to spawn after an 80-year absence. Waterfowl and eagles now reside on the new island (Figure 2).

Figure 2 The new habitat island at South East False Creek. Herring have returned to spawn in this marine environment after an 80-year absence.

ECONOMICS

It should be noted that 'engineered ecology' and integrated design measures cover a range of disciplines and therefore an inter-disciplinary team approach is required to promote urban design that integrates best practice water planning and management measures with attractive streetscapes and open spaces. This integration can create attractive and sustainable urban landscapes that can provide developers with a marketing advantage (Corkery *et al.* 2004). A radical shift that aligns sustainability with decisions that make economic sense (financially viable and affordable) can be achieved by moving from 'cost based' evaluation to 'value based' evaluation. 'Cost based' design seeks outcomes at *least cost* and historically has been the dominant paradigm for infrastructure selection and design. 'Value based' approaches, in contrast, seek to design systems to yield the *most value*. Such a shift in perspective has the potential to radically change how infrastructure decisions are made. It also aligns with global shifts in valuation practices (for example, the Vancouver Valuation Accord) that seek to bring in 'green' values, and begins to open up opportunities for capturing increases in value associated with new infrastructure provision (Mitchell *et al.* 2008; O'Riordan *et al.*, 2008).

Valuation of Ecosystem Services

The natural world and its processes provide numerous free ecological goods and services to the human populations that live within them. "Ecosystem goods [such as food] and services [such as water filtration] represent the benefits human populations derive, directly or indirectly, from ecosystem functions" (Costanza *et al*. 1997). Human societies rely on these ecological services, which may also be described as Nature's infrastructure. Without adequate water quality and quantity, most of these services cannot exist. When natural infrastructure is disturbed by human activities such as urbanization, the functions this infrastructure provides often need to be replaced. The way in which the infrastructure is designed becomes important for long-term management efficiency, cost and liability concerns.

Water infrastructure has evolved over the past century or so from a conventional engineering approach of ditches and pipes, to an environmental engineering approach such as rain(storm)water ponds (Figure 3). When taken to its logical extent, we find "Engineered Ecology" which uses Nature's design principles to create natural systems that are ecologically functional and capable of replacing or at least significantly augmenting, conventional infrastructure. Engineered Ecology involves using Nature's principles as the design criteria in order to build an infrastructure element that is resilient and adaptive, is low maintenance due to its emphasis on natural function, and is connected to its surrounding watershed.

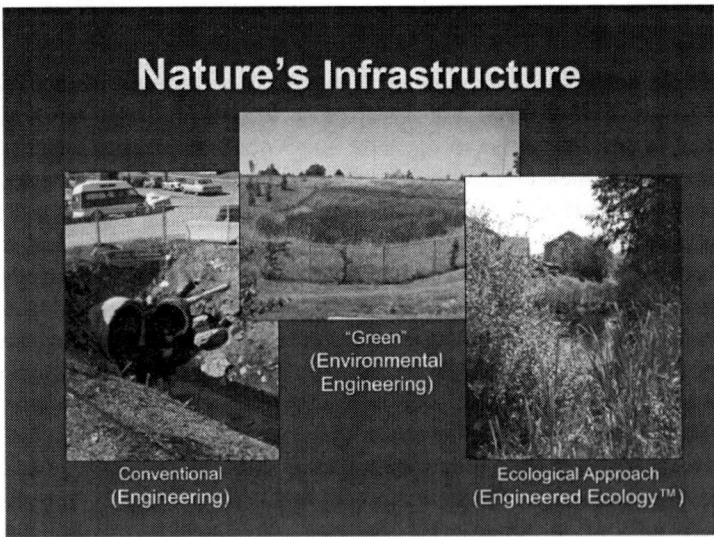

Figure 3 The evolution of stormwater management from conventional buried pipes (left) to ecologically functional wetlands and watercourses (right)

The natural infrastructure described above, supports human populations in urban and rural areas and reduces the amount of money required to create and maintain infrastructure. For example, by rehabilitating and conserving healthy freshwater ecosystems, communities can save money in drinking water and stormwater infrastructure costs by "deferring or eliminating the need for conventional pipes and valves" (Aqua-Tex 2005). Furthermore, healthy freshwater ecosystems regulate hydrological flows, store and retain water, provide pollution treatment via filtration and microbiotic organisms, and regulate local climate. In this way, the freshwater ecosystem provides a revenue source via cost savings in community infrastructure thereby providing a revenue stream (termed Nature's Revenue Streams™ by Aqua-Tex).

The valuation of ecosystem services has been the subject of considerable study by high profile agencies such as the United Nations (Millennium Ecosystem Assessment 2003) and the World Business Council. These agencies have stressed the need to monetize ecosystem services in order for ecosystems to be recognized, and thus protected, by business and management decision-making processes. So-called "green infrastructure" is a key component of urban ecosystems, and building infrastructure to mimic nature's processes is one method of replacing lost ecosystem services within the urban environment. While the ecological value of green infrastructure is readily apparent in the form of green trees, healthy streams and abundant wildlife, its economic significance is less obvious (Barraclough and Hegg 2008a). The following case study illustrates this point.

Case Study: From Ditch to Functional Creek

Blenkinsop Creek is the major tributary to Swan Lake, an urban lake surrounded by a nature sanctuary in the middle of the municipality of Saanich, BC on Southern Vancouver Island. Blenkinsop Creek has been severely channelized by agriculture and urban development throughout the watershed and is the receiving waterway for urban storm drainage. Approximately 700m of Upper Blenkinsop Creek was rehabilitated by relocating and restructuring the channel within a farmer's field, replanting native riparian vegetation and allowing the creek to access its floodplain (Figure 4). Roads on either side of the former ditch were eliminated, three bridges were removed and the ditch was rebuilt as a functional stream along one edge of the field, adjacent to an existing strip of mature vegetation (rails to trails public walkway). As a direct result of the stream realignment, the farmer agreed to permit a municipal commuter trail to be built along the border of his property, thus connecting two sides of the valley via a major pedestrian and cycling route.

An overview of the benefits accrued from using Nature's design principles is presented in Table 1.

Table 1 Benefits of stream realignment and regeneration

Pre-Existing Condition	Post-Realignment Condition
Three roads	One road
Three bridges	No bridges
One long narrow field isolated from the main field	One large field with 7% more arable land due to reclamation of roads
Two irrigation systems	One irrigation system
Poor stream water quality	Improved water quality[1]
Low biodiversity	Enhanced biodiversity, particularly birds and amphibians
Trespassing and vandalism	Trespassing and vandalism eliminated; new connector trail constructed
Minimal flood storage	Significant flood storage
Weekly pesticide use	Integrated Pest Management (IPM program and virtual elimination of pesticide use)
Poor riparian health	Riparian health restored to 700 m of channel
Barrier to local non-motorized transportation	New connector trail provides key link in the municipal trail network

[1]Barraclough and Hegg, 2008b

Figure 4 Blenkinsop ditch prior to realignment and restoration as a functional creek (a). The field prior to creek reconstruction (b) and the finished creek 5 years after construction (c).

Utilizing Engineered Ecology ideas resulted in an accrued economic benefit to both the landowner and the municipality within which the farm is located. Under conventional practices, the cost to the farmer was \$1,421,226 (Present Value over 25 years with a discount rate of 5%). There was no financial benefit to the community. Under the sustainable approach, the present value to the farmer was \$179,598 for a net benefit of \$1.6 M. There was an additional benefit to the community of \$3.5 M (PV) for such things as the trail connector, carbon sequestration and avoidance of flooding (Barraclough and Hegg 2008a).

SUMMARY

Nature's design principles of utilizing wastes as a resource and ensuring that each design element serves multiple functions can be incorporated into a wide range of urban designs. Buildings, sites, neighbourhoods, agricultural fields and even whole cities can benefit from this approach. As cities become more crowded, ecologically functional greenspaces will not be a luxury, but a necessity to ensure preservation of ecosystem services (clean water, nutrient recycling) and the good health of our urban citizens. As demonstrated in these case studies, this sustainable approach can also be very economically viable, but realizing the benefit requires a shift from 'cost-based' to 'value-based' economic assessment and a system boundary that considers all the benefits to the community.

ACKNOWLEDGEMENTS

The authors wish to extend their appreciation to Dr. Xaodi Hao and the entire organizing team for the invitation to present at SWIF 2009 in Beijing. We also extend our thanks to Nicholas Arsenault and Tracy Motyer for their contributions and thoughtful insights in the preparation of this chapter.

REFERENCES

Akbari H., Pomerantz M. and Taha H. (2001). Cool surfaces and shade trees reduce energy and improve air quality in urban areas. *Solar Energy* **70**(3), 295–310.

Aqua-Tex (2005). *Environment as Infrastructure-Nature's Revenue Streams: Project Profiles*. Colquitz River Watershed, Saanich B.C. Aqua-Tex. 19 pp.

Barraclough C. L., Hegg D. A. and Aqua-Tex Scientific Consulting Ltd. (2008a). *Nature's Revenue Streams: Five Ecological Value Case Studies*. Prepared for the District of Saanich and Canada Mortgage and Housing Corporation.

Barraclough C. L., Hegg D. A. and Aqua-Tex Scientific Consulting Ltd. (2008b). *Nature's Revenue Streams: Assessment of Stormwater Treatment via Engineered Ecology™ Treatment Systems and Stream Restoration*. Prepared for the District of Saanich and Canada Mortgage and Housing Corporation.

Coombs P. (2006). *Integrated Water Cycle Management: Improved Water Security and Ecological Health in an Uncertain World.* Keynote Presentation: Water in the City Conference, September 17th – 20th, 2006. Victoria, British Columbia.

Corkery N., Kielniacz A and Chubb D. (2004). *Water Sensitive Urban Design: Technical Guidelines for Western Sydney.* URS Australia Pty Ltd.

Costanza R., d'Arge R., de Groot R., Farber S., Grasso M., Hannon B., Limburg K., Naeem S., O'Neill R. V., Paruelo J., Raskin R. G., Sutton P. and van den Belt M. (1997). The value of the world's ecosystem services and natural capital. *Nature*, **387**, 253–260.

Dockside Green Website. http://www.docksidegreen.com (accessed December, 2009).

Lucey, W.P. (2009). Resources from waste: creating resilient cities in uncertain times. *News for the Decentralized Wastewater Industry*, **18** (1).

Millenium Ecosystem Assessment (MEA). (2003). *Ecosystems and their Services. Ecosystems and Human Well-being: A Framework for Assessment*, pp. 49–70. http://www.millenniumassessment.org/en/Framework.aspx.

Mitchell, C., Abeysuriya, K. and Fam D. (2008). *Development of Qualitative Decentralized Systems Concepts for the 2009 Metropolitan Sewerage Strategy*, vol. 1 – Synthesis Report, for Melbourne Waters. Institute for Sustainable Futures, p. 7.

O'Riordan J., Lucey W. P., Barraclough C. L. and Corps, C. G. (2008). Resources from waste: An integrated approach to managing municipal water and waste systems. *Industrial Biotechnology*, **4**(3), 238–245.

Parkes M. W. and Horwitz P. (2009). Water, ecology, and health: ecosystems as settings for promoting health and sustainability. *Health Promotion International* **24**(1), 94–102.

Postel S. L. (2008). The forgotten infrastructure: safeguarding freshwater ecosystems. *Journal of International Affairs* **61**(2), 75–90.

Smith K. R., Brown B. B., Yamada I., Kowaleski-Jones L., Zick C. D. and Fan J. X. (2008). Walkability and Body Mass Index: density, design, and new diversity measures. *American Journal of Preventative Medicine* **35**(3), 237–244.

Sullivan W. C. and Kuo F. E. (1996). *Do Trees Strengthen Urban Communities, Reduce Domestic Violence?* Urban and Community Forestry Assistance Program Technology Bulletin No.4. USDA forest Service, Southern Region, Atlanta, GA.

Taylor A. F., Kuo F. E. and Sullivan W. C. (2008). Coping with ADD: The surprising connection to green play settings. *Environment and Behaviour*, **33**(1), 54–77.

The Challenge Series. *Water + Building Landscape* (2009b). Millennium Water: The Southeast False Creek Olympic Village. Vancouver, Canada. http://www.thechallengeseries.ca/chapter-06/

The Challenge Series. *History + Policy* (2009b). Millennium Water: The Southeast False Creek Olympic Village. Vancouver, Canada. http://www.thechallengeseries.ca/chapter-01/

Todd, N. and Todd, J. (1993). *From Eco-Cities to Living Machines: Principles of Ecological Design*. North Atlantic Books. Berkley California, p.197.

Optimized Distribution and Sustainable Utilization of Water Resources in Jinan Municipality

Zhang Shoubin and Qiu Liping

School of Civil & Architectural Engineering, University of Jinan, Jinan 250022, China
E-mail: zhangshoubin1980@yahoo.com.cn

Abstract Jinan, suffers from some of the most severe water shortages in China. Serious contamination of surface water has lead to a series of environmental, ecological and economic problems such as over-exploitation of underground water, land subsidence and spring malfunction. The problem of water resources in China has led to a bottleneck situation in the development of Jinan. When analyzing the existing problems and utilization of water resources in Jinan, corresponding measures to optimize distribution of water resources and to realize sustainable utilization were proposed.

Keywords Jinan, water resources, optimized distribution, sustainable utilization

INTRODUCTION

Currently, China is experiencing rapid urbanization. With the rapid development of the economy and the sharp increase of population, the conflicts between population, resources and environment are becoming more problematic. As the political, economic and cultural centre of the Shandong province, the development of Jinan is closely linked with reasonable exploitation and sustainable utilization of water resources. To sustain positive development, building an ecological city, utilizing water resources reasonably and distributing water resources optimally appears to be particularly important in Jinan. At present, the consumption of water resources per capita (350 m^3 per person) in Jinan is far lower than the world average. In addition, urban water pollution is becoming more severe, exacerbating the reduction of quality water resources. Therefore, in order to achieve sustainable development in Jinan, it is important to study the status quo of the use of water resources and its existing problems. The study of present conditions can also provide a powerful assurance for establishing the optimal distribution and utilization of water resources in Jinan.

STATUS QUO OF WATER RESOURCES OF JINAN

Total Volume of Water Resources in Jinan

The water resources in Jinan include atmospheric precipitation and passing river water. Atmospheric precipitation forms surface and underground water. Passing river water refers to Yellow River. The average annual precipitation in Jinan is about 636 mm, equal to 5.201 billion m³ of water. The total volume of surface water is 0.860 billion m³ and that of underground water is 1.163 billion m³. Deducting the overlapped 0.482 billion m³, the total volume of the water resources in Jinan is 1.539 billion m³.

The Status Quo of Water Resources in Jinan

Limited recharge capacity of water resources

Table 1 Comparison Between Rainfall and Evaporation in Jinan

	mm							
item	1985	1990	1995	1997	1998	1999	2000	2001
annual rainfall	593.2	844.9	636.2	568.0	742.7	448.3	635.1	496.9
annual evaporation	1717.4	1739.7	2034.4	2159.6	1906.5	1872.6	1870.3	1971.5

Jinan is in the temperate zone with a continental monsoon climate. The four seasons differentiate conspicuously and the rainfall is intensified to some extent. The annual evaporation capacity is much more than the rainfall. The comparison between rainfall and evaporation (Shandong Statistics Bureau, 2002) is listed in Table 1.

As shown in the above table, the average annual rainfall, calculated in the 8 yr period, is only 620.7 mm. However, the average annual evaporation volume is 1909 mm. Limited rainfall provides little recharge capacity for water resources in Jinan.

Imbalanced spatio-temporal distribution of recharge capacity of water resources

Influenced by monsoons and topography, the spatio-temporal distribution of atmospheric precipitation in Jinan is obviously imbalanced. The regional layout of rainfall is that there is more rain in the south than in the north. It is also more concentrated in the middle than in the east or west. The average annual rainfall in the south mountain area is 100 mm more than that in north plain area. The temporal distribution of rainfall in Jinan is also imbalanced. Rainfall between different seasons and different years changes greatly. Firstly, rainfall among different seasons can vary significantly. Every year, the flood season lasts from June to September. In this period, there is about 467 mm Rainfall, which is 75.7% of the annual rainfall (Jinan Statistics Bureau, 2002). Secondly, rainfall between different years can change. According to fifty years of records, taken between 1951 and 2001, the most rainfall was recorded in 1962 as 1145 mm. This is about 3.4 times more than what the smallest rainfall recorded in 1968 as 336 mm.

Low water resources per capita

The total volume of water resources in Jinan is no more than 1.539 billion m^3. The water resources per capita is 266 m^3 which accounts for 11% of the country's total. Where the water resources per capita in China accounts for 1/4 of the world average, it is seen that Jinan is severely short of water.

THE EXISTING PROBLEMS IN WATER UTILIZATION IN JINAN

The Gap Between Supply and Demand is Wide, and the Bottleneck Influence is Higher

If the present water supply guarantee rate of Jinan is 95%, the total volume that can be utilized is about 1.643 billion m^3. However, the gross water requirement is 1.881 billion m^3. So the shortage of water resources is about 0.238 billion m^3. The gap between supply and demand is wide, and the bottleneck influence is becoming more and more obvious. To ensure the supply of domestic water, industrial water utilization has to be monitored. Due to water shortages, some enterprises have to cut down their output, which can lead to huge economical losses. The present water supply capacity can only meet domestic and industrial requirements. As far as agricultural, it can only be supplied by wastewater. At the end of 2001, there were 0.332 million ha of cultivated land in Jinan. However, only 72.32% of this cultivated land can be irrigated effectively.

Water Utilization Structure is Unreasonable

Water consumption of different departments of Jinan is shown in Table 2.

Table 2 Comparison of water consumption among different departments of Jinan

Unit	Total intake	Irrigation water	Country domestic water	Forestry, stockbreeding, fishery water	Industrial water	Town domestic water
×10⁸m³	17.84	11.10	1.04	1.00	3.05	1.65
proportion%	100	62.22	5.83	5.60	17.10	9.25

Table 2 illustrates that agriculture, using 67.82% of Jinan's total water, plays the biggest part of water consumption in Jinan. As far as agricultural water is concerned, irrigation water occupies 92.74% and forestry, stockbreeding and fishery water occupies the other 7.26%. At the same time, the utilization ratio of irrigation water in Jinan is very low. Its effective utilization coefficient is only 0.5. Industrial water occupies 17.10% of the total water consumption of Jinan. Metallurgy industry, machine building industry, petrochemical industry, light textile industry, building materials industry, paper making industry, are the main industries of Jinan. These types of industry traditionally have a very large value of water consumption per unit of output. Domestic water only occupies 15.08% of the total water consumption of Jinan (WU Xingbo et al., 2004).

The Problem of Water Pollution is Getting Serious

Since the 1990s, the problem of water pollution in Jinan is becoming more and more serious. In 2001, Industrial wastewater discharge in Jinan amounted to 63 million tons. Amongst this discharged industrial wastewater, 93.20% meets emission standards. Though the proportion of industrial wastewater discharge meeting these standards is very high, there is still a large volume of industrial wastewater discharged directly into water. The treatment rate of domestic sewage is only 34.89%. Thus domestic sewage is also an important source of surface water pollution. Lots of pollutants in domestic sewage discharges into waters and causing the water quality deteriorate (PENG Yuanxin et al., 2003).

The Utilization Ratio of Surface Water

Out of the available water resources of Jinan, the proportion of underground water is larger than surface water. A considerable degree of surface water is wasted. Its utilization ratio is very low. Compared with other cities in China's Shandong province, the utilization ratio of underground water in Jinan is relatively high. As far as annual rainfall is concerned, Jinan ranks tenth in the Shandong province. The volume of runoff is 10.4 billion m³. But the utilization ratio of surface water in Jinan, is only 20%. This is lower than the average level of whole country which is 25%.

Over-Exploitation of Underground Water has Resulted in a Series of Problems

Because of the shortage of surface water resources, for a long time, underground water has been used as the main water resource in Jinan. In 2001, the exploitation volume of underground water in Jinan was about 1.063 billion m^3, occupying 59.58% of the total water consumption of that year. Out of the 1.063 billion m^3 of underground water, the exploitation volume of deep layer underground water was about 0.402 billion m^3. This occupied 22.53% of the total water consumption of that year. The exploitation volume of underground water accounts for 91.3% of the average annual volume of groundwater resources. The exploitation of underground water resources has caused the groundwater table to decline greatly. In turn, this has resulted in a series of environmental problems such as land subsidence, spring malfunction and ground water contamination etc (MAO Xiaoping, 2002).

COUNTERMEASURES FOR OPTIMIZED DISTRIBUTION AND SUSTAINABLE UTILIZATION OF WATER RESOURCES OF JINAN

Accelerating the Development of Foreign Water Resources and Utilizing Cisborder Water Resources Reasonably

Water resources in Jinan consist of cisborder water resources and foreign water resources. Cisborder water resources can be divided into surface water and ground water. Foreign water resources mainly refer to water from Yellow River water. The use of Cisborder water resources is regarded as the root for water saving and spring protection. Under this guiding ideology, 400 wells in city of Jinan have been shut down. At present, only eastern suburbs still employ ground water as a water source. Through the use of Yellow River and water transfer projects such as QueShan reservoir and YuQing reservoir, the city zone, western suburbs and southern suburbs of Jinan have established a joint supply of surface water, ground water and foreign water. In order to realize the goal of water saving and spring protection, Jinan should focus its attention on long term development, setting up projects to use new water resources. The East Lake reservoir is part of the South-to-North Water Transfer Project in China and should be regarded as the main water source for industry in Jinan henceforth.

Adjusting Industrial Structure and Building up an Economic Structure for Water-Saving

As part of a traditional industrial structure, the industrial water consumption of Jinan is relatively large. The water saving potential of industrial water is also relatively big. So the industrial layout should be adjusted reasonably. In areas where the water supply is insufficient, industries that consume large amount of water should be restricted. Projects where the water consumption is particularly high should also be controlled. At the same time, the development of some hi-tech industries such as electronic industry should be accelerated. Big enterprises can be set up to encourage industries to multi-use industrial wastewater. To realize the multi-using of industrial water, factories can be concentrated and constructed in the eastern industrial zone of Jinan.

Establishing Water Price Reasonably

Experiences of water utilization at home and abroad indicate that increasing the price of water is the most effective way to save water. The present water price of Jinan is not integrated, which is the main reason for water price staying on the low side. The implementation system is not reasonable and cannot reflect the change of cost and conditions in the development of water resources. The current institution of water price is a single measurement price, that is to say, water price is equal to the water unit price multiplied by water consumption. This kind of measurement method of water price can not play a role in satisfying the balance of supply and demand and accelerating water saving. So, implementing a ladder measure for water price is suggested. The ladder measure for water price is as follows. The radix of water consumption should be determined at first. When water consumption is below the radix, a basic water price is used. Based on the proportion that is over the radix, the price of water consumption that exceeds the radix will increase in a form of ladder. The more water consumption to exceed the radix, the higher price is (ZHANG Lina, 2002). In addition, establishing a seasonal water price is needed. The price of water should be increased in dry seasons and be decreased in wet seasons.

Accelerating the Cyclic Utilization Rate of Water

Firstly, the construction of wastewater treatment plants and matching pipe networks should be accelerated. The technologies for wastewater treatment should also be improved. Currently, the processing rate of domestic sewage can not achieve 50%. Accelerating the construction of wastewater treatment plants, improving the technologies for wastewater treatment and reducing the costs of wastewater treatment are effective measures to realize the target of water saving. Other effective measures to save water are as follows: establishing separate sewerage systems gradually, supplying water according to water quality, determining water price on the basis of water quality. Using reclaimed water for industrial and environmental water use is a particularly effective way to save water. Projects of reclaimed water should be actively encouraged and developed in order to enhance the reuse rate of water. To improve and develop methods of water saving, technologies of reclaimed water treatment should be popularized and implicated widely. According to the 2000(3) command of Jinan government, hotels and comprehensive service buildings with a building area beyond 20000 m^2, must build up matching reclaimed water facilities. This also applies to organs, scientific research institutions, colleges and universities that have a building area beyond 30000 m^2 and also high-rise buildings.

Strengthen Water Conservation and Reinforce Tree Planting and Afforestation in Southern Mountain Areas

The Southern mountain area of Jinan is the recharge area of groundwater. The WoHuShan reservoir and JinXiuChuan reservoir are also located in southern mountain area of Jinan. The 1000 km^2 of forest land in the southern mountain area can store up to 50% of rainfall. According to predications of the forestry department, when the forest coverage of southern mountain area is 30%, precipitation can increase by 0.15 billion m^3. Meanwhile, the runoff volume can decrease by 0.16 billion m^3. The sum of above two values can increase water volume by 0.31 billion m^3 every year (LIU Qingyong 1999). Compared to the city zone of Jinan, the air humidity of the southern mountain area is higher. Under suitable climatic conditions, artificial rainfall can be implemented to enhance precipitation. This is a way to increase water resources in Jinan. Thus, playing a leading role in the protection of the ecological environment, the development of the southern mountain area of Jinan should take construction. Large scale building and ground hardening should be decreased. The management of institutions in forest regions must be improved. In order to form a benign circle of water resources in Jinan, the above measures must be enforced strictly.

CONCLUSIONS

Jinan is characterised by its organic integration of mountain, spring, lake, river and city. As far as urban planning is concerned in Jinan, the concept of Jinan as a 'Spring City' can not deviate from the need to optimise the distribution of water resources. Realizing sustainable development of the city and building ecological city can not deviate from the reasonable exploitation and utilization of water resources too. Thus in the process of urbanization, with the principle of 'intercepting and storing natural water fully, utilizing Yellow River water actively and protecting ground water reasonably', Jinan should overall arrange the spatial layout of water resources, establish regional water supply and drainage system, realize sharing and reasonable utilization of water resources. The ultimate aim is to realize the sustainable development of urban water resources.

REFERENCES

Jinan Statistics Bureau (2002) *Statistical Yearbook of Jinan-2002*. China Statistics Press, Beijing, pp. 30–31.

LIU Qingyong (1999). Study on the strategies of sustainableutilization of ground water resources. *Haihe Water Resources*, 7(3), 23–26.

MAO Xiaoping. (2002). Analysis on main factors influence spewing of famous spring in Jinan. *Journal of Shandong Meteorology*, 2(6), 6–8.

PENG Yuanxin, LONG Xuemei and XU Yuetong (2003). The water pollution of Jinan City and its treatment. *Territory & Natural Resources Stuty*, 11(1), 5–6.

Shandong Statistics Bureau (2002). *Statistical Yearbook of Shandong Province-2002*. China Statistics Press, Beijing, pp. 5–7.

WU Xingbo, SONG Xingyuan, NIU Jingtao et al. (2004). Study on water resources countermeasure for water supply and spring protection. *Water Saving Irrigation*. **12** (3), 2–3.

ZHANG Lina (2002). A Study of water resources price in Jinan city. *Territory & Natural Resources Stuty*, 7(4), 62–63.

A New Strategy for Water Supply Systems in Local Cities and Towns in Developing Country

Kiyoshi Yamada

Dept. of Environmental Systems Eng., Ritsumeikan University, Noji-higashi 1, Kusatsu City, Shiga 525–8577, Japan
E-mail: yamada-k@se.ritsumei.ac.jp

Abstract In developing countries, despite the implementation of water supply systems, many people are still unable to get safe drinking water. This is due to water shortages and insufficient pipeline management. In this paper, a new strategy for water supply in a local cities and towns is introduced. A new hybrid system is based on a feasibility study. Results from surveys in three developing countries showed that safe drinking water could be guaranteed for all residents by employing the hybrid system and an appropriate charge. At the same time, the money balance for waterworks companies could be improved. Profits from waterworks companies can then be used to maintain and rehabilitate existing water supply systems. In addition, the hybrid system could contribute to the realization of a low carbon society. Thus, the proposed hybrid system results in a WIN-WIN situation for residents, waterworks and the global environment.

Keywords Water supply systems, drinking water, hybrid system, waterworks management

BACKGROUND AND NEW STRATEGY

Although water supply systems exist and are in use, there are a lot of cities and towns in developing countries that cannot get access to safe drinking water. In order to obtain safe drinking water, a maintenance system is required to repair leakages and pipe erosion; a continuous supply of water needs to be provided at a stable pressure. However because of high maintenance costs, quality maintenance systems are rarely established in the cities or towns of developing countries.

People have obtained drinking water, supplied through boiling for years. However, as fuel and time are required to boil water, people often drink or wash their hands with the water directly, without boiling it first. Having done this, they become susceptible to water-born diseases. Recently, some people have started to buy water purification equipment or even bottled water. However, whilst this is a satisfactory strategy for rich people, it is not a feasible option for poorer communities. In this way, there is indeed a survival gap.

Every waterworks company has to provide safe drinking water. People have a right to get this as a basic human right. In this paper, a hybrid water supply system is proposed. Safe drinking water is delivered, having been bottled at a treatment plant. The remaining water is sent by the existing pipeline. People have to pay for the both bottled water and water supplied by pipeline. The bottled water is fairly cheap, as large 20 litre bottles are used and delivered to customers. By employment of this system, water-born diseases can be decreased and the money balance of waterworks can be improved. Profits can then be used for the rehabilitation or renewal of their water supply systems

RESEARCH OBJECTIVES AND OUTLINE OF SURVEY

Research Objectives

Access to a safe drinking water supply constitutes one of the most basic human needs. In this research, it is proposed that until safe piped water supply systems are established in developing countries, like those in developed countries, a two way system of piped and bottled water should be employed. This is the so called Hybrid System. Both should be provided by the same public waterworks company. The concept of the Hybrid Water Supply System is shown in Figure 1.

In the proposed Hybrid System, part of the water treated in water treatment plants is packed in bottles of 10 to 20 litres. It is then distributed by the waterworks companies directly to consumers as drinking water (and partly as water for cooking, dish washing and hand washing). Used bottles are returned to the water treatment plants and reused after rinsing. Additional treatment is done if needed. For all other uses other than drinking, most of the water for household use is provided to consumers via pipeline. As the piped water does not need to be used for drinking, just a little chlorine may be added to ensure safety.

Every consumer has to pay a charge for both water supplies. If waterworks companies provide the bottled water in big bottles (10 – 20 litres) by themselves, consumers can get it at a relatively lower price. Furthermore, some government subsidies could be provided to those of very poor communities. As a result, all people, both rich and poor, can use bottled water for drinking. The objective of this research is to clarify the feasibility of the proposed Hybrid System from financial, welfare and environmental points of view.

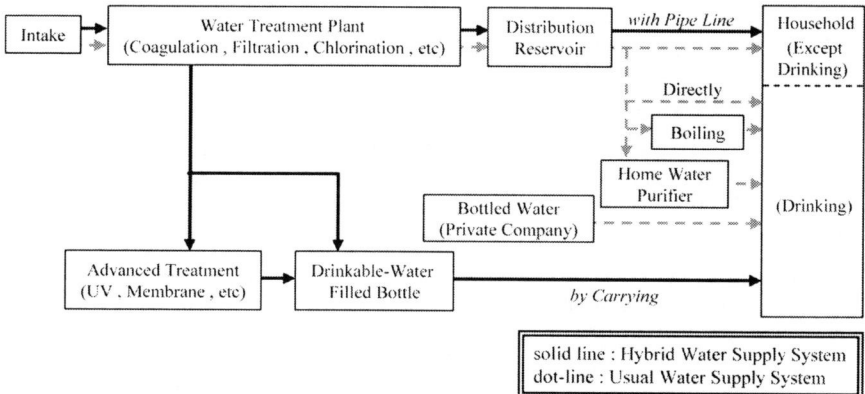

Figure 1 Hybrid Water Supply System

Outline of the Survey

Field surveys were carried out at water supply systems in 9 areas in Laos, the Philippines and Vietnam (Nishida *et al.*, 2008). All water supply systems surveyed are connected to houses. Laos is a country where the use of bottled water is popularized by the public sector. In the Philippines, various alternatives of water sources are possible. However, Vietnam is one of the countries where the provision of the Hybrid System by the public sector has been implemented.

The basic data for current water usage in urban households, in large cities, and in villages of rural areas, was collected using questionnaire surveys. The questionnaires covered items such as household type, water charge, metered water volume, water quality, service time, satisfaction with water supply, and volume/price of purchased bottled water, etc. as shown in Table 1, the number of collected questionnaire samples came to more than 1000.

Table 1 Outline of survey

Country	Survey area		Scale of the area	Survey Period	Sample of Questionnaire
	Code	City name			
Philippines	Pu2	Angeles	Provincial city	30 Aug–	90
	Pu3	Batangas	Provincial city	10 Sep,	131
	Pu4	Tarlac	Provincial city	2006	62
	Pu5	Magdalena	City (semi-urban)		126
Laos	Lu1	Vientiane	City (urban)	13–25	177
	Lu2	Luangprabang	City (semi-urban)	Sep, 2007	115
Vietnam	Vu1	Gia Lam, Bode,Trau Quy,Sai Duong	Part of capital city; Hanoi		124
	Vu2	An Duong,Binh Hai,Van My, Ha Ly	Third biggest city in Vietnam	9–18 Sep, 2008	149
	Vu3	Thanh Binh (Hai Duong)	Third biggest city in north part of Vietnam		71

RESULT AND DISCUSSION

Unit Water Consumption

Table 2 shows the metered unit water consumption on a yearly basis. In some areas, such as Laos, unit water consumption was high and close to that in developed countries. In the Philippines, because of the availability of different alternatives of water sources, unit water consumption varied widely amongst regions.

Table 2 Water consumption

Country		Philippines				Laos		Vietnam		
Area code		Pu2	Pu3	Pu4	Pu5	Lu1	Lu2	Vu1	Vu2	Vu3
Total samples		84	130	60	120	148	111	121	144	43
Water consump-tion (L/cap/ day)	Average	194.1	160.9	136.1	85.9	194.4	211.0	123.0	127.8	120.7
	84% value	333.3	243.0	186.9	134.9	293.2	335.8	170.0	180.3	203.5
	50% value	157.6	125.8	113.3	60.6	164.1	170.8	113.1	104.3	88.9
	16% value	83.7	73.7	77.0	31.4	94.3	88.9	64.9	69.3	49.8

Usage of Sold Water or Other Water Sources

The rate of the usage of sold water (or other sources in addition to house connection water) is shown in Table 3. Sold water consists of mostly bottled water. The rate of the usage of sold water is very high in Laos and in two areas in the Philippines. However, it is not so high in Vietnam where the Hybrid System has been employed. The rate of the usage of other sources, in addition to house connection water, is fairly high in some areas.

Table 3 Percentage of using sold water or other source water

Country	Philippines				Laos		Vietnam		
Area code	Pu2	Pu3	Pu4	Pu5	Lu1	Lu2	Vu1	Vu2	Vu3
Number of samples	90	131	62	126	177	115	124	149	71
Solid water (%)	27.8	15.3	61.3	68.3	75.1	93.9	1.6	7.4	18.3
Other source (%)	28.9	23.1	59.7	38.9	4.0	3.5	16.3	13.4	49.3

Sold Water Consumption

The water consumption of sold water users is shown in Table 4. The average usage of sold water is more than 3L/c/d in some areas, and is less than 2L/c/d in other areas. If people could purchase cheaper options of sold water, the number of users and amount of water used would increase.

Table 4 Sold water consumption

Country		Philippines				Laos	
Area code		Pu2	Pu3	Pu4	Pu5	Lu1	Lu2
Total samples		25	18	38	85	133	108
Sold Water	Average	0.39	1.72	1.12	4.03	2.74	3.88
Consumption	84% value	0.64	3.81	1.90	5.71	3.81	4.76
(L/cap/day)	50% value	0.34	0.44	0.79	2.29	2.14	2.45
	16% value	0.07	0.03	0.36	0.95	0.95	1.14

Reasons for Using Sold Water

The reasons for using sold water are shown in Table 5. The major reason for purchasing sold water is due to the poor quality of house connection water. Other reasons are insufficient amount of water or limited time of supply for the house connection water.

Household Income

The situation of household incomes is shown in Table 6. The average income in the Philippines is slightly higher than in Laos. As income on the questionnaire sheet was categorized in 5 stages, there are some errors inherent in calculating the values in Table 6. However, it may be said that the income difference is about 2 times for the areas surveyed and 3 to 5 times between the higher 16% and the lower 16%.

Table 5 Reasons for using sold water

Country		Philippines				Laos	
Area code		Pu2	Pu3	Pu4	Pu5	Lu1	Lu2
Total samples of Using Solid Water		14 (25)	5 (20)	12 (38)	41 (86)	133	108
Reasons for Using Sold Water	Insufficient amount of tap water	0 (0)	0 (2)	0 (0)	0 (0)	4	20
	Poor quality of tap water	11 (15)	3 (3)	12 (34)	40 (48)	114	76
	High cost of tap water	0 (7)	0 (2)	0 (8)	1 (7)	6	1
	Limited time for which tap water is available	0 (0)	0 (4)	0 (3)	0 (0)	3	7

() is including households using other source

Table 6 Household income

Country		Philippines				Laos	
Area code		Pu2	Pu3	Pu4	Pu5	Lu1	Lu2
Total samples		59	131	59	119	173	115
Income (US$/HH/Month)	Average	249.8	334.6	278.1	180.3	140.8	241.4
	84% value	334.8	558.0	446.4	267.9	200.0	300.0
	50% value	223.2	223.2	223.2	111.6	120.0	150.0
	16% value	133.9	111.6	111.6	67.0	40.0	90.0

Philippines; 1US$ = 44.8peso, Laos; 1US$=10,000kip

Water Charge

Water charge is shown in Table 7. In spite of the much higher water consumption in Laos, the water charge is lowest in Laos.

Table 7 Water charge

Country		Philippines				Laos	
Area code		Pu2	Pu3	Pu4	Pu5	Lu1	Lu2
Total samples		88	130	62	126	166	115
Water Charge	Average	12.0	8.8	10.5	5.2	3.9	3.8
(US$/HH/Month)	84% value	20.1	15.6	15.6	6.9	6.7	6.4
	50% value	8.9	7.1	8.9	3.3	2.2	2.7
	16% value	3.8	3.8	3.6	2.3	1.1	1.4

Philippines; 1US$ = 44.8peso, Laos; 1US$ = 10,000kip

Sold Water Cost

The amount of money paid for sold water use is shown in Table 8. It is clear that people pay a lot of money for sold water.

Table 8 Sold water cost

Country		Philippines				Laos	
Area code		Pu2	Pu3	Pu4	Pu5	Lu1	Lu2
Total samples		25	18	38	86	133	108
Sold Water Cost	Average	7.4	23.5	7.6	6.4	4.5	6.1
(US$/HH/Month)	84% value	11.6	25.0	16.1	8.9	7.2	7.2
	50% value	3.6	9.8	3.8	3.2	4.0	4.1
	16% value	2.7	3.6	2.7	1.4	2.0	2.4

Philippines; 1US$ = 44.8peso, Laos; 1US$=10,000kip

Ratio of Water Charge and Sold Water Cost Relative to Income

People spend from 2 to 5% of their income on house connection water that is supplied by waterworks. Those who use sold water pay an additional 2 to 4% of their income. It is clear that people are paying in excess of 5% of their income to get sold water for drinking.

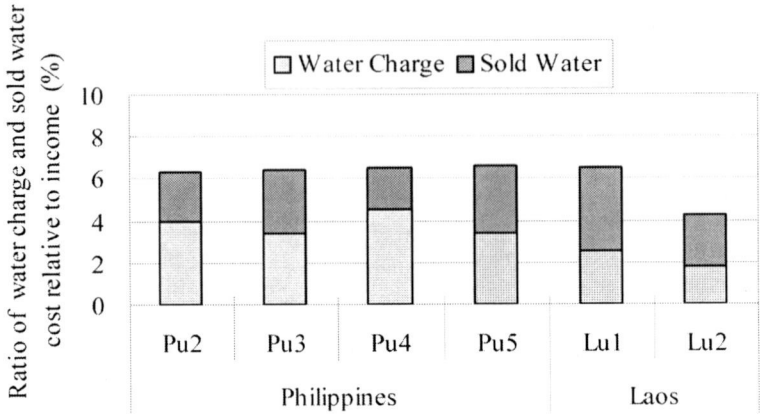

Figure 2 Ratio of water charge and sold water cost relative to income

Relationship Between Household Income and Water Charge

Figure 3 shows the relationship between water charge and income. Water charge seems to be independent of household income. In the case of poor people (low income), the water charge is in excess of 5% of the income for many households, and in excess of 10% of the income for some households.

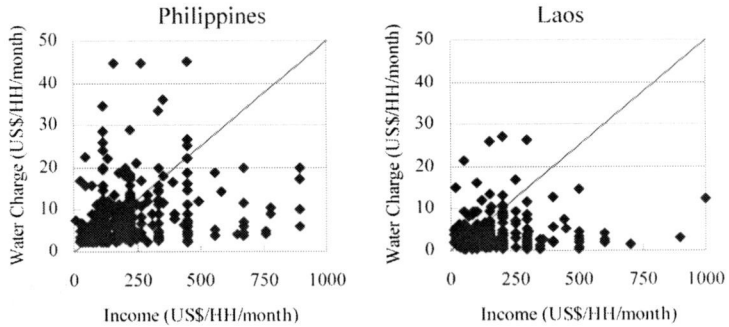

Figure 3 Relationship between income and water charge

Relationship Between Household Income and Sold Water Cost

Figure 4 shows the relationship between sold water cost and income. As for house connection water charge, sold water cost seems to be independent of household income in Laos. However, for the Philippines, there seems to be little correlation between the two. For some households, the cost of sold water is in excess of 5% of the household income.

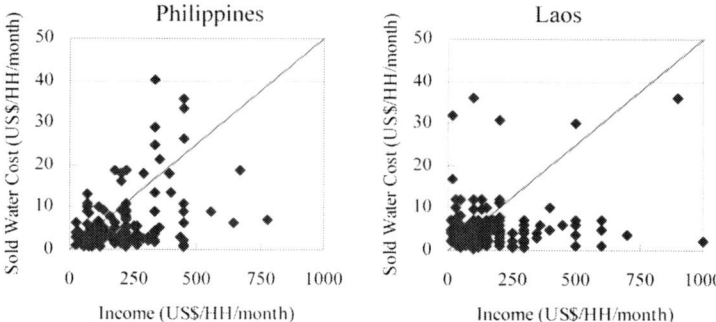

Figure 4 Relationship between income and water charge

CONCLUSION

From the above results, the use of the Hybrid Systems in developing countries has been proposed. Some feasible conditions can be assumed, as follows:

- Bottled water is to be provided to all of users.
- The amount of bottled water to be supplied is about 5 L/c/d. In addition to drinking, this amount should cover water for cooking, dish washing and hand washing.
- Bottles of 20 L are used in place of the commonly used 0.5 L bottles and should be delivered to users directly by each public waterworks company.
- Used bottles are to be collected and reused after cleaning.
- The cost of bottled water will be reduced to less than one-tenth of the present cost.
- Users pays 3.5% of their income for water charges and 1.5% of their income for bottled water.

Results simulated under the above conditions, for the areas surveyed showed that users can get more bottled water than ever at a cheaper price. At the same time, the waterworks companies can get a lot of profits from the sale of bottled water. They can then use the profits for the rehabilitation or renewal of their water supply systems. This could provide a promising sustainable supply system for developing countries.

Furthermore, the large reduction of small sized disposable bottles and the reuse of large sized bottles could contribute to the realization of an energy saving, low carbon society. Employment of the Hybrid system should create a WIN-WIN situation for three key stakeholders, namely, residents, public waterworks companies and the global environment.

REFERENCE

Nishida T., Yamada K., Muhandiki V. S., Shimizu T. and Murakami S. (2008). *Basic Study on Introduction of Hybrid Water Supply System in Developing Countries*. The 59th Annual Conference and Symposium of the Japan Water Works Association, Sendai, Japan, 28–30 May, pp. 96–97 (in Japanese).

Evaluation of Community-owned Water Resources Based on Water Quality Labeling System

Hiroaki Furumai, Michio Murakami and Tushara Chaminda G.G

The University of Tokyo, 7-3-1, Hongo, Bunkyo-ku, Tokyo 113-8656, Japan
E.mail: furumai@env.t.u-tokyo.ac.jp

Abstract Global freshwater resources are in growing demand, owing to rapid urbanization and climate change. This is particularly evident in Asian developing countries. As a result, the use of alternative water resources, such as rainwater and reclaimed water, are being considered. However, due to concerns regarding water quality, the general public is still hesitant to use these alternative resources. It is important, therefore, to develop evaluation methodologies which overcome public uncertainties over the use of community-owned water resources. This paper examines the use of water quality evaluation methodologies in Japan. The proposed method of water quality evaluation is 'the water quality labeling system'. This method provides a good representation of the water quality in different resources. Another method used is the evaluation of CO_2 emissions through water production which provides a clear comparison between existing water supply systems and community-owned water resources. By adopting these evaluation methodologies, the public can understand the benefits and risks of using alternative water resources.

Keywords Water reuse, water quality evaluation, water resource, rainwater, reclaimed water, energy consumption, CO_2 emission

INTRODUCTION

Climate Change and its Impacts on Water Resources

Climate change is one of the greatest environmental, social and economic threats facing the world. The Second Climate Conference, in Geneva 1990, established that the effects of climate change on the hydrological cycle and water management should be stressed as the most important impact of climate change (Haman and Brown, 1994). As a result, climate change and its potential impacts on water resources has become a common concern amongst engineers, managers in water sectors and scientists in the field of global warming.

Over the past several decades, it has been observed that climate change is closely linked with changes in the hydrological cycle and systems i.e. changing precipitation patterns, widespread melting of snow and ice, increased atmospheric water vapor, increased evaporation, and changes in soil moisture and runoff (Bates, 2008, Trenberth, 2007, Brekke, 2009). As climate change has the potential to alter hydrological cycles, it has the ability to affect and alter the quantitative and qualitative status of fresh water resources. This has wide-ranging consequences for human societies and ecosystems. The second assessment report of the IPCC (1995) discusses the uncertainties of water availability due to the impacts of climate change. Table 1 depicts the water per capita in some Asian countries as of 1990 and forecasts for 2050 under the present climate condition (Frederick and Major 1997).

   Climate change, population growth, economic development, land use change and aging infrastructure bring the problem of water scarcity to the forefront. Both water engineers and water users have to address these issues in the development of sustainable water resource management. To achieve sustainability, it is necessary to ensure a long-term supply of adequate quality water, for all required purposes, thus, minimizing adverse economic, social and ecological impacts. The implementation of sustainable water use will require the implementation of policy and regulation, the development of technology and the participation of citizens (Furumai, 2008). This paper looks at how to assess the availability of rainwater and reclaimed water in an urban area. These water resources, termed "community-owned water resources" are evaluated in terms of water quality and sustainable water use. Focusing on a case study in Japan that is deeply dependant on surface water resources, this paper describes how a water quality labeling approach is used.

Table 1 Forecast for water availability for some Asian countries (m³/yr/per capita) under the present climate condition

Country	Water availability in 1990	Forecast for 2050 (present climate condition)
China	2,500	1,630
India	1,930	1,050
Japan	3,210	3,060
Sri Lanka	2,500	1,520
Thailand	3,380	2,220
Vietnam	6,880	2,970

Source: Adapted from the second assessment report of the IPCC (1995) cited in Frederick K. D. and Major D. C. (1997)

Urbanization and Increased Water Demand

Population growth, industrial development and the degradation of available resources has put a growing pressure on global freshwater resources. With the onset of climate change, it is anticipated that the situation will only get worse. Insufficient water availability may lead to the degradation in human health, ecosystems, agricultural, and industrial output. It might also increase the potential of water-related conflict. The increasing urban population and the growth of mega cities will put an additional strain on existing public services. With current economic realities as they are, this could result in chaos throughout many towns and cities in the developing world (Vairavamoorthy, *et al.*, 2008).

As shown in the Figure 1, it is expected that developing countries will experience the most urban population growth. This is particularly evident in Asia and Africa. In Asia, it is anticipated that after 2025, its urban population will dominate over its rural population. With urban populations on the increase, we can expect an increased demand for water and a greater water consumption per-capita in these regions. This demonstrates how rapid urbanization processes directly affect water availability and quality. With increased water demands and accompanying wastewater discharge, this has implications at both city and regional levels. The change of land use, from rural to urban, can affect water use and water intake patterns; as urbanization develops, water abstraction from surface water and ground water inevitably increases to support human activities. The change in water intake patterns can cause significant impacts on fresh water sources such as the instability of river flows, the drying up of spring water and the deterioration of ecosystems. Thus, urban water consumption can have a major impact on both the natural water cycle and the aquatic environment (Furumai, 2007).

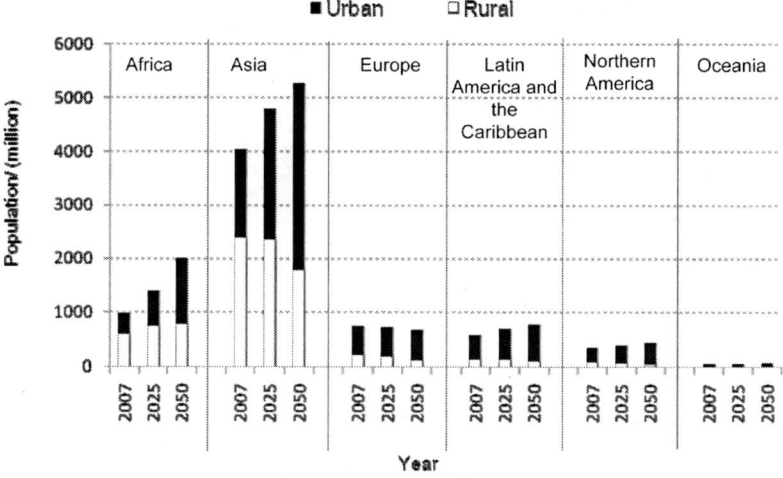

Figure 1 Urban and rural population in the world. (United Nation Population division, 2007)

COMMUNITY-OWNED WATER RESOURCES FOR SUSTAINABLE WATER USE

To meet growing water demands, new strategies must be applied to find alternative water resources. One solution is to use water resources, which can be found within a community; for many countries, a community resource can increase water availability and militate against future water scarcity. Identifying new water sources is a very economical way to meet increased water demands for a community. Meeting non-potable needs, with alternative waters available in the community, can substantially reduce the consumption of high quality potable water. To achieve this objective, roof runoff, road runoff, reclaimed water and groundwater should be within reach of any house or community (see Figure 2). Rainwater and reclaimed water have been receiving growing interest as complementary water resources for the community.

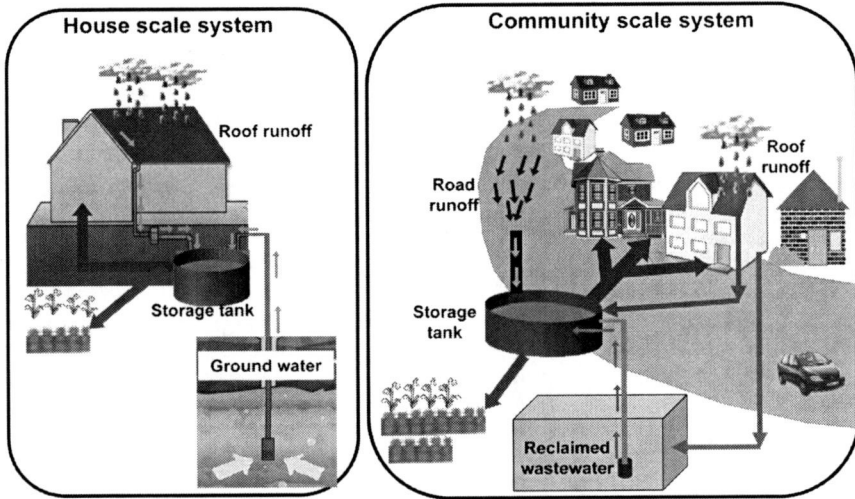

Figure 2 Community-owned water resources available for house scale and community scale levels

Reclaimed Water

Reclaimed water use has been practiced in various ways for decades. Reclaimed water use has developed from the direct use of wastewater, without any treatment, in rural regions, to the more sophisticated use of wastewater, after advanced treatment, in urban regions. The rapid development of new technologies in wastewater treatment, such as membrane technology, has made it possible to deal with contaminants like pathogenic organisms and viruses present in wastewater. Clearly, the emphasis of these new technologies lies in developing reclaimed water as a water resource. Reclaimed water use has several benefits: it can serve as a more dependable water resource; it reduces extensive exploitation of natural water; it can reduce pollutant load discharge, subsequently conserving water and the environment; it is economically efficient since it minimize the water transporting cost and energy from distant sources (UNEP, 2005; Furumai, 2007).

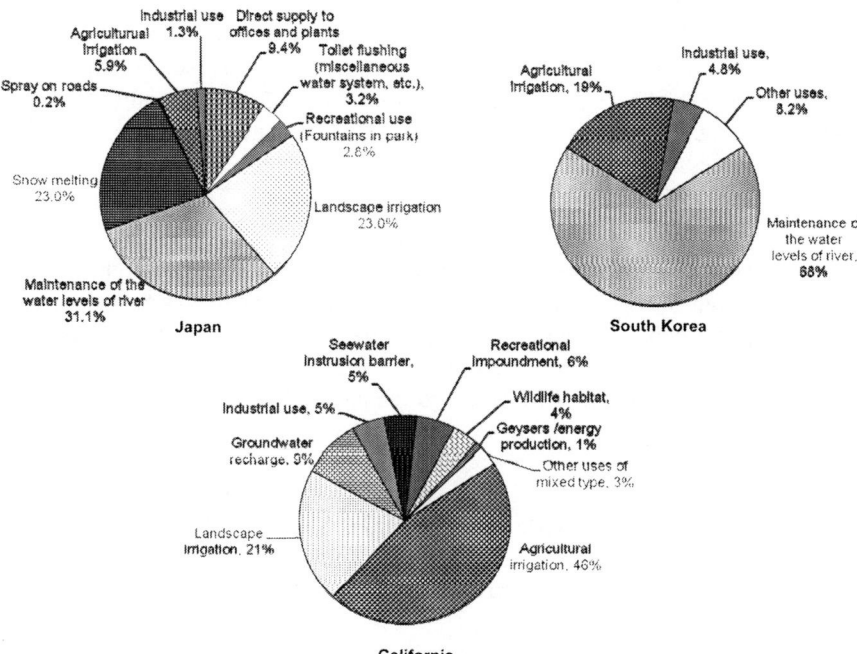

Figure 3 Percentage and purpose of reclaimed water use in Japan, Korea and California

Figure 3 provides a breakdown of water reuse in Japan, Korea and California. Although the water use categories are different among the three countries, wastewater reuse can be applied to agriculture, industry, maintenance of river water level, groundwater recharge, and urban usage (including toilet flushing, and landscape irrigation). In Japan, reclaimed water is mainly used for the maintenance of water levels of rivers, landscaping and snow melting. The total amount of wastewater treated in the wastewater treatment plant (WWTP) is 14.1 billion m^3 per year in Japan, in year 2004. However, the reuse rate is only 1.4% (0.19 billion m^3). In Korea, the total amount of wastewater treated is 6.6 billion m^3 per year (Kim, 2009). The rate of re-use is 4.7% (0.31 billion m^3). The reclaimed water is mainly used to maintain the water level of rivers, similar to Japan. However, the rate of reclaimed water used for agriculture is higher than that of Japan.

It is known that the reclaimed water use for agriculture is remarkably high in the State of California as shown in Figure 3. By the end of 2001, reclaimed water use in California had reached over 0.65 billion m^3/yr (Metcalf & Eddy, 2007). Compared to Japan and Korea, agricultural and landscape irrigation is the dominant use of reclaimed water in California. This accounts for about two thirds of the total volume of water reuse. Artificial ground water recharge is also practiced successfully in California.

Rainwater

The rainwater has also been considered as an alternative community water resource. As rainwater is usually of better quality than treated wastewater, there is generally better public acceptance for its utilization. Rainwater utilizing systems vary, from the small and simple, to the large and complex. Where a suitable harvesting surface exists, such as a rooftop or even a road, the collection of rainwater can provide a sufficient and low cost source of water for non-portable usage. Roof runoff for miscellaneous usage, such as toilet flushing and water-cooling, is employed, successfully, in Japan. It is used at both an individual and large-scale level. Although road runoff contains higher pollutant concentrations than roof runoffs, there is still a trend to use of the relatively polluted rainwater drained from streets and courtyard surfaces for service water in some countries. (Nolde, 2007).

Rainwater collection can be successfully practiced at different scales in both urban and rural areas. An early attempt of large-scale harvesting in Japan can be seen at Kokugikan as shown in Figure 4(a). This system was introduced in 1984 and utilizes a tank with an effective capacity of 750 m^3 to store rainwater collected from the roof with an area of 8400 m^2. The collected water is used for cooling system and toilet flushing. Other early examples are the City Office Building in Sumida ward, Tokyo in 1988, and the Tokyo Dome Stadium in 1988.

Figure 4(b) shows the system at the Tokyo Dome Stadium having a storage tank with a capacity of 3000 m³. The maximum capacity for storage is controlled at 2000 m³. This means that one third of its capacity is used for flood control (this capacity is left empty for the case), another 1000 m³ is used for miscellaneous water use and the remaining 1000 m³ is used for water use such as earthquake and fire events.

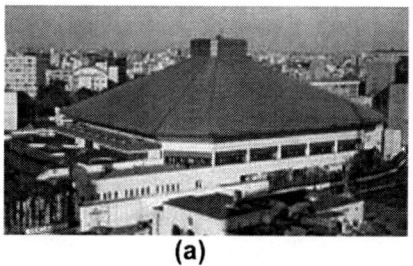 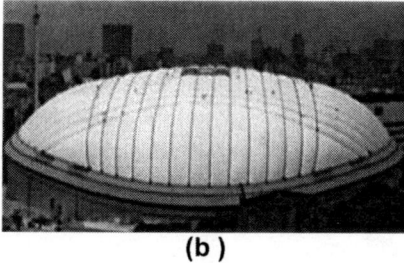

(a)   **(b)**

Figure 4 Rainwater harvesting at (a) Kokugikan for Sumo wrestling and (b) Tokyo Dome Stadium

The water retention pavement, used as a countermeasure against heat, is one of the new types of rainwater use adopted in Tokyo. The Rainwater, falling on each square block, flows into a side ditch and is then stored in an underground tank (155 m³) as shown in Figure 5. Water from the tank is then supplied to paving blocks via sprinkling pipes. Unlike the conventional asphalt pavement, data show that there has been a reduction in surface temperature of 13°C or less.

However, there are limitations in the use of rainwater as a stable and sustainable water source. The amount being collected deeply relies on the variability of weather and climate. Rainwater should be used, therefore, in conjunction with other water resources such as groundwater or reclaimed water (as shown in the Figure 2).

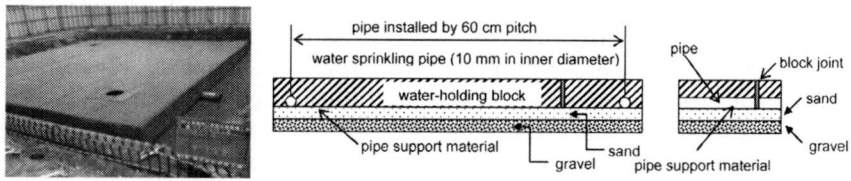

Figure 5 Rainwater use for water-holding block pavement in Tokyo

WATER QUALITY EVALUATION FOR PUBLIC UNDERSTANDING

Although using reclaimed water as an alternative water resource is becoming more common, there are still concerns from the general public about adopting this. Generally, there is a misconception about the quality of reclaimed water. Since municipal wastewater is discharged to the drain from toilets, bathrooms, kitchens, and laundries, there is frequently a psychological objection to having close contact with it (Baumann, 1983; Asano, 1998). The origin of reclaimed water always leads to concerns from citizens, even after reclamation. Unlike reclaimed water, it is thought that rainwater is relatively free from impurities. However, the rainwater and runoff quality in urban areas can often deteriorate. During harvesting and long storage, rainwater picks up pollutants from the atmosphere. It then picks up deposits on the harvesting surface. This is particularly noticeable in first flush runoff, which carries most contaminants, such as micropollutants and pathogens.

The general public often has a poor understanding of water quality in alternative resources. Many communities, for example, have a negative perception of reclaimed water and urban rainwater. The complexities in recognizing suitable water quality in these resources, causes many citizens to hesitate before using them.

One of the most difficult tasks in water evaluation is translating the scientific, complex data into comprehensive and accessible information that is useful to, both, the general public and technical teams. It is important, therefore, to develop a simple and effective way to represent the differences in water quality using alternative resources. The objective of this section is to propose a methodology to develop a water quality information platform. The information platform provides a criteria reference for water quality, and applies it to different waters using a water quality labeling and ranking system. The identification of water quality is crucial when evaluating the water quality level in different community water resources.

The general practice is to describe water quality according to the official water quality guidelines, on the basis of different water quality parameters. As these traditional practices consist of 'bunch' data, the general public has neither the inclination nor the training to study the results. The proposed water quality information platform reduces the multivariate nature of water quality data by providing a scoring system that will logically combine all water quality measures; it provides a simple and comprehensive description of different waters. The proposed water quality evaluation method consists of five steps;

(1) Data Gathering: Collecting water quality data of water sources and water quality standard values
(2) Water Quality Scoring: Determining 5-level scores based on scientific evidences such as drinking and ambient water quality standards
(3) Water Quality Characterization: Characterizing water quality using multiple scores based on scientific evidence
(4) Water Use Ranking: Ranking by water availability and water quality characterizing information for several water usages
(5) Water Quality Labeling and User judgment: Judging the type of water usage such as gardening, toilet flushing, car washing, landscape, etc.

   This framework can evaluate and summarize water quality data in a way that should be scientifically accepted, and easily understood, by both professionals and the general public. The relative evaluation of water quality, in this case, is beneficial to using absolute values and parameters of water quality. If the evaluation of water quality is conducted using familiar community water resources such as well-known river waters, it provides citizens with a better understanding of water quality levels in general. The development of a water quality information platform, therefore, might help change the perceptions of the general public towards the use of reclaimed water. In addition to this, the platform will provide a useful tool for describing the water quality of different waters and ranking their suitability for different usages. Using a case study in Japan, the next section will provide a detailed description of each step of this methodology.

CHARACTERIZING AND RANKING DIFFERENT WATER RESOURCES - A CASE STUDY IN JAPAN

The water quality information platform provides a water quality reference for the community to evaluate reclaimed waters (or any alternative water sources). The water quality information platform should integrate the significant physico-chemical and biological constituents of different water resources available to the community. It should then present them in a simple and scientifically defensible way. The following case study in Japan describes five steps to develop a nation-wide water quality information platform that indicate the ranking and labeling of water resources. The steps described in this case study can be used as a guide for water quality evaluation in any community or a region, depending on the needs or objectives of that particular community. In this case study, the water quality of advanced treated wastewater, rainwater, runoff waters and groundwater were evaluated and compared with the first-class rivers in Japan.

Data Gathering

To develop the water quality information platform, an appropriate assessment of different water quality data is essential. The first step is to make a water quality reference that can be used to compare the water quality level of different water sources. In this study river waters were selected as the reference water resource. This is because, in Japan, surface water is the major source of water supply. Out of the 109 rivers throughout Japan, designated as the "first-class" rivers by the River Bureau, Ministry of Land Infrastructure Transport and Tourism, thirty seven rivers were selected to develop the reference profile (Harada et al., 2006). Water samples were collected during the dry season from the most downstream section of the river.

To confirm the selection was a fair representation, catchment areas, population density, concentrations of nutriments of the selected rivers, were compared to the annual average of the total 109 rivers. This was based on monthly monitoring (The River Bureau, Ministry of Land Infrastructure Transport and Tourism Japan, 2004). The histograms in Figure 6 show that the distributions of catchment areas, population density, T-N and T-P concentrations in the selected 37 rivers were in agreement with 109 first-class rivers. The selected rivers, therefore, were regarded as representative and could then be used as a water quality criteria reference.

The first objective was to select the water quality parameters that would be used to evaluate the water resources. In order to develop the water quality information platform in this study, the data related to several chemical analyses (Managaki et al., 2005; Harada et al., 2006; Komatsu et al., 2007; Murakami et al., 2008a; Nakada et al., 2008).

- Organic matter (TOC, COD, BOD)
- Nitrogen (T-N, NO_2-N, NO_3-N, NH_4-N)
- Phosphorus (T-P, PO_4)
- Metals (Zn, Cu, Mn, Ni, Pb, Cd, As, etc)
- Endocrine disrupting chemicals (E1, α-E2, β-E2, E3, BPA, NP, OP)
- Polycyclic aromatic hydrocarbons (benzo(a)pyrene etc.)
- Perfluorinated surfactants (PFOS, PFOA etc)
- Pharmaceuticals and personal care products (ibuprofen, carbamazepine etc)
- Synthetic detergents (LAS, SPC etc)

Asides from the typical chemical water quality parameters of organic matters nitrogen and phosphorus, other micro pollutants such as heavy metals, endocrine disrupting chemicals (EDCs), polycyclic aromatic hydrocarbons (PAHs), perfluorinated surfactants (PFSs), pharmaceuticals and personal care products (PPCPs) and synthetic detergents were also taken into account for developing the water quality information platform. Whilst only the chemical data were used in this paper, in a more advanced assessment study, it is better to use multiple bioassay data along with chemical data. Bioassays are important for the effective assessment of water quality. This is because, chemical analysis gives only quantitative statements for a limited number of chemicals. For an integrated assessment of the effects or toxic potentials of chemical substances, bioassays such as estrogenic activities, algal growth inhibition, algal growth potential, Microtox, and mutagen formation potential (MFP) are required. However, as bioassay data are not easily available, this case study will be described with the use of chemical data only.

The selecting of the water quality parameters to be evaluated depends on the intended water usage. If the water usage is only limited to purposes such as toilet flushing, the conventional water parameters, relating to coloring and appearance, may be used to develop the water quality information platform. However, additional parameters are recommended when the water quality information platform is being used as a decision tool for more advanced water usage such as amenity purposes involving human contact.

The water quality references of each parameter were represented as percentile distributions. Percentiles are values that divide a set of observations into 100 equal parts. The percentile rank is the proportion of values in a distribution that a specific value is greater than or equal to. Therefore, the lower percentile rankings, in this case, indicate the better water quality while the higher percentile rankings indicate poorer water quality. Thus, the 0th and 100th percentiles describe the best and worst of the water quality distribution respectively, while the median or 50th percentile describes the water quality value at the middle of the distribution. These percentile distributions can be used as a water quality criteria reference for other water resources such as reclaimed waters. In this study, different types of water resources such as sub-urban rainwater (Shimamura et al., 2007; Ichiki et al., 2008; Murakami et al., 2009a); infiltrated water of road runoff (Murakami et al., 2008b); secondary treated effluent (Nakada et al., 2007a); advanced treated wastewater by sand filtration (Nakada et al., 2007a); infiltrated water of secondary treated effluent (Shinohara et al., 2006; Nakada et al., 2007b); advanced treated wastewater by ozonation (Nakada et al., 2007a); and ground water (Nakada et al., 2008; Murakami et al., 2009b) were compared with the river water quality references.

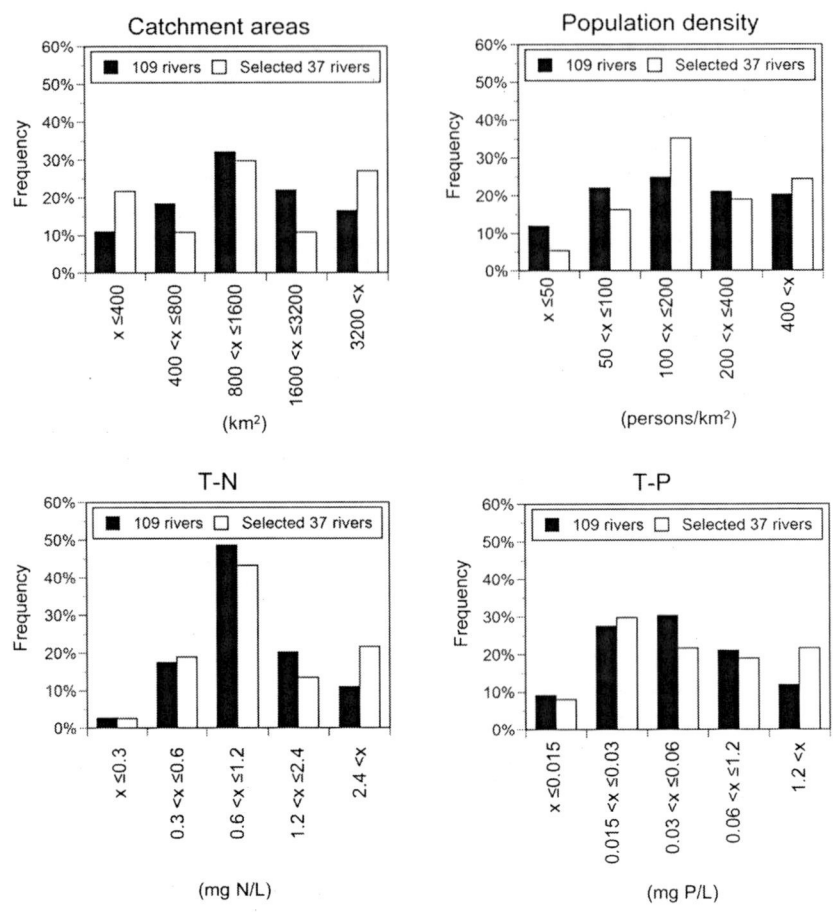

Figure 6 Comparison of catchment areas, population density, T-N concentrations, and T-P concentrations between 109 rivers and selected 37 rivers

Figure 7 shows the percentile distributions of TOC, NO_2-N +NO_3-N and Mn concentration of 37 first class rivers in Japan. The figure also compares the rainwater, infiltrated water of road runoff, secondary effluent, sand filtered effluents and ozonated effluent with the reference river water quality. As in this example of Figure 7, the Mn concentration in rainwater is equal to 10% of the percentile distribution. The 10th percentile describes a value (or 0.0015 mg/L of Mn) for which around 10% of selected rivers in the distribution are equal to, or lower to the Mn concentration in the rainwater. In the case of TOC, the percentile rank of the rainwater is 20%, indicating 80% of the river water contains a higher concentration of TOC than the rainwater. There is no significant difference in NO_2-N +NO_3-N among the three types of treated wastewater effluent. The value is equal to the 100th percentile rank of the selected rivers. Since there is no water quality standard set for emerging contaminants such as PPCP, this kind of comparison with known river waters can be used to reflect the water quality condition of the water source.

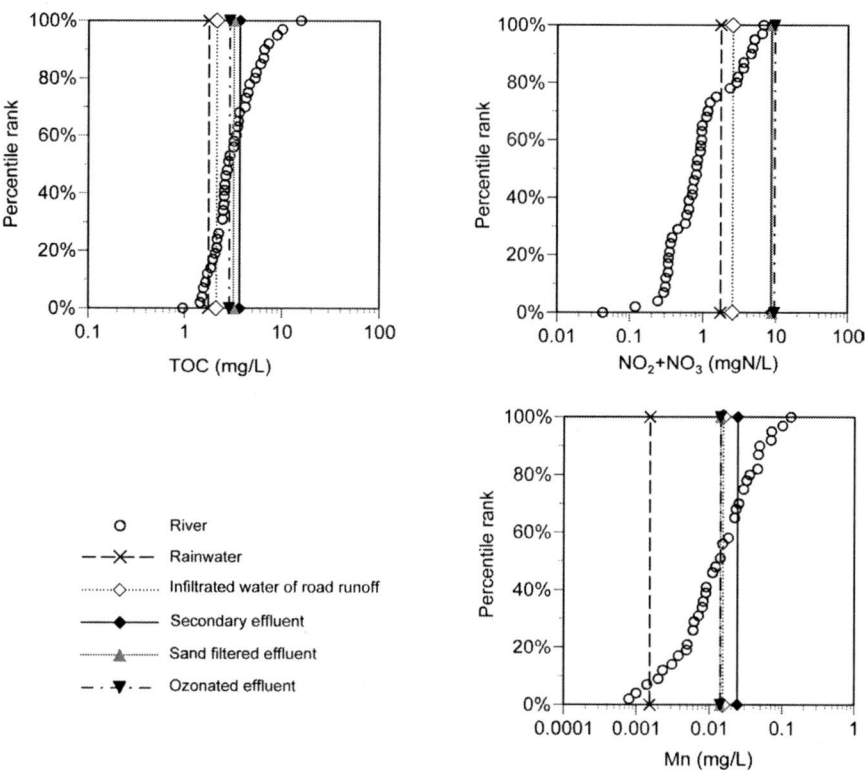

Figure 7 Comparison of the TOC, NO_2-N +NO_3-N and Mn concentration in rainwater, infiltrated water of road runoff, secondary effluent, sand filtered effluents and ozonated effluent with those of percentile distribution of reference river waters

Water Quality Scoring

Once the water quality references have been developed as the percentile distribution, the next step is the scoring process. The first stage of this step is to divide the water quality percentile distribution of rivers in to 5 factors according to the criteria of each water quality parameters. This describes the state of the water quality based on scientific evidence such as drinking and ambient water quality standards. The other water resources were compared with the reference scale to determine their scores.

All the chemical water quality parameters were classified into the following five different water quality factors.

1. Coloring and appearance
2. Eutrophication
3. Conservation of aquatic ecosystem
4. Chronic toxicity to human
5. Acute toxicity to human

These five factors are combined to form the score matrix. Although the five different factors were used in this study, the number of factors may depend on the availability of data and the purpose of the water quality evaluation. The factors based on chemical analysis were scored according to the water quality standard. The 5-level scores index, which gives a number between "0" and "4" was used, with "0" being the best and "4" indicating the poorest water quality within the distribution. The score "0" implies the water is safe with a virtual absence of threat or impairment. The score "1" implies there is a minor degree of threat or impairment. Continuous monitoring is needed for that particular water. The score "2" to "4" means the water is threatened or impaired in terms of that particular water quality parameter and some action is needed to improve the water quality. Depending on the degree of action level, scores were given as "2", "3" or "4".

Figure 8 demonstrates the scores for the distribution of Mn concentration in the 1st class rivers of Japan. Within the same percentile ranking, it considers the factor of "coloring and appearance". Since 0.05 mg/L of Mn is the guideline value for coloring and appearance in Japan, less than 0.005 mg/L of Mn in waters is regarded as score "0", 0.005–0.05 mg/L as score "1", 0.05–0.5 mg/L as score "2", 0.5–5 mg/L as score "3", and more than 5 mg/L as score "4". The waters of score "3" show that the water, even after 10 times dilution, is affected by coloring and appearance. Similarly, scores can be given to all other water quality parameters.

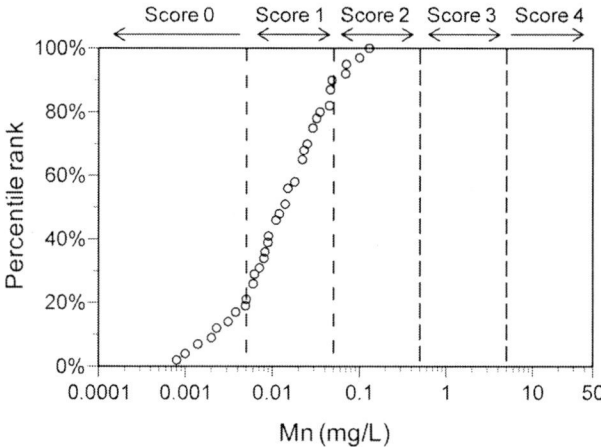

Figure 8 Distribution of Mn concentrations and their scores in the 1st class rivers based on the water quality factor for coloring and appearance

Table 2 is an example of the water quality threshold values for scoring. Once all the five factors being evaluated had been divided into the five level index in the reference distributions, the scores of the different types of water resources could be determined. Table 3 shows the final score matrix of the different water resources.

Table 2 Example of water quality thresholds value for scoring

	Unit	Score 0	Score 1	Score 2	Score 3	Score 4
Coloring appearance						
pH	-	5.8–8.6	4.8–5.8, 8.6–9.6	3.8–4.8, 9.6–10.6	2.8–3.8, 10.6–11.6	2.8, 11.6
TOC	mg/L	<0.5	0.5–5	5–50	50–500	500
Cu	mg/L	<0.1	0.1–1	1–10	10–100	100
Zn	mg/L	<0.1	0.1–1	1–10	10–100	100
Mn	mg/L	<0.005	0.005–0.05	0.05–0.5	0.5–5	5
...						
Acute toxicity to human						
NO_2+NO_3	mgN/L	<1	1–10	10–100	100–1000	1000
As	mg/L	<10	10–100	100–1000	1000–10000	10000
...						
Conservation of aquatic ecosystem						
NH_4	mgN/L	<0.02	0.02–0.2	0.2–2	2–20	20
Zn	mg/L	<0.003	0.003–0.03	0.03–0.3	0.3–3	3
...						

Table 3 Example of scores in different water resources

	100% percentile rank river	50% percentile rank river	0% percentile rank river	Groundwater	Infiltrated water of road runoff	Rainwater	Infiltrated water of secondary wastewater	Wastewater treated by sand filtration	Wastewater treated by ozonation
Coloring and appearance									
pH	1	0	0	0	0	2	0	0	0
TOC	2	2	1	2	1	1	1	1	1
Cu	0	0	0	0	0	0	0	0	0
Zn	1	0	0	0	0	0	0	0	0
Mn	2	1	0	0	1	0	1	1	1
...									
Acute toxicity to human									
NO_2+NO_3	1	0	0	1	1	1	1	1	1
As	0	0	0	0	0	0	0	0	0
...									
Conservation of aquatic ecosystem									
NH_4	2	1	0	0	0	1	1	1	1
Zn	2	1	0	0	1	1	1	2	2
...									

Water Quality Characterization

In this section, the scores given to each resource are summarized and the water quality is characterized. Characterization works in a similar way to a water quality label on a drinking water bottle. As several water quality parameters (nitrogen, fluoride, etc) are seen in a water bottle to show the quality of water, five water quality water quality factors (coloring, eutrophication, acute toxicity, etc) were used in this study to characterize the different water resources including reclaimed waters. Based on the scores in Table 3, characterizing indexes were calculated. In the case of the water quality factor, "Acute toxicity to human", the characterizing indexes were determined by taking the maximum score value, while the index values of all the other 4 factors might be determined by taking the average of each score under a particular water quality factor. It is also possible to give weighting coefficients to the water quality parameters of each factor.

In order to understand the water quality characterizing indexes easily, radar charts (also known as spider charts) were developed to express water quality graphically. In the radar charts, each axis represents the water quality factor developed at step 2 of water quality Scoring. The five factors are plotted along the chart axes from the center of the chart. For a given water quality factor, a score of "4" is located in the center and "0" at the edge. Lines connect the axes at the position of each data object, forming a spiral around the center. The larger the area of the spiral, the better the water quality is of a particular water source.

Figure 9 shows the radar chart related to the 0th, 50th and 100th percentile rank of 1st class rivers in Japan, as well as groundwater, ozonated wastewater effluent and rainwater. According to the graphical representation, the overall water quality of groundwater is roughly equal to the 50th percentile rank of the 1st class rivers. The radar charts of ozonated wastewater effluent and rainwater depict that their overall water quality is between the 50th and the 100th percentile rank of the 1st class rivers.

However, each water quality factor should be carefully evaluated in terms of their water usage purpose. For example, ozonated wastewater has a poorer water quality score for eutrophication than the 100% percentile rank of a river. This indicates that the ozonated wastewater, in this case, is not suitable for water usage such as landscaping pond water, in which the eutrophication would be a possible impairment factor.

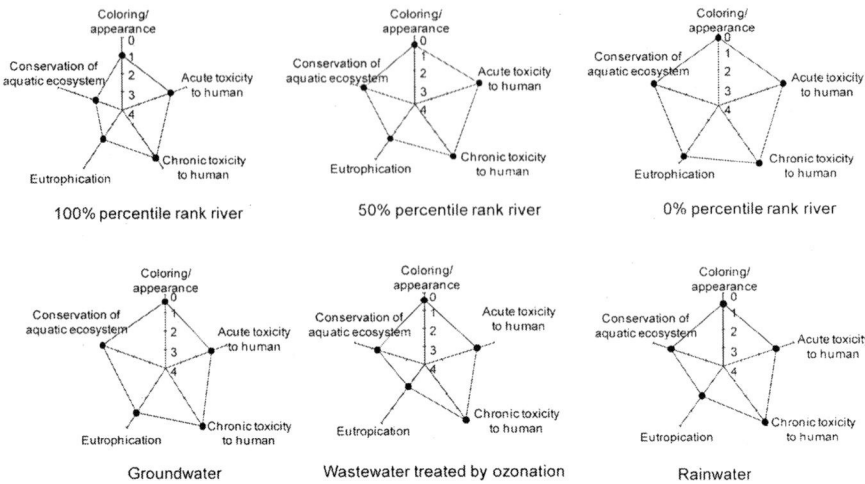

Figure 9 Example of water quality characterization of several water resources

Water Use Ranking

The aim of this section is to integrate the characterized water resources with the objectives of different water usages. Water used for amenity, landscaping, toilet flushing and fire protection all have different water quality requirements. The suitability of the water resource, therefore, depends on the usage for which it is intended. In this step of the methodology, the values of the water quality factors were used to provide a water use ranking as "good", "fair" or "not applicable", for different water usages. Table 4 describes the criteria used to judge the ranking for different water usages. When considering this, it is not necessary to consider all the factors for each water usage. As in the Table 4, for amenity water usage, only four factors were considered whilst five factors were considered in water for landscaping usage.

Table 5 shows a ranking of the different water usages of ozonated wastewater effluent in terms of its water quality factors. When considering the water quality factor, "acute toxicity to human", ozonated wastewater effluents can be ranked as "good" for most water usage. However, when used as water for emergencies, in which household water use is involved, it is only ranked "fair".

Table 4 Criteria for several water applications

	Coloration/ appearance	Acute toxicity to human	Chronic toxicity to human	Eutrophication	Conservation of aquatic ecosystem	...
Amenity water	≥2: n.a. 1–2: fair <1: good	≥4: n.a. 3–4: fair <3: good	≥3: n.a. 2–3: fair <2: good	≥3: n.a. 2–3: fair <2: good	-	
Water sprinkling and car wash	≥3: n.a. 2–3: fair <2: good	≥4: n.a. 3–4: fair <3: good	≥4: n.a. 3–4: fair <3: good	-	-	
Toilet water	≥3: n.a. 2–3: fair <2: good	≥4: n.a. 3–4: fair <3: good	≥4: n.a. 3–4: fair <3: good	-	-	
Landscaping water	≥3: n.a. 2–3: fair <2: good	≥4: n.a. 3–4: fair <3: good	≥4: n.a. 3–4: fair <3: good	≥3: n.a. 2–3: fair <2: good	≥3: n.a. 2–3: fair <2: good	
Water for emergency use	≥2: n.a. 1–2: fair <1: good	≥2: n.a. 1–2: fair <1: good	≥3: n.a. 2–3: fair <2: good	-	-	
Water for fire protection	≥4: n.a. 3–4: fair <3: good	≥4: n.a. 3–4: fair <3: good	≥4: n.a. 3–4: fair <3: good	-	-	

n.a.: not applicable   -: not considered

Table 5 User judgment of wastewater treated by ozonation for several water applications

	Coloration/ appearance	Acute toxicity to human	Chronic toxicity to human	Eutrophication	Conservation of aquatic ecosystem	...	Example of user judgement
Amenity water	good	good	good	fair	-		fair
Water sprinkling and car wash	good	good	good	-	-		good
Toilet water	good	good	good	-	-		good
Landscaping water	good	good	good	fair	good		fair
Water for emergency use	good	fair	good	-	-		fair
Water for fire protection	good	good	good	-	-		good

Water Quality Labeling and User Judgment

The last, but most significant, step is to look at the user judgment of the various water resources and their different usages. In the previous step, each water resource was ranked with respect to factors such as "coloration", "acute toxicity" etc. The aim of this step is to decide the overall judgment for water usage of each water resources. As shown in Table 5, ozonated wastewater use for amenity purposes has been ranked as "good" for the water quality categories of coloration, acute toxicity to human, chronic toxicity to human. However, it has been ranked as "fair" for the water quality factor of eutrophication. The overall water quality rank of ozonated wastewater for amenity purposes was, therefore, labeled "fair", taking the lowest rank in each factor. All of the water resources subjected to evaluation have been ranked and labeled in Table 6.

As summarized in Table 6, water usage labeling evaluates each water resource by simply describing "how good or how bad" they are for different water usages. This labeling information provides an easy way to compare the water quality of different water resources directly with those in the river water quality reference. It can help, therefore, to overcome public uncertainty surrounding the use of alternative water resources (reclaimed water, runoff, etc). The concept and process described in this case study can be used as a guide for water quality evaluation for all community-based water resources. However, it is worth keeping in mind that the required water quality may vary, depending on the needs and objectives of that particular community. It must also be combined with effective management and monitoring system.

ENERGY CONSUMPTION FOR COMMUNITY-OWNED WATER RESOURCES

Although the water quality labeling framework can indicate the relative water quality of a particular water resource, the water quality labeling system should never stand alone when considering water usage. In order to optimize community-owned water use, it is important to analyze both the quality and the quantity of the available water resources and their potential reuse applications. Moreover, the energy consumption for water intake, the purification and the distribution of water should be a further consideration.

Figure 10 shows the percentage of water demand for different domestic applications in Tokyo. The daily water consumption in the Tokyo, per capita is around 245 liters (Japan Water Work Agency, 2003). As the most common application, 28% of this is being used for "toilet flushing" followed by "water for bathing" (23%), and "water for kitchen" (23%). "Water for washing" (17%) is next, with the

Table 6 Example of labeling of several water resources

	Amenity water	Water sprinkling and car wash	Toilet water	Landscaping water	Water for emergency use	Water for fire protection
100% percentile rank river	fair	good	good	not applicable	not applicable	good
50% percentile rank river	good	good	good	fair	fair	good
0% percentile rank river	good	good	good	good	good	good
Groundwater	good	good	good	good	fair	good
Rainwater	fair	good	good	fair	good	good
Infiltrated water of road runoff	good	good	good	fair	fair	good
Infiltrated water of wastewater	fair	good	good	fair	not applicable	good
Wastewater treated by sand filtration	fair	good	good	fair	fair	good
Wastewater treated by ozonation	fair	good	good	fair	fair	good

remaining 8% of water being used for "sprinkling and car washing". As discussed in the previous section, alternative water sources such as untreated groundwater, rainwater, road runoff or reclaimed water, available in the community, can be adopted to meet water demands for different purposes. These resources, therefore, can contribute to the conservation of freshwater resources. A large percentage of water used for domestic activities does not need to be of the same high quality as drinking water or water used in kitchen.

Untreated ground water or rainwater is of lower quality than tap water or treated ground water. However, it is generally of higher quality than reclaimed water or water from road runoff. Table 7 demonstrates the applicability of different water resources for domestic usages. Untreated ground water or rainwater can be used for washing, bathing, sprinkling, car washing and toilet flushing. This can account for up to 77 % of the total per capita water demand. Toilets flushing, sprinkling and car washing consume about 36% of the domestic water usage. Reclaimed water or road runoff can be used to cover this water application and can then conserve the high quality water by 36%. This example in Tokyo shows how using alternative water resources for non-potable domestic usage could reduce demand for traditional potable water supplies by as much as 77%.

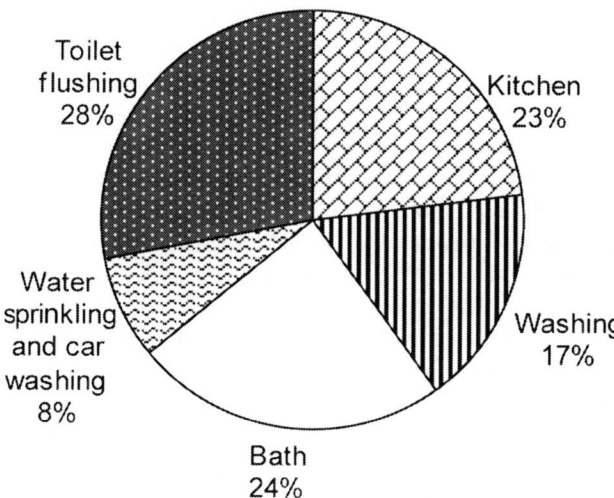

Figure 10 Percentage of water demand for different domestic applications in Tokyo

Water Infrastructure for Sustainable Communities

Table 7 Applicability of different water resources for domestic use

Water resources	Water application				
	Kitchen	Washing	Bath	Water sprinkling and car washing	Toilet flushing
Tap water	Yes	Yes	Yes	Yes	Yes
Treated ground water	Yes	Yes	Yes	Yes	Yes
Untreated ground water	No	Yes	Yes	Yes	Yes
Rain water	No	Yes	Yes	Yes	Yes
Road runoff	No	No	No	Yes	Yes
Reclaimed wastewater	No	No	No	Yes	Yes

In addition to the quality and quantity assessments of community-owned water resources, the energy consumption is also an important factor to consider when discussing sustainable water usage. Figure 11 compares the carbon dioxide (CO_2) emissions for the production of tap water, reclaimed water, rainwater and treated and untreated ground water in Tokyo. The CO_2 level can be used as a universal unit of measurement to assess global warming potential. The energy requirements for the production of different water resources can be estimated as a part of the treatment process configuration.

In the case of treated groundwater, energy requirements for pumping, membrane treatment, disinfection and water supply have all been taken into account in when calculating the total energy requirement per unit volume ($1 m^3$). In this example, the total energy requirement for the production of treated groundwater has been calculated as 1.51 kwh/m^3. It considers the operating time of all pumps, water supply units and control units, etc for an average household in Tokyo. The energy requirement (kWh) can be transformed to a value of CO_2 emission (kg-CO_2 equivalent) by multiplying it with the carbon conversion factor, that is 0.332 (kg-CO_2/kWh). This carbon conversion factor is specific to Tokyo in 2008. This value can vary between other cities and countries, depending on power generation method, fuel type for electricity generation and emission permit transactions.

As shown in Figure 11, CO_2 emissions from the production of treated groundwater can be estimated as 0.50 g/L. The CO_2 emissions from the production of tap water (0.34 g-CO_2/L), reclaimed water (1.20 g-CO_2/L), rainwater (0.18 g-CO_2/L) and untreated groundwater (0.14 g-CO_2/L) have been estimated in a similar way (see Figure 11). Rainwater and untreated groundwater production emits less CO_2 than of tap water, because these waters only require energy for pumping and supply control. Thus, rainwater and untreated groundwater can be considered as environmentally friendly water resources, as well as attractive solutions for water shortages. If one household in Tokyo, comprising of four people, were to use around 1 m³ of water per day, but they used rainwater for washing, bathing, sprinkling, car washing and toilet flushing, it could be possible to decrease CO_2 emissions by up to 123 g-CO_2/day. In summary, rainwater usage for non-portable water application could reduce the demand for traditional potable water supply by 77% while cutting down the emission of greenhouse gas by as much as 36%. Similarly, the usage of untreated groundwater, which could account for 77% of total water needs, could might down CO_2 emissions by 45%.

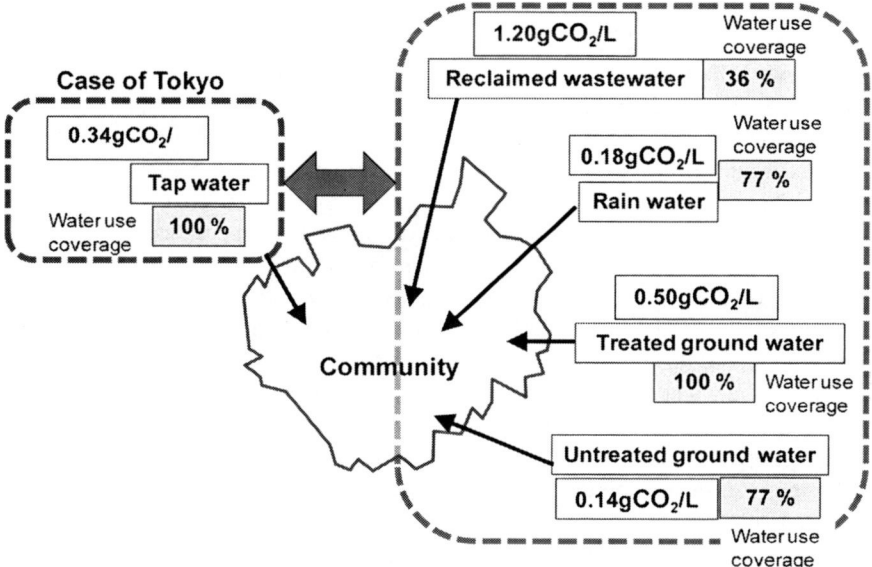

Figure 11 Comparison of CO_2 emission for different water production in Tokyo

However, the CO_2 emissions for reclaimed water are as high as 1.2 g-CO_2/L. It must be noted that this value may underestimate the potential benefits of reclaimed water. This value includes the total energy consumption for the biological treatment process as well as the energy consumption for advanced treatment. If reclaimed water production from secondary effluent, by sand filtration and disinfection, is considered, the CO_2 emission is estimated as 0.17 g-CO_2/L. Also, if the reclaimed water is made available to local water users only, there will be less water distribution energy required.

CONCLUSION

Global freshwater resources are in growing demand, owing to rapid urbanization and climate change. This urbanization is particularly evident in developing countries in Asia and Africa. In Asian developing countries, it is predicted that the water availability per capita has drastically decreased from 1990 to 2050. This is due to a combination of factors such as climate change and increased water demand through population and economical growth.

The use of alternative community water resources is a promising option for many countries needing to meet increasing water demands under decreasing water availability conditions. The usage of reclaimed water and rainwater has received particular interest. Although reclaimed water and rainwater have been applied in some cases, the rate of reuse is still limited to just Japan; the general public still fears adopting these alternative water resources. There are still concerns relating to water quality.

It is important to develop evaluation methodologies of community-owned water resources based on water quality and energy consumption. The proposed evaluation of water quality composes of five steps: (1) data gathering, (2) water quality scoring, (3) water quality characterization, (4) water use ranking, and (5) water quality labeling and user judgment. This water quality labeling system evaluates the applicability of different water resources for different applications. The outcome of this system can provide an easy representation of water quality. It compares the water quality of different water resources directly with water quality reference of well-known rivers. The water resources are also evaluated based on the energy consumption for their water production. This evaluation provides a comparison of the energy consumption of community-owned water resources and that of the existing water supply system. This way, the public can see the benefits of using community-owned water resources. The proposed evaluation methodologies are expected to overcome public uncertainties relating to the use of alternative water resources.

ACKNOWLEDGEMENTS

The authors acknowledge the CREST project, 'Risk-based Management of Self-regulated Urban Water Recycle and Reuse System' (granted by JST), for the concept of water quality ranking and labeling.

REFERENCES

Asano T. (1998). *Wastewater Reclamation and Reuse*, 1st edn, CRC Press.

Bates B. C., Z. W. Kundzewicz, S. Wu and J. P. Palutikof (eds.) (2008). *Climate Change and Water*. Technical Paper of the Intergovernmental Panel on Climate Change, IPCC Secretariat, Geneva, 210 pp.

Baumann D. D. (1983). Social acceptance of water reuse. *Applied Geography*, 3(1), 79–84.

Brekke L. D., Kiang J. E., Olsen J. R., Pulwarty R. S., Raff D. A., Turnipseed D. P., Webb R. S. and White K. D. (2009). Climate change and water resources management—A federal perspective: U.S. Geological Survey Circular 1331, 65 p. (Also available online at http://pubs.usgs.gov/circ/1331/.)

Frederick K. D. and Major D. C. (1997). *Climatic Change*, 37. Kluwer Academic Publishers, Printed in the Netherlands, pp. 7–23.

Furumai H. (2007). Reclaimed stormwater and wastewater and factors affecting their reuse. *Cities of the Future Towards Intergrated Sustainable Water and Landscape*. IWA publishing, pp. 219–235.

Furumai H. (2008). Rainwater and reclaimed wastewater for sustainable urban water use. *Physics and Chemistry of the Earth*, 33, 340–346.

Haman D. Z. and Brown D. A (1994). AE250, Agricultural and Biological Engineering Department, Florida Cooperative Extension Service, Institute of Food and Agricultural Sciences, University of Florida. Original publication date April, 1994. Reviewed July, 2002. Visit the EDIS Web Site at http://edis.ifas.ufl.edu.

Harada A., Nakada N., Yamashita N., Sato N., Ito M., Suzuki Y., Tanaka H. and Furumai H. (2006). Relative evaluation of reclaimed wastewater in urban area in comparison with the distribution of river water quality. *Environmental Engineering Research*, 43, 501–508. (in Japanese)

Ichiki A., Ido F. and Minami T. (2008). Runoff characteristics of highway pollutants based on a long-term survey through a year. *Water Science and Technology*, 57(11), 1769–1776.

Japan Sewage Works Association (2005). http://www.jswa.jp/05_arekore/dataroom/05/riyou/shorisui.html.

Japan Water Works Association (2003). Japan Water Supply Data Report 2003. http://www.jwwa.or.jp/english/water_en/water-eindex.html.

Kim K. (2009). The Present State and Policy Plan of Sewerage in Korea, Special Seminar at Dept. of Urban Engineering, the University of Tokyo.

Komatsu T., Manabe Y., Himeno S., Harada A., Murakami M. and Furumai H. (2007). Distribution of mutagen formation potential in nationwide river waters in Japan and relationship with general water quality or river basin characteristics. *Journal Japan Society on Water Environment*, 30(8), 433–440.

Managaki S., Kojima S., Harada A., Nakada N., Tanaka H. and Takada H. (2005). Development of analytical method of linear alkylbenzenesulfonates and their degradation intermediates by high performance liquid chromatography equipped with 28 Water Infrastructure for Sustainable Communities: China and the World tandem mass spectrometry and its application to major Japanese rivers. *Journal Japan Society on Water Environment*, 28(10), 621–628. (in Japanese).

Metcalf and Eddy (2007). *Water Reuse: Issues, Technologies, and Applications, Inc.* An AECOM Company.

Murakami M., Imamura E., Shinohara H., Kiri K., Muramatsu Y., Harada A. and Takada H. (2008a). Occurrence and sources of perfluorinated surfactants in rivers in Japan. *Environmental Science & Technology*, 42(17), 6566–6572.

Murakami M., Kuroda K., Sato N., Fukushi T., Takizawa S. and Takada H. (2009b). Groundwater pollution by perfluorinated surfactants in Tokyo. *Environmental Science & Technology*, **43**(10), 3480–3486.

Murakami M., Sato N., Anegawa A., Nakada N., Harada A., Komatsu T., Takada H., Tanaka H., Ono Y. and Furumai H. (2008b). Multiple evaluations of the removal of pollutants in road runoff by soil infiltration. *Water Resources*, **42**(10–11), 2745–2755.

Murakami M., Shinohara H. and Takada H. (2009a). Evaluation of wastewater and street runoff as sources of perfluorinated surfactants (PFSs). *Chemosphere*, **74**(4), 487–493.

Nakada N., Kiri K., Shinohara H., Harada A., Kuroda K., Takizawa S. and Takada H. (2008). Evaluation of pharmaceuticals and personal care products as water-soluble molecular markers of sewage. *Environmental Science & Technology*, **42**(17), 6347–6253.

Nakada N., Shinohara H., Murata A., Kiri K., Managaki S., Sato N. and Takada H. (2007a). Removal of selected pharmaceuticals and personal care products (PPCPs) and endocrine-disrupting chemicals (EDCs) during sand filtration and ozonation at a municipal sewage treatment plant. *Water Resources*, **41**(19), 4373–4382.

Nakada N., Yamashita N., Miyajima K., Suzuki Y., Tanaka H., Shinohara H., Takada H., Sato N., Suzuki M., Ito M., Nakajima F. and Furumai H. (2007b). Multiple evaluation of soil aquifer treatment for water reclamation using instrumental analysis and bioassay. In: *Southeast Asian Water Environment*, H. Furumai, F. Kurisu, H. Katayama, H. Satoh, S. Ohgaki and N.C. Thanh (eds.), 2, pp. 303–310.

Nolde, E., (2007). Possibilities of rainwater utilisation in densely populated areas including precipitation runoffs from traffic surfaces. *Desalination*, **215**, 1–11.

Shimamura T., Iwashita M., Iijima S., Shintani M. and Takaku Y. (2007). Major to ultra trace elements in rainfall collected in suburban Tokyo. *Atmospheric Environment*, **41**(33), 6999–7010.

Shinohara H., Murakami M., Managaki S., Kojima S., Takada H., Sato N., Suzuki Y. and Nakada N. (2006). Removal of water-soluble organic micro-pollutants by soil infiltration. *Environmental Science (Japan)*, **19**(5), 435–444. (in Japanese).

The River Bureau, Ministry of Land Infrastructure Transport and Tourism Japan (2004). River water quality yearbook 2002. (Kasen suishitsu nenkan 2002); River Association Japan ed., (in Japanese).

Trenberth K. E., Jones P. D., Ambenje P., Bojariu R., Easterling D., Klein T. A., Parker D., Rahimzadeh F., Renwick J. A., Rusticucci M., Soden B. and Zhai P., (2007). Observations: Surface and Atmospheric Climate Change. In: *Climate Change 2007: The Physical Science Basis. Contribution of Working Group I to the Fourth Assessment Report of the Intergovernmental Panel on Climate Change*, S. Solomon, D. Qin, M. Manning, Z. Chen, M. Marquis, K. B. Averyt, M. Tignor and H. L. Miller (eds.). Cambridge University Press, Cambridge, United Kingdom and New York, NY, USA.

UNEP (2005). Water and Wastewater Reuse, An Environmentally Sound Approach for Sustainable Urban Water Management, United Nations Environment Programme (UNEP), Division of Technology, Industry and Economics - International Environmental Technology Centre. http://www.unep.or.jp/.

United Nation Population Division (2007). http://www.un.org/esa/population/publications/wup2007/.

Vairavamoorthy K., Gorantiwar S. D. and Pathirana A. (2008). Managing urban water supplies in developing countries – Climate change and water scarcity scenarios. *Physics and Chemistry of the Earth*, **33**, 330–339.

PART TWO

Precipitation, Stormwater Drainage and Hydrologic Cycle

Main Cause of Climate Change: Decline in the Small Water Cycle

M. Schmidt

Technical University of Berlin, WaterGy Group, A 59, Strasse des 17.Juni152, 10623 Berlin, Germany
E-mail: marco.schmidt@tu-berlin.de

Abstract While we may all agree that global warming is caused by man-made activities, it is no longer acceptable to attribute this to the increase in greenhouse gas emissions. Rather, global changes in land use, in particular urbanization, deforestation and desertification, are responsible for increased temperatures. Problems of global climate and the water cycle are largely due to unsustainable land use.

Urbanization results in a huge impact on the natural water cycle. Reduced evaporation rates are mainly responsible for the urban heat island effect and translate in a reduction in overall precipitation. A sustainable urban development requests a new waterparadigm. Instead of shifting rainwater into sewer systems it could be used for irrigation and cooling. Roofs should be greened and wastewater should be used after being pre-treated. Not a single drop of water should leave urban surfaces, simply to be funnelled into sewer systems.

Keywords climate change, land use, water cycle, evaporation, urban heat island effect, new waterparadigm

INTRODUCTION

Evaporation of water is the largest and most important hydrological component on earth. Only water that has been evaporated will result in rainfall. The effects of deforestation and growing urbanization are causing significant losses to global evapotranspiration. Reduced evapotranspiration, locally and globally, means that more short-wave global solar radiation is converted into long-wave thermal emissions and sensible heat. Reduced evaporation causes higher surface temperatures, this is the main contributor to the urban heat island effect. On a global scale, it is largely responsible for local climate change and overall global warming.

A sustainable approach concerning water requires a new water paradigm (Kravčík *et al.*, 2007; www.waterparadigm.org). Until recently, evaporation has always been defined and understood as a *loss*. In fact, evaporation is the very *source* of precipitation. Drought is conventionally expressed as a result of rising global temperatures, but if we take this new perspective then increased aridity is the cause, not the result, of the global warming. Our intensive land use patterns are causing the planet to dry out (Ripl, Pokorny, Scheer, 2007; Kravčík *et al.*, 2007).

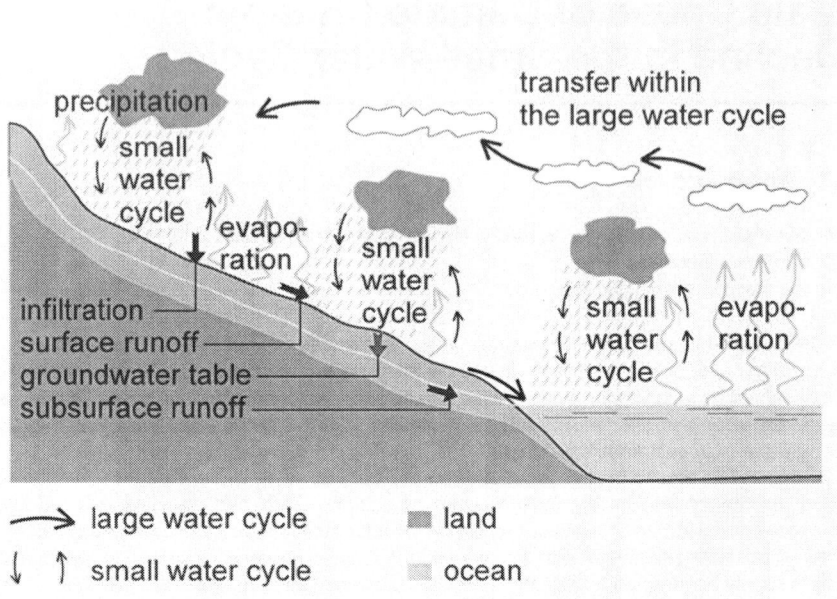

Figure 1 The large and small water cycles on land (Kravčík *et al.*, 2007)

IMPACT ON THE NATURAL WATER CYCLE

Urban areas have a huge impact on the natural water cycle. Contrary to public opinion, the largest influence on the hydrologic cycle is not the reduction in groundwater recharge, but the reduction in evaporation. As a mean, about 75% of the precipitation on land in Europe is redirected to evaporation (DVWK 2004). Once evaporation declines due to urbanization, this translates to a reduction in overall precipitation, effecting a further reduction in evapotranspiration, thus creating a "snowball-effect". Landscapes start to dry out, and short wave global radiation is converted into long-wave thermal emissions and sensible heat. All components in which global radiation is converted on the Earth's surface are illustrated in Figure 4. There is a mean energy flux of one square meter per day. Of this, 7.3% of the incoming solar radiation is reflected, and 38% is directly converted to thermal radiation due to the increase of surface temperatures. Net radiation can be either converted into sensible heat (575 Wh/ (m²d)) or consumed by evaporation, a conversion into latent heat. With 1888 Wh/ (m²d), the energy conversion by evaporation is the largest component of all. It is even more dominant than the thermal radiation converted from incoming short-wave radiation. Additionally, evaporation reduces the long-wave thermal radiation due to the decrease in surface temperatures. Having looked at Figure 4, it is clear that the entire global radiation balance is dominated by evaporation and condensation. Urbanization results in a huge change to the small water cycle. Additionally, hard materials and surfaces in urban areas absorb and re-radiate solar irradiation and increase that area's heat capacity. Fundamentally, the main driving factor for the urban heat island effect is the lack of vegetation and absence of unpaved soils (Figures 2 and 3). Impermeable surfaces like roofs and streets influence urban microclimates through a change in radiation components. Due to the changes in radiation, air temperatures inside buildings also rises and leads to discomfort and/ or greater energy consumption through climate management. For an example of radiation changes in urban areas, Figure 5 illustrates the radiation balance of a black asphalt roof. Compared to Figure 4, most of the net radiation from the urban settings is converted to sensible heat rather than evaporation. Higher surface temperatures also increase the thermal radiation. Greening buildings is a logical solution to create more comfortable air temperatures in cities and to improve the microclimate around buildings.

Figure 2 and 3 An urbanized landscape (Rio de Janeiro, left) in contrast to natural landscapes (Germany, right) significantly alter a region's patterns of radiation and hydrology

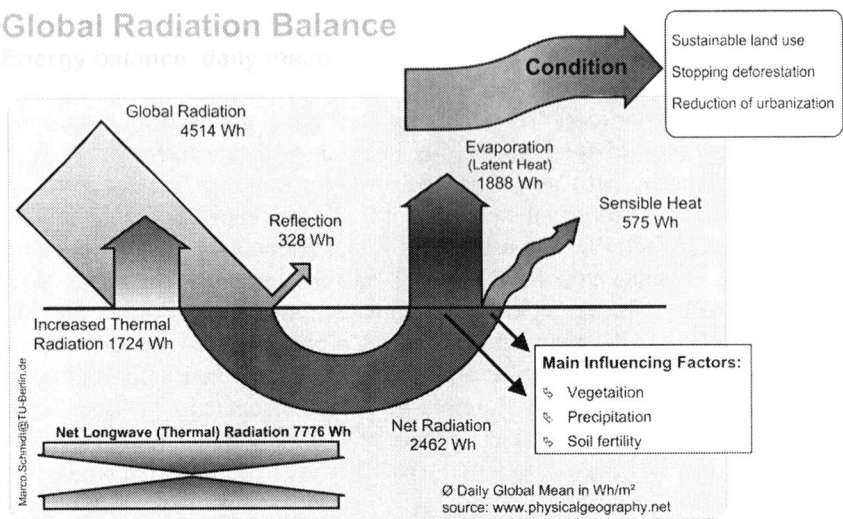

Figure 4 Global daily radiation balance as annual mean (Schmidt *et al.*, 2007). Energy data based on www.physicalgeography.net

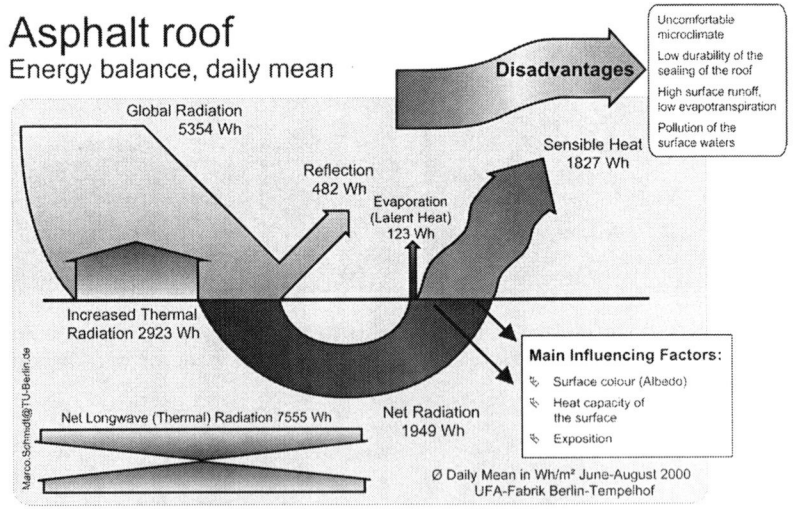

Figure 5 Radiation balance of a black asphalt roof as an example for urban radiation changes (Schmidt, 2005)

Sustainable urbanization needs to reflect the small water cycle of evaporation, precipitation and condensation. While water is evaporated and redirected into the atmosphere by plants in photosynthesis, it cools the landscape. This missing process in urban areas could be substituted. Instead of shifting rainwater into sewer systems it could be harvested and used in air conditioners using evaporative exhaust air cooling techniques. Another cheap and reliable measure is to green façades and roofs. According to measurements taken at the UFA Fabrik in Berlin, a greened vegetated roof covered with 8 cm of soil, can transfer 58% of the net incident radiation into evapotranspiration during the summer months (Figure 6). The annual average energy consumption is 81%, the resultant cooling-rates are 302 kWh/ (m²*a) with a net radiation of 372 kWh/ (m²*a) (Schmidt, 2005). The roof in Figure 6 was monitored in parallel to the asphalt roof in Figure 5.

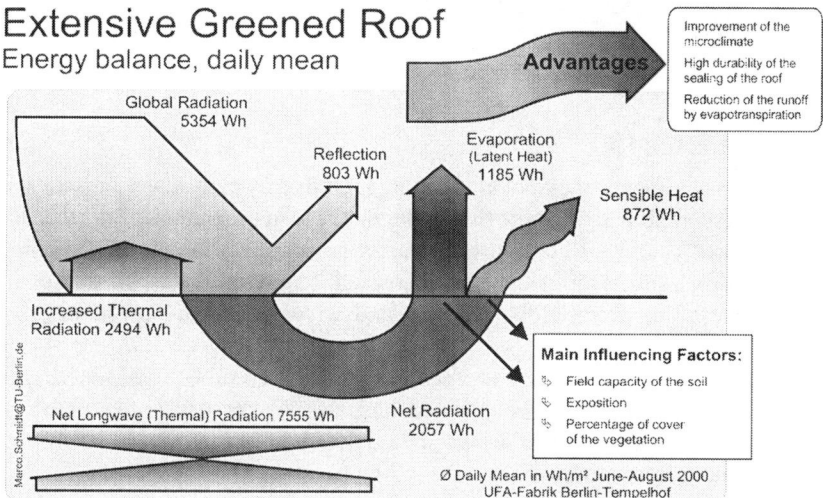

Figure 6 Extensive green roofs transfer 58% of net radiation into evapotranspiration during the summer months, UFA Fabrik in Berlin, Germany (Schmidt, 2005)

Figure 7 Extensive green roof in Berlin- Kreuzberg, greened in 1984

CONCLUSIONS

While we may all agree that global warming is caused by man-made activities, it is no longer acceptable to attribute this to the increase in greenhouse gas emissions. Rather, global changes in land use, in particular urbanization, deforestation and desertification, are responsible for increased temperatures. Our tremendous problems resulting from global climate and water issues are owed largely to unsustainable land use.

Instead of shifting rainwater into sewer systems it could be used for irrigation and cooling. Roofs should be greened and wastewater used after being pre-treated. Not a single drop of water should leave urban surfaces simply to be funnelled into sewer systems.

REFERENCES

Kravčík M., Pokorný J., Kohutiar J., Kováč M. and Tóth E. (2007). "Water for the Recovery of the Climate - A New Water Paradigm". Publisher Municipalia. http://www.waterparadigm.org/

Ripl W., Pokorny J. and Scheer H. (2007). Memorandum on Climate Change. *The Necessary Reforms of Society to Stabilize the Climate and Solve the Energy Issues*. Berlin. http://www.aquaterra-berlin.de.

Schmidt M. (2003). Energy saving strategies through the greening of buildings. *Proc. Rio3, World Energy and Climate Event*. Rio de Janeiro, Brasil. http://www.rio3.com.

Schmidt M. (2005). The interaction between water and energy of greened roofs. *Proceedings World Green Roof Congress*. Basel, Switzerland.
Schmidt M., Diestel H., Heinzmann B. and Nobis-Wicherding H. (2005) Surface runoff and groundwater recharge measured on semi-permeable surfaces. In: *Recharge Systems for Protecting and Enhancing Groundwater Resources. Proceedings of the 5th International Symposium on Management of Aquifer Recharge ISMAR5.* Berlin, Germany, 11–16 June 2005, pp. 190–193. http://unesdoc.unesco.org/images/0014/001492/149210E. pdf.

Sustainable Solutions to Our Water Crisis: Generating Freshwater from "Thin Air"

D. Milani, A. Abbas and D. Al Bakri

School of Chemical and Biomolecular Engineering J01, the University of Sydney, NSW 2006, Australia
E-mail: dia.milani@sydney.edu.au

Abstract As a result of increasing population and climate change, many regions in the world are facing serious water crises. Some countries have endured severe shortage of good quality drinking water for decades. Conventional alternative solutions such as recycling sewage and wastewater, desalination plants, and rainwater harvesting tanks (RHTs) have not been a panacea due to serious economic, environmental and social constraints. Growing water demands coupled with the community's environmental awareness and aspiration for sustainability provide a strong impetus for the water industry to develop clean and sustainable technology for the production and management of drinking water supply. Within this context, air moisture may provide the answer. The atmosphere potentially contains over 12.9×10^{12} m³ of renewable water. The amount of generated freshwater is highly dependent on climate conditions, energy source and the characters of the adopted technology. This study investigates the feasibility of using thermoelectric coolers (TECs) in a proposed dehumidification system. The system can be linked to RHTs to store generated water and programmed to respond to the specific need of individual households. Generated water's quantity and quality can easily satisfy normal household freshwater demands. The main attractions of this approach include compact, safe and noiseless technology that can provide independent and clean water supply to end-users. Intensive energy requirement may be eased by integrating this technique with latest condensation, gas separation and cooling technologies to develop a hybrid dehumidification system that can generate freshwater at competitive prices. This approach, if successfully adopted, will facilitate a thriving water industry, cut hefty rates and reduce the need for building more treatment plants, pipelines and related unwieldy infrastructures.

Keywords Air dehumidification; rainwater harvesting tanks; thermo-electric coolers; psychrometric; water vapor

INTRODUCTION

Two thirds of our planet's surface is covered with water. But, 97% of the world's water is salt water in the oceans, and only 3% is fresh water. Of this tiny fraction, only one-third of 1% is usable by humans- the rest is locked in polar icecaps, or is deep underground (Moore 2003). Freshwater is our most valuable and precious resource which must be handled conservatively. One of the traditional methods to preserve freshwater is to collect rainwater in large tanks instead of discharge it to receiving waterways as stormwater that typically contains pollution. Rainwater harvesting tanks (RHT) used to be a familiar water supply solution in rural regions, arid and semi arid regions worldwide. Recently, with the impact of drought that is stretching to big cities, urban people are also increasingly turning to RHT as an extra tool in aid of rapidly depleting water resources. But the move was not optimal due to the unpredictability and uncertainty of rain events along with high cost of produced freshwater compared to other competitive sources. This study analyses the likelihood of extending the productivity of RHTs beyond rain events via condensing water vapour that naturally presents in the atmosphere to standard freshwater via a dehumidification system (Figure 1).

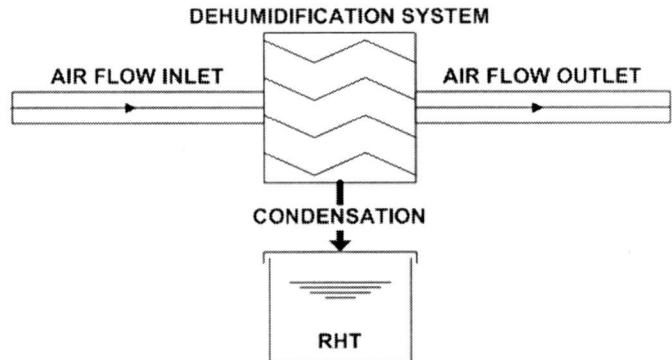

Figure 1 Sketch of the air dehumidification process

REFRIGERATION TECHNIQUES EVALUATION

Vapor compression refrigeration VCR systems are bulky, heavy and require high compression of hazardous refrigerants (CFCs) which is typically achieved via consuming substantial electrical energy (Zubair 1994). Vapor absorption refrigeration VAR systems are using toxic or flammable refrigerant and they are heavier, more expensive and less efficient than VCR systems (Butler 2001). Metal Hydride and Zeolite systems have very poor thermal conductivity (Feng *et al.* 2007) and design difficulties (Lang *et al.* 1999). Active magnetic regenerator AMR systems are excessively heavy and cost-prohibitive for building applications (Butler 2001), while thermoacoustic refrigeration TAR systems have a low cooling capacities and demand additional heating energy source (Zoontjens *et al.* 2005). Thermoelectric coolers TECs are in solid state, noiseless and have no moving parts or refrigerant. TECs have compact size, light weight, precise temperature control and can be oriented in any direction inside the dehumidification system. TECs are also having poor thermal efficiency and power density; however, that can be simply skipped by using more units at divergent locations in the system since they are small, cheap and easy to replace. Figure 2 provides a comparison of various refrigeration systems in terms of relative cost to manufacture and total equivalent warming impact (TEWI). Figure 2 indicates that TEC device cost is reasonably attractive and is leading the group in terms of environmental sustainability with very little effect of total equivalent warming impact (Bhatti 1999a).

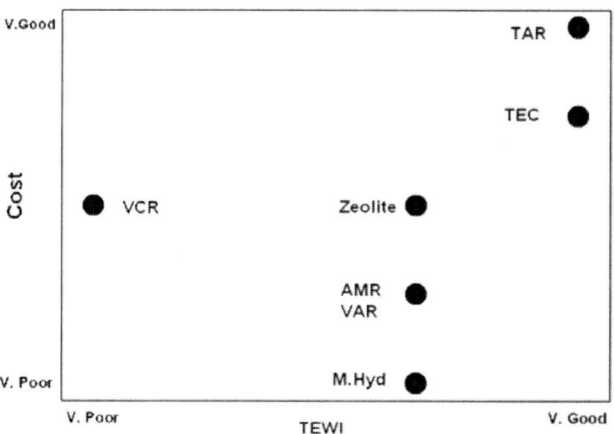

Figure 2 Cost and TEWI comparison of different refrigeration systems. (Modified after Zoontjens *et al.* 2005)

PSYCHROMETRY

Psychrometry is concerned with the measurement or determination of atmospheric attributes that describe the thermodynamic behaviour of moist air (ASHRAE 2002). It is necessary to obtain psychrometric properties of critical point and cooling target point from psychrometric charts (Shallcross 1997) at sea level atmospheric pressure. Moisture condensation occurs when moist air is cooled to a temperature below its initial dew point. Although water can be condensed at various temperatures ranging from the initial dew point down to 0 °C temperatures or beyond, it is assumed that condensed water is cooled to the final air temperature

t_2 before it drains from the system. Equation (1) derives the amount of generated water in the dehumidification system (ASHRAE 2002).

$$m_w = m_{da1}\omega_1 - m_{da2}\omega_2 \qquad (1)$$

$$m_{da1} = \frac{F}{v_1} \qquad (2)$$

$$m_{da2} = \frac{E}{v_2} \qquad (3)$$

From (1), (2) and (3)

$$m_w = F\left(\frac{\omega_1}{v_1} - \frac{\omega_2}{v_2}\right) \qquad (4)$$

Where; m_w - the mass of generated water (Kg_w/s); F- air flow rate (m^3/s); m_{da1}, m_{da2} - are the mass of dry air at entrance and exit respectively (kg_{da}/s); ω_1, ω_2 -are the absolute humidity at entrance and exit respectively (kg_w/kg_{da}); v_1, v_2 -specific volume of the moist air at entrance and exit respectively (m^3/kg_{da}). ASHRAE (2002) plots the energy balance inside the system without heat exchange with surroundings in equation (5):

$$m_{da1}h_1 = m_{da2}h_2 + m_w h_{w2} + Q_2 \qquad (5)$$

h_1, h_2 -Enthalpy of the moist air at entrance and exit respectively (kJ/kg_{da}); h_{w2} - Specific enthalpy of generated water (kJ/kg_w). Then, the energy consumed inside the system can be calculated as Q_2

$$Q_2 = m_{da1}h_1 - m_{da2}h_2 - m_w h_{w2} \qquad (6)$$

Exchanging (2 to 4) into (6)

$$Q_2 = F\left[\frac{h_1}{v_1} - \frac{h_2}{v_2} - h_{w2}\left(\frac{\omega_1}{v_1} - \frac{\omega_2}{v_2}\right)\right] \qquad (7)$$

The total energy Q_t needed for the system is the sum of required energy for cooling and condensation process Q_2 plus required energy to confront heat exchange with surroundings Q_3

$$Q_t = Q_2 + Q_3 \qquad (8)$$

If good insulation materials and proper wrapping were carefully chosen, Q_3 can be reduced to a very tiny fraction of the total energy. However, the air flow F

is the most influencing attribute of Q_3 which is functional to the stream intensity and frequency, therefore combining (7) and (8)

$$Q_t = F\left[\frac{h_1}{v_1} - \frac{h_2}{v_2} - h_{w2}\left(\frac{\omega_1}{v_1} - \frac{\omega_2}{v_2}\right) + \lambda\right]$$

(9)

Where λ- is assumed to be a functionality factor of Q_3 to F and can be experimentally derived.

RESULTS

Water vapor density in lower atmosphere varies from 4 to 25 g/m³ of air (Wahlgren 2001) and can be easily converted to liquid water via a dehumidification system. Water productivity is a function of climate conditions, the processed air flow rate and the efficiency of the dehumidification system. A temperatures range of (10–50) °C and relative humidity between (30%–100%) will be monitored and analysed. If the temperature of 1 m³ of an air parcel can be dropped to 10 °C in any cooling system, the air parcel can not hold more than 9.4 gram of moisture when it leaves; the rest of the moisture would be condensed to liquid water in a process which is typically associated with heat release. Obviously cooling the air parcel further would decrease its moisture holding capability and consequently generate more water, but that would be associated with higher energy consumption. Initial analysis of local climate range and the cost of required energy to run the system were conducted. This analysis suggests that any cooling beyond 10 °C will be economically unwise because it would not generate enough water even with significant increase of energy consumption. Therefore this study will claim this point (10 °C) as a cooling target and calculate accordingly.

Water Productivity

The design of the system should maintain sufficient surface contact between cooled surfaces and passing air parcel to optimize heat exchange and condensation process. These surfaces, coils or fins, must be adequately designed to form, gather and collect droplets and prevent any water loss or re-evaporation which may reduce the efficiency of the system. The efficiency of a dehumidification system can be defined as the ratio of the amount of water extracted to the total moisture content of the air sucked into the plant per unit of time (Hellström 1969). The efficiency E should be included in the equations governing the process of generating water from ambient air (eq. 10).

$$L_\omega = \frac{H_\chi^\phi - H_{10}^{100\%\,\phi}}{100} \times E$$

(10)

where; L_ω - Generated liquid water (g/m³); H_x - Absolute humidity at any point (g/m³); H_{10} - Absolute saturated humidity at 10 °C (g/m³); E- Efficiency of the dehumidification system and n - Relative Humidity at any point. Equation 11 is employed to calculate the volume of generated water per day D presuming same ambient weather conditions and fixed rate of air flow

$$D = C_1 \ L_\omega \ F \tag{11}$$

where; D- Daily production (L/day); F- Air flow rate (m³/s); C_1- Conversion constant. The efficiency of dehumidification process can be further enhanced to strip more moisture from exiting air to detain more water. However for more conservative approach, exiting air is assumed to be left unmodified and fully saturated with maximum moisture holding capacity. From this prospect, generated liquid water and daily production are calculated for an ambient air temperature range from 10 °C to 50 °C and n>30% as shown in Table 1. Daily water generation is variable and linearly related to ambient air temperature, relative humidity, air flow rate and system's efficiency. At very hot saturated air (50 °C and 100% n) generated freshwater exceeds 3000 litre per day equivalent to 4 times Sydney's typical household daily water consumption which is about 304 kL/yr (Troy *et al.* 2005) and that is even achievable at 50% efficient dehumidification system from an inlet rate of only $F = 1$ m³/s.

Energy Requirements

A dehumidification system design must cope with latent heat or heat of condensation released whenever water changes phase from gas to liquid. This heat must be dissipated to prevent liquid water from re-evaporating before storage. Dissipating the heat would be associated with significant energy consumption that comes from three types of energy loads and must be considered in the design calculation (eq. 12):

$$Q_t = Q_L + Q_{\Delta t} + Q_i \tag{12}$$

Where; Q_t- total energy load (kw); Q_L- condensation load (kw); $Q_{\Delta t}$- cooling load (kw) and Q_i- insulation load (kw).

Condensation Load (Q_L)

Latent heat is the heat quantity absorbed or released by a substance during a change of state (Perrot 1998). Every gram of water to be condensed would release 2450 J of heat at STP conditions (Wahlgren 2001) which must be dissipated from the system via proper heat exchange with cooled fins. The governing equation is:

$$Q_L = \frac{L_w.h_w.F}{1000} \tag{13}$$

Cooling Load ($Q_{\Delta t}$)

Cooling the air from ambient conditions to targeted temperature would add a significant sensible load that should be considered. The governing equation for cooling is:

$$Q_{\Delta t} = \frac{F\, \rho_{air} C_\rho \Delta t}{1000} \tag{14}$$

Where; C_ρ- specific heat capacity of the air (J/kg.°C); Δt – temperature differential $(t_x - t_{10})$(°C); ρ_{air}- The density of air (kg/m³). The moist air density ρ_{air} and specific heat capacity C_ρ at standard atmospheric pressure and 20 °C is equal to 1.205 (kg/m³) and 1.005 (J/kg.°C) respectively (ASHRAE 2002).

Table 1 Water production from 1m³/s air parcel that is being refrigerated through the system to 10 °C and maximum n of 100%

[°C]	ϕ	30%	40%	50%	60%	70%	80%	90%	100%
50	L_ω	15.5	23.8	32.1	40.4	48.7	57	65.3	73.6
	D	1339	2056	2773	3491	4208	4925	5642	6359
45	L_ω	10.2	16.8	23.3	29.9	36.4	43	49.5	56
	D	881	1452	2013	2583	3145	3715	4277	4838
40	L_ω	5.9	11.1	16.2	21.3	26.4	31.5	36.6	41.7
	D	510	959	1400	1840	2281	2722	3162	3603
35	L_ω	2.5	6.4	10.4	14.4	18.3	22.3	26.2	30.2
	D	216	553	899	1244	1581	1927	2264	2609
30	L_ω	0	2.7	5.8	8.8	11.9	14.9	17.9	21
	D	0	233	501	760	1028	1287	1547	1814
25	L_ω	0	0	2.1	4.4	6.7	9	11.3	13.6
	D	0	0	181	380	579	778	976	1175
20	L_ω	0	0	0	1	2.7	4.4	6.2	7.9
	D	0	0	0	86	233	380	536	683
15	L_ω	0	0	0	0	0	0.9	2.1	3.4
	D	0	0	0	0	0	78	181	294
10	L_ω	0	0	0	0	0	0	0	0
	D	0	0	0	0	0	0	0	0

Insulation Load (Qi)

An insulation analysis must be carried out to minimize heat exchange with surroundings and save energy. Using materials with lowest thermal conductivity to insulate the system will employ most of consumed cooling energy for internal processes and increase the efficiency. Therefore the escaped energy out of the system will be a very small fraction of total consumed energy for cooling plus condensation and consequently can be neglected.

Design Calculations

It is widely acknowledged that TECs reliability, efficiency and durability are products of excellent design, proper mounting, adequate temperature range and operational stability. However, TECs are limited in terms of cooling capacity, size and operational temperature range. The largest available cooling capacities obtained from three different suppliers are presented in Table 2 (Milani 2008). Initial calculation of TECs design and the total cost of dehumidification system will be based on (TB-199-2.0- 0.9) from Kryotherm. The combination of influential factors such as price, duration and maximum cooling capacity Q_{max} plus largest size and lowest electrical resistance of (TB-199-2.0-0.9) clearly prevail other products. To determine the required number of TECs, the total heat load at a specific efficiency will be divided on the maximum cooling capacity of the (TB-199-2.0-0.9) from Kryotherm (Table 2).

Table 2 Largest TECs capacities and other properties from three different suppliers

TEC Name	Company	I_{max} A	Q_{max} W	U_{max} V	Δt_{max} C	Rac Ω	L mm	W mm	H mm	Price $	Duration years
TEC1-28719T125	Huimao/ China	19	410	34	67	1.5	50	50	4	17.5	5–7
TECA 980-127	TECA/ USA	8.5	83.2	15.4	72.2	1.63	15.7	15.7	1.3	93	23–24
TB-199-2.0-0.9	Kryotherm Russia	20.6	310	24.6	69	0.87	62	62	3.2	39.5	23–24

$$N_{TEC} = \frac{Q_t}{Q_{max}} \tag{15}$$

Then, the cost of total TECs is:

$$C_{TEC} = N_{TEC} \times P_{TEC} \qquad (16)$$

where; P_{TEC} - the price of every TEC (\$); C_{TEC} - the total cost of all TECs (\$); N_{TEC} - the required number of TECs and Q_{max} - Maximum cooling capacity of a thermoelectric cooler (kW). The total cost of the system consists of the sum of required TECs, materials, construction of the system, maintenance and total energy consumption during the lifespan of the system based mainly on TECs duration. To estimate the total required energy, the maximum required number of TECs at optimum efficiency and worst case scenario will be multiplied by the duration of TECs and the price of kWhr energy assuming that these conditions will govern the process for the entire lifetime of the system for conservative approach and analysis dignity.

$$C_{en} = C_2.Q_t.P_{en}.\psi \qquad (17)$$

where; C_{en} -Energy consumption cost (\$); P_{en}-the price of energy (\$/kWhr); ψ- Duration of the system (yr); C_2- Conversion constant. At last, the total cost;

$$C_t = C_{TEC} + C_{con} + C_{main} + C_{en} \qquad (18)$$

C_t-Total cost (\$); C_{TEC}-TECs cost (\$); C_{con}-Construction plus materials cost (\$); C_{main}-Maintenance cost (\$). Dividing the total cost on the total water production in the entire system's lifetime:

$$price = \frac{C_t}{D_{ent}} \qquad (19)$$

Where; Price- the price of generated water (\$/kL); D_{ent} - total production for the entire lifetime of the system (kL).

Assumptions

To simplify the calculation of total cost of the dehumidification system and the price of produced water (\$/kL), the following assumptions were used in the system design.

1. Air flow rate F is constant and equal to 1 m³/s at all times.
2. Pressure variation inside the system is negligible.
3. Ambient conditions are constant at the same level during the day.

4. Ventilation and heat exchange at the TEC's heat sink is optimum and dose not cause any drop in Q_{max}.
5. Drop of cooling capacity Q_{max} at higher temperature deferential Δt_{max} is negligible.
6. Maintenance is low and equal to $ 2000 in all cases.
7. Materials and construction cost are similar and equal to $ 8000 in all cases.
8. Price of energy is fixed at $P_{en} = 0.1$ ($/kWhr) (Wahlgren 2001).

Critical thermal point

Maximal thermal point represents the highest temperature and relative humidity that a particular location can record. In Sydney, the highest recorded temperature is 45 °C at saturation level will be adopted for design calculations. Minimal thermal point determines the lowest climatic conditions to generate a satisfactory amount of water at reasonable price. The lowest production point (worst production outcome and highest water prices) at 20 °C 60% has been chosen because it is more common in Sydney's weather conditions. At a specific air flow rate, water productivity is directly related to system's efficiency. However, increasing the efficiency does not necessarily improve the outcome substantially because of higher power consumption and operational complexity which adversely affect the economical and environmental sustainability of the system. Table 3 illustrates how the efficiency of a system influences each parameter at maximal thermal point and minimal thermal point. Raising the system's efficiency far above 50% will significantly increase TECs number and energy consumption which causes the system to be impractical.

Controlling variables

To determine the main variables controlling the price of generated water, equations (4), (18) and (19) are integrated in equation 20

$$price = \frac{C_{TEC} + C_{con} + C_{main} + C_{en}}{C_3 \psi\, m_w} \tag{20}$$

Where; ψ- Active duration of the system (yr); C_3- Conversion constant [$C_1 \times 365$ day/yr = 31536 kL.s/(kg.yr)]

Table 3 TECs numbers, daily water production, total energy consumption and water price in response of efficiency variation at maximal thermal point (45 °C & 100% n) and minimal thermal point (20 °C & 60% n)

E	N_{TEC}	Maximal Thermal conditions			Minimal Thermal conditions		
		D L/day	Q_t kW	Price $/kL	D L/day	Q_t kW	Price $/kL
10%	44	484	13.8	70.7	8.6	0.26	233.8
20%	89	968	27.5	69.3	17.3	0.50	162.2
30%	133	1452	41.2	68.9	25.9	0.75	138.9
40%	177	1935	54.9	68.7	34.6	0.99	126.8
50%	221	2419	68.6	68.5	43.2	1.24	119.9
60%	266	2903	82.4	68.4	51.8	1.48	115.3
70%	310	3387	96.1	68.4	60.5	1.73	111.8
80%	354	3871	109.8	68.3	69.1	1.97	109.3
90%	398	4355	123.5	68.3	77.8	2.22	107.3
100%	443	4838	137.2	68.2	86.4	2.46	105.8

Substituting equations (16) and (17) in (20) beside assuming that the cost of construction and maintenance of the system are functional to the total quantity of TECs and accordingly will be functional to Q_t

$$price = \frac{N_{TEC}P_{TEC} + \alpha N_{TEC} + \beta N_{TEC} + C_4 \psi Q_t P_{en}}{C_3 \psi m_w} \qquad (21)$$

Where; α,β- are directly related to the number of TEC devices to be installed and can be derived experimentally. From (15) replacing N_{TEC} with Q_t

$$price = \frac{Q_t \left[\dfrac{P_{TEC} + \alpha + \beta}{Q_{max}} + C_4 \psi P_{en} \right]}{C_3 \psi m_w} \qquad (22)$$

Then, from (4) and (9) substituting Q_t and m_w into (22)

$$price = \frac{F \left[\dfrac{h_1}{v_1} - \dfrac{h_2}{v_2} - h_{w2} \left(\dfrac{\omega_1}{v_1} - \dfrac{\omega_2}{v_2} \right) + \lambda \right] \left[\dfrac{P_{TEC} + \alpha + \beta}{Q_{max}} + C_4 \psi P_{en} \right]}{C_3 \psi F \left(\dfrac{\omega_1}{v_1} - \dfrac{\omega_2}{v_2} \right)} \qquad (23)$$

Simplifying (23) by substituting $(C_3 = 3.6 \times C_2)$ and eliminating F

$$price = \frac{\left[\dfrac{h_1}{v_1} - \dfrac{h_2}{v_2} - h_{w2}\left(\dfrac{\omega_1}{v_1} - \dfrac{\omega_2}{v_2}\right) + \lambda\right]\left[\dfrac{P_{TEC} + \alpha + \beta}{Q_{max}\psi} + C_4 P_{en}\right]}{C_3\left(\dfrac{\omega_1}{v_1} - \dfrac{\omega_2}{v_2}\right)} \qquad (24)$$

Surprisingly, equation (24) indicates that the price of generated water ($/kL) is only functional to the psychrometric attributes of the entrance and exit points of the system and the physical properties of TEC devices. The air flow F intensity and frequency have no influence on the price of generated water. TEC main physical properties that influence the price are the cooling capacity Q_{max}, lifetime and the cost P_{TEC}. Energy cost P_{en} as expected is a major contributor which requires involving more sustainable technologies such as renewable energies to reduce the energy cost and sustain the system.

CONCLUSIONS

Using TEC devices in dehumidification process to refrigerate ambient air and generate renewable freshwater is clean, quite and achievable. To further enhance its economic and environmental sustainability, TECs dehumidification system can be integrated with RHTs to store generated water and programmed to respond to the specific water demand of individual households. Such systems can reduce the stress on the precious freshwater resources and alleviate pressure from aquatic ecosystems. Furthermore, TECs dehumidification system could be used by scattered settlements where full scale freshwater infrastructures are cost prohibitive. While more than 90% of water price in TEC dehumidification system is generated from energy consumption rather from capital cost, this enhances the necessity of facilitating renewable energies (e.g. solar and wind) to reduce the energy cost and further sustain the system. Water price from dehumidification is relatively higher compared to conventional water sources. However, in conventional water pricing, huge capital costs of dam construction, treatment plants, pipelines and other infrastructures were usually subsidised, the value of natural assets were bypassed and environmental degradations were shamelessly ignored. In long term prospective, this technology could help to transform the cumbersome management system of water industry to a more efficient, sustainable and decentralised community enterprise system where households have a direct control on their individual water supplies. This approach, if successfully adopted, will facilitate privatization of the water industry, cut hefty rates and stop building more treatment plants, pipelines and other water unwieldy infrastructures. The main limitation of TECs dehumidification system is their low capacities which require allocating high number of TECs to meet extreme weather conditions. More investigations required to optimise the system using gas separetion techniques via hydrophilic membranes and recycling the energy. TECs can play a crucial rule in hybrid processes by integrating latest condensation, gas separation and cooling technologies supported by renewable energy sources. Advances in the design and development of TECs technology could lead to more powerful and applicable devices that might dictate dehumidification and refrigeration industry in the future.

REFERENCES

ASHRAE (2002). *The ASHRAE Handbook CD*, American Society of Heating Refrigerating Air-conditioning Engineers, Atlanta.

Bhatti M. S. (1999a). Enhancement of R-134a Automotive Air-conditioning System. *SAE International*, no. 1999-01-0870.

Butler D. (2001). *Life after CFCs and HCFCs*, Technical report, Building Research Establishment (BRE), Watford, UK.

Feng Q., Jiang-ping C., Wen-feng Z. and Zhi-jiu C. (2007). Metal hydride work pair development and its application on automobile air-conditioning systems. *Journal of Zhejiang University- Science A*, **8**(2), 197–204.

Hellström B. (1969). Potable water extracted from the air- report on laboratory experiments. *Journal of Hydrology*, **9**(1), 1–19.

Lang R., Roth M., Stricker M. and Westerfeld T. (1999). Development of a modular zeolite-water heat pump. *Heat and Mass Transfer*, **35**(3), 229–234.

Milani D. (2008). Product's Manuals and Direct Correspondence, Faculty of Agriculture, Food and Natural Resources, A04 - R.D. Watt, The University of Sydney, NSW 2006 Australia.

Moore M. (2003). A World Without Walls: Freedom, Development, Free Trade and Global Governance, Cambridge University.

Perrot P. (1998). *A to Z of Thermodynamics*, Oxford University Press, Oxford.

Shallcross D. C. (1997). *Handbook of Psychrometric Charts: Humidity Diagrams for Engineers*, Blackie Academic Professional, London.

Troy P. N., Randolph W. and Holloway D. (2005). *Water Use and the Built Environment: Patterns of Water Consumption in Sydney*, City Futures Research Center, University of New South Wales, Sydney.

Wahlgren R. V. (2001). Atmospheric water vapor processor designs for potable water production: A review. *Water Research*, **35**(1), 1–22.

Zoontjens L., Howard C., Zander A. and Cazzolato B. (2005). Feasibility study of an automotive thermoacoustic refrigerator. In: *Proceedings of Acoustics*, Busselton, Western Australia, School of Mechanical Engineering, The University of Adelaide, South Australia, pp. 363–371.

Zubair S. M. (1994). Thermodynamics of a vapor-compression refrigeration cycle with mechanical subcooling, *Energy*, **19**(6), 707–715.

Studies and Practices of Urban Rainwater Harvest and Runoff Pollution Control in Beijing

J. Q. Li, W. L. Wang and W. Che

Key Laboratory of Urban Stormwater System and Water Environment (Beijing University of Civil Engineering and Architecture), Ministry of Education, Beijing, China, 100044
E-mail: jqli6711@vip.163.com

Abstract Based on the investigation and analysis of the effectiveness of RWH projects, the development of studies and practices about urban rainwater harvest (RWH) and runoff pollution control (RPC) in Beijing was reviewed. The typical technical process, storage volume calculation methods, first flush regular and control facilities were described in this article. The characteristics of urban runoff pollution and its proportion for the water environment in Beijing were analyzed. The laboratory and practice test results of porous pavements and rain gardens were discussed. As a result, it has been proved that urban RWH and RPC systems will play a more and more important role in Beijing, and will have more significant developments in the near future.

Keywords Urban rainwater harvest, runoff pollution control, first flush, porous pavement, rain garden, Beijing

INTRODUCTION

Since the 1990s, the wastewater industry has developed rapidly in Beijing. At present the treatment rate has reached 90%. But the lakes and rivers' water quality has not been significantly improved. Eutrophication frequently occurs in rainy seasons. One of the main reasons is the non-point source pollution of the urban runoff. In addition, Beijing is facing a crisis of available water due to the scarce amount of water resources lower than 300 m³/ca. Modern urban rainwater harvest (RWH) and runoff pollution control (RPC) is a system of multi-objective integrated technology, to achieve water conservation, control urban soil erosion and groundwater depression cone, reduce water pollution and improve the urban ecological environment goals. The research and engineering practices were rapidly developed. Urban RWH and RPC systems played a more and more important role in water resource supply, control of runoff pollution and flood hazards, etc. The policies of the RWH and RPC perform an important function. The policies, regulations and codes are multilevel in coordination, which includes law and by-law, economy, government, technology propaganda and education etc.

The increasingly perfect frame of policies have established a good base of RWH and RPC systems in Beijing. It can be predicted that Beijing will make more significant development of RWH and RPC in the near future.

URBAN RAINWATER HARVEST

The Situation of Urban RWH

Based on 10 years of research and engineering practices, the modern urban RWH technology system could be classified into direct harvesting, infiltration and integrated utilization. Taking Beijing as an example: until the rainy season of 2009, more than 700 RWH projects had been built. Investigation and analysis on the effectiveness of 274 RWH projects was completed in 2009. Fig. 1 is the statistical chart according the project numbers of different models of RWH, and Fig. 2 is the statistical chart showing the total water quantities of different models of RWH.

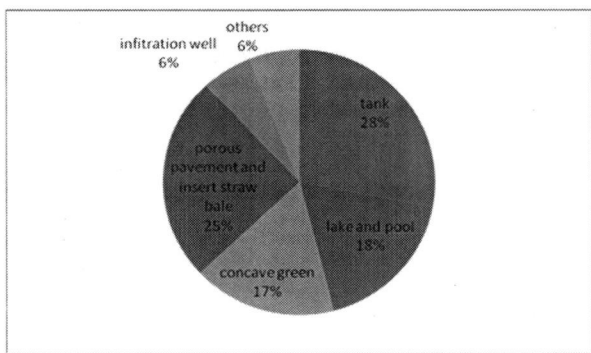

Figure 1 Statistical chart showing the project numbers of different models of RWH

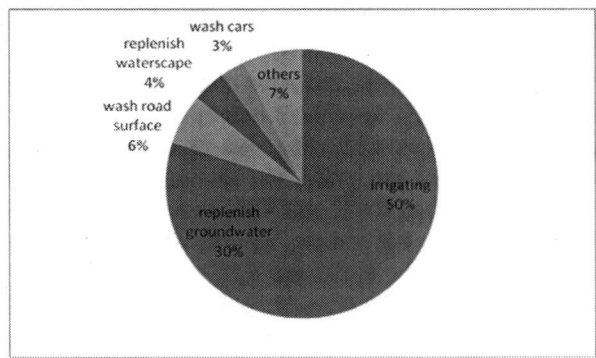

Figure 2 Statistical chart showing the total water quantities of different models of RWH

The Typical Technical Process of Rainwater Purification

The direct harvesting project mainly concentrates on green belt irrigation. Only a small fraction of the water is used to replenish water bodies, wash cars and road surface, flush toilets, and perform other tasks.

To explore some efficient and cost effective methods to purify rainwater quality, laboratory tests and practical applications of vertical flow artificial soil filters (VFASF) were carried out from 1999 to 2002. The results indicate that VFASF is an efficient and cost effective technique for rainwater purification. During the rainwater soil-plants filtration process, different filtration media had different level of removal efficiency. Regarding the inlet water, COD concentration varied from 60~200mg/L, and the inlet water load was in the range of 10~100mg COD/m²s. COD removal efficiency was related to soil quality and depth. Removal efficiency of artificial soil (slag: soil 1:1; depth 1.0m) was 60%~85%. Natural soil (depth 0.5m) removal efficiency was 30%~40%; Natural soil (depth 1.0m) removal efficiency was 45%~80%. General appraisal was that the purification effects sequence was: Depth 1m artificial soil (slag: soil 1:1) >Depth 1m natural soil >Depth 0.5m artificial soil (slag: soil 1:1) >Depth 0.5m natural soil. The results could be used to direct the filter design (Li et al., 2009).

The Storage Volume of RWH

The fundamental standard for the storage volume of RWH should be subject to water balance analysis and improving water circle system or amending the environment. The calculation methods of rainwater storage volume differ, according to the purposes, types, outlets, infiltration capacities and so on. The rainwater cistern types and calculation principles for storage volume are reviewed, practical calculation methods are proposed and analyzed, including rainfall estimating method, rainfall intensity-duration curve method, and statistical rainfall frequency accumulative method (Li et al., 2005). In Beijing, when water consumption is greater than the harvested quantity, the harvesting volume can be determined with a 24h rainfall, and the rainfall return period should not be greater than 0.5 years, or design rainfall should not be larger than 50mm, or using water quality volume for design rainfall instead. When runoff in a practical project is sufficient with small daily water consumption, the harvesting storage volume could be set at water quantity for 3~5 days water consumption. For those projects with artificial water bodies with high levels of evaporation and storage capacity, the rainwater storage volume should be confirmed by water balance and flooding control analysis.

According to the procedure of storage facilities and cost-benefit theory, a method of designing the optimum volume of rainwater basins was put forward by analyzing relationships among storage volume, average daily precipitation and the number of full volume, and then several critical questions were discussed, such as the number of full volume, available harvesting volume, investment and cost, comprehensive benefits, time value of money, and so on. This method can be used generally and practically for optimizing rainwater projects in Chinese cities (Li *et al.*, 2005).

First Flush Control

Based on theoretical analysis and actual monitored results, it was affirmed that the first flush regularity of runoff existed in small catchment surfaces despite the possibility that it might appear differently under various local conditions. The results show that actual conditions, such as the character of the catchment or sewer, the treatment operations or rainwater utilization technology should be taken into consideration to determine the first flush control volume. The first flush control methods include volumetric and small pipe split-flow methods. Highly efficient first flush control equipment was invented by Beijing University of Civil Engineering and Architecture, which has the advantage of volumetric and small pipe split-flow methods and a high investment benefit in the projects of pollution control and sewer rainwater, as well as rainwater utilization (Che and Li, 2006).

The analysis indicates that 2mm split of first flush for runoff from roof would control above 60% of the pollutant load from such a catchment's surface during most storm events. The study also suggests that treating 6mm first flush from road surface would treat above 60% of the pollutant load. Consequently, 2~3mm controlled quantity of first flush from roof and 6~8mm from road surfaces are suggested in many Chinese cities (Che and Li, 2006).

URBAN RUNOFF POLLUTION CONTROL

The Situation of Urban Runoff Pollution

With rapid development in the effective control of point-source pollutants, the proportion of pollutant from runoff into receiving waters is rising. Urban stormwater impact constitutes a complicated environmental problem in cities which undoubtedly needs to be considered in environmental protection programs. It is important to establish an effective strategy of runoff pollution control to solve the urban water environmental problem.

Take Beijing as an example: the planning urban area is 1042 km² in 2010. According to 585 mm as many years' average rainfall and the value of runoff pollution, the COD of the stormwater runoff, which accounts for the total gross, is shown in Table 1.

Table 1 The COD of the stormwater runoff account for the total gross

Year	COD of the runoff, municipal and industry/ (10^3 kg)	Thereinto			Rate of runoff account for total gross/ (%)
		Municipal/ (10^3 kg)	Industry/ (10^3 kg)	Runoff/ (10^3 kg)	
2000	233,600	157,000	21,500		23.59
2001	225,500	152,300	18,100		24.43
2002	207,800	138,500	14,200		26.52
2003	189,100	136,000	10,400	55,100	29.14
2004	184,800	118,400	11,300		29.82
2005	171,100	105,000	11,000		32.20
2006	165,000				33.39

The total runoff is an average of $3.54 \times 10^8 m^3/a$, COD emissions are $5.51 \times 10^7 kg/a$, the unit area load is 577.4 kg/($10^4 \cdot m^2 \cdot a$). When the urban sewage treatment rate was 40.6 % in 2000, the runoff COD emissions accounted for the total COD of the runoff, municipal, industry wastewater is 23.59%. Until 2006, with increasing the sewage treatment rate to 90%, when the proportion was raised to 33.39%. Compared to urban point source pollution, not only is the number considerable, but also the runoff in the rainy season is greatly intensified. As the further sewage treatment and standard rate will be rising continuously and sewage treatment processes will improve, the proportion of the runoff COD emission accounted for the total COD will also rise. Runoff pollution will be rising continuously, which has become an important issue for pollution decreasing.

The Characteristics of Urban Runoff Pollution

During the period from 1998–2002, the runoff qualities of more than 100 rainfall events from roofs and roads were observed in Beijing, and major factors which affected their quality were analysed. The results showed that the qualities of runoff are relatively poor, especially the first flushes of runoff, which are seriously polluted, even more highly than raw sewage, in most cases. The major pollutants of urban runoff include organic substances (COD, TOC and BOD), suspended solids (SS), nutrients (TN, TP), oil, and some heavy metals etc. COD is well related with SS and it also has some relation with TOC, ammonia-nitrogen, TN and TP in certain cases. The degradation rule fits the "first order" flush model (Che and Li, 2006). Table 2 shows the stormwater runoff concentrations in Beijing urban areas.

Table 2 The runoff contaminations concentration in Beijing urban areas (mg/L)

Types of runoff		COD	SS	TN	TP
Asphalt roof	First flush	350–2800	400–2400	7.9–39.2	3.28–4.1
	After first flush	164–656	68–272	7.8–14.7	0.75–0.94
	Average	328	136	9.8	0.94
Tile roof	First flush	100–800	400–2400	-	-
	After first flush	60–246	68–272	-	-
	Average	123	136	-	-
Road	First flush	610–3660	967–5802	6.5–65	2.8–11.2
	After first flush	291–1164	367–1468	5.6–22.4	0.87–3.48
	Average	582	734	11.2	1.74

The most important affecting factors are roof material, road type and rubbish on the surfaces. Other factors such as air pollution, rainfall intensity and volume, interval of rainfall events, season, air temperature and so on impact on the runoff quality also. Generally speaking, green belt runoff quality is better than roof runoff, which in turn is better than road runoff. From the roof runoffs, tile is better than asphalt; the runoff in the uptown is better than municipal district; the runoff in new uptowns, parks or development zones in good environmental conditions is much better than in the city center. In good environmental conditions, the runoff of more rainfall and short intervals of events is good. In high temperature, the asphalt roof material decomposes noticeably.

Urban Runoff Pollution Control Practices

Based on the scientific evaluation of runoff pollution in the Beijing urban area, a control and management practice system was proposed which emphasized the combination of the structure and non-structure technique, and the concept of quantitative and source control. Since 1998, some Low Impact Development/Best Management Practices (LID/BMP) technical facilities have been researched in the laboratory and were used in the urban areas of many cities, mainly including sunken greenbelts, permeable pavements, green roofs, rain gardens, infiltration wells and so on.

The sunken greenbelt and permeable pavements are typical measures in rainwater infiltration projects. In some places, the grassed swales, infiltration wells, infiltration pipelines and trenches are also adopted. These decentralized facilities are mainly used in residential areas, parks, roads, factory districts, and so on. The surface soil in urban Beijing is mainly mixed powdery and sandy types, the infiltration rate of which is between 10^{-5} m/s and 10^{-6} m/s, and therefore suitable for infiltration. The ground water level is descending 1 m every year because of excess use. The infiltration technology is simple, easily constructed and requires less investment. It can also supply and reimburse some losses of the groundwater resources, and therefore refresh the environment, alleviate surface descent, reduce flooding, and so on. Rainwater infiltration is the dominant type of stormwater management in Beijing and other similar cities. But the runoff pollution must be controlled in order to avoid groundwater being polluted.

Permeable pavements

Besides the infiltration capacity, the pollutant removal efficiencies for five types of porous pavements were tested. The surface of porous pavement is permeable brick, but the five sub-based structures from the lower to upper layers are different. The first sub-based structure is made of graded aggregate, geotextile and medium sand. The second is graded aggregate and medium sand. The third is gravel and hollow concrete. The forth is lime-soil. The fifth is lime-soil, graded aggregate, geotextile and medium sand. The results of sub-based structures of porous pavement experiments indicates that the removal efficiency of five sub-based structures for SS, TP, COD, dissolved COD and TN are 42.3% ~ 95.5%, 56.4% ~ 95.6%, 11.7% ~ 84.9%, 5.3% ~ 71.3%, 11.3% ~ 45.1% respectively. In conclusion, the bigger the material interior surface, the higher the removal ratios for pollutants. The ratios from high to low level for SS is the structure V, IV, II, I and III successively, however, the permeability capacity from high level to low is the structure III, II , I,V and IV successively. The third sub-based structure is best for permeability capacity. With the increase of SS in the sub-based structures, permeability coefficient reduces quickly.

Rain gardens

Two types of rain garden for building and road runoff pollution control were built in 2006. The slag and local sand soil as medium materials were installed under the rain gardens. The continuous monitoring of rainfall events was tested during the period 2006~2008 and the results indicated that: the removal for building roof runoff of SS, COD, heavy metal (Pb, Zn, Cu, Fe) and turbidity is excellent, except for TP, phosphate phosphorus (PO_4-P), TN, and nitrate nitrogen (NO_3-N). The removal efficiencies of road rain gardens for SS, TP, COD are notable, and can remove a part of dissolved COD too. The removal efficiency of TN has presented fluctuations. Sometime it appeared that effluent concentration was higher than initial concentrations of road runoff. As to two types of rain gardens, removal performances can recovery gradually after duration of two events and continuing sunshine. Obviously, in urban land use, using decentralized rain gardens is effective to control runoff pollution and reduce the runoff volume of small catchment areas. Table 3 is the main pollutants removal efficiencies of rain gardens for roof and road runoff (Xiang *et al.*, 2008).

Table 3 The removal efficiencies of rain gardens for roof and road runoff in Beijing (%)

	Rain garden for roof runoff	Rain garden for road runoff
SS	≥90.0	75.8~87.9
COD	35.0~91.4	59.9~69.3
Dissolved COD	/	36.0~48.7
TN	22.0~45.4	−34.8~31.4
NO_3-N	−9.0~20.2	/
TP	−86.3~76.0	52.9~67.9
PO_4-P	−574.2~78.9	/
Fe	30.0~90.0	/
Pb, Zn, Cu	≥90.0	

CONCLUSIONS

Some conclusions are summarized as follows:

Beijing faces a critical contradiction of rapid development of urbanization with water resource shortages and water pollution. Beijing will have more significant developments in urban rainwater harvest (RWH) and runoff pollution control (RPC) in the near future. The urban rainwater harvest project for reusing mainly concentrates on green belt irrigation; only a small fraction of the water is used to replenish water bodies, wash cars and road surface, flush toilets, and for other activities.

Vertical flow artificial soil filter (VFASF) is an efficient and cost effective technique for rainwater purification.

There are many methods for storage volume calculation, including the rainfall estimating method, the rainfall intensity-duration curve method, and the statistical rainfall frequency accumulative method, etc. The optimum volume of rainwater basins should be decided by analyzing relationships among storage volume, average daily precipitation, the number of full volume and the greater benefit/ cost.

The first flush regularity of runoff existed in small catchment surfaces. When the controlled quantity of the first flush is analyzed, 2–3 mm from roofs and 6–8 mm from road surfaces are appropriate, and this can remove most urban pollutants.

With the rapid development in the effective control of point-source pollutants, the proportion of pollutants from runoff into receiving waters is rising. By 2006, the proportion of runoff pollution had been raised to 33.39% in Beijing.

Rainwater infiltration is the dominant type of stormwater management in Beijing and other similar cities in China. Sunken greenbelts and permeable pavements are typical measures for rainwater infiltration. The infiltration technology is simple, easily constructed and does not require too much investment.

Rain gardens could be effective in controlling runoff pollution for small catchment surfaces such as roofs and roads.

ACKNOWLEDGEMENTS

These studies were financially supported by the Projects of Water Pollution Control and Treatment Key Technology Program (2010ZX07317-011 and 2008ZX07314-007).

REFERENCES

Che W. and Li J. Q. (2006). Urban rainwater harvest technology and management. China Architecture & Building Press. Beijing. (in Chinese)

Che W., Liu Y. and Li J. Q. (2003a). Flush model of runoff on urban non-point source pollutants and analysis. In: *Water and Environmental Management Series, Water in China*, P. A. Wilderer, J. Zhu and N. Schwarzenbeck (eds.), IWA Publishing, pp. 143–150.

Che W., Liu H. and Wang H. Z. (2001). Study on roof-rainwater pollution and harvesting in Beijing. *China Water & Wastewater*, 17(6), 57–61. (in Chinese)

Li J. Q., Che W., Meng G. H., Wang H. L. (2001). Preliminary plan and economic analysis of urban rainwater utilization. *Water & Wastewater Engineering*, 27 (12), 25–28. (in Chinese)

Li J. Q., Xiang L. L., Mao K. Che W. and Li H. Y. (2009). Using rain garden to control roof runoff: the case study of Beijing. *Journal of Nanchang University*, 31, 217–222.

Li J. Q., Yu P., Che W. *et al.* (2005). The study of economic scale of rainwater harvesting and utilization projects in cities. *China Water & Wastewater*, 21(3), 49–52. (in Chinese)

Xiang L. L., Li J. Q., *et al.* (2003). Discussion on the design method of rain garden. *Water & Wastewater Engineering*, 34(6), 47–51. (in Chinese)

Efficiency and Economy of a New Agricultural Rainwater Harvesting System

JI Wen-hua[a], CAI Jian-ming[a] and Marinus van Veenhuizen[b]

[a]Institute of Geographic Sciences and Natural Resources Research, Chinese Academy of Sciences, Beijing 100101, China
[b]ETC-Urban Agriculture, Leusden 3830 AB, Netherlands
E-mail: jiwh.07b@igsnrr.ac.cn

Abstract Shortage of water is the key limiting factor for agricultural development of Beijing. Rainwater harvesting (RWH) could provide an alternative water source for greenhouse agriculture, but local natural and socioeconomic conditions challenge the application of the technology. This paper analyses the advantages and disadvantages of different types of greenhouse RWH in Beijing, and describes a new greenhouse RWH system demonstrated in 2008 in Huairou, a suburb district of Beijing. It analyses the efficiency, cost-benefit ratios and limiting factors of the new system. The results show that with the new system, rainwater harvesting efficiency can be as high as 66% (of total rainfall) and the rainwater usage rate can reach 69% of total water usage. The ratio of benefit to cost of government investment can be 1.84, and the ratio of benefit to cost of a farmer's investment could be 1.68 provided the project is designed to save water and also increase income. This paper also suggests a number of measures to increase the efficiency of the system in order to apply it on a large scale.

Keywords Beijing, greenhouse agriculture, rainwater harvesting, cost-benefit

INTRODUCTION

Rapid population growth and frequent droughts in recent decades have widened the gap in many parts of the world between the need for water and its availability. The situation is worst in arid and semiarid countries and regions. Rainwater harvesting (RWH) for agricultural production has been practiced for thousands of years (Evenari *et al.*, 1961, Wesemael, 1998, Scott *et al.*, 2001). Nowadays, it is applied widely as one of the key measures to relieve water shortages in agriculture.

The annual per capita water resource of Beijing is below 300 m³, which is only about 1/30 of the world average (Liu Baoqin *et al.*, 2003). A sharp reduction in surface water has forced farmers to rely on groundwater. In 2003, the agriculture sector used 1.38 billion m³ water and 90% was groundwater. Groundwater has been increasingly overexploited as a means to fulfill the city's water demand. As a result, the groundwater table dropped by about 1 m annually from 2000 to 2007.

The shortage of water threatens the sustainability of agriculture around Beijing, in that a) the cost of agriculture rises along with diminishing availability of water, and b) farmers increasingly face water shortages during the peak water use periods of spring and summer. Thus RWH could be a solution to the farmers' problems.

RWH systems can be divided into two general types according to the collecting surface, i.e. natural or artificial slopes. To capture rainfall, farmers need to use either hills or roofs. In mountainous areas, famers use natural slopes for RWH, whereas in flat areas it is difficult to practice RWH without access to roofs. RWH using greenhouses has become an important way to ensure a diversity of sources for water in water-scarce regions.

The development of greenhouse agriculture in Beijing has created an important opportunity for farmers. According to an investigation carried out jointly by Urban Development and Reform Commission, Municipal Bureau of Statistics, Water Authority and Municipal Bureau of Agriculture in June 2005, Beijing had a total of about 20,666 ha of greenhouse agriculture, which was close to 10% of total arable land. Greenhouse agriculture will develop quickly in the future due to strong support from the government.

Four basic types of greenhouse RWH systems are used in Beijing (Table 1). Types A, C and D save more water than type B, because they have closed storage pools and thus lose less water through evaporation. The water quality of type B is also poor due to pollution and algal bloom. Type B is thus not used extensively in Beijing.

Table 1 Types of greenhouse RWH systems in Beijing

Type	Water from	Description of SP	Cost	Function of SP
A	One GH	One, closed, small	High	Storing water
B	Cluster of GHs	One, open, big	Low	Storing water
C	Cluster of GHs	One, closed, big	High	Storing water
D	Cluster of GHs	One, closed, big	High	Multiple

Notes: GH = greenhouse and SP = storage pool.

Researches show that economies of scale exist for storage pools (Zhao, 1996). Obviously it is more economical to build big pool than small pools (see Figure 1). Furthermore, it is easier to maintain one big pool than several small ones. The main challenge in implementing types A, B and C is the high cost and low economic incentives. Type D, which combines water collection and agricultural production, could therefore be a better choice.

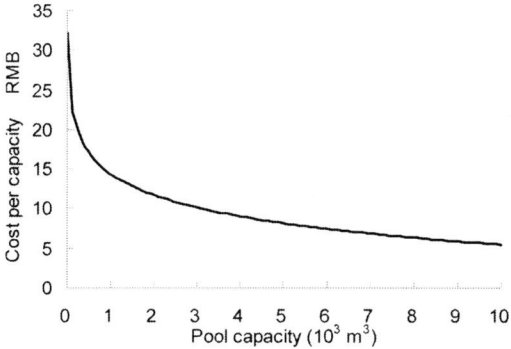

Figure 1 Relation between pool capacity and cost per capacity

This paper presents the results of a first year of analysis of a type D RWH demonstration. It focuses on the efficiency of water harvesting and the benefits versus the costs. It also discusses the system's potential and upscaling mechanisms.

MATERIALS AND METHODS

Description of the Demonstration

The RWH demonstration is located in Angezhuang Village, Beifang Town, Huairou District, Beijing. It covers 1.33 ha with a total of five greenhouses (640 m² per greenhouse). Available water resources are confined to groundwater and rainwater here. Recent years, the groundwater level has decreased rapidly in Huairou district. It even dropped to nearly 40 m around the demonstration in 2008.

The 2008 rainfall data come from a weather observation station near the demonstration area. It rained 22 times from April to September that year, amounting to a total of 583 mm (Figure 2). It rained most in August (151 mm) and the heaviest rain fell on August 10th with 56.5 mm.

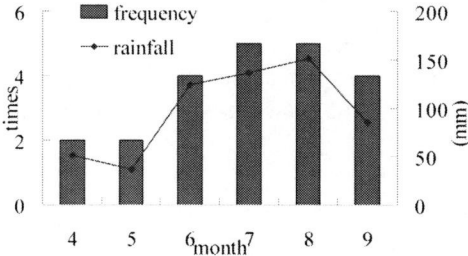

Figure 2 Rainfall around demonstration area in 2008

Altered Rwh Mode

Construction of a RWH project began in May 2007 and was completed by early 2008. The material of the greenhouses' surface is plastic film. Runoff slots and one deposit pool were built with bricks and concrete. Reinforced concrete and anti-leakage material were used for the storage pool. To prevent water loss, a 60 cm of soil is laid up to the storage pool. It then is covered by a concrete layer.

The pool is divided into four cisterns with a volume of 125 m³ each. One of the cisterns, which is connected directly to the deposit pool and stores rainwater all the time, is the 'permanent cistern'. Three 'spare cisterns' are interconnected by doors. During the rainy season, all four cisterns can be used to store rainwater. In the dry season, the spare cisterns gradually dry up and then be reused to plant crops or feed animals.

Rainwater Harvesting Efficiency

Rainwater harvesting efficiency is the rate of the rainwater harvested against the total rainfall on the surface of the greenhouses, expressed as:

$$RWHE = RWH/TR*100 \qquad (1)$$

In the equation, RWHE stands for rainwater harvesting efficiency (in %); RWH stands for rainwater harvested in storage pool (in m³) and TR stands for total effective rainfall on the surface of the greenhouses (in m³). Before and after each rain, the water level in the storage pool was recorded. Thus the amount of water collected from each rain could be calculated.

Substitution Effect

The rainwater substitution rate (SR, in %) refers to the proportion of rainwater used for irrigation as part of total water use. The SR can be expressed as,

$$SR = RWU/(RWU + GWU)*100 \qquad (2)$$

In the equation, RWU stands for rainwater used (in m³) and GWU for groundwater used (in m³). The rainwater used can be calculated by subtracting the rest rainwater in the storage pool at the end of the year from the total rainwater harvested, provided there is no water leakage or evaporation from the storage pool.

Income Effect

Studies show that the economy of a RWH system relies mostly on the size of the storage pool (Fewkes, 1999). In the case of a fixed runoff area, the bigger the storage pool is, the higher the cost will be, and the water can be harvested up to a certain maximum depending on rainfall. So the optimal scale of storage pool should be carefully designed.

Water for irrigation is free in Beijing. Incentives are needed and therefore the multiple functions of the RWH system are important in this respect. Additional income can thus be obtained by growing production in the storage pool during dry seasons. Using rainwater can also save electricity.

Calculation of Cost

The total fixed investment of the RWH and reuse system was 125,000 RMB, excluding the cost for greenhouses and the drip irrigation system. Financial support for 70% of the total investment, i.e. 87,500 RMB, was provided by the government.

The electricity cost of pumping rainwater was 0.015 RMB per m^3 according to power consumption in 2008. Maintenance costs, increased a little with the new system. However, with or without RWH, farmers have to pay for the greenhouse's plastic surface, so the cost of replacing it should not be taken into account when calculating the cost of the system. Generally, 6 days each year is enough for a farm to keep the system clean, so the annual cleaning cost is about 240 RMB (labor cost is calculated at 40 RMB per day). Farmers need to buy new equipment to produce mushrooms. Which cost 100,000 RMB and could be expected to use for 10 years. The running cost of mushroom production includes the costs for materials, labor and transportation, which amounted to 72,000 RMB in 2008.

In 2006, farmers built a new 100-metre-deep well. The total investment was 80,000 RMB. The electricity needed to pump groundwater costs about 0.15 RMB/ m^3. The maintenance cost of the well, including a pump replacement in 2008, reached 1,200 RMB.

Cost-Effective Analysis

As farmers cannot afford to invest in the RWH system all by themselves, part of the costs are subsidized by the government. Thus, the cost-benefit analysis of the RWH system has to be done from both the government's and the farmers' perspective.

Cost-Benefit Ratio for the Government

The RWH project can benefit the city by reducing a) investments in the irrigation infrastructure, and b) economic losses caused by groundwater overexploitation.

a) The Beijing government invests a lot each year in the city's irrigation infrastructure in order to resolve water problems in agriculture. In 2007 alone, this amounted to 1.42 billion RMB. The average cost of a greenhouse (667 m²) was about 408.4 RMB. Expansion of the RWH project could reduce annual investment in irrigation infrastructure.

b) Groundwater overexploitation could cause a series of environmental problems, such as land subsidence and surface collapse, which can have a huge economic impact. Ni Hongzhen estimated the economic loss in 2000 to be 9.4 RMB per m³ of overexploited groundwater (Ni Hongzhen *et al.*, 2006). According to Li's model (Li Xiaokai *et al.*, 2003), the economic loss of overexploitation was 8.6 RMB per m³ in 2007. Based on these analyses, we assume that the economic loss of water overexploitation is 9 RMB/m³.

Generally, a system is considered economically feasible if its dynamic cost-benefit ratio exceeds 1 (Che Wu *et al.*, 2006). Otherwise, the project should be rejected from an economic point of view. The following formulas are used to calculate the ratio.

$$EV = B \times \frac{(1+i)^n - 1}{i(1+i)^n} \tag{3}$$

$$PV = I \tag{4}$$

$$\alpha = \frac{EV}{PV} \tag{5}$$

In the formulas, EV is the present value of total benefits (in RMB); B is the average annual benefits (in RMB/y). i is the discount rate (7 %) according to the Engineering and Technical Norms of Rainwater Utilization. n is the service life of the RWH (expected at 30 years). PV is the present value of total cost (in RMB). I is the fixed investment of RWH (in RMB). α is the dynamic cost-benefit ratio.

Cost-Benefit Analysis for Farmers

The same methodology mentioned above can be used to calculate the benefit coefficient from the farmers' perspective.

$$EV = nB \tag{6}$$

$$PV = I + nC \tag{7}$$

In the formulas, B is annual present income (in RMB/y). I is fixed investment. And C is the annual present running costs (in RMB).

RESULTS AND DISCUSSION

Rainwater Harvesting Efficiency

The sum of rainfall on greenhouses was 1,865 m³. The measured harvested water in that same period was 1,233 m³. The rainwater harvesting efficiency was thus 66.1%. This was not high, for two reasons: 1) In Beijing, average temperatures are high in the summer period (when rainfall is also high). So farmers have to uncover greenhouses to keep down the air temperature inside. This can result in water loss if the farmers do not put the covers back on the greenhouses in time for the next rain. 2) A design mistake was made when building the greenhouse. The plastic film was not connected directly with the runoff slot, so part of the rainwater infiltrated into the soil. Of course, these problems can be resolved by: a) providing timely weather information to farmers, and b) improving design of RWH system. The rainwater harvesting efficiency could thus be improved significantly.

Substitution Effect

Water consumption for agriculture in the demonstration was 1,050 m³ in 2008. 850 m³ out of 1,233 m³ collected rainwater was used for grape, tomato, cabbage and mushroom production 200 m³ of groundwater was used for mushroom mycelium production. The substitution rate was thus 81%. Groundwater was used for the production of mushroom mycelium because it requires a high water quality.

Greenhouse RWH could possibly prevent the recharging of groundwater. However, the RWH system is designed to recharge groundwater through an infiltration well. Rainwater overflow from the storage pool could be turned into groundwater.

Income Generation Effect

Farmers earned a gross income of 140,000 RMB through mushroom production in the storage pool in 2008. Profits can be expected in the long term whereas the high investment in the short term.

The electricity cost for irrigation is about 158 RMB (1,050*0.15), provided that all of the water used is groundwater. And farmers could save about 115 RMB (850*(0.15–0.015)) of electricity when using rainwater.

Cost-Benefit Analysis for Government

In the tested case in which 70% of total fixed investment is subsidized by the government, the cost-benefit ratio of the RWH project increases year by year over the duration of its service life (Figure 3). The investment itself could be returned in ten years (the cost-benefit ratio is then 1.06). This ratio could reach 1.86 in the end, which means that the project has a high economic return. However, the water supply and demand balance in Beijing in future will influence the economic evaluation of the project. For example, in rainy years, the RWH system could harvest more water, meaning that the project would have a higher economic return in those years than in a normal year. The economic loss caused by the shortage of water would then decrease due to the increase in available water resources. This would decrease the economic value of the project. Further economic evaluation of the project, taking fluctuations in rainfall into account, is thus needed to better understand its potential economic impact.

Cost-Benefit Analysis for Farmers

The new system, which combines rainwater harvesting with agricultural production, could return farmer's investment in three years (see Figure 3). The cost-benefit ratio of the project grows rapidly in the first ten years, reaching 1.63 in the tenth year. Additional investment is needed in the eleventh and twenty-first years to replace equipment. Thus the cost-benefit ratio of these two years decreases a little compared to the prior years', but it is still at a high level. The cost-benefit ratio increases at a slower rate after the eleventh year. At most, it could reach 1.68, which indicates that the new method is economically feasible if its potential multi-functionality is fully realized. This also indicates that it is feasible to upscale the system.

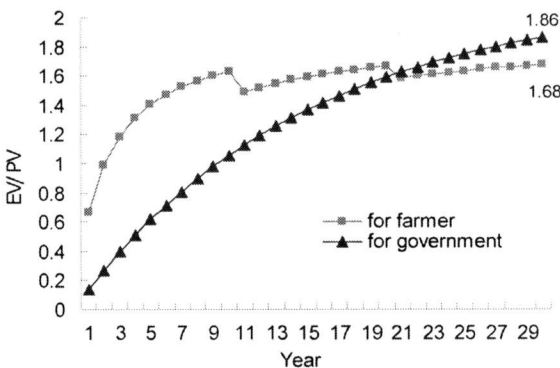

Figure 3 Cost-benefit ratio of the RWH project

As shown in Figure 3, the project's cost-benefit ratio is higher for the government than for farmers. Therefore, more financial support could be given to farmers. If a RWH system is intended only for water saving and not designed to be multifunctional, like most existing RWH systems in Beijing, then the economic incentive is insufficient. This is in large part why some RWH systems are not in active use, especially in water-rich areas. This problem can be resolved by making RWH multifunctional, increasing financial support or setting a water price.

CONCLUSIONS

Agricultural greenhouse RWH could be an alternative water source for greenhouse agriculture in Beijing. This paper focuses on the efficiency and feasibility of a new type of greenhouse RWH demonstrated in Beijing. The RWH system had ordinary RWH efficiency due to imperfect design and negligent management that could be easily resolved. The total rainwater harvested in the first year of operation was more than the amount of water consumed thanks in part to the high annual rainfall, large storage pool and drip irrigation technology used. However, the system's economic efficiency was reduced by the excessive size of the storage pool.

The cost-benefit analysis of government investment reveals that the cost-benefit ratio of the RWH project is high over the duration of its service life. However, the economic value of the project is strongly related to the water balance of the city, which is determined by the urbanization process and fluctuations in rainfall. The cost-benefit ratio of the farmer's investment in the RWH system is slightly lower than that for government investment, but it can still be considered high as long as the system is used in a multifunctional way.

Several factors are crucial for improving the efficiency of the RWH system and upscaling this new method. First, the RWH system should be built in water-scarce locations, where the RWH system would have a high economic value and be easily accepted by farmers. Second, the RWH system must be optimally designed to lower cost and maximize the volume of harvested water. Third, the operation and management skills of farmers should be improved through education and publicity activities. Fourth, more financial support from local governments should be given to farmers to help them overcome the high investment cost of the new method. Finally, the price system for irrigation water should be improved.

ACKNOWLEDGEMENTS

This work is financially supported by the SWITCH Project (018530), which was launched by UNESCO-IHE (Institute for Water Education) in 2006.

REFERENCES

Baoqin Liu, Zhijun Yao and Yingchun Gao (2003). Analysis on structural change of water use and driving force in Beijing. *Resources Science*, **2**, 38–43.
Che Wu and Junqi Li (2006). Technologies and management of rainwater use in city. *China Architecture and Building Press*, 155–157.

Evenari M., Shanan L. and Tadmor N. (1961). Ancient agriculture in the Negev. *Science*, **133**(3457), 979–996.

Fewkes A. (1999). The use of rainwater for WC flushing: the field testing of a collection system. *Building and Environment*, **34**, 765–772.

Hongzhen Ni, Hao Wang and Dangxian Wang (2006). Water price of Beijing based on green accounting method. *Journal of Hydraulic Engineering*, **37**(2), 210–217.

Scott C. A. and Silva-Ochoa P. (2001). Collective action for water harvesting irrigation in the Lerma-Chapala Basin, Mexico. *Water Policy* **3**, 555–572.

Songling Zao (1996). *An introduction to RWH agriculture*, Shaanxi Science and Technology Press, Xi'an.

Wesemael, B. V. (1998). Collection and storage of runoff from hillslopes in a semi-arid environment: geomorphic and hydrologic aspects of the aljibe system in Almeria Province, Spain. *Journal of Arid Environments*, **40**, 1–14.

Xiaokai Li and Haifeng Shi (2003). Calculation method of economic loss by groundwater overexploitation and its application in typical region. *Advances in Science and Technology of Water Resources*, **23**(6), 13–20.

Outline of Some Stormwater Management and LID Projects in Chinese Urban Area

W. Che, W. Zhang, J. Q. Li, H. Y. Li and J. L. Wang

Key Laboratory of Urban Stormwater System and Water Environment (Beijing University of Civil Engineering and Architecture), Ministry of Education, Beijing, China, 100044
E-mail: chewu812@163.com

Abstract Urbanization significantly impacts water environments with increased runoff and the degradation of water quality. With the fast development of urban expansion, water environment problems have become increasingly serious in Chinese urban areas in recent years. The main causes of these problems are discussed from a strategic point of view. Furthermore, the importance and necessity of urban stormwater management and water environmental protection are put forward. Over 82 projects have been undertaken by our research team in recent years, consisting of about 35 research projects and 47 engineering projects related to rainwater harvest, landscape and water environment, stormwater management, runoff pollution control and proprietary technology development. Four typical projects are briefly introduced subsequently. The design of urban stormwater management systems associated with low impact development (LID) and waterscape planning were implemented successfully and creatively in these four projects. Based on projects undertaken by the Key Laboratory of Urban Stormwater System and Water Environment (Beijing University of Civil Engineering and Architecture), the Ministry of Education (KLUSSWE) and other Chinese scholars in the past 10 years, the development of urban stormwater management and water environment eco-remediation in China are summarized. Finally, the prospects of future potential urban stormwater management and water environment eco-remediation in China are discussed.

Keywords Stormwater Management, Rainwater Harvest, LID, Projects, Chinese Urban Area

INTRODUCTION

Urban expansion transforms local environments and, in the context of effective urban resource planning and management, the recognition of the impacts of urbanization on the water environment is among the most crucial. This significance stems from the fact that water environments are greatly valued in urban areas as environmental, aesthetic and recreational resources, and hence are important community assets (Goonetilleke *et al.*, 2005). As an essential element of the water cycle, stormwater plays an important role in an urban water environment. Unfortunately, according to traditional stormwater management methods in Chinese urban regional planning and some relevant planning, stormwater has been regarded as "waste" and "disaster source". Many traditional measures of stormwater management were only employed to control peak flows and provide local drainage during wet weather flow. Due to the disadvantages of the traditional ways, some obvious problems of urban water environment are inevitable, including:

- Serious non-point source pollution caused by urban runoff;
- Increasing urban impervious surfaces, resulting in serious flood disaster;
- Water resource loss and serious water crises brought on by direct discharges of more and more runoff;
- Excessive exploitation of groundwater and decreasing infiltration of rainwater, resulting in the decline of groundwater levels and large-scale "groundwater depression" in Chinese urban areas;
- The destruction of urban beneficial hydrologic cycles and ecosystem deterioration, etc.

There is no doubt that the stormwater runoff problems in Chinese urban areas mentioned above are partly caused by natural hazards. However, unsustainable urban development and traditional conceptions of urban stormwater management are the main reasons for this. In any case, the stormwater runoff problems described above in Chinese urban areas are simulatenously both natural and man-made disasters.

Along with the problems of urban runoff and water environment, which have gradually been gathering more and more attention in China, rainwater harvest & stormwater management have been focal issues in many Chinese scholars' researches (Yin *et al.*, 2009; Che and Li, 2006).

AN OVERVIEW OF THE PROJECTS

To solve the remarkable problems mentioned above, systemic researches and engineering projects have been implemented successfully by KLUSSWE over the past ten years. Wide cross researches and engineering projects related to rainwater harvesting, stormwater management and water environment were accomplished by KLUSSWE, with over 82 projects being completed, consisting of about 35 research projects and 47 engineering projects, mainly distributed in Beijing, Ningbo, Hangzhou, Tianjin, and many other cities, as illustrated in Figure 1 and Figure 2. These projects are classified to rainwater harvest, landscape and water environment, rainwater harvest and stormwater management, runoff pollution control and proprietary technology development (Figure 3 and Figure 4). According to different land use and scale, these engineering projects were mainly applied to public buildings, parks, open spaces, schools, government agencies, residential areas, and even regional stormwater management planning (Figure 5).

Reciprocal combinations and promotion between researches and engineering projects is the key principle and should persist. There is no obvious boundary between researches and engineering ones.

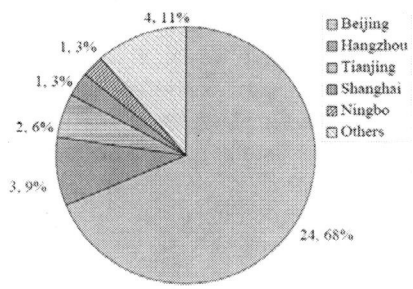

Figure 1 Research projects by city located in

Figure 2 Engineering projects by city located in

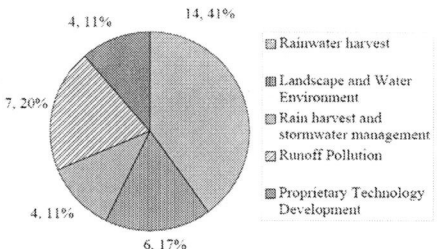
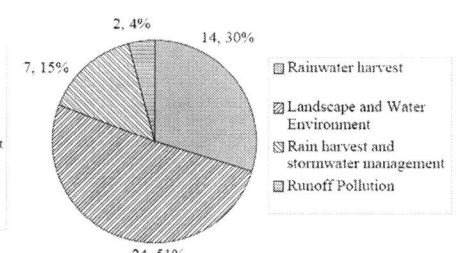

Figure 3 Research projects by type   **Figure 4** Engineering projects by type

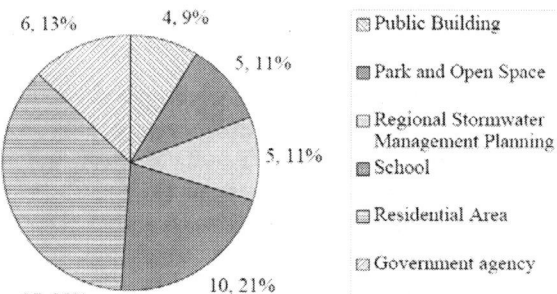

Figure 5 Engineering projects by land use and scale

OUTLINE OF SEVERAL TYPICAL PROJECTS

Project 1 Master Planning of Stormwater Management & LID in a New Urban Area, Ningbo

The project is located in the east of Ningbo, a city in southern China. The total area of the development site is about 8 km², which is a developing new centre of Ningbo. As a result of a dense waterways network existing in this area, the maintenance of urban beneficial water environment is a key goal in the new city planning and development. According to local natural conditions and main water environment problems in this district, the master planning of stormwater management is aimed at protecting local eco-system and creating a sustainable water environment.

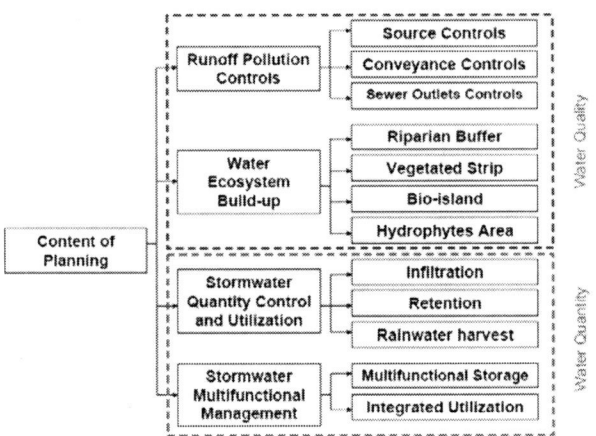

Figure 6 Framework of the integrated planning

The stormwater management planning included several important components, such as stormwater runoff pollution control, water quality improvement and water ecosystem build-up, stormwater quantity control and utilization, and stormwater runoff multifunctional management (Figure 6). Landscape design and LID measures were incorporated into this integrated planning according to different land uses, including residential areas, roadways, parking lots, plazas, shopping centres, public buildings, culture and education areas, surface water, eco-corridors, open spaces and other areas. Furthermore, a series of stormwater management goals in the future and relevant implementary measures were made under regional general planning. Stormwater management practices employed in this integrated planning are shown in Figure 7.

Planning was not limited only to structural practices mentioned in Figure 7. Nonstructural practices were also considered in this planning, such as limiting pesticide use or retaining rainwater on residential lots, public education, public involvement and participation, illegal connection control, etc. As different functions of sub-districts, the district stormwater planning was made according to different sub-district land use, which was intended to control runoff pollution and volume, to utilize rainwater, and to maintain or enhance existing aquatic habitats, etc. Consequently, different stormwater management practices can be implemented in centralized or decentralized ways.

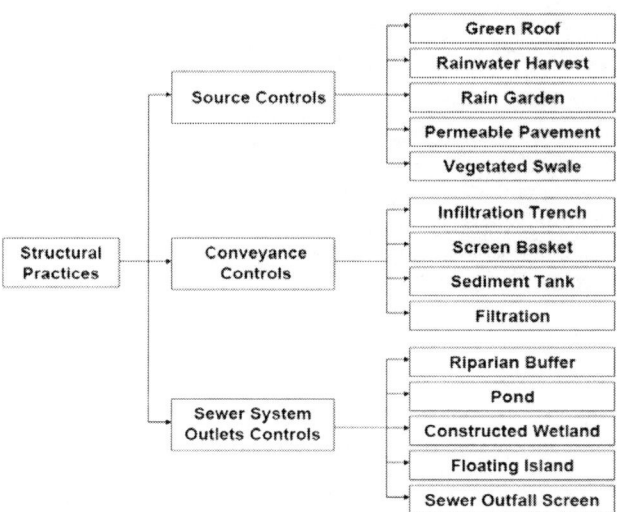

Figure 7 Stormwater management practices (BMP/LID) in the master planning

Project 2 LID Application in Wanke Centre in Shenzhen Aimed at the LEED Platinum Certification

This project is located in Shenzhen, a modern city in southern China. Figure 8 shows the master plan of the project. The total catchment area is about 6.2 ha, which consists of about 1.61 ha of building area (all green roof), 0.46 ha of water surface and less than 50% of impervious areas, etc.

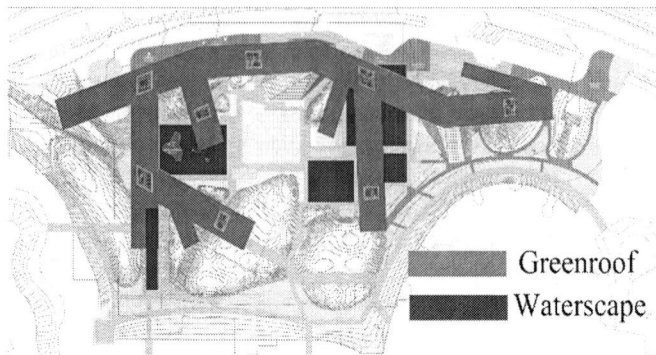

Greenroof
Waterscape

Figure 8 Master plan of Wanke Centre

Based on the basic requirements of the project and datum, several main targets of the water environment planning are summarized as follows:

- Reasonably utilize rainwater and grey water to save tap water;
- Achieve the post-development runoff coefficient not exceeding the pre-development one (on the basis of the site natural state) through infiltration, harvesting and retention facilities to reduce runoff discharge;
- Control stormwater pollution and reduce the discharge of contaminants;
- Safeguard water quality in order to obtain pleasant waterscape and improve the environment;
- Optimize investment and operation costs of the water system.

The water system in this project is a complicated system which includes some sub-systems, such as stormwater management, rainwater harvest, waterscape and landscape, grey water and constructed wetland which is designed for recycled water purification. Among the sub-systems, the stormwater management for quantity and quality control is the core. The main targets are to prevent the post-development peak discharge rate and quantity from exceeding the pre-development ones for the one and two-year 24-hour design storms, and to treat runoff from 90% of the average annual rainfall events by the integrated stormwater management.

Based on systematic water-balance analysis, an integrated water management plan was put forward (Figure 9). It involves grey water and rainwater utilization reasonably, wetland and waterscape designing, BMPs & LID for quantity and quality control of excess runoff, etc.

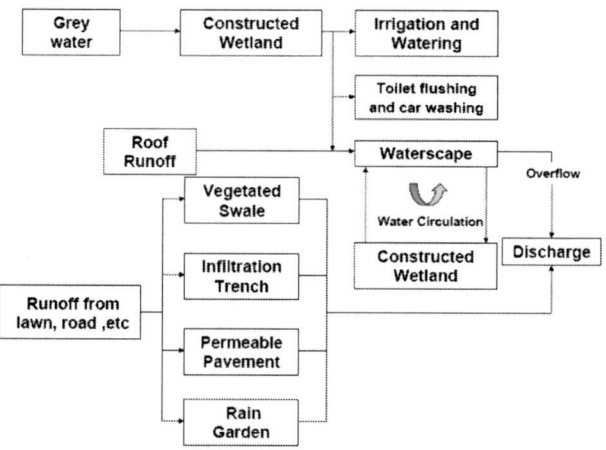

Figure 9 The integrated water management planning of Wanke Centre

Water quality is a very important component which benefits water system sustainability. Several issues should be considered in safeguarding water quality in waterscapes, maintaining a good environment around waterscapes, designing a healthy water ecosystem, water circulation with purification of the constructed wetland.

Rainwater resources are convenient for harvesting and utilization, because rainwater is abundant in this area. A series of practices were employed in this project, such as green roof, vegetated swale, permeable pavements, bio-retention and other infiltration facilities, to control the quantity and quality of stormwater. Runoff water quality from the green roof is acceptable. A proportion of infiltrated rainwater can also be collected for reusing.

Project 3 Arcadia Project of Stormwater Wetland and Landscape Lake

This project is located in Beijing. The community has a total area of 11.66 ha with a 1.2 ha water surface and 4,800 m³ storage capacity. About 40,660 m³ of rainwater on average every year could be harvested into the artificial lake. Based on the basic conditions of this project, several main targets of the water environment planning are summarized as follows.

- Saving water;
- Reducing runoff discharge;
- Controlling flood and runoff pollution;
- Designing ecological landscape.

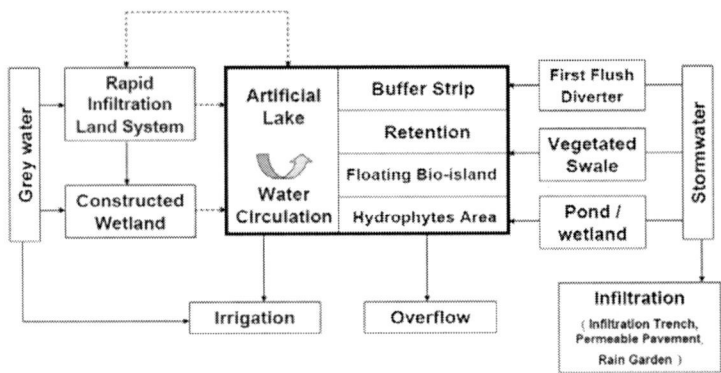

Figure 10 The integrated water management planning of Arcadia

As the main targets mentioned above, water management planning of this project should be integrated and multifunctional, as is shown in Figure 10. Based on systematic water-balance analysis, systematic and multifunctional design included runoff pollution control, water quality improvement, well-liking water ecosystem build-up, flood control, rainwater efficient utilization and multifunctional stormwater runoff management. Among all the above practices and measures, an artificial lake with functions of eco-remediation, retention, detention, flood control and water quality improvement is the key practice to maintain and enhance sustainability of the water system.

Rainwater harvested efficiently is the principal water replenishment for the artificial lake. During periods of dry weather, a little replenishment of purified grey water can also be considered. Meanwhile, buffer strips, floating bio-islands, hydrophytes areas and larger storage volume are necessary to achieve the target of flood control and water quality improvement. Overflow is also important when the stormwater volume harvested exceeds the lake storage capacity. These ecological structures also provide a beneficial habitat for hydrophytes and birds and aesthetic environment for inhabitants.

(a) Wild water birds

(b) Autumn scenery

(c) Wetland and abundant hydrophytes

Figure 11 Various waterscape in the Arcadia project

On average, about 40,000 m³ tap water every year has been saved over the latest two years. Meanwhile, drainage design recurrence period and flood control standard are also enhanced. Since rainwater harvest, flood control, runoff pollution control and waterscape design are integrated into this project design successfully, the artificial lake and stormwater management system has run successfully since the projects construction (Figure 11).

Project 4 Rainwater Harvest & Stormwater Management System for a Large Residential Area in Tianjing, China

The waters system in Dongli Lake Uptown in Tianjin is a successful case of decentralized stormwater management with rainwater harvest, waterscape design and water quality maintenance, etc.

The community in Tianjin had a planned total area of 255 ha, in which waterscape were about 38 ha. The project was implemented in four phases, and about 10 ha waterscape was constructed in the first phase. The integrated land use of the project is summarized in Table 1. The master plan of the project is shown in Figure 12.

Table 1 The land use of the overall project

Land use	Area (ha)	Proportion (%)
Roof and road	133	52.14
Lawn	84	32.87
Waterscape surface	38	14.99
Total	255	100
Nature reed wetland	17	
Total Area	272	

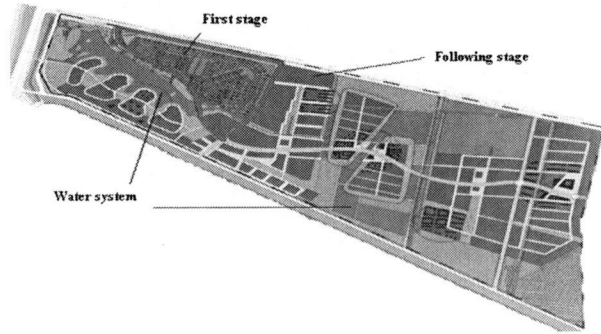

Figure 12 Master plan of the project

This community lies in an arid zone, where rainfall is about 560 mm/a, centralized from May to September, and much less than evaporation. Due to evaporation, the large waterscape results in huge amounts of water being lost. A sustainable system of water replenishment to the waterscape is very important in this project. Special consideration is also needed in water quality maintenance. According to the basic information of the project, the water environment planning and stormwater management program are summarized as follows:

- Harvesting rainwater as fully as possible for the waterscape replenishment, reducing the runoff discharge and reducing consumption of tap water;
- Proper design of the stormwater retention or detention facilities combined with the waterscape design and flood control;
- Runoff pollution control in stormwater conveying process;
- Water quality and good ecosystem maintenance in the waterscape.

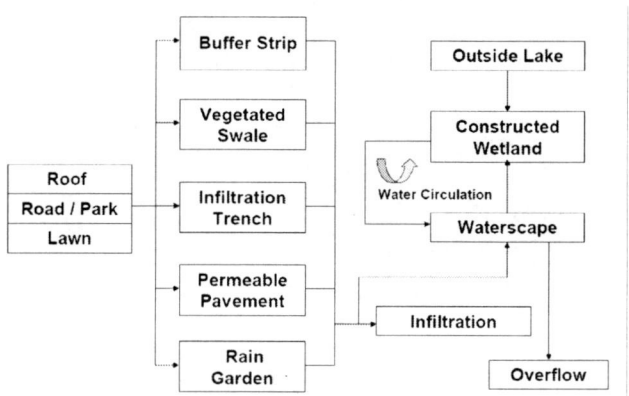

Figure 13 The integrated stormwater management planning

The stormwater management strategy is shown in Figure 13. Rainwater flowed into the lake along suitable routes in different catchment areas according to the land use planning and the ground elevation. The stormwater runoff was purified by vegetated swale, screen basket, first flush diverter, bio-retention and buffer strips, etc. Storage capacity of the water body was utilized to harvest rainwater, to reduce the volume of stormwater discharge, and to control flood with the concept of multifunctional storage design.

Based on water balance analysis, the storage capacity of the waterscape was designed properly. Therefore rainwater utilization was improved. The water body was designed as an ecological structure. Meanwhile, a wetland water environment was proposed. An efficient vertical flow constructed wetland with 8,000 m² was built to purify the water from waterscape or Dongli Lake nearby.

Decentralized stormwater management and integrated rainwater utilization is quite a useful strategy, especially in a project suffering from water shortages and with large-scale waterscape environment. The final phase of this project is currently under construction. Good operational results were achieved during the operation of the initial three phases in recent years (Figure 14). Huge amounts of rainwater were harvested efficiently. Meanwhile, a large quantity of water resource and investment costs has been saved. Furthermore, floods have also been controlled successfully. It is proved that rainwater harvesting aimed at maintaining the waterscape's operation can be achieved by combining integrated planning and rational designing of the waterscape and stormwater system.

(a) Swale

(b) Buffer strip

Figure 14 Several stormwater management practices implemented in this project

CONCLUSION

Development continues at a rapid pace throughout the urban area in China. Runoff from the impervious surfaces in these watersheds continues to be a major cause of degradation to freshwater bodies, estuaries, and flood disasters. Low impact development and stormwater master planning techniques have been recommended to reduce these impacts. The outline of this article showed rapid development in this research field. All the research and engineering projects undertaken by KLUSSWE and other Chinese scholars would provide good experiences for other Chinese cities and projects and by now, more and more projects have been implemented in many other Chinese cities. Based on the outstanding problems and challenges of urban stormwater and the water environment in China, a higher demand and greater development in the next decades can be predicted.

ACKNOWLEDGEMENTS

These studies were financially supported by the Projects of Water Pollution Control and Treatment Key Technology Program (2010ZX07317–011) and (2008ZX07314-007).

REFERENCES

Chen H. P., Che W., Li J. Q. and Zhang W. (2007). Water resources comprehensive application in urban residential building district. *Building Science*, **23**(2), 96–100. (in Chinese)

Che W., Cheng W. J. and Li H. Y. (2006). Application of rainwater utilization and water balance analysis in design of waterscape for urban residential subdivisions. *Chinese Landscape Architecture*, **12**, 62–69. (in Chinese)

Che W., Li J. Q. and Chen H. P. (2004). "Water resources comprehensive utilization and the water environment improvement case analysis for urban residence zone." *Water And Development (water and development international seminar anthology)*, Geological Publishing House, 9, pp. 146–153.

Che W., Li J. Q. and Liu Y. (2003). First flush control for urban rainwater harvest systems. *11th International Conference on Rainwater Catchment Systems. Proceedings*, August 25–29, Mexico.

Che W. and Li J. Q. (2006). *Rainwater Harvest and Management*, China Architecture & Building Press. Beijing. (in Chinese)

Che W., Liu Y. and Li J. Q. (2003). Flush model of runoff on urban non-point source pollutants and analysis. In: *Water and Environmental Management Series, Water in China*, P. A. Wilderer, J. Zhu and N. Schwarzenbeck (eds.). IWA Publishing, pp. 143–150.

Che W., Wang H. Z., Li J. Q., et al. (2001). The quality and major influencing factors of runoff in Beijing's urban area. *10th international Conference on Rainwater Catchment Systems*, Fakt and IRCSA/Europe, Mannheim (Germany), pp. 13–16.

Che W. and Zhou X. B. (2008). New stormwater management and utilization system in urban landscape architecture. *Chinese Landscape Architecture*, **11**, 52–56. (in Chinese)

Li H. Y., Che W., Huang Y., et al. (2007). Case study of environmental planning in urban residence zone. *Water & Wastewater Engineering*, **33**(2), 80–83. (in Chinese)

Li J. Q., Li B. H., Zhang J. and Che W. (2006). Rainwater utilization and water quality control design for landscape water body in residential area. *China Water & Wastewater*, **22**(24), 57–60. (in Chinese)

Li J. Q., Meng G. H. and Che W. (2007). On regulation, reservation and practical calculation in urban rainwater utilization. *Water & Wastewater Engineering*, **33**(2), 42–46. (in Chinese)

Goonetilleke, A., Thomas E., Ginn S. and Gilbert D. (2005). Understanding the role of land use in urban stormwater quality management. *Journal of Environmental Management*, **74**(1), 31–42.

Yin C. Q., et al. (2009). *Urban Diffuse Pollution Control Principles & Technologies*, China Architecture & Building Press, Beijing. (in Chinese)

Survey of Storm Sewer Sediments in Beijing

Li Haiyan[1,*,a], Xu Boping[a], Huang Yan[b], Liu Dake[c] and Li Xiaoheng[c]

[a]Key Laboratory of Urban Stormwater System and Water Environment, Beijing University of Civil Engineering and Architecture, Ministry of Education, Beijing, China, 100044
[b]Beijing Environmental Protection & Monitoring Center, Beijing, China, 100044
[c]Department of sewage network management, Beijing Drainage Group, Beijing, China, 100044
E-mail: lihaiyan@bucea.edu.cn

Abstract Accumulation of solids in storm sewers results in losing flow-carrying capacity, which might block flow and cause an upstream surcharge, local flooding, and enhanced solids deposition. Sewer-solid accumulation in urban drainage systems also creates septic conditions that could create such problems as odour, health hazards, and corrosion. Moreover, sewer sediment from prior storms could contribute a significant amount of pollutants to receiving waters. Continuous in-situ survey was carried out on separated sewers in a district of Beijing with this study for investigating the sediment accumulation state in sewers and the corresponding impacting factors. It would also be helpful to the management of storm sewers network and useful for urban communities in developing better plans reducing risks caused by storm sewer sediments.

Keywords Storm sewer, sediment, survey, pollution

INTRODUCTION

Storm sewers (also called stormwater lines or pipes) are designed to collect rain-water and melting snow, and carry them to a nearby water body. However, runoff pollution is a serious problem worldwide, because such pollutants as motor oil from roads and parking lots, pesticides from lawns and gardens, and loose soil from construction sites can be washed into the receiving waters. Part of flushed sediment-solids deposit by natural settling, and hydro-resistance inevitably.

Sewer sediment is one of major sources of pollutants in urban wet-weather flow (WWF) discharges that include combined-sewer overflow (CSO), separate sanitary-sewer overflow (SSO), and runoff. In most cases, re-suspended sewer sediments are carried by CSO and generate a highly concentrated pollutant load sometimes associated with the "first-flush" phenomenon (Saget *et al.*, 1996; Arthur and Ashley, 1998). And in most cases, the re-suspended sewer sediments are carried into local waterways by CSO. It has been verified that sewer sediment deposited during dry weather flow (DWF) contributes about 30% to 80% of pollutants into receiving waters (Ahyerre and Chebbo, 2002).

Storm sewers are designed to convey designed discharge, but their requested diameter seems too large. The oversized and mildly sloping storm sewer segments possess a potential for sedimentation during low storm flow intensity. During wet weather, the accumulated solids would be re-suspended into receiving waters. During low-flow periods, solids deposit and accumulate in storm sewers because the flow velocity is usually less than the particle-settling velocity. The accumulated sediments in storm sewers would also result in a loss of flow-carrying capacity that may block flow and cause an upstream surcharge, local flooding, and enhanced solids deposition. Sewer-solid accumulation in storm sewers also creates septic conditions that pose odor, health hazard, and corrosion problems. It has been verified that solids deposited from 5% to 30% of the daily SS pollution loading (Cheebo *et al.*, 2003) in sewers. In Europe, average deposition rates have been measured as ranging from 30 to 500 g/m/d (Ashley *et al.*, 2003). A review (Heaney *et al.*, 1999) showed that impervious areas directly connected with sewers contributed a high pollutant loading in separate storm sewers. The flushing of garbage deposited in the stormwater inlet is another important sewer sediment source (Che *et al.*, 2003).

As a result, it is essential to survey the deposition and characteristics of the storm sewer sediment. It would also provide insights to understand the relationship between the storm sewer sediment and the receiving water body pollution control.

SURVEY AREA

Xicheng District is located in the west of Beijing, China. The city was equipped with a hybrid drainage system. A combined drainage system was applied in the traditional region while the separate one was distributed in the majority of the city. The experimental site is about 1,500 m wide and 3,000 m long separate sewer reach located in the densely urbanised centre of Xicheng District. The study area extends to Fuwai Avenue in the south, the Xizhimenwai Street in the north, the Xinjiekounan Avenue in the east, the Xisanhuanbei Road in the west, and a part of the Beijing Financial Street. Old and new residential sundivisions, the campus area and commercial districts were included. The manholes on the sidewalk or subsidiary roads were checked, and 72 lactations were involved, excluding the deserted ones.

METHODS

The sediment depth, materials and inner diameter of the pipes and the corresponding characteristics of catchments were recorded. The sediment depth along the sewer reach was measured with a graduated metal measuring rod, with an accuracy to within 1–2 mm.

RESULTS AND DISCUSSION

The sediment depths were measured in 72 investigated locations. The sediments were found in about 80% of the investigated storm sewers, and deposited in different depths of about 10% to 50% of the pipe diameters, the maximum ratio even reaching 66.7%. These details are illustrated in Table 1.

The Sediment State

As illustrated in Figure 1, the sediment depths account for about 10% to 50% of the diameter in 45.84% of the investigated pipes. In almost half of the pipes, the sediment depths were below 10% of the diameter. Only in 4.16% pipes was the ratio over 50%.

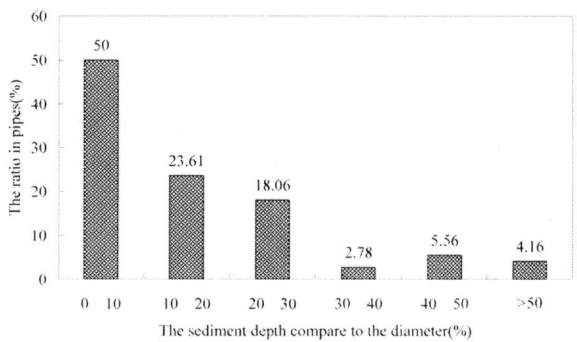

Figure 1 The depth ratio in different storm sewers

The Sediment State in Sewers From Different Runoff Catchments

Runoff coefficients for every catchment in the investigated field could be regarded as similar in this study, and the sewer gradients were between 2% to 3%. The sediment depths in the investigated pipes (the diameters are 300 mm, 400 mm and 500 mm, respectively) were varied, and the details are illustrated in Figure 2. It is necessary to point out that the data of storm sewers with 500 mm diameter in the campus area and those with 300 mm diameter in the business district were not recorded, due to traffic problems and discarded manholes.

Table 1 The sediment depth in storm sewers

Land use function	Inner diameter of the pipe (mm)	Sediment depth (mm)	Ratio (sediment depth/inner diameter of the pipe)	Land use function	Inner diameter of the pipe (mm)	Sediment depth (mm)	Ratio (sediment depth/inner diameter of the pipe)
old residential area	200	30	0.150	commercial district	800	75	0.094
	200	35	0.175		800	100	0.125
	200	65	0.325		800	110	0.138
	200	45	0.225		1000	110	0.110
	200	55	0.275		1500	20	0.013
	300	60	0.200		1600	25	0.016
	300	33	0.110		1600	25	0.016
	300	185	0.617		1600	45	0.028
	300	40	0.133	new residential area	300	15	0.050
	300	170	0.567		300	10	0.033
	300	200	0.667		300	20	0.067
	300	75	0.250		300	25	0.083
	300	50	0.167		400	90	0.225
	300	120	0.400		400*	55	0.138
	300	35	0.117		500	20	0.040
	300	135	0.450		500	45	0.090

Size	Depth	Ratio	Location	Size	Depth	Ratio	Location
300	45	0.150		500	50	0.100	
300	30	0.100		800	20	0.025	
300	85	0.283	commercial district	900	30	0.033	
400	15	0.038		1000	55	0.055	
400	100	0.250		1000	65	0.065	
400	75	0.188		1100	65	0.059	
400	100	0.250		1100	85	0.077	
500	60	0.120		1100	90	0.082	
300	100	0.333		200	10	0.050	campus area
300	120	0.400		300	20	0.067	
300	15	0.050		300	20	0.067	
400	35	0.088		400	10	0.025	
400	25	0.063		400	10	0.025	
400	100	0.250		700	25	0.036	
500	140	0.280		700	20	0.029	
500	100	0.200		400	15	0.038	business district
500	45	0.090		400	15	0.038	
500*	20	0.040		500	200	0.400	
500*	50	0.100		500	20	0.040	
500*	30	0.060		500	10	0.020	

Note: The majority of pipes were made of reinforced concrete; the cast iron storm sewers are labeled as * .

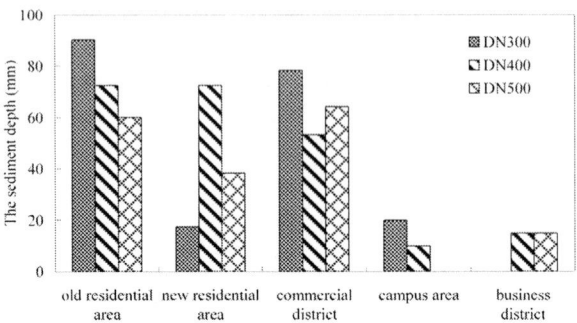

Figure 2 The sediment depths in sewers from different runoff catchments

Distinct differences of the sediment deposition depths were presented in different runoff catchments, and the results are as follows: old residential area > commercial district > new residential area > campus area > business district. The average sediment depths in the old residential area and the commercial district are larger than those in other catchments with the same diameter sewers. It was deduced that less maintenance for the storm inlet and storm sewers is perhaps the main reason for their being more sediments in the old residential subdivision. In the investigation, a large amount of garbage was found in these storm inlets, as it would enter the storm sewers during storm events. And it was also a fact that waste management was ineffective in these areas. However, the sediment depths in such runoff catchments as new residential areas, campus areas and business districts are less due to the relatively effective management of the wastes and good infrastructures.

The Sediment Depths in the Same Diameter Storm Sewers

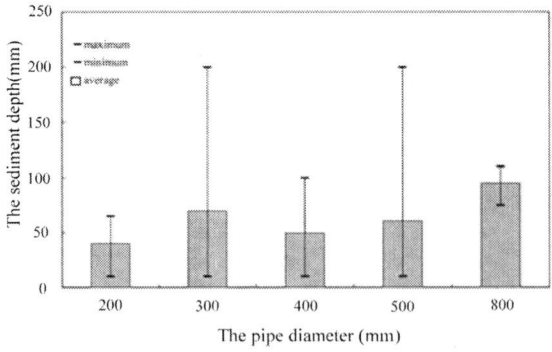

Figure 3 The sediment depths in the same inner diameter storm sewers

The sediment depths in sewers with the same inner diameters were analyzed in this study. As shown in Figure 3, the maximum sediment depths were investigated in storm sewers with 300 mm and 800 mm diameters. Apart from those sewers with 300 mm diameter, the larger the sewer diameters are, the thicker deposition depth. The sediment depths in two sewers located in the same catchment are depicted in Figure 4. From the analysis with SPSS (Zhang, 2002), it was discovered that there is a positive correlation between the sediment depth and sewer diameter, and the correlation coefficient is 0.989 and 0.983 with two-tailed significance of 0.097 and 0.017, respectively.

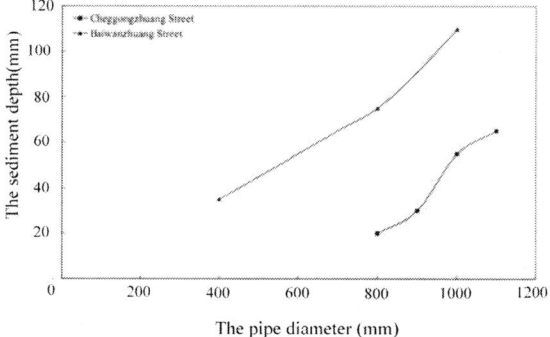

Figure 4 The trend of deposition in same diameter pipes

It was verified that in runoff accumulation period, pollutant concentration flux increases initially, then reduces with the flow decreasing after the peak value until approaches zero. Liu *et al*. determined that runoff mostly reached the peak flow when the pollutant flux was the maximum (Liu, 2005). During low-flow periods or in storm events with low recurrences, the flow velocities are typically inadequate to maintain settleable solids in suspension; and so the particles will deposit. Generally, runoff flow velocity reduces with the sewer diameter increasing. And it could be concluded that the deposition of suspended solids would be easily influenced by sewer diameter in storm events with low recurrences. The larger the sewer diameter, the thicker the sediments. Pan *et al*. analyzed the rainfall data from 1977 to 2006 in Beijing and found that about 58.34% of rainfalls were those with low recurrences (rainfall in 24h being between 2mm to 10mm), and the solids in runoff would be more easily deposited in storm sewers in Beijing (Pan, 2008).

The Sediment Depths in Sewers with Different Materials

The roughness of the storm sewer would influence the sediments deposition. Storm sewers investigated in this study were mainly made of cast iron or reinforced concrete, with the majority being reinforced concrete. Comparison was practised between those pipes with different materials, but the same diameter (500 mm) in three locations in business districts, the results are shown in Figure 5.

It was verified that the sediment depths in reinforced concrete pipes were thicker than those in cast iron pipes. Under the same flow and gradient slope, the rougher pipes are, the more hydro-resistance they have, and the slower the flow velocity is, which would increase the possibility of sediment deposition. As a result, cast iron pipes are perhaps the better choice to avoid sediment's aggradation, but a more expensive choice than reinforced concrete ones.

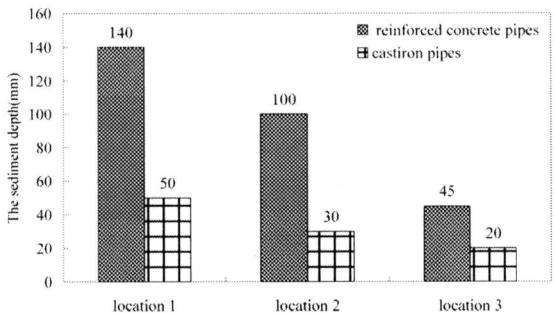

Figure 5 The sediment depths in the same inner diameter pipes with different materials

CONCLUSIONS

The investigations focused on sediment deposition in sewers were applied in this study. The relationships between sediment depth and such impacting factors as the material of the pipes and the characteristics of runoff catchments were put forward. These would be useful for urban water pollution control and storm sewer management. The main conclusions are as follows:

(1) Sediment was found in about 80% storm sewers in Xicheng District, Beijing, China. The sediment depths in different storm sewers range from 10% to 50% of the pipe diameter, and the ratio of above 65% were also found in several pipes.

(2) The sediment depths were distinctly different in sewers from different runoff catchments. The details were as follows: old residential area > commercial district > new residential area > campus area > business district.
(3) Sediment depths became larger with the sewer diameters increasing in a majority of sewers.
(4) Solids deposited more easily in the reinforced concrete pipes than those in the cast iron ones.

ACKNOWLEDGEMENTS

The work is supported by the Natural Science Foundation of China (No.50808009), the Beijing Education Committee Science and Technology Plan Project (No. KM200810016012) and Funding Project for Academic Human Resources Development in Institutions of Higher Learning Under the Jurisdiction of Beijing Municipality (PHR201008379).

REFERENCES

Ahyerre M. and Chebbo G. (2002). Identification of in-sewer sources of organic solids contributing to combined sewer overflows. *Environmental Technology*, **23**(9), 1063–1073.

Arthur S. and Ashley R. M. (1998). The influence of near bed solids transport on first foul flush in combined sewers. *Water Science and Technology*, **37**(1), 131–138.

Ashley R., Crabtree B. and Frase A. (2003). European research into sewer sediments and associated pollutants and processes. *Journal of Hydraulic Engineering*, **129**(4), 267–275.

Che W., Liu Y. and Li J. Q. (2003). Investigation on pollution in urban gutters. *Urban Environments & Urban Ecology*, **16**(6), 153–155.

Cheebo G., Ashley R. and Gromaire M. C. (2003). The nature and pollutant role of solids at the water sediment interface in combined sewer networks. *Water Science and Technology*, **47**(4), 1–10.

Heaney J. P., Wright L., Sample D., Field R. and Fan C. Y. (1999). Innovative wet-weather flow collection, control, and treatment systems for newly urbanizing areas in the 21st century. In: *Sustaining Urban Water Resources in the 21st Century*, C. Rowney, P. Stahre, and L. A. Roesner (eds.) ASCE. ISBN 0-78440-0424-0.

Liu C. Y. (2005). *Research on Pollution and Control of Municipal Drainage System*. M.Sc. thesis, Beijing Univ. of Civil Eng. and Architecture, Zhanlanguan Road 1, Beijing, China.

Pan G. Q. (2008). *Study on Pollution Load and Control Measures for Different Drainage System*. M.Sc. thesis, Beijing Univ. of Civil Eng. and Architecture, Zhanlanguan Road 1, Beijing, China.

Saget A., Chebbo G. and Bertrand J. L. (1996). The first-flush sewer systems. *Water Science and Technology*, **33**(9), 101–108.

Zhang W. T. (2002). SPSS 11 Advanced Statistical Analysis (first edition). Beijing Hope Electronic press.

Innovative Stormwater Management of C10 Road on Guangzhou International Bioisland

Jian Liu[a], Nian She[b] and Xinli Yang[c]

[a]PhD, PMP, Associate Professor, College of Civil Engineering, Shenzhen University, Nanshan District, Shenzhen 518060, China
Tel & Fax: +86-755-2673-2827; E-mail: liujian@szu.edu.cn
[b]PhD, Visiting Professor, Shenzhen University; Seattle Public Utilities, Seattle Municipal Tower, 700 Fifth Avenue, Suite 4900, PO Box 34018, Seattle, WA 98124-4018, USA
Tel: +1-206-684-5164; E-mail: nian.she@seattle.gov
[c]Master graduate, College of Civil Engineering, Shenzhen University, Nanshan District, Shenzhen 518060, China; Tel: +86-755-2673-2827; E-mail: yangxinli200888@163.com

Abstract This paper describes an innovative approach to stormwater management for the C10 Road on Guangzhou International Bioisland in China. Instead of using conventional storm drains to convey storm runoff to downstream drainage systems, a natural drainage system consisting of bioswales and rain gardens was designed to infiltrate, retain, and convey the excess runoff. The costs between a conventional convey system and a natural system were compared.

Keywords Low impact development, Stormwater management, Natural drainage system, Bioswale, Rain garden

INTRODUCTION

Even though it has abundant rainfall, Guangzhou is one of the many cities in China that suffers from water "shortage" problems. The problems are caused by pollution discharged from point and nonpoint sources to its rivers and lakes. In addition, during the wet season, many areas of the city are flooded due to a rapid increase in impervious area. Therefore, efficiently managing the city's stormwater is critical to solving the water "shortage" and water "abundance" problems in Guangzhou. The city government has realized that rainfall water is a valuable resource that can be utilized to effectively improve the city's ecological systems, and is looking into some successful examples of stormwater management using low impact development (LID) techniques in the United States, Japan, Australia, and European countries. LID has proven to be a scientifically sound approach to many urban environmental problems (Akinyemi, 2008; Albano et al., 2008;

Easton and Ansen, 2008; Schreier and Marsalek, 2008), and can be applied to Guangzhou and elsewhere in China.

LID is an innovative stormwater management approach with a basic principle of managing rainfall at the source using decentralized micro-scale engineering controls. The goal of LID is to mimic a site's predevelopment hydrology by using design techniques that infiltrate, filter, store, evaporate, and detain runoff close to its source. LID techniques are based on the premise that stormwater management should not be seen as stormwater disposal. Instead of conveying and managing stormwater in large, costly end-of-pipe facilities located at the bottom of drainage areas, LID addresses stormwater through small, cost-effective landscape features located at the lot level. LID techniques can reduce project costs and improve environmental performance. The LID principles originate from Prince George's County, Maryland, in the mid-1980s where the concept of bioretention was first introduced. The technique helped the County to address the growing economic and environmental limitations of conventional stormwater management practices. LID allows for greater development potential with less environmental impacts through the use of smarter designs and advanced technologies that achieve a better balance between conservation, growth, ecosystem protection, and public health or quality of life. Today, bioretention is just one of many LID techniques. Other techniques include permeable pavement, rain gardens, vegetated swales, ecoroofs, tree box planters, and disconnected downspouts. These techniques are useful for controlling nonpoint pollutants, reducing peak flow and runoff volumes, recharging ground water, and addressing a number of social and environmental concerns (EPA, 2007; Akinbobola, 2007). Many examples have shown that LID is a versatile approach that can be applied equally well to new development, urban retrofits, and redevelopment projects (LID Center, 2009). LID benefits are classified into three categories: 1) environmental benefits, which include reductions in pollutants, protection of downstream water resources, ground water recharging, reductions in pollutant treatment costs, reductions in frequency and severity of combined sewer overflow (CSO), and habitat improvements; 2) land value benefits, which include reductions in downstream flooding and property damage, increases in real estate value, increased parcel lot yield, increased aesthetic value, and improvement of quality of life by providing open space for recreation; and 3) compliance incentives (EPA, 2007).

Since 1980s, numerous LID projects have been constructed and operated in the United States. These projects have demonstrated that the cost of applying LID in urban watershed is generally lower than that of the traditional approach to achieve the same environmental benefits. Based on the statistics collected by US EPA the total capital cost savings ranged from 15% to 80% (EPA, 2007). As a case study, the cost of the 2nd Avenue Street Edge Alternative (SEA) Street project in Seattle, Washington State is listed in Table 1. The SEA Street project was a pilot project undertaken by Seattle Public Utilities to redesign an entire 201.17 m block with a number of LID techniques. The goals were to reduce stormwater runoff and to provide a more livable community. Throughout the design and construction process, Seattle Public Utilities worked collaboratively with street residents to develop the final street design. The design reduced imperviousness, included retrofits of bioswales to treat and manage stormwater, and added 100 evergreen trees and 1,100 shrubs. Conventional curbs and gutters were replaced with bioswales in the rights-of-way on both sides of the street, and the street width was reduced from 7.62 m to 4.27 m. The final constructed design reduced imperviousness by more than 18%. An estimate for the final total project cost was $651,548. A significant amount of community outreach was involved, which raised the level of community acceptance. Community input is important for any project, but because this was a pilot study, much more was spent on communication and redesign than what would be spent for a typical project. Managing stormwater with LID techniques resulted in a cost savings of 29%. Also, the reduction in street width and sidewalks reduced paving costs by 49%. The avoided cost for stormwater infrastructure and reduced cost for site paving accounted for much of the overall cost savings. The nature of the design, which included extensive use of bioswales and vegetation, contributed to the increased cost for site preparation and landscaping. For this site, the environmental performance improvement has been even more significant than the cost savings. Hydrologic monitoring of the project indicates a 99% reduction in total potential surface runoff, and runoff has not been recorded at the site since December 2002, a period that included the highest-ever 24-hour recorded rainfall at Seattle-Tacoma Airport. The site retains more than the original design estimate of 19mm of rain. A modelling analysis indicates that if a conventional curb-and-gutter system had been installed along 2nd Avenue instead of the SEA Street design, 98 times more stormwater would have been discharged from the site (EPA, 2007).

Table 1 Cost comparison for 2nd Avenue SEA Street

Item	Conventional development cost ($)	SEA Street cost ($)	Cost savings ($)	Savings (%)	Total savings (%)
Site preparation	65,084	88,173	−23,089	−35%	−11%
Stormwater management	372,988	264,212	108,776	29%	50%
Site paving and sidewalks	287,646	147,368	140,278	49%	65%
Landscaping	78,729	113,034	−34,305	−44%	−16%
Misc.	64,356	38,761	25,595	40%	12%
Total	868,803	651,548	217,255	25%	—

Though there are a number of rainwater reclamation projects being implemented or under construction in Beijing, Hong Kong and Taiwan, there are very few LID applications to be found in China. The rainwater reclamation of the Longhua No. 2 Line Project in Shenzhen is one of a few projects that only applied LID techniques in the planning level. But the rainwater reclamation works in the Longhua No. 2 Line Project have not been constructed due to disharmonious opinions of governmental departments, and the passive attitude of design professionals (Liu et al., 2008a, 2008b, 2009). At present, stormwater management and rainwater reclamation issues have not been considered thoroughly during the planning of the new urban developments in China mainly due to a lack of relevant laws, regulations and standards (Liu et al., 2009).

Low impact development and rainwater reclamation are an effective means of sustainable development and environmental improvement in terms of reducing disasters caused by storm events, creating a healthy aqua-landscape, and strengthening the general public's water conservation awareness. It has also been proven that LID can reduce the local heat island effects, recharge groundwater, and prevent the pollutants carried by runoffs to receiving waters (Liu and She, 2009). The object of this paper is to describe the use of the LID approach such as bioswales and rain gardens in place of a conventional concrete storm drain system in the design of C10 road on Guangzhou International Bioisland in China.

This research was conducted through literature review, site reconnaissance, and discussions with the developers and design professionals.

OUTLINE OF C10 ROAD

C10 Road is on the Guangzhou International Bioisland (originally called as Guanzhou Island) located on the Pearl River in southeast Guangzhou, Guangdong province. The island has been developing into a biotech hub to host multinational biotech companies and research institutes.

The gross land area of the catchment under the study is 40,000 m², and the net land area is 30,045 m². The total construction area in the catchment is approximately 58,000 m², of which 48,000 m² are buildings on the ground, and 10,000 m² are underground structures. The widths of the main roads and branch roads are 30 m and 15 m, respectively. A 60 meters wide sidewalk along the island dike called the Milky Way and several 30 meters wide main roads intertwine to form a spatial pattern resembling DNA double helixes. The size of the compound under construction for research, business development and residence is a square approximately 200 meters long by 200 meters wide.

According to the discussions with the Developer, Guangzhou International Bioisland Construction Office, the east section of C10 Road between Gangtou

Road and C7 Road was selected as the demonstration project for application of LID techniques to stormwater management (see Figure 1). Total length of C10 Road is 942.84 meters. The demonstration project covers about 400 meters in length.

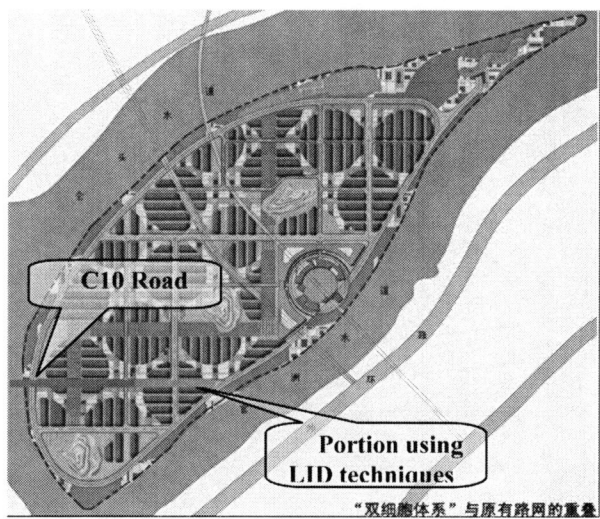

Figure 1 Location of C10 Road and ecological park

STORMWATER MANAGEMENT MEASURES

The original design of C10 Road was carried out according to the current laws, regulations and codes. The runoff from the catchment of C10 Road is directly conveyed into concrete storm drains under the road. The design did not consider stormwater control and pollution reduction. An alternative design is to apply LID in the catchment of C10 Road that will serve as a conveyer of stormwater runoff with less peak flows and smaller volume, but have additional environmental benefit such as reducing nutrients discharged directly to Pearl River. According to literature review and site reconnaissance, the following five LID components were proposed for C10 Road design and construction.

Construction of Rain Gardens

By using the space at the intersections of the Gangtou Road and C10 Road, and C7 Road and C10 Road in the Gangtou Hilly Park, two rain gardens were designed to detain the stormwater and control pollutants in the drainage area. The locations of the rain gardens are shown in Figure 2. The radius of each rain garden was set to be equal to 10 m. The design method of the rain gardens is similar to the natural drainage system described hereafter.

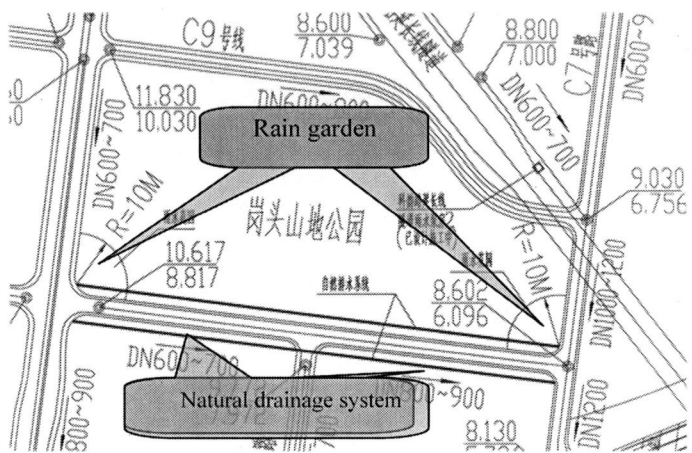

Figure 2 Location of rain gardens and natural drainage system

Replacing the Original Green Belts with Natural Drainage System

The original green belts on both sides of C10 Road were replaced with natural drainage systems (see Figure 2). Generally, natural drainage systems consist of bioswale and vegetation. The size of a natural drainage system is determined by its drainage area and stormwater intensity. The stormwater intensity design standard for a LID project in USA is either 25-year or 50-year rainfall. There are no long term rainfall records on or near the Bioisland. The original storm drain system was designed based on the Guangdong provincial rainfall intensity formula. The rainfall intensity calculated by the formula is 116 mm/h which corresponds to a 200-year storm. On the other hand, the maximum recorded rainfall in Guangzhou City in recent years is the event on March 29, 2009 with an intensity of 53 mm/h, which is equivalent to 50-year storm. The larger value between the recorded rainfall and calculated storm was selected as design storm for the natural drainage systems of the demonstration portion of C10 Road. As shown in Figure 3, the cross section of the natural drainage system on the side of the Gangtou Hilly Park is 3 meters wide by 2 meters deep, and the size of the natural drainage system on the other side is 2 meters wide by 2 meters deep. The lengths of the two natural drainage systems are 400 meters. The ponding depth of each natural drainage system is 30 centimeters. The side slope of the mixing soil is about 1 (height): 4 (width). Native evergreen trees and shrubs will be planted in the bioswales. In order to reduce stormwater overflow from the bioswales, a PVC pipe that is laid 10 centimeters underground connects to the storm drain under C7 Road. The conventional green belts and storm drains were replaced by the natural drainage systems and rain gardens.

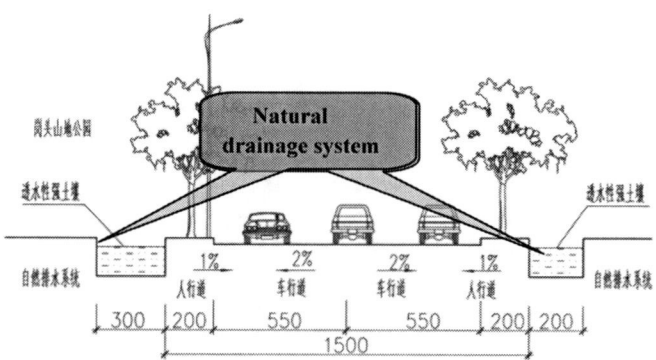

Figure 3 Cross section of natural drainage systems

Modelling Hydrology of Natural Drainage System

Recall that the original conveyer system was designed based on rainfall intensity of 116 mm/h. The proposed natural drainage system should perform equal or better than that of the original design in reducing peak flows and runoff volumes. To insure that the proposed design is appropriate we use SWMM5-LID beta version developed by US EPA (Lewis Rossman, personal communication, October, 2009) to compare the simulated runoffs from the catchment of original design and proposed design. 5 minute intervals of a 24-hour synthetic storm event with a maximum intensity of 432 mm/h were used in the simulations. This would be an extreme event that would only occur with a powerful typhoon sweeping across Southern China's coastal cities. Figure 4 shows the hydrograph of the runoffs for both systems. It can be seen that the natural drainage system eliminates the first peak flow from the storm runoff and reduces the consequent peak flows and volumes. The reduction of peak flow is about 54%, and volume is about 45%.

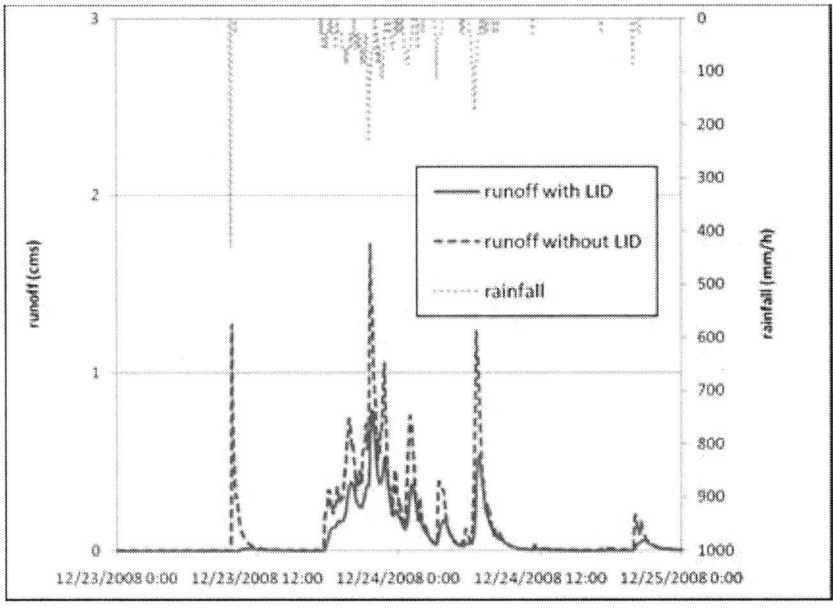

Figure 4 Comparison of simulated runoff for conventional convey system vs. natural drainage system using a synthetic 5-minutes interval of 24-hour storm event

To see a more realistic case we used a rainfall event collected from University of Auckland's green roof. The rainfall was collected at 5-minutes interval from February 20, 2009 to February 22, 2009. The maximum intensity is 62.4 mm/h. The simulations for both conventional conveyer and natural drainage system are shown in Figure 5. This event is similar to a large event that occurred on March 29, 2009 in Guangzhou, which caused severe flooding in the city street. The simulations showed that the peak flow is reduced by 68%, and runoff volume is reduced by 56% by the natural drainage system.

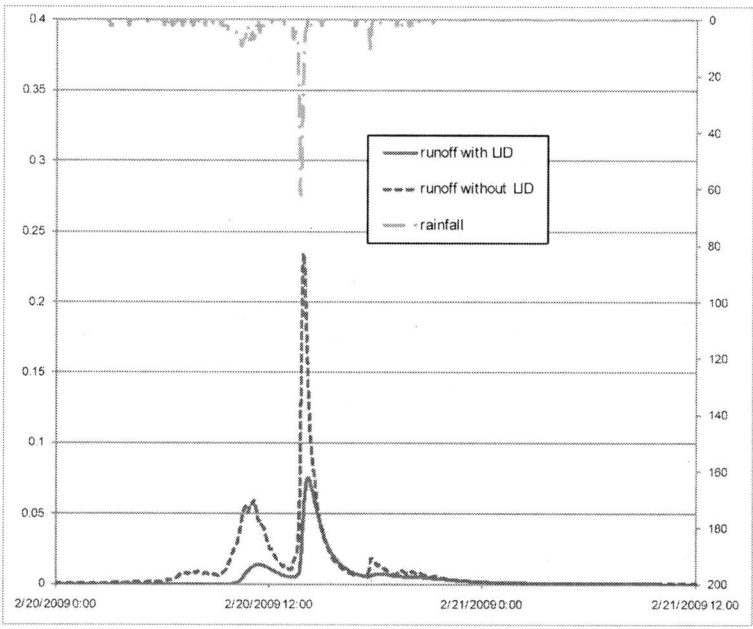

Figure 5 Comparison of simulated runoff for conventional convey system vs. natural drainage system using University of Auckland green roof rainfall data

Replacing the Original Interlocking Bricks on Sidewalks with Permeable Pavements

In order to increase rainfall infiltration, permeable pavement of sidewalks has been used in some countries such as Japan for more than 20 years. In addition, the permeable pavement sidewalks are also adopted by Beijing Olympic Stadium. The technical requirements for the permeable pavement of the sidewalks of C10 Road were recommended as follows:

a) The strength on 28th day is larger than 15 MPa;

b) Effective porosity is larger than 15%, and permeability coefficient is 0.2 cm/s;

c) The depth of the permeable concrete is determined by the maximum value of the depth calculated by road stability and permeability for 2-year storm intensity;

d) The permeability coefficient of gravel base is larger than the one of the permeable concrete; and

e) The permeability coefficient of soil base is not smaller than 0.0001 cm/s.

Replacing Original Casting-iron Rain Well Lids and Sewage Covers with Reactive Powder Concrete Covers

Cast-iron manhole lids and sewage covers have been used for many years in China and elsewhere in the world. Not only are the costs of these kinds of lids high, but they are also easily stolen, and cause many traffic accidents. In order to reduce costs and accidents, we proposed to replace the cast-iron covers within the catchment of C10 Road with high-tech reactive powder concrete (RPC) covers. The RPC covers have low permeability, long durability and high strength.

COST COMPARISON

Table 2 shows the cost comparison of LID approach to C10 Road vs. the conventional approach. The construction cost and design fee of the conventional approach are 0.7 and 0.2 million Yuan, respectively for the pilot project while the construction cost and design fee of LID approach are 0.6 and 0.55 million Yuan. The project cost of LID approach is higher than that of conventional approach in the pilot study because of additional expenses on communication, redesign, and a monitoring program that would not have been spent for a typical project. However, if the overall cost includes 50-year maintenance, then LID approach is much cheaper than that of conventional approach. Managing stormwater using LID approach resulted in a cost savings of 14%. The savings from conventional drainage infrastructure (i.e., concrete drainage pipes and green belts), site pavement, and maintenance cost account for much of the overall cost reduction.

Table 2 Cost comparison of the pilot project of C10 Road (Unit: 1,000 Yuan)

Item	Conventional development cost	LID techniques	Cost savings	Savings (%)
Construction cost	70	60	10	14
Design fee or consulting fee	25	55	−30	−120
Monitoring equipment	0	10	−10	−
50-year maintenance fee	100	20	80	80
Expense at completion	95	125	−20	−21
Expense after 50 years	195	145	50	26

CONCLUSION

This paper reports an innovative approach to stormwater management for road construction. A natural drainage system design was proposed to replace conventional concrete drainage pipes in the catchment of C10 Road on Guangzhou International Bioisland in China. Porous pavement is also proposed for the sidewalks. Using SWMM5-LID beta version we have demonstrated that the natural drainage system proposed would reduce more than 50% of peak flows, and reduce more than 45% of the total volumes of the large events compared with conventional drainage system. Therefore, adopting the natural drainage system would significantly reduce the risk of street flooding. In addition to flood prevention, the proposed natural drainage system design has water quality benefits in terms of biofiltration to reduce nutrient load directly to Pearl River. The construction of the demonstration project is planned to be commenced at the end of 2009. If the maintenance is included, the cost of applying LID approach is cheaper than the one by conventional system.

Future research opportunities exist to analyze the infiltration mechanism of the natural drainage system and study the environmental benefits after the monitoring data are collected.

ACKNOWLEDGMENTS

The research was supported by the Shenzhen Science & Technology Program (No. 200739), Fund of Land Investment & Development Centre of Shenzhen Works Bureau for Rainfall Usage Study of the Longhua No. 2 Line Project (Contract No. 2007139) and Fund of the Guangzhou International Bioisland Construction Office for Application of Low Impact Development Techniques to C10 Road of the Guangzhou International Bioisland. The authors wish to thank for Lewis Rossman of US EPA to provide SWMM5-LID beta version for this work, and acknowledge Dr. Jian Zuo of University of South Australia for his comments to improve the quality of this paper.

REFERENCES

Akinbobola C. (2007). Low impact development urban retrofit program in Prince George's County, Maryland. *Sustainable Water Management and River Development*, 1, Sichuan University Press, pp. 304–320.
Akinyemi E. (2008). International experiences with low impact development. *Proceedings of the 2008 International Low Impact Development Conference*, ASCE, Seattle, Washington, Nov. 16–19.
Albano C., Carlone C., Cattaneo S. and Medina D. E. (2008). Sustainable stormwater management: Implementation of pilot low impact development stormwater controls

at US Department of Defense Installation in Europe. *Proceedings of the 2008 International Low Impact Development Conference*, ASCE, Seattle, Washington, Nov. 16–19.

Easton H. and Ansen J.-A. (2008). Growth of low impact design in the Auckland region (New Zealand) through an innovative grants programme. *Proceedings of the 2008 International Low Impact Development Conference*, ASCE, Seattle, Washington, Nov. 16–19.

Environmental Protection Agency (EPA). (2007). *Reducing Stormwater Costs through Low Impact Development (LID) Strategies and Practices*. December, http://www. epa.gov/nps/lid.

Liu J. and Chen S.-C. (2007). Study on rainfall usage of the Longhua No. 2 Line project in Shenzhen. *The 11th Cross-Strait Water Resources Conference*, Changchun, China, September, 361–371.

Liu J., Mo L.-K. and Nian, S. (2008a). *Evaluation Report of the Stormwater Utilization in China and Abroad for the Longhua No. 2 Line project*, Shenzhen University, April (in Chinese).

Liu J., Mo L.-K. and Nian S. (2008b). *Proposal for Improving the Jianshang Road Design*, Shenzhen University, August (in Chinese).

Liu J., Mo L.-K., Nian S., Chen S.-C. and Yang X.-L. (2009). Application of Low Impact Development Techniques to Rainwater Utilization Planning of the Longhua No. 2 Line Project. *34th Australasian Universities Building Educators Conference (AUBEA 2009)*, Barossa Valley, SA, Australia, July 7–10.

Liu, J. and Nian, S. (2009). Apply low impact development techniques to solve flooding, nonpoint source and desertification problems in urban development. *The Proceedings of 13th Cross Strait Water Resources Conference*. Taizhong, Taiwan, Nov. 23–24.

Low Impact Development (LID) Center. (2009). Introduction to LID, http://www.lid-stormwater.net/background.htm#What_is_LID#What_is_LID[accessed June 2009].

Schreier, H. and Marsalek J. (2008) Innovative stormwater management in Canada. *Proceedings of the 2008 International Low Impact Development Conference*, ASCE, Seattle, Washington, Nov. 16–19.

PART THREE

Used Water Source Separation,
Decentralized Systems

Water Metabolism Concept and its Application in Designing Decentralized Urban Water Systems with Wastewater Recycling and Reuse

X. C. Wang and R. Chen

Key Lab of Northwest Water Resource, Environment and Ecology, MOE, Xi'an University of Architecture and Technology, Xi'an 710055 China
E-mial: xcwang@xauat.edu.cn; chenrong@xauat.edu.cn

Abstract In order to reconsider the configuration of an urban water system to meet the needs for sustainable water use and water environmental improvement in our water stringent world, the concept of water metabolism which stresses the harmony of the artificial water cycle with the natural hydrological cycle is discussed. By using the Second Law of Thermodynamics as a theoretical tool, the natural water cycle with minor human disturbance is considered to be a pseudo-reversible system with minimum entropy change from endogenous contribution. The minimization of entropy increase corresponds to the maximization of the metabolic capacity of a system. Two baselines can thus be proposed for the design of an urban water system: one is to decrease the entropy increase from human disturbance and another is to make the artificial or engineering part of the water system as close to the nature as possible. These principles are applied in two model cases of decentralized urban water systems that demonstrate a harmonic integration of water supply, sewerage, water reuse, and local water environment within one framework. Sound water environment is well sustained with minimized freshwater supply. The water metabolism concept and its application may direct a new paradigm for urban water system design towards the future.

Keywords Water metabolism, thermodynamics, urban area, decentralization, water reuse

INTRODUCTION

The total renewable water resource in the world amounts to about 55000 km^3 (World Resources Institute, 2005). Taking into account a total population of about 6.67 billion (World Resources Institute 2008), the per capita water resource can be calculated as more than 8200 m^3/person. However, due to uneven distribution of both the water resource and population, in different area of the world the availability of water resource differs from each other. The per capita water resource can be as high as 45000 m^3/person or more in South America and Oceania, while it can be as low as about 1400 m^3/person in the Middle East and North Africa (World Resources Institute, 2005). The distribution of water resource is also uneven within one area or one country. For example, in China the average per capita water resource based on the 2008 data (Ministry of Water Resources, 2010a; National Bureau of Statistics of China, 2009) is 2066 m³/person, while of the 10 major river basins those in the northern China such as the Liaohe River, Yellow River, Huaihe River, and Haihe River, the per capita water resources are only 711 m³/person, 510 m³/person, 443 m³/person, and 164 m³/person, respectively. Due to low availability of water resources, over withdrawal of surface water or groundwater is commonly practiced in these basins. The direct result of over withdrawal is the decrease of water flow in the river channel and/or the decline of groundwater table. This results in deteriorated water quality because of insufficient water quantity for diluting pollutants.

In China surface water quality has been categorized into five classes according to its suitability for drinking water supply (Class I to Class III), industrial water use (Class IV), and agricultural water use (Class V). By a calculation using the national river water quality monitoring data of 2008 for all the major river basins (Ministry of Water Resources 2010b), Figure 1 can be obtained to show the relationship between the river water quality and per capita water resource. There is an apparent tendency that the % of polluted water (water quality worse than Class V) decreases with the increase of per capita water resource, indicating that water pollution often occurs simultaneously with water shortage.

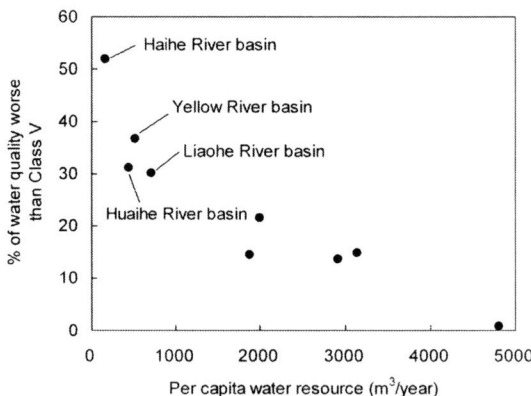

Figure 1 Relation between per capita water resource and % of water quality worse than Class V in major river basins in China (calculated according to 2008 surface water quality monitoring data)

In China there are more than 400 cities, most in the northern region, suffering from both water shortage and serious water pollution (Wang and Jin, 2006).

Although various actions have been taken recently such as development of new water resources, long distance water transfer and so on, it is widely recognized that in most cases water reuse is the most feasible option for mitigating urban water shortage.

When treated wastewater becomes part of the usable water resource in an urban area, how to design the urban water system may become a newly encountered problem because the traditional philosophy of system design for urban water supply is no longer completely apt to the current circumstance. Firstly, as water supply will be through at least two qualitatively different sources, i.e. freshwater and reclaimed water, the water demand has to be accounted both quantitatively and qualitatively. Secondly, as water shortage often occurs simultaneously with water pollution, restoration of a sound water environment should be one of the main objectives of water reuse. Thirdly, sustainable utilization of water resources should be the sole principle of urban water system design.

A detailed discussion on the above mentioned issues may need the development of an innovative philosophy by looking at the natural behaviour of water in the world. It thus becomes the topic of this chapter to introduce the concept of water metabolism into urban water system design. Examples of applying this concept to the design of decentralized urban water system with wastewater recycling and reuse are also presented.

CONCEPT OF WATER METABOLISM

Metabolism and Metabolic Capacity of Natural Waters

Let us consider what happens in the natural hydrologic cycle which is the cycling of water through the environment following a simple pattern. Moisture in the atmosphere condenses into droplets that fall to the Earth as rain or snow. Water, flowing over the Earth as surface water or through the soil as groundwater, returns to the oceans, where it evaporates back into the atmosphere to begin the cycle again. Such a water cycle is important for keeping a worldwide or regional circulation of water in various water bodies such as rivers, lakes, and groundwater aquifers. On the other hand, the water cycle is also a process of water purification that ensures the provision of fresh water resources in the cycle by a series of physical, chemical, and biological reactions. As a result, various water bodies can be kept "healthy" to perform their environmental functions well.

These processes in the hydrological cycle can be considered metaphorically as "metabolism" which, by definition, is the set of chemical reactions that occur in living organisms to maintain life (Smith and Morowitz, 2004). Such a metaphor was first used by Wolman (1965) in his famous paper "The metabolism of cities", and then by Tambo (2004) who proposed the "urban water district" as a water metabolic space within the hydrological cycle. We may thus give "water metabolism" a terminological definition as the set of natural purification reactions to maintain a water system in a living condition. The capability of a water system to perform natural purification may be called "metabolic capacity" which is its capacity of natural purification to maintain a healthy condition.

Human Disturbance on Natural Water Cycle

As human beings depend on natural water for sustaining life, the scale of human disturbance on the natural hydrological cycle became larger and larger with urban development. From ancient time people found traditional ways to take water from various water bodies for daily use and then discharge the used water which goes back to the water bodies through various routes. Because the scale of the traditional water use is very small, the disturbance on the natural water cycle is minor. However, in a modern city the human disturbance is no longer negligible and a large scale artificial water cycle is added to the natural water system (Figure 2). The pollutant loading from the artificial cycle to the natural cycle may be beyond the metabolic capacity for the water bodies to maintain "healthy". For these reasons, human beings have nothing to do but to take engineering means to "protect" or "cure" the natural water bodies, such as to practice water purification and wastewater treatment.

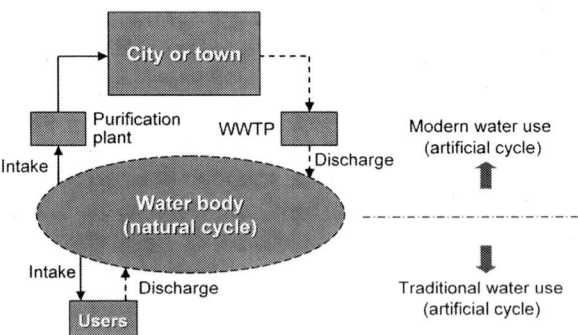

Figure 2 Artificial cycles added to the natural hydrological cycle (lower part: small artificial cycle by traditional water use; upper part: large artificial cycle by modern water use)

Concept of Water Metabolism Relating to Urban Water System Design

A thermodynamic consideration

The thermodynamic principles have been widely used for evaluating aquatic ecosystems (Ludovisi and Poletti, 2003; Aoki, 2006; Aoki, 2008), agro-ecosystems (Steinborn and Svirezhev, 2000), and water resources availability (Kawachi *et al.*, 2001; Maruyama *et al.*, 2005). In order to evaluate an urban water system in a similar way, we can consider the water system shown in Figure 2 to be principally an ecosystem. According to the Second Law of Thermodynamics, the entropy increase in the ecosystem can be written as

$$\Delta S = \oint_B \frac{\partial q}{T} \tag{2.1}$$

where ΔS is entropy increase, B is the system boundary, ∂q is any small change of energy or heat, and T is absolute temperature.

For an isolated system, it is considered to be reversible if

$$\Delta S = 0 \tag{2.2}$$

or it is considered to be irreversible if

$$\Delta S > 0 \tag{2.3}$$

However, since no ecosystem could ever exist as an isolated system, the second law of thermodynamics cannot be applied without adaptation. One prevailing method is to consider that the change in entropy for a non-isolated ecosystem is composed of two parts: an external contribution from outside as $\Delta_e S$ and an endogenous contribution due to the internal processes as $\Delta_i S$ (Ludovisi and Poletti, 2003).

From a worldwide viewpoint, all the natural processes in the hydrological cycle can be considered as internal processes that bring about endogenous contribution to changes in entropy, i.e. $\Delta_i S$ within the large natural aquatic ecosystem, while the external contribution of $\Delta_e S$ is considered to be from only human disturbances. Strictly speaking, any natural process can only progress in a direction which results in an entropy increase (Ludovisi and Poletti, 2003). However, it may be reasonable to assume that the natural hydrological cycle as discussed in 2.2.1 is a pseudo-reversible process by its nature of self maintenance of water and materials balance. Of course, such an assumption should be restricted to a comparatively short time span (e.g. the time scale of human life) but not a long time span (e.g. the time scale of natural evolution). In this way, we assume that the following condition almost holds for the natural hydrological system:

$$\Delta_i S \rightarrow 0 \qquad (2.4)$$

We can thus bridge between the concepts of water metabolism and the thermodynamics a simple relationship as:

$$\textit{"Maximized metabolic capacity"} = \textit{"Minimized entropy increase"} \qquad (2.5)$$

Theoretical strategy of urban water system planning

An urban water system is closely related to the natural hydrological cycle (Figure 2). From what discussed above, there would be two baselines we have to follow in urban water planning. The first baseline is to decrease as far as possible $\Delta_e S$ which is the entropy change resulted from human disturbance on the natural hydrological cycle, and the second baseline is to make the artificial part of the urban water system as close to the natural part as possible so that the nature of $\Delta_e S$ can be modified in the way as we discussed for $\Delta_i S$. From the former, the envisaged strategy is to protect the natural watershed or water bodies such as lakes and streams and decrease human disturbance on them. This coincides with the principle of low impact development (LID) for local water system design (van Roon 2007) and for combined sewer overflows control (Montalto et al., 2007). From the latter, the strategy is to learn more from the nature and try to build the artificial or engineering components of the urban water system in a natural manner.

Here we have to say that the conventional ways of urban water planning are often against the abovementioned principles. Under the human-centric consideration of supply of high quality drinking water, collection of sewage and storm water and advanced treatment of the collected water prior to discharge into natural water bodies to meet the needs of the public health, industrial growth and prosperity of the society (Wilderer, 2001), centralized water supply and sewerage networks become the central part of the urban water system almost in every city or metropolitan in the world. For water supply, "to meet the demand of water use" is the basic philosophy and the water supply network should cover every corner of the city to ensure water provision. For the sewerage work, "to collect and discharge the used water quickly and smoothly" is the basic philosophy and a sewerage network covering the whole water supplied area should also be provided in the city. In such an urban water system, water is in fact taking a journey through the artificial networks in a way as shown in Figure 3.

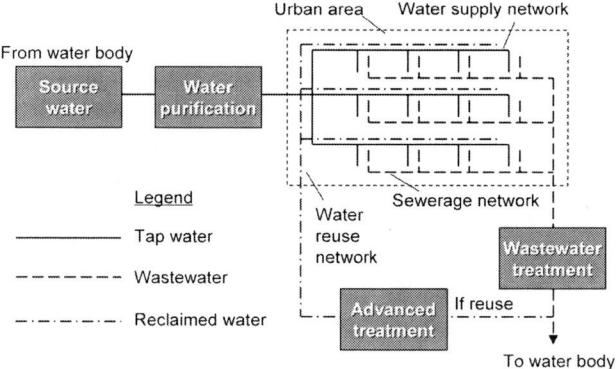

Figure 3 A conventional urban water system which composes of water supply network, sewerage network, and additional water reuse network as treated water reuse is practiced

As the source water is usually at a discrete location upstream of the city while the wastewater treatment plant is at another discrete location downstream of the city, long distance water and wastewater transfer pipelines have to be constructed for the water to take a long journey. When treated water reuse is to be practiced, a third network shown in Figure 3 will have to be constructed for bringing the reclaimed water back to the city area again for various purposes of reuse. The discrepancies between different parts of the artificial urban water system result in large amount of energy consumption which inevitably brings about additional increase of entropy as $\Delta_e S$. This is against the abovementioned first baseline. Another thing noticeable in Figure 3 is that such an artificial urban water system is related to the natural water only at the beginning and end points of the system, i.e. the source water which locates at upstream of the city and the water body which locates at downstream of the city to receive urban discharge. This is against the abovementioned second baseline.

Configuration of a Local Water System Under the Concept of Water Metabolism

Now we consider the configuration of an urban water system under the concept of water metabolism. Following the strategies discussed above, the basic policies for configuring such a system are (i) it should be a harmonic integration of the subsystems of water supply, sewerage, water reuse, and urban water environment; (ii) decentralization should be an important option for system selection; (iii) the system should be as close to the nature as possible; and (iv) the principle of ecological design should be applied.

A conceptual configuration of a local water system can then be proposed as shown in Figure 4. Comparing with a conventional urban water system, this system has the following features:

- It is an enclosed water system with minimized supply of fresh water and minimized discharge of wastes across its boundary.
- The primary objective of wastewater treatment is for water reuse. Therefore, as long as economically and technologically feasible, non-potable water use and environmental water use should be covered by reclaimed water.
- Where applicable, natural or artificial water bodies, such as lakes, ponds, streams, can be introduced into the system. They use the reclaimed water as source for water replenishment, and meanwhile play the functions of regulation basin and water quality polishing before being used for miscellaneous purposes.

Although at this stage there is yet a mathematic tool available for quantitative thermodynamic calculation of the system, it conceptually follows the two baselines we discussed, i.e. human disturbance on the nature being minimized, and the system itself being close to the nature.

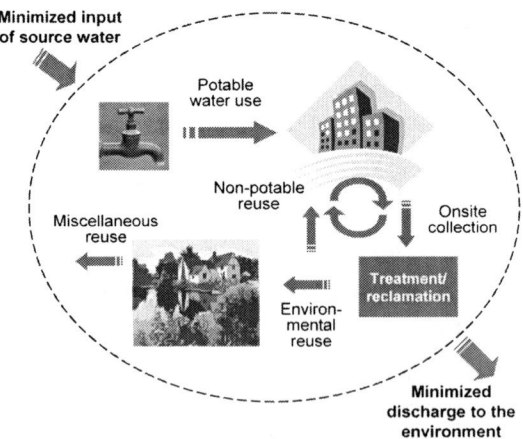

Figure 4 Conceptual configuration of a local water system under the concept of water metabolism

MODEL CASES FOR THE APPLICATION OF THE CONCEPT OF WATER METABOLISM

A Decentralized Water Environmental System with Grey Water Reuse

This model case is a decentralized water environmental system in a newly developed residential community in Xi'an, China. Environmental reuse of the treated grey water, including replenishment of an artificial pond and green belt gardening, is the main purpose. The total households in the project area are 400 and the total population is about 1600 people. The green belt covers 6400 m^2 and the artificial pond is with a water surface of 6500 m^2 and an average water depth of 0.5 m.

Figure 5 shows the system provided for the residential community. In the 6 residential buildings, dual pipe collection system is installed for separate collection of black water and grey water. The black water is treated by a septic tank system while the grey water is treated for local environmental reuse. The treated grey water is led to the artificial pond for water replenishment. The pond also performs the function of a regulation tank for other reuse purposes. In order to control the pond water quality, part of the stored water is circulated. The average retention time of the pond water is about 15 days.

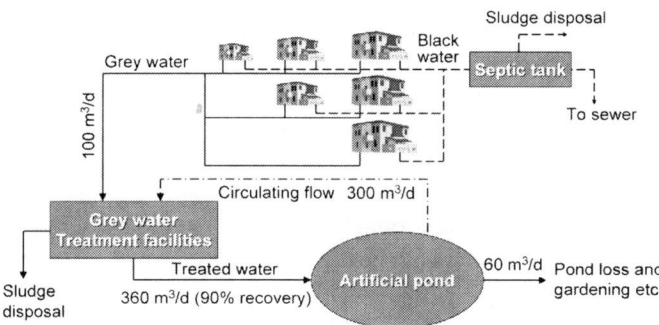

Figure 5 System composition of the decentralized water environmental system with grey water reuse

The grey water is treated by a process combining enhanced primary treatment with ozone enhanced flotation. The enhanced primary treatment is performed by a fluidised pellet bed bioreactor which is a specially designed wastewater treatment device for onsite wastewater treatment and can perform chemical coagulation, biological degradation, particle pelletization and separation in one unit (Wang et al., 2007). The ozone enhanced flotation is performed by a dispersed-ozone flotation separator which is a compact device combining coagulation, ozonation and flotation in an integrated unit (Jin et al., 2006). As for the circulated pond water, it only enters the ozone enhanced flotation unit for treatment.

Decentralized Water and Wastewater System
Serving a College Campus

This model case is a project in a college located in the southeast suburban area of Xi'an, China. The campus is on top of a hill covering an area of about 87 hectares of which 45 hectares are green belts. About 30 thousands students are living in the campus. The college is away from the centralised urban water supply system and urban drainage system. Available water source is only several groundwater wells with a maximum water supply capacity of 3000 m^3/d.

Figure 6 shows the system composition of the decentralized water and wastewater system serving the college campus. As the fresh water source is unable to cover the total water demand for various uses, it is decided that the available groundwater should only be used for potable water consumption and all water for non-potable consumption should be covered by wastewater treatment and reclamation. In order to meet the high demand for toilet flushing (1200 m^3/d), lake replenishment (650 m^3/d), and gardening and road washing (1800 m^3/d) which much surpassed the reclaimable quantity based on the freshwater supply, several measures are taken in this system for increasing the recovery ratio of water reclamation, such as 100% recycling the toilet flushing water and partially escalated water use between lake replenishment and gardening. Another feature of this system is the practice of dual-quality reclaimed water supply to meet the requirement for different water uses, i.e. high quality for indoor toilet flushing and lake water replenishment, and normal quality for gardening and road washing. The water budget of the whole system is also shown in Figure 6. It shows that the total water consumption of more than 6000 m^3/d is well covered by the freshwater supply of merely 3000 m^3/d through this system.

Wastewater treatment and reclamation in this system is through two units. In the first unit with a treatment capacity of 1500 m^3/d, an anaerobic-anoxic-oxic (A2O) process is employed to produce the normal quality reclaimed water to meet the need for gardening and road washing, while in the second unit with a treatment capacity of 2000 m^3/d, a hybrid process of A2O combined with membrane bioreactor (MBR) is employed to produce the high quality reclaimed water to meet the need for toilet flushing and lake water replenishment.

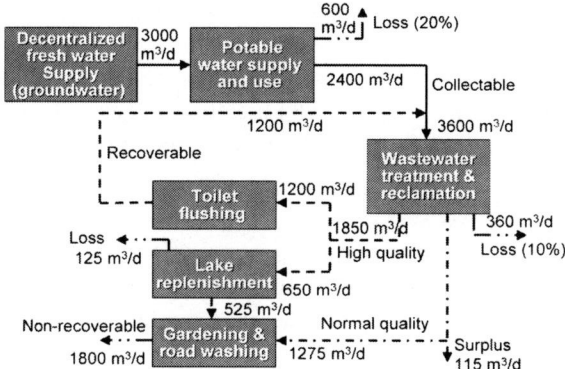

Figure 6 System composition of the decentralized water and wastewater system serving the college campus

CONCLUDING REMARKS

Water metabolism concept has been discussed in this paper. It stresses the harmony of the artificial water cycle with the natural hydrological cycle in the urban area. The Second Law of Thermodynamics can be used as a theoretical tool for analysing either a natural water system or an artificial water system. Under an assumption that the change in entropy for a water system is composed of an external contribution from outside and an endogenous contribution due to the internal processes, the natural water cycle with minor human disturbance can be taken as a pseudo-reversible system due to its nature of self maintenance of water and materials balance, which is a condition equivalent to $\Delta_i S \rightarrow 0$. It can be bridged between the water metabolism and thermodynamic concepts a relationship that minimization of entropy increase equals to the maximization of metabolic capacity. Two baselines thus have to be followed for the design of an urban water system: firstly to decrease the entropy increase from human disturbance and secondly to make the artificial part of the system as close to the nature as possible.

Harmonic integration of water supply, sewerage, water reuse, and urban water environment should be taken as the basic policy for urban water system design. These principles are applied in two model cases of decentralized urban water systems. The first case of grey water treatment and environmental reuse in the residential area is characterized by the application of treated grey water for replenishment of artificial pond and gardening in the residential area, and the maintenance of a sound local water environment. The second case of decentralized water and wastewater system serving a college campus is characterized by a harmonic integration of local water supply and water environmental system in which dual-quality water reclamation is practiced to meet the requirement for different reuse purposes. It is made possible to use the limited available freshwater resource to sustain the water demand of doubled scale. Zero discharge of wastewater is thus realized. The water metabolism concept and its application may direct a new paradigm for urban water system design in the future.

ACKNOWLEDGEMENT

This study was supported by the National Natural Science Foundation of China (50838005), the National Program of Water Pollution Control (2008ZX07317-004), and the Program for Changjiang Scholars and Innovative Research Team in University (IRT0853).

REFERENCES

Aoki I. (2006). Ecological pyramid of dissipation function and entropy production in aquatic ecosystems. *Ecological Complexity*, **3**, 104–108.

Aoki I. (2008). Entropy law in aquatic communities and the general entropy principle for the development of living systems. *Ecological Modelling*, **215**, 89–92.

Jin P. K., Wang X. C. and Hu G. (2006). A dispersed-ozone flotation (DOF) separator for tertiary wastewater treatment. *Water Science and Technology*, **53**(9), 151–157.

Kawachi T., Maruyama T. and Singh V. P. (2001). Rainfall entropy for delineation of water resources zones in Japan. *Journal of Hydrology*, **246**, 36–44.

Ludovisi A. and Poletti. A. (2003). Use of thermodynamic indices as ecological indicators of the development state of lake ecosystems. 1. Entropy production indices. *Ecological Modelling*, **159**(2), 203–222.

Maruyama T., Kawachi T. and Singh V. P. (2005). Entropy-based assessment and clustering of potential water resources availability. *Journal of Hydrology*, **309**, 104–113.

Ministry of Water Resources (2010a). *Bulletin of water resources in China 2008* (in Chinese). The Ministry of Water Resources of the People's Republic of China.

Ministry of Water Resources (2010b). *Bulletin of water resource quality in China 2008* (in Chinese). The Ministry of Water Resources of the People's Republic of China.

Montalto F., Behr C., Alfredo K., Wolf M., Arye M. and Walsh M. (2007). Rapid assessment of the cost-effectiveness of low impact development for CSO control. *Landscape and Urban Planning*, **82**(3), 117–131.

National Bureau of Statistics of China (2009). *Statistic bulletin of national economy and social development 2008* (in Chinese). National Bureau of Statistics of China.

Smith E. and Morowitz H. (2004). Universality in intermediary metabolism. *Proceedings of the National Academy of Sciences of the United States of America*, **101**(36), 13168–13173.

Steinborn W. and Svirezhev Y. (2000). Entropy as an indicator of sustainability in agroecosystems: North Germany case study. *Ecological Modelling*, **133**, 247–257.

Tambo N. (2004). Urban metabolic system of water for the 21st century. *WST: Water Supply* **4**(1), 1–5.

van Roon M. (2007). Water localisation and reclamation: Steps towards low impact urban design and development. *Journal of Environmental Management*, **83**(4), 437–447.

Wang X. C. and Jin P. K. (2006). Water shortage and needs for wastewater re-use in the north China. *Water Science and Technology*, **53**(9), 35–44.

Wang X. C., Yuan H. L., Liu Y. J. and Jin P. K. (2007). Fluidised pellet bed bioreactor: a promising technology for onsite wastewater treatment and reuse. *Water Science and Technology*, **55**(1–2), 59–67.

Wilderer P. A. (2001). Decentralised versus centralised wastewater management. In: *Decentralised Sanitation and Reuse*, IWA Publishing, London, UK, pp. 39–54.

Wolman A. (1965). The metabolism of cities. *Scientific American*, **213**, 179–190.

World Resources Institute (2005). Earth trends data tables: *Freshwater resources 2005*. http://www.earthtrends.wri.org/pdf_library/data_tables/wat2_2005.pdf.

World Resources Institute (2008). Earth trends data tables: *Population and human wellbeing*. http://www.earthtrends.wri.org/pdf_library/data_tables/population_2008.pdf.

Urine Separation for Sustainable Urban Water Management

Thorsten Schuetze[a] and
Mark M. C. van Loosdrecht[b]

[a]Dr.-Ing. Assistant Professor, Delft University of Technology, Faculty of Architecture, Julianalaan 134, 2628 BL Delft, The Netherlands
E-mail: t.schuetze@tudelft.nl
[b]Prof. Dr., Delft University of Technology, Faculty of Applied Sciences, Environmental Biotechnology, Julianalaan 67, 2628BC Delft
E-mail: M.C.M.vanLoosdrecht@tudelft.nl

Abstract In many cultures worldwide human faeces and urine were traditionally not mixed with water but collected separately. This was either related to an absence of a sewer system or because recovery of the nutrients in the human waste was an economical beneficial activity until the early 20th century. With the introduction of centralized water supply and water born sewer systems in industrialized countries in the 19th century, wastewater from households, industries and rainwater runoff from urban areas is mixed and (with or without primary treatment) discharged in nearby surface water bodies. When the effects of the resulting severe river pollution became a real concern, advanced treatment of wastewater was introduced by biological treatment and tertiary treatment to remove nutrients and reduce eutrophication of the receiving water bodies. These three steps now represent the present state-of-the-art in wastewater treatment. Sewer born systems have improved the public health situation but they have also caused severe problems, like polluted fresh water resources, broken nutrient cycles, impoverished soils, and high monetary cost. The separated collection of urine offers manifold possibilities to enhance existing and new urban sanitation and wastewater management systems. Energy savings, pollution control and nutrient recovery can be achieved in centralized, collective or decentralized sanitation systems. This paper discusses urine separation in urban developments and wastewater management systems using research results, good practice examples and sustainability criteria.

Keywords Urine separation, nutrient recovery, pollution control, wastewater management, sustainability

INTRODUCTION – HISTORICAL PERSPECTIVE

With the introduction of permanent settlements around 10000 B.C.E., solutions for the management of solid and liquid wastes, particularly organic wastes, excreta and used water that accumulated in the built environment, were developed. Many ancient agricultural societies recognised the value of excreta for soil fertility and practiced its collection and use together with organic waste, ash or charcoal, either directly or after treatment, e.g. in form of urine or compost. This enabled them to live for centuries with sanitation systems, which were based on the principle of closed loop recycling. In China (McGarry, 1978), Japan and Korea as well as in South America (Heckenberger, 2008), soil fertility was maintained over millennia, despite high population densities and different climates.

In Japan for instance, a disciplined use of excreta in agriculture was practiced, with it being applied at rates of up to 4 t/ha on fields in an environment that was considerably more urbanised than that of China. King in 1911 reported seeing night soil transported out of Yokohama and Tokyo "carried on the shoulders of men and on the backs of animals, but most commonly on strong carts drawn by men bearing six to ten tightly covered wooden containers holding forty, sixty or more pounds each" (Brown, 2003). Statistics from the Japanese Bureau of Agriculture for 1908 show that almost 24 million tons of excreta had been used on around 13.5 million hectares of arable land. Public toilets were provided with the aim of collecting excreta for use in agriculture. Urine was particularly regarded as a useful fertiliser and was collected separately for direct use (Matsui, 1997).

In Korea it was also well known that urine and faeces have different properties and that they can be used as a fertilizer and for enhancement of the fertility of land. It was recognized that faeces had to be handled in a safe way as they could cause illness. Due to the microbiological contamination of fresh faeces, the application of fresh and poorly composted faeces was only allowed in early spring or in autumn after the harvest. During the vegetation period only the use of fully composted matter was allowed. Until 1900, a graded pricing system existed for the marketing of faeces with different qualities. Excreta from city-households were evacuated and transported to agricultural land outside the cities. The high value of composted excreta for food production is reflected in the old Korean proverb: "You can always give away a bowl of rice, but never a bag of compost." However, with the gradual introduction of centralized water supply, water borne sanitation and sewer systems in the 20th century, the traditional sanitation system disappeared (Lee, 2000). The use of urine and faeces was prohibited in the 1960s, when chemical fertilizer production began and centralized water supply and sewer systems were in wide use. However limited amounts of human urine were still collected and used for the production of pharmaceuticals until the end of the 1990s (Schuetze, 2005).

Figure 1 Traditional manual transport of excreta in Korea. On the left side urine is transported in an earthen container. A stick is used for supporting the lifting and safe walking. At the right side faeces are transported in pots, which are fixed on a horizontal carry-stick (Lee, 2000).

The introduction of piped domestic water supplies in cities in the nineteenth century made water flushed sewerage systems possible. The installation of the centralized water supply was generally combined with the construction of mixed sewer systems. Sewage was flushed away from homes and the build up areas in cities into nearby rivers. This greatly increased the volume of sewage, at the same time diluting the nutrients, making it virtually impossible for them to be recovered and reused as they were previously. Urban areas became cleaner, healthier places to live, but city pollution became river pollution and downstream communities suffered. The concept of the water-borne sewer system became the standard approach for urban areas of industrialised countries during the second half of the 19th century through the 20th (van Zon, 1986; Lange and Otterpohl, 2000; Lange 2002) and into the 21st century.

However, since the introduction of the centralized sewer systems in the 1840s there is a critical and ongoing discussion about the benefits and risks of water borne sanitation systems. In 1866 Justus van Liebig (known as the "father of the chemical fertilizer industry") wrote to the Frankfurt based Physician Georg Varrentrapp that it would be crucial for the means of agriculture that the sewer content would not be discharged in rivers but used for the irrigation and fertilization of agricultural land. He was convinced that the marketing of sewage for agricultural purpose would be a significant source of income for the cities (Hapke, 1997).

The critical discussion about the centralized sewers was further ratchet by publications by Georg Varrentrapp in 1868 ("About drainage of cities, about value and degradation of water closets") and Justus van Liebig in 1876 ("On composting of urban waste to serve fertilization"). They wrote that the introduction of water closets in most English cities would destroy the basic conditions for the reproduction of food for 3.5 million people irrecoverable. The whole immense amount of fertilizer, which was imported yearly to England would be discharged to the most biggest part back to the rivers and the sea. The products produced there would not be sufficient to feed the growing population. The worst would be that this process of self-destruction takes place in all European cities, however not in such a big scale as in England.

Despite the doubts and critical discussions by scientists finally the water-borne sanitation prevailed. The nutrient demand of farmers was met by cheap chemical fertilisers, such as nitrogen produced with the Haber-Bosch process, which was available in Europe from the beinning of the 20th century, making any efforts to recover and reuse the nutrients and organic material from the large volumes of sewage completely obsolete.

This basic outcome is still leading for the international development of urban sanitation and drainage systems, the so-called "end of pipe technology" with all its impacts on the environment. For advertising the water born sanitation and canalization systems the concept of "self purification of rivers" was invented. Even though it was not true, it was stated that the pollution of rivers would not matter because light, air and bacteria in the river water would clean the pollutants already after a short distance (Maquet, 1903). When the effects of the resulting severe river pollution became obvious, mechanical treatment of wastewater was introduced, followed in time by biological treatment for the degradation of organic substances, and tertiary treatment to remove nutrients and reduce eutrophication of the receiving water bodies. These three steps now represent the present state-of-the-art in wastewater treatment. Although these conventional sewer systems have improved the public health situation in areas that can afford the installation, operation and maintenance cost, they have also significant disadvantages. Amongst others these are investment and operation costs, the mixing of sewage streams with different characteristics, and the discharge of nutrients, which are on the one hand polluting receiving water bodies and are on the other hand lost for the regeneration of impoverished soils.

For a big part of almost half of the world's population, the estimated 2.6 billion people who do not have access to adequate sanitation today, conventional centralized sewer born sanitation systems with wastewater treatment plants are both unaffordable and inappropriate. Statistics from India show that only 17 of 3700 cities and large towns have any kind of primary sewage treatment (Davis, 2006). Other countries report similarly low treatment rates, for example Argentina reports treating 10% of its sewage and Colombia only 5%, while only 2% of cities in sub-Saharan Africa have sewage treatment, and only 30% of these are operating satisfactorily (UNEP, 2002). It is estimated that more than 90% of sewage in the developing world is discharged directly into rivers, lakes, and coastal waters without treatment of any kind. However, proper disposal of human waste remains a challenge even in the "developed" countries. Until 2007 only 349 out of the 571 big cities of Europe (population greater than 150,000) complied with the treatment requirements of the Urban Waste Water Treatment (UWWT) Directive. In fact, 17 of these cities had no treatment at all. In southeastern European countries (Turkey, Bulgaria, Romania) approximately 40% of the population is connected to wastewater treatment facilities (Commission of the European Communities, 2007).

With increased population pressure and increasing pressure on freshwater and other resources (such as nutrients), this human waste disposal system is no longer able to meet the pressing global needs. Around the globe, a range of systems has been developed, based on traditions and recycling principals of the past and using new and innovative technological and systems approaches. (Luethi *et al.*, 2009) The separation of sewage streams with different properties is a basic condition for the appropriate treatment of the different streams and the reuse of resources. Particularly the separated collection of urine offers a wide range of possibilities for the improvement of different new or existing sanitation systems.

URINE AND DOMESTIC WASTEWATER

So-called human waste and domestic wastewater can be separated in different flow streams. In conventional water born sanitation systems generally all sewage streams are mixed with each other. However they can be separated in different flow streams, such as rainwater, greywater (from kitchens and bathrooms) and so called blackwater from toilets (consisting of flush water, urine, faeces and wet or dry anal cleansing material). Generally rainwater or stormwater is the flow stream with the lowest pollution load, followed by greywater (from washing food, clothes, dishware and bathing), which may contain traces of excreta and pathogens, but contains only 3 to 20 percent of the nitrogen in blackwater. The blackwater from toilets has the highest load of nutrients and so called micro pollutants. In case of urine separation at the source (in the toilet or the so called user interface) the flow streams can be separated in yellowwater (urine, which is more or less mixed with flush water) and the remaining brown water (faeces which are more or less mixed with water).

The average amount of urine produced by humans is (depending on the diet) approx. 1,4 litres per person and day. The total amount of 500 litres per person and year contains 2 to 4 kg Nitrogen. This is similar to approx. 90% of the total nitrogen excreted by humans and up to 87% of the nitrogen in domestic wastewater. Urine contains approx. 60% of the phosphorous excreted by humans and 50% in domestic wastewater (in case that detergents with phosphorous are used the percentage may be lower). Generally urine is sterile when it leaves the body. The average amount of faeces produced by humans is approx. 140 ml per person and day. The total amount of 50 litres per person and year contains approx. 10% of the Nitrogen, 30 to 40% of the Phosphorous and 12 to 20% of the potassium excreted by humans.

The biggest part of the Nitrogen, Phosphorous and Potassium load in domestic sewage can be separated from the wastewater flow, by the separation of urine alone. The volume of 1.4 litres urine per person and day is similar to only 1% of the total sewage from a household with an average water consumption of 140 litres per person and day. While the total amount of wastewater discharged by households may very significantly, dependent on cultural habits and the related water consumption, the amount of excreted urine is comparable similar.

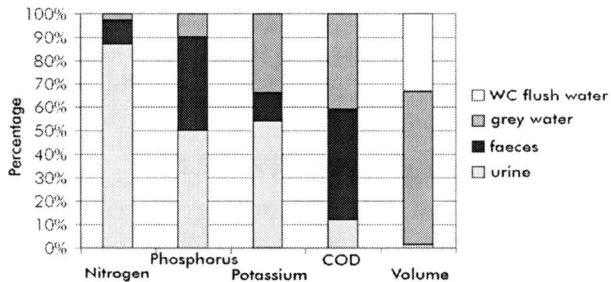

Figure 2 Portions of WC flush water, greywater, faeces and urine in relation to the total volume of domestic wastewater as well as the specific nutrient flows and the Chemical Oxygen Demand (COD). The percentages are calculated based on the average water consumption of 117 and nutrient discharge per person and day in households in Hamburg/Germany (Otterpohl, 2001; Schuetze, 2005)

Due to the small volume and the high nutrient and pollution load, the separation of urine has a big potential to solve the current water quality problems, which are related to the discharge of more or less treated domestic sewage. This is true for areas in which enhanced sewage treatment plants are already operating but also for areas in which this is not the case. Generally it can be stated that urine separation is a highly promising option for all sanitation systems and scales, from centralized, over collective to individual systems.

The increasing water quality problems caused by the discharge of sewage with high pollution and nutrient loads require the effective elimination of nitrogen in wastewater. With growing urban populations, quantities of wastewater and rising water temperatures (amongst others caused by increased cooling for power plants and climate change) the eutrophication of surface water bodies is expected to become more problematic and it is likely that emission limits will be tightened in the future. From a global view point municipal wastewater treatment plants will not be able to tackle the problem of nutrient overload efficiently in the short term, but with urine separation it may be possible to attain that goal (Eawag, 2007). In municipal sewage treatment plants, a considerable amount of energy is required to provide the oxygen necessary to convert ammonia to nitrate (i.e. NH_4 to NO_3). Phosphorus can be removed either biologically, or it can be removed by adding metal salts. Keeping nitrogen and phosphorus out of the wastewater would be positive in terms of saving energy and chemical additives. Optimally treated wastewater has to be nutrient free before discharge. However, all biological processes require a specific (stoichiometric) ratio of carbon, nitrogen and phosphorus as well as micronutrients, oxygen and moisture. Wastewater treatment requires some nitrogen, but too much nitrogen is an energy burden.

Based on research results (Wilsenach and Loosdrecht, 2003[1]) urine diversion could reduce aeration energy and chemical costs in municipal wastewater treatment plants. However there is an optimal level of urine diversion beyond which the wastewater would become deficient in terms of having enough nitrogen and phosphorus to process the degradable carbon present (Hellebust, 2006). For the Dutch situation and dependent on the type of sewage treatment this portion is 60–80%, according to Wilsenach. In areas which are equipped with wastewater treatment plants, a comparable easy and satisfying solution (respectively first step towards source separation) would be the introduction of faeces and urine separation where most easily practiced, e.g. by waterless urinals in men's washrooms only. Significant amounts of urine could be collected e.g. in highly frequented public toilets, in commercial and office buildings as well in sport stadiums. Such solutions would satisfy the requirement for nutrients in biological treatment and facilitate the reuse of urine as a fertilizer in agriculture.

The water pollution control authorities in the Netherlands are investigating whether urine source separation would enable nutrient emission limits to be complied with at a lower cost than is possible with conventional wastewater treatment technologies. Around 20 pilot projects are being planned or have already been launched, which are a direct consequence of the stringent Dutch limits on nutrient emissions (STOWA, 2009). Pilot projects in other European countries such as Switzerland, Sweden, Germany and Austria as well as in fast industrializing countries outside Europe, such as China, highlight the recent developments that urine separation is increasingly regarded as a realistic alternative to conventional nutrient elimination in wastewater treatment plants.

The source separation of urine would also be beneficial for water pollution control because the ecotoxicological hazard posed by human medicines in domestic wastewater (so called micro pollutants) could be reduced by approximately 50% (Larsen and Lienert, 2007). Furthermore the separated collection of urine could turn a wastewater treatment plant from an energy consumer into an energy producer: Instead of 11 watt per person being consumed, 2 watt of primary energy per person could be generated, as the energy efficiency of many processes could be increased and the energy in wastewater could be better exploited. (Wilsenach and Loosdrecht, 2006).

SEPERATED COLLECTION AND REUSE OF URINE

Urine can be processed to products which can be used in agriculture or industrial processes. Research results show that urine can be stabilized by biologically processes, Phosphorous can be precipitated chemically (to magnesium ammonium phosphate, MAP) and physically (with membrane technologies) micro pollutants, bacteria and viruses can be separated. Compared with diluted wastewater the separated treatment of urine facilitates the easy and safe reuse of nutrients (Larsen and Lienert, 2007).

In case of area wide application of urine separation from municipal wastewater systems, portions of the collected urine could be used to optimize the biologically treatment process in wastewater treatment plants (in case that the amount of available nitrogen in the wastewater would be too small). However, experiences with urine separation toilets show that the total separation of urine from remaining wastewater can hardly be achieved (Larsen and Lienert, 2007). Particularly in already existing urbanized areas, which are equipped with water born sanitation systems without urine separation, the introduction of urine separating systems would only allow the separation of comparable small portions of the totally discharged urine at the beginning. In the framework of an area wide application the portion of separated urine could be constantly increased.

The optimizing effect of urine separation in different portions on the centralized treatment of sewage is illustrated using an example from Kunming in China. Kunming, the capital of Yunnan Province in Southwest China has a population of about 2.4 million. The city lies on Lake Dianchi, a shallow water body with excessive phosphorus levels. Although six modern wastewater treatment plants have been built in recent decades, the water quality in Lake Dianchi has not improved because only about 25 % of all urban wastewater is treated. The remainder is discharged into the lake untreated. One of the main reasons is that ground and river water infiltrates the sewers, so that wastewater is diluted in a ratio of approximately 1:1. As a result, around 1600 of a total of 1960 tons of phosphorus are discharged with urban wastewater into the lake each year. The ambitious goal of the local authorities is to restore the water quality to the levels of 1960. To reach this goal only 30 tons of phosphorus per year may be discharged into the lake, which is similar to a reduction of Phosphorous discharge by 98%. Even with state of the art sewer systems as well as enhanced treatment technologies for the wastewater, with first, secondary and tertiary treatment, the targets set for water quality would not be attainable. At least 56 tons of phosphorus per year would still be discharged into the lake. If additionally two thirds of all urine would be separated, and this source control measure would combined with 'ideal' wastewater treatment technology, the goal would only be narrowly missed, with phosphorus inflows of 39 tons per year. The real challenge lies therefore in finding a combination of rational measures, which are both, realistic and will make it possible to achieve the ambitious goals. Different solutions will be required for urban areas, rural regions, and older parts of downtown Kunming as well as for new developments, which are not equipped with infrastructures yet. Household centered demand driven approaches for the installation of dry urine separating sanitation systems have been for example very successfully implemented in peri-urban and rural areas of Kunming (Larsen, 2007).

The goal to reduce the phosphorous discharge in the lake Dianchi by 98% could be achieved with only primary and secondary treatment of sewage in the centralized treatment plants and decentralized urine separation of 60%. According to findings of Wilsenach & Loosdrecht (2003[2]), which are the basis for the subsequent calculations, the partly (60%) urine separation has significant advantages regarding nutrient removal in the conventional wastewater treatment. Remaining N & P could be removed with waste sludge and no 3rd treatment would be required. By anaerobic digestion of sludge after thickening 100% of slowly biodegradable COD can be transformed into biogas. After wastewater treatment in an aerobic reactor, the sludge could be digested (for biogas production) thickened, dewatered and incinerated for energy production. The supernatant and centrate could be collected and mixed with separated collected urine for Struvit (MAP - Magnesuim-Ammonium-Phosphate) production. The MAP is free from Micro pollutants and a single solid product, which is well-established slow-acting multi-component fertilizer. In this way 98% P and 80% of N could be recovered by precipitation. The removal rate from the effluent would be 98% P & 99% N (19% is emitted in the atmosphere in form of nitrogen gas). By application of such a system the wastewater treatment process could be turned from energy consumer to energy producer: Instead of a consumption of 6 W per person and day, 1 W per person and day could be produced; A complete nutrient removal could be achieved with 75% urine separation (Wilsenach and Loosdrecht, 2003[2]).

The most direct and practically proven way to reuse separately collected urine is the application in agriculture as a liquid fertilizer, either diluted or undiluted and either before planting or during plant growth. Undiluted urine provides a harsher environment for microorganisms, increases the die-off rate of pathogens and prevents the breeding of mosquitoes. Urine contains few disease-producing organisms (while faeces may contain many). Before urine can be used as liquid fertilizer it is therefore advised to store the urine for a specific period in which organisms are decomposed. However urine can be used directly at the household level, where crops are intended for the own consumption. The storage of undiluted urine for one month will render urine safe for use in agriculture. When urine is collected from many urban households and transported for re-use in agriculture, the recommended storage time at temperatures of 4–20 °C varies between 1 and 6 months depending on the type of crop to be fertilized. The nitrogen concentrations in urine are varying between 3–7 grams per litre. The amount of urine needed per crop, can follow the recommendations given for chemical nitrogen fertilizers. As a rule of thumb the urine collected from one person in a year is sufficient to fertilize 300–400 square metres of crop (Rosemarin, 2008). In average the urine collected from households could substitute 2.5 kg nitrogen and 0.25 kg phosphorous from artificial fertilizers. The fertilizing effect of urine applied on spring crops is approximately 80 to 90% of artificial fertilizer (Johansson and Nykvist, 2001). Regarding the deteriorating quantity and quality of artificial phosphate fertilizers (remaining mineral resources of phosphorus have a comparable high heavy-metal content) – the recycling of relatively pure phosphorus from urine seems to be worthwhile (Larsen *et al.*, 2007). In the case of nitrogen recovery from urine, the key considerations concern the energy efficiency of nitrogen from urine and the quality of the fertilizer produced. The energy demand for the transport of liquid urine fertilizer compared with chemical fertilizer is lower if the total distance is less than 100 km (Johansson, 2001). The transport effort (amount of required trucks in case of motorized transport) for urine in urban areas is significantly lower than for solid household waste, due to the comparable low volume (Schuetze, 2005).

For the application of urine separating systems in the built environment there are already many examples. Famous pilot projects in Europe are amongst others the housing estates and residential areas in Understenshjöden and Palsternackan (Sweden) and in the Solar City Linz (Austria). Non-residential examples can be e.g. found in the GTZ main office building in Germany or the Forum Chriesbach in Switzerland.

Contemporary urine separating flush toilets were developed in the 1990 in Sweden, where also the first projects were realised. The toilets and collection systems can be installed easily in the framework of the construction of new buildings or renovation of existing buildings. The additional space demand for the drainage pipelines is low and the storage tanks can be either installed in the basement of buildings or underground, outside of the buildings. The additional installation costs for the drainage and storage systems are also low but the investment costs for the available urine separating flush toilets are comparable high. With a rising demand it can however be expected that the costs will decline and that a wide variety of comfortable separation toilets will be developed, which are functioning well and offer the same comfort as conventional toilets.

Finally it can be concluded that the separation of urine at the source offers an immense potential for the optimization of existing wastewater management systems as well as for the design of new sustainable water and sanitation systems, which are based on the principles of Integrated Resource Management and a Closed Loop Recycling Economy. For the desired widespread application of urine separating sanitation systems, the proper construction, operation and management as well as the user-acceptance of these systems will be crucial. On the one hand the decentralized approach for the collection of urine at the source is challenging for operators of centralized wastewater management systems, and requires the development of new business areas for operation and maintenance of such decentralized systems. On the other hand, in areas with collective and individual operation and mangement structures for sanitation systems, the introduction of urine separating systems is comparable easy to realize and offers the possibility to generate new sources of income for the local stakeholders. The national and regional policy plays an important role for the implementation of measures at the source because they can facilitate the sepaerated collection and reuse of urine by an appropriate legal and instutional frameworks. Education and awareness rising amongst the population are essential for the development of demand driven sanitation systems and entrepeneurship.

REFERENCES

Brown A. D. (2003). *Feed or Feedback: Agriculture, Population Dynamics and the State the Planet*, International Books Utrecht, the Netherlands.
Commission of the European Communities (2007). *4th Commission Report (Executive Summary) on Implementation of the Urban Waste Water Treatment Directive*, Brussels, Belgium, 2007.
Davis M. (2006). Planet of Slums. Verso, USA.
Eawag (2007). *Mix or NoMix – a closer look at Urine Source Separation*, Eawag News 63, March 2007, Duebendorf, Switzerland.
Hapke T. (1997). Stadthygiene und Abwasserreinigung Ende des 19. Jahrhunderts. Pictures and texts of an exhibition 5; TUHH, Hamburg, Germany.
Heckenberger M. J., Russell J. C., Fausto C., Toney J. R., Schmidt M. J., Prereira E., Franchetto B., and Kuikuro A. (2008). Pre-Columbian urbanism, anthropogenic landscapes, and the future of the Amazon. *Science*, **321**, 1214–1217.

Hellebust A. (2006). *Wastewater Reuse in Residential*, Institutional and Commercial Buildings in Canada: Current Motivations, Future Scenarios and Initiatives. Water Soft Path Lexicon, Vol. 1, 2007, Canada.

Johansson M. (2001). Urine Separation. VERNA Ecology, Stockholm, Sweden.

Johansson M. and Nykvist M. (2001). Closing the nutrient cycle, Stockholm, Sweden.

Lange J. (2002). *Zur Geschichte des Gewässerschutzes am Ober- und Hochrhein*. PhD thesis, Albert-Ludwigs-Universitaet, Freiburg, Germany.

Lange J. and Otterpohl R. (2000). Abwasser - Handbuch zu einer zukunftsfaehigen Wasserwirtschaft. Mall-Beton-Verlag, Donaueschingen-Pfohren, Germany.

Larsen T. A., Lienert J. (2007). Novaquatis final report. NoMix – A new approach to urban water management. Eawag, 8600 Duebendorf, Switzerland.

Larsen T. A., Maurer M., Udert K. M. and Lienert J. (2007). Nutrient cycles and resource management: Implications for the choice of wastewater treatment technology. Accepted for presentation at IWA Advanced Sanitation Conference, Aachen, 2.–13.3.2007, submitted to Water Science and Technology.

Lee D.–B. (2000). The dream of the nature-orientated toilette Dickan. Seoul, South Korea.

Luethi C, McConville J., Norström, Panesar A., Ingle R., Saywell D. and Schuetze T. (2009). Rethinking Sustainable Sanitation for the Urban Environment. *Proceedings of the 4th International Conference of the International Forum on Urbanism (IFoU)*, The New Urban Question – Urbanism beyond Neo-Liberalism, Amsterdam/Delft, The Netherlands.

Maquet C. (1903). Das Abfuhrsystem. In: *Sammlung von Abhandlungen ueber Staedtereinigung*, Bd I, F. Leineweber, Leipzig, Germany.

Matsui S. (1997). Nightsoil collection and treatment in Japan. In: *Ecological Alternatives in Sanitation*. Proceedings from Sida Sanitation Workshop, 6th-9th August, J.-O. Drangert, J. Bew and U. Winblad (eds.), Balingsholm Swedish International Development Cooperation Agency Stockholm, Sweden.

McGarry M. and Stainforth J. (1978). Compost, fertilizer and biogas production from human and farm wastes in the People's Republic of China. IDRC Ottawa IDRC-TS8e.

Otterpohl R. (2001). Design of highly efficient source control sanitation and practical experiences. In: *Decentralised Sanitation and Reuse*, P. Lens, G. Zeemann and G. Lettinga (eds.), IWA Publ. pp. 164–179.

Rosemarin A. (2008). Dry Sanitation, Urine Management, Containment Methods and Reuse Potential in Agriculture. *International Symposium Coupling Sustainable Sanitation & Groundwater Protection*, Hannover October 14–17, 2008, Lecture.

Schuetze T. (2005). Dezentrale Wassersysteme im Wohnungsbau internationaler Grossstaedte am Beipiel der Staedte Hamburg in Deutschland und Seoul in Sued-Korea. Dissertation, Leibniz University Hannover, Department Landscape and Architecture, Germany.

STOWA, Foundation for Applied Water Research (2009). Available at the Internet: http://www.stowa.nl.

UNEP - United Nations Environment Programme (2002). Environmentally Sound Technologies for Wastewater and Stormwater Management: An International Source Book. United Nations Environment Programme: Technical Publication Series, No.15.

Wilsenach1 J. and van Loosdrecht M. C. M. (2003). Impact of separate urine collection on wastewater treatment systems. *Water Science and Technology*, **48**(1), 103–110.

Wilsenach2 J. and van Loosdrecht M. C. M. (2003). From waste treatment to integrated resource management. *Water Science and Technology*, **48**(1), 1–9.

Wilsenach J. A. and van Loosdrecht, M. C. M. (2006). Integration of processes to treat wastewater and source-separated urine. *Journal of Environmental Engineering*, **132**(3), 331–341.

Zon H. v. (1986). Een zeer onfrisse geschiedenis, studies over niet-industriële vervuiling in Nederland, 1850–1920. Proefschrift Rijksuniversiteit Groningen, faculteit der letteren (A rather filthy history: studies into non-industrial pollution in the Netherlands, 1850–1920). The Netherlands.

Evaluation of Technologies for Decentralized Wastewater Treatment in China

Meixue Chen, Junxin Liu*, Min Yang, Xuesong Guo and Hanwen Liang

State Key Laboratory of Environmental Aquatic Chemistry, Research Center for Eco-Environmental Sciences, Chinese Academy of Sciences, Beijing 100085, P.O. Box 2871, China
E-mail: jxliu@rcees.ac.cn

Abstract In China, decentralized wastewater treatment is used to solve the problem of rural sewage. In this study, the primary types of decentralized processes applied to rural wastewater treatment are reviewed using typical case studies in China. In the past several years, as a result of recent policy directions and various economic factors, ecological technologies such as constructed wetlands and soil treatments have been developed. However, due to the lack of professional construction and appropriate management in rural areas, it has been difficult to establish a suitable system of effluent quality, technology selection and construction. Recently, the combination of biological and ecological processes in wastewater treatment has been brought into focus. It is urgent that a standard system of effluent quality, technology selection and construction is established. In addition to this, routing management is a significant factor to consider.

Keywords Decentralized wastewater treatment, rural sewage, technologies, biological treatment, ecological treatment

INTRODUCTION

Rural sewage in China comprises of: domestic wastewater, toilet sewage, surface runoff and a small amount of livestock manure. This sewage contains pathogens, suspended solids, nutrients (nitrogen and phosphorus) and other organic pollutants. China currently has more than 600,000 administrative villages. 2,500,000 of these are 'natural' villages with a total population of about 760,000,000 (Li, 2010). According to an investigation conducted by The Ministry of Housing and The Urban-rural Development of the People's Republic of China, the total amount of sewage produced by villages and towns in China is 9.2×10^8 t per year (Table 1). Additionally, the total amount of COD discharged from rural communities, annually, is as high as 8.0×10^6 t per year. This is practically equal to the amount of COD discharged from cities. Moreover, the amount of nitrogen discharged from towns and villages was 1.6×10^6 t per year. This is 1.6 times greater than domestic wastewater production. Despite these figures, less than 4% spray drainage and wastewater treatment has been constructed to date (Li, 2010). Providing reliable and affordable wastewater treatment in rural areas is currently a real challenge for China.

Table 1 Wastewater and the concentration of contaminants discharged from towns, villages and cities annually

Items	Town	Village	Town and village	City
Sewage volume ($10^8 m^3/a$)	3.6	5.6	9.2	33.0
COD ($10^6 t/a$)	2.6	5.4	8.0	8.6
N ($10^6 t/a$)	0.5	1.1	1.6	0.97

Centralized wastewater collection and treatment systems are costly to build and operate. This is particularly evident in areas with low population densities and dispersed households. Due to the geography and level of economic development in some areas, decentralized wastewater treatment systems such as onsite and/or cluster systems have been implemented in many rural areas in China. To satisfy the demand for public health and water quality goals, it is necessary to apply decentralized rural sewage treatment systems (Massound et al., 2009; Gikas et al., 2009). Not only do these systems meet the demands for health, but they also meet economic requirements; they are mostly easy to operate, manage and maintain. It is important, however, to consider how the treated wastewater generated by these systems will be used. It is particularly important, if the water will be provided to livestock, used for irrigation or directly/indirectly discharged into water bodies. Therefore, combined biological and ecological technologies have been applied to the decentralized treatment systems in China. Many on site wastewater treatment systems have been constructed and operated. Although such facilities do not provide a long-term solution for rural areas, the technologies used are reliable and cost effective.

In this study, the primary types of decentralized processes applied to rural wastewater treatment are reviewed and illustrated through typical case studies in China. The advantages and disadvantages of each technology are also reviewed to asses their practical suitability. Finally, the barriers obstructing the development of decentralized technology in China are discussed.

TYPICAL DECENTRALIZED WASTEWATER TREATMENT APPLIED IN CHINA

China's rural population is scattered; therefore, the choice of treatment used should be based on the following: (1) the operation should enable stable treatment to give water that meets all discharge standards and effluent so that it can be utilized as miscellaneous water or irrigation water; (2) the required capital investment costs should be relatively low; (3) the operating costs and operating costs should be affordable to the local residents in the town or village; (4) the system should be able to accommodate the local natural and geographical conditions, such as landscape characteristics; (5) the operation and management should be relatively simple. The local towns and villages should also recruit trained technicians for the day to day maintenance and operation of the system.

The septic tank is the most common primary treatment technology that is applied to rural areas (Gill *et al.*, 2009). Although ecological technologies such as constructed wetlands and soil treatment have been encouraged in the past, the lack of professional construction and management in rural areas has meant that it has been difficult to ensure quality construction. Therefore, most communities have focused on the development of systems that combine biological and ecological processes. In such systems, COD and ammonia are removed effectively by oxidation processes, after which, excellent effluent quality can be achieved.

Septic Tank

The three-grid septic tank shown in Figure 1 is most widely applied in rural areas in China (Ministry of agriculture of PRC, 2008). The septic tank removes most solids, functioning as an anaerobic bioreactor that promotes partial digestion of organic matter. The benefits of adopting this system are that it is inexpensive and simple to operate. However, poor maintenance and inappropriate design often leads to failure. If the septic tank is not properly managed, there can be an overflow of wastewater into the surrounding localities, causing detrimental health impacts (Carroll *et al.*, 2006). In addition to this, septic systems do not effectively remove nitrate and phosphorous compounds or reduce the levels of pathogenic organisms. It has also been discovered that some systems have been built without water-proofing and have not been properly maintained. This can lead to failure and subsequent surface and groundwater contamination. Septic tanks should be applied, therefore, as primary treatment units that are combined with other treatment methods.

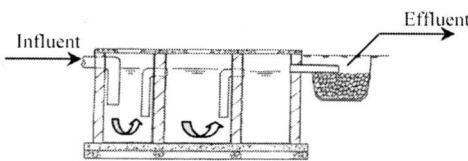

Figure 1 Typical septic tank in China

Biological Treatment

Biological treatment is a useful method for both centralized and decentralized wastewater treatment systems. It enables the removal of contaminants such as COD, SS, NH_4–N, P as well as pathogenic organisms. Biological treatment in rural, decentralized systems differs from biological treatment in urban, centralized systems. They must satisfy more stringent economic and availability requirements, while being simple to operate and manage.

Anaerobic digester

In the past few decades, anaerobic treatment systems have become widespread in China. According to the 2006–2010 National Rural Biogas Construction Plan (Ministry of Agriculture of PRC, 2007), there will be 40,000,000 households that use biogas by 2010. This is amounts to nearly 30% of rural households. A typical biogas system is shown in Fig. 2. In such systems, the waste and wastewater are collected from toilets and livestock. The produced biogas is then used in the kitchens and bathrooms. However, such systems are often poorly maintained and do not effectively remove nutrients from wastewater.

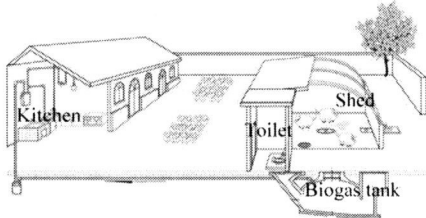

Figure 2 A typical biogas water treatment system

Aerobic treatment

The anticipated tightening of environmental legislation in developed rural areas such as the Yangzi Delta and Zhujiang Delta, has lead to the implementation of electric aerobic treatment facilities. This is to enhance the COD, BOD_5 and ammonia removal processes of existing systems. Fig. 3 shows an aerobic sequencing batch reactor. Not only is it useful for wastewater treatment in rural areas, but it is easy to control automatically (Li, 2010). The treatment ability of the systems depicted in Fig. 3 was 1m³, 2m³, 5m³, 10m³ and 15m³/day, while the effluent COD and BOD_5 levels were less than 60 mg/L and 20 mg/L, respectively. However, this method is still expensive for single families and the management of such systems is relatively complex because they involve suspended growth. Bio-contact oxidation technology has also been applied to rural wastewater treatment (Ji *et al.*, 2006). In addition to this, high removal efficiencies were obtained by biofilters with ceramic carriers (Tong *et al.*, 2009). Removals of 83.6–98.3% for ammonia and 84.4–95.2% for COD were observed at a hydraulic retention time (HRT) of 2.15–5.7h under hydraulic loading of $2.8–7.5m^3m^{-2}d^{-1}$. When compared with activated sludge systems, the technology is relatively easy to manage.

Figure 3 SBR for rural wastewater treatment

Enhanced nutrient removal process

As a result of increased water pollution, rural sewage treatment is a major concern in some areas. The rural sewage pollution in Taihu Lake accounts for 35.35% of the nitrogen in the system and 59.65% of the phosphorus (Li *et al.*, 2006). As a result, we can say that rural wastewater has become one of the primary factors leading to the eutrophication process in Taihu Lake. To enhance nutrient removal in wastewater treatment, it is necessary to increase the efficiency of many of the current systems. The most appropriate technology for enhancing nutrient removal is the application of Anoxic-oxic (A/O) processes. Fig. 4 shows a case study operating in Taihu basin (Li, 2010). The actual sewage treatment ability of the system shown is about 5m^3. d^{-1}. The system employs alternating aerobic and anaerobic conditions in a sequencing tank. Overall, the system has removal rates of COD, 87.2%, ammonia, 95.4%, TN, 73.8%, and TP, 97.4%, respectively. This technology is particularly suitable in rural mountainous areas, where the use of existing geographical features can enable the operating costs to be reduced. In these areas, the use of technology, such as this, can significantly reduce the power consumption associated with the process.

The use membrane bioreactors (MBR) (Schories, 2008) have also been implemented in some rural decentralized wastewater treatment systems. These systems generally produce effluent levels of COD, 50 mg/L, BOD$_5$, 10mg/L, and ammonia nitrogen 5 mg/L, respectively. This allows water to be reused on site for irrigation purposes (Shen *et al.*, 2009). However, the maintenance of membrane bioreactors is complex. The operation and management of these systems is also expensive. Such technology, therefore, is not appropriate for many rural areas in China.

Figure 4 Ladder style A/O treatment

Ecological Treatment

In the past several years, ecological technologies such as constructed wetlands and soil treatment systems have been encouraged due to various policy directions. However, many large scale wetlands have been constructed for the direct treatment of rural domestic wastewater without any pre- or post treatment. In rural areas, Due to the lack of professional construction and operation management, it is often difficult to ensure quality of these constructions.

Constructed wetlands

Constructed wetlands have been widely applied in rural areas. The advantages of implementing constructed wetlands, is that they are inexpensive and are can accommodate different land uses. Fig. 5 shows a small scale constructed wetland, capable of treating $0.2 - 1$ t/d of the gray wastewater from two neighbouring families. The water treated by this system has COD, SS and ammonia nitrogen levels of less than 150 mg/L, 150 mg/L and 12mg/L, respectively. This is acceptable for irrigation (Li, 2010).

Figure 5 Small constructed wetland for gray wastewater treatment

However, due to the lack of knowledge regarding the management of such systems, deteriorating water quality and the clogging of the constructed wetlands commonly occurs.

Soil treatment systems

Soil treatment systems have low construction costs. They employ a series of pipelines that can be easily introduced to homes in urban areas for the collection of wastewater. Soil treatment systems can remove organic pollutants and nitrogen. This is because the nitrification and denitrification reactions occur through existing natural systems, in the aerobic and anaerobic zones in the soil treatment system. Fig. 6 demonstrates how a system in Jiangxi Province (Li, 2010) was used to treat the wastewater collected from a residential septic tank. During the primary periods of operation, the system could perform with zero maintenance. However, in order to maintain the performance during all seasons and long-term operation, it is necessary to improve the system.

Figure 6 Soil treatment system

Lagoon

Lagoons are popular sewage disposal methods in rural areas in China. Lagoon treatment systems usually employ a natural, or a constructed, pond built with an impermeable layer. The advantage to using this technology is that it is low cost. Also, due to its high nutrient levels, there is the opportunity to use the effluent for irrigation. Furthermore, eco-lagoons have been reported to have COD, ammonia, TN and TP removal rates of 55%, 70%, 80% and 75%, respectively (Ji *et al.*, 2007).

Combined Bio-Eco treatment

Combined biological and ecological treatment technologies have been developed for economic reasons and to improve the water quality of effluent. In the Jiangsu Province, a combination of anaerobic, drop aeration and constructed wetland technologies have been used (Wu *et al.*, 2008). In this system, the nitrogen removal occurs through anaerobic and aerobic processes, while part of the phosphorous removal occurs through plants in the constructed wetland. Fig. 7 shows a combined bio-eco treatment system operated in a rural tourism area in Yunnan Province (Liang *et al.*, 2009). In this combined system, the bio-technique unit is an integrated oxidation ditch with a vertical circle (IODVC). The eco-technique unit consists of a wetland eco-filter, a surface infiltration area and an eco-ditch. The concentration of COD, ammonia nitrogen and total phosphorous in effluent was less than 10, 1 and 0.1 mg/L, respectively. The effluent of the full-scale wastewater treatment station was used as the source water for a local landscape ecological wetland.

Figure 7 Bio-Eco system

Recently, the use of systems containing a combination of biological and ecological processes for decentralized wastewater treatment has generated a great deal of discussion. In these systems, COD and ammonia are removed effectively by biological processes. After this, an excellent effluent quality can be achieved by treatment using subsequent ecological techniques.

SUSTAINABILITY OF DECENTRALIZED RURAL WASTEWATER TREATMENT IN CHINA
Choice of Appropriate Technology

Currently, most sewage treatment technologies operating in rural areas are based on the municipal sewage treatment technology model. However, due to the lack of professional construction and operation management, in rural areas, it is difficult to ensure quality construction. China's economic situation in different rural regions varies greatly. The main driving forces behind the selection of treatment technologies in different areas are: the performance requirements, the site conditions and the wastewater characterization (Massoud et al., 2009). For decentralized rural sewage treatment technology, it is necessary to achieve a balance between, improving general living conditions, and protecting the environmental. Other factors to consider are: the economic suitability of processing technologies; local social and environment conditions and the ease of operation, maintenance and management. In Table 2, the primary advantages and disadvantages of the most common technologies are compared.

Table 2 Advantages and disadvantages of the most common wastewater treatment technologies

Technology	Main advantages	Main disadvantages	Application
Septic tank	1. Inexpensive; 2. Simple to maintain	1. Does not effectively remove ammonia, phosphorus and pathogenic organisms; 2. Potential source of groundwater pollution	Primary sediment tank or anaerobic unit
Anaerobic digester	1. Low cost; 2. Energy use	1. Low removal rate; 2. Maintenance is necessary	Biogas use
Activated sludge	1. Flexible and effective for COD and ammonia removal 2. Automatic control	1. Expensive for single families 2. Management is relatively complex	Second treatment cluster system for a village
Biofilm	1. Effective for COD and ammonia removal 2. Routine maintenance is relatively easy when compared to activated sludge	1. Low temperature will affect the removal rate of contamination in northern areas	Second treatment for on-site or cluster system
Lagoon	1. Low coat 2. Effluent for irrigation	1. Not effective for the removal of pollutants 2. Land use 3. Poor sanitation	Large area of land
MBR	1. High effluent quality	1. Expensive in capital and operation 2. Complex membrane maintenance	Sensitive area
Soil treatment	1 Simple construction and operation 2. Low cost	1. Pollution of groundwater 2. Poor quality of effluent	Large area of land
Con-structed wetland	1. Constructed cost 2. Flexible land use	1. Low removal rate 2. Management	Gray wastewater or enhancing process
Bio-Eco treatment	1. High effluent quality 2. Balance of capital and operation cost	1. Climate effect	Almost occasions for protecting environmental quality

Thus, single families should adopt small-scale equipment, constructed wetlands and land treatment methods to ensure that environmental standards are met for living purposes. However, combined biological treatment and ecological approach systems can be used for multi-family units. When there are many households, a cluster system consisting of combined biological and ecological processes is recommended. In such a system, biological treatment removes most of the COD and ammonia nitrogen, after which further processing using an ecological based system improves the water quality.

Management

In the last decade, non-power technologies have been applied to wastewater treatment in rural areas. This is due to the need for low investment requirements and low-cost operation. Non-power techniques seem simple and do not require expert management. However, during actual operation, processing cannot achieve the desired results due to the unstable effluent quality. Accordingly, the management of these treatment facilities must be improved. From a technical point of view, standardization of operating procedures and operator training is very important. As most technical operators are local farmers, treatment systems should employ appropriate technologies that are easy to operate. Centralized management of decentralized wastewater treatment systems is one method that can help ensure that these systems are inspected and maintained regularly (Massound et al., 2009).

CONCLUSIONS

In recent years, rural sewage treatment has attracted a great deal of attention. Decentralized wastewater treatment was actually included in the Chinese government's "new countryside" construction plan. However, lack of appropriate environmental standards and inadequate long-term operation management has had a negative impact on the application of sewage treatment facilities in practice. When considering the development of decentralized wastewater treatment systems in China, it is necessary to establish a standard system of effluent quality, technology selection and construction. In addition, management is very important. Furthermore, the choice of technology should meet the environmental protection standards and improve the local quality of life. Based on the actual situation in different regions, the combined bio-eco technology should be considered in environmentally sensitive areas to achieve good water quality, while the appropriate technology for other areas should be chosen based on whether they will accommodate single-families, multiple-families or villages.

REFERENCES

Carroll S., Goonetilleke A., Thomas E., Hargreaves M., Fros R. and Dawes L. (2006). Integrated risk framework for onsite wastewater treatment system. *Environmental Management*, **38**(2), 286–303.

Gikas P. and Tchobanoglous G. (2009). The role of satellite and decentralized strategies in water resources management. *Journal of Environmental Management*, **90**, 142–152.

Gill L. W., O'Luanaigh N., Johnston P. M., Misstear B. D. R. and O'suilleabhain C. (2009). Nutrient loading on subsoils from on site wastewater effluent, comparing septic and secondary treatment systems. *Water Resources*, **43**, 1739–1749.

Ji Z. and Lv X. (2006). Contact oxidation method for medium and small sized domestic sewage treatment. *Water Purification Technology*, **25**(3), 43–45.

Ji Z., Li T., Lv X. and Jin Q. (2007). The effect analysis of sewage treatment by ecotypic bed. *Jiangsu Environmental Science and Technology*, **20**(5), 39–41.

Li B. (2010). *Case Sets of Rural Sewage Treatment Technology*, 1st edn. China Building Industry Press.

Li X., Lv X., Kong H., Luo X. and Li X. (2006). Research on rural sewage treatment techniques and application in demonstration projects. *China Water Resource*, **17**, 19–21.

Liang H., Liu J., Guo X., Shan B., Zhao J., Yu L., Li L. and Liu J. (2009). A novel bioeco technology combined system for rural domestic wastewater treatment in a tourism area: a full-scale study. *Environmental Engineering Science*, **26**(9), 1419–1427.

Massoud M. A., Tarhini A. and Nasr J. A. (2009). Decentralized approaches to wastewater treatment and management: applicability in developing countries. *Journal of Environmental Management*, **90**, 652–659.

Ministry of Agriculture of Peoples Republic of China (2007). National Rural Biogas Construction Plan.

Ministry of Agriculture of Peoples Republic of China. (2008). Earthquake-Stricken Areas of Agricultural Ecological Environment Quiz.

Schories G. (2008). IWAPIL-innovative wastewater treatment applications for isolated locations. *Desalination*, **224**, 183–185.

Shen Y., Liao R., Huang Y. and Gu H. (2009). Case analysis and evaluation of suitability for treatment technologies of rural domestic sewage. *China Water and Wastewater*, **25**(18), 19–26.

Tong J., Ji G., Zhou Y. and Xie C. (2009). A high efficient multi-function ceramic bio-filter for treatment rural domestic sewage. *Journal of Agro-Environment Science*, **28**(9), 1924–1931.

Wu C., Xiang S. and Lu X. (2008). Combined technology of biological contact oxidization and constructed wetland for processing rural domestic sewage. *Hubei Agricultural Sciences*, **47**(1), 44–46.

Investigation of Domestic Wastewater Separately Discharging and Treating in China

Chen Hong-bin*, Yu Feng, Ruan Jiu-li, Qian Liang, Wang Shao-yong and He Qun-biao

State Key Laboratory of Pollution Control and Resource Reuse, College of Environment Science and Engineering, Tongji University No. 588, Miyun Rd., Shanghai, P.R.China, 200092
Tel: 0086-21-65984569; E-mail: bhctxc@ tongji.edu.cn

Abstract In this article, a 12-month investigation of domestic wastewater separately discharging was conducted in a district in Shanghai to provide the technical support for domestic wastewater source separating, treating and reusing. The results show that the characteristics of blackwater, greywater, kitchen water and shower water are obvious and have changing tendencies at different time of day and seasons. Blackwater has the highest concentrations of major pollutants, kitchen water has the second, and shower water has the lowest pollutants concentrations. Except for shower water, other wastewaters have good biodegradability, of which the ratio of B/C is about 0.4 or higher. It is suggested that kitchen water can be sorted into greywater in north China, which can be discharged and treated together with washing water, shower water and laundry water; however, in south China, kitchen water can be discharged into the blackwater pipeline for treating and recovering nutrients.

Keywords greywater, black water, kitchen water, source separating, nutrients

INTRODUCTION

There are two typical sewer systems at present in the world which perform the key functions of transporting and treating municipal wastewater. One is the centralized treatment system, and the other is the decentralized treatment system. With more and more serious environment pollution, and resources gradually becoming exhausted, both systems have shortcomings: it is difficult to reuse and recycle the materials, energy and reclaimed water, because of one-way, not closed-loop flows, used by centralized treatment systems, municipal sewage is transported to wastewater treatment plants (WWTPs) far away from the city, which needs a lot of energy to remove contaminants, such as carbon nutrients, nitrogen and phosphorus, which can be utilized for agricultural interests. If the treated water is to be reused in the city, a second pipeline, pump stations and power are indispensable to re-transport to the end user (Otterpohl, 1997). As to a decentralized treatment system, it is advantageous for the simplicity of the devices, sewers and low costs of investing and running the operation. However, it is limited by the unstable effluent quality and low capacity of loading rock resistance (Lundun, 2004). With the recent development of new conceptions of environmental protection and a recycling economy, new ideas or conceptions of water supplying, domestic wastewater source separating, and treating and reusing are more and more prominent and studied; "Ecological Sanitary Systems (ECOSAN)" and "Semi-centralized treatment system" (Meisheng, 2004) are examples. Their characteristics are: drinking water and reclaimed water were separately supplied and consumed by user; greywater is treated and reused; the nutrients of blackwater and kitchen biological residual solids are treated to be fertilizer or greening soil.

The key to domestic wastewater treatment and reuse is source separating and collecting. It is feasible that domestic wastewater can be divided into several portions according to their origination (Li, 2009; Henze, 1997; Jefferson, 1999; Lu, 2003; Ayres, 1996; Larsen, 1996; Lazarova, 2003):

1. Blackwater: includes manure, urine and flushing water;
2. Yellow water: toilet urine and flushing water only.
3. Greywater: coming from washing water, shower water and cleaning water.
4. Brown water: kitchen water, sometimes including broken kitchen biological solids.

Some scientists think that domestic wastewater can be composed of blackwater and greywater: the former being comprised of manure, urine and flushing water which devote the most nutrient, nitrogen and phosphorus to domestic wastewater; the latter is the mixture of (3) and (4), in which the contaminants strength is low and easy to remove. Some previous studies proved that it is useful to lower energy consumption, reduce the amount of discharged pollutants and improve the recycling rate of wastewater by separately collecting, treating and reusing domestic wastewater (Li, 2009).

Small successful projects, which were applied by separately treating and recycling domestic sewage with the size of a single family or one building, have been reported at home and abroad (Weihui, 2008; Friedler, 2006). However, there is no study or practice on applying new conceptions to communities or one part of a town, which serve the equivalent of several hundreds to dozens of thousands. One of the key reasons is a lack of basic data to support the theoretical study and technique development. In this article, the contaminants categories and concentration range of greywater, blackwater, shower water and kitchen water through 12-month analyses has been investigated in a community in Shanghai, P.R. China. The objectives of the study are to obtain fundamental data on different domestic kinds of sewage, catch the properties and seasonal characteristics of them, and propose suitable domestic source separating modes for north and south China.

MATERIALS AND METHODS

The Source of Water Samples and the Sampling Way

There are two kinds of sewer systems in Shanghai. One is a separate sewer system which was constructed by most new communities to transport wastewater and rain water respectively; the other is combined sewer systems built in many old communities over the last century, few of which have separate pipes for transporting greywater and blackwater from the buildings to the yard. After flowing through their own check well, different wastewater flows combine in the branch sewer, then the rain water conduit and domestic wastewater pipe join in the main sewer.

The greywater and blackwater samples were obtained from the check well in a community in Yangpu district, Shanghai. The kitchen water sample was gathered from another community in the same region. As there was no shower water check well, a public bathing room in the campus of Tongji university was selected to be the source of the shower water samples. If the sampling day was rainy, then the next day after rain was used for sampling date.

(1)   Greywater sampling. Four or five check wells were chosen to gather greywater samples which included shower water, washing water, cleaning water and kitchen water. To investigate the changing tendencies of different times of day, samples were captured every two hours from 7:00 to 23:00. All the samples from check wells at the same time were blended as one sample. The samples were kept in cold temperatures and analyzed in the Laboratory. Such sampling continued for four days. To know the monthly changing tendencies of greywater, the same date of every month was chosen to obtain samples.

(2)   Blackwater sampling. The sampling frequencies, time and methods for blackwater were similar to those of greywater. After that, the same day was used to sample and analyze blackwater in the next six months.

(3)   Kitchen water sampling. The sampling method is that equal amounts of kitchen water from three check wells were mixed to make one sample. As kitchen water is discharged intermittently, the sampling time was three or four hours before and after diner. Kitchen water sampling and analyses lasted for 12 months.

(4)   Shower water sampling. Two or three hours after the public bathing room opened, the sampling procedure was carried out, with a sample obtained from the shower water check well. There was a total of six day's samplings from December to the following March.

Analyzing Method

The pollution parameters, including pH, SS, VSS, TSS, COD_T, COD_s, BOD_5, NH_3-N, NO_2-N, NO_3-N, TN,TP,PO_4-P, TOC,TC,IC, Alkaline, the sum of E.coli. and bacteria, Linear Alkyl Sulfonate (LAS), were analysed, the method were described by literature. *(Water and Wastewater Standard Analysing Method (IV) M.).*

RESULTS AND DISCUSSION

Characteristics of Greywater, Kitchen Water, Shower Water and Blackwater

Based on the results of the investigation over 12 months, the range and average concentration as well as the ratio value between different parameters for major pollutants in greywater, kitchen water, shower water and blackwater are shown in Table 1.

It can be seen from Table 1 that, except for blackwater which is weakly alkaline, all three other wastewater types are weakly acidic; the concentration of COD_{Cr}, BOD_5, TN, TP and NH_3-N of blackwater are the highest, the kitchen wastewater is the second highest, and the shower water is the lowest. The concentrations of pollutants in greywater vary from those of kitchen water to shower water. Greywater varies from kitchen water to shower water, in which most of the nutrient substances especially nitrogen and phosphorus, come from kitchen water. As most LAS derives from domestic detergent, the highest and lowest concentrations of LAS are found in shower water and kitchen water. Similarly, there is no LAS detected from blackwater, which is in accordance with expectations. One can find that the concentration of organic pollutants in black water is 2.5~3.5 times larger than those in grey water; the content of nitrogen and phosphorus of blackwater are 9.6 and 15.7 times as large as those n grey water. In the view of variation of pollutant content, black water and kitchen wastewater have broad ranges which are concerned with suspended solids, whereas shower water has the most stable contaminant concentrations.

Further analyses are discussed from Table 1. It is obvious that kitchen water has the highest ratio of B/C and the best biodegradability; shower water is the worst, with a ratio of B/C below 0.3. Except for shower water, B/C values for all other three kinds of wastewater were higher than 0.4, which is to say, they had good biodegradability. The values of BOD_5:N:P of greywater and shower water are 135:10:1 and 92:17.6:1 respectively, which is slightly lower than the growth requirements of microbes. On the other hand, blackwater is very rich in nitrogen and phosphorus, with an average ratio of BOD_5:N:P of 24.4:6.21:1, which will lead to carbon nutrients shortage if nitrogen and phosphorus are removed from

Table 1 Pollutants concentration in greywater, kitchen water, shower water and blackwater (Unit: mg/L)

Parameter	Greywater*		Kitchen water		Shower water		Blackwater	
	Range	Average	Range	Average	Range	Average	Range	Average
pH	5.7–6.6	6.3±0.3	4.6–8.3	5.9±0.8	6.8–7.7	7.3±0.3	6.7–8.9	7.8±0.9
COD_T	202–639	432±123	183–1776	792±347	73–463	201±126	432–3856	1213±1028
COD_S	94–497	165±105	108–607	314±163	–			676
BOD_5	88–271	186±64	88–858	370±182	15–69	46±20	170–1249	458±322
TC	35.07–168.5	89.2±41.5	57.5–281.1	143.8±72.5	31.7–140.5	67.6±36.7	136.6–447	310.8±130.6
TOC	11.2–145.5	65.3±47.5	32.6–264.3	110.7±80.6	5.3–98.8	37.6±32.2	38.8–265.2	141.6±80.4
IC	4.38–39.7	23.9±11.8	10.7–90.4	33.1±72.5	18.1–46.1	30±29	90.4–376.7	169.2±91.3
TN	5.7–25.0	13.7±5.7	11.3–196.9	43.5±43.0	10–32.6	18.4±8.8	113.9–376.7	204.3±81.3
TNs	2.99–15.6	7.48±4.0	1.35–89.8	15.2±19.4	4.9–12.3	8.8±2.7	19.3–196.9	116.7±62.3
NH_3-N	0.86–5.65	3.28±1.4	0.7–62.1	6.2±13.3	2.3–6.1	3.2±1.3	42.1–174.3	122±46.0
NO_3-N	0.13–2.19	0.89±0.5	0.08–5.36	1.72±1.3	1.5–2.8	2.5±0.7	2.0–6.10	5.2±1.5

(Continued on next page)

Table 1 Pollutants concentration in greywater, kitchen water, shower water and blackwater (Unit: mg/L) (Continued)

Parameter	Greywater*		Kitchen water		Shower water		Blackwater	
	Range	Average	Range	Average	Range	Average	Range	Average
TP	0.4–2.93	1.18±0.7	1.0–10.1	3.6±2.3	0.3–0.7	0.5±0.2	5.7–54.2	18.8±14.3
PO_4-P	0.03–1.13	0.31±0.3	0.1–6.8	1.7±1.6	0.03–0.4	0.2±0.1	3.0–43.6	12.6±12.3
LAS	1.06–15.49	5.95±3.7	1.65–12	3.6±3.2	6.0–15.2	9.9±3.5	N.D	N.D
Alkaline ($CaCO_3$)	32.53–122.6	100.9±24.6	46–339	113±57.5	82–127	98±81.7	246–808	491±187.2
TSS	335–2656	810±583.9	325–2258	1100.9±528	570–779	715±72.0	726–2754	1300±670.8
SS	53–576	285±146.9	29–2130	522.8±562	32–359	136±114	47–2174	508±673
VSS	32–514	184±122.4	2–1360	259.2±333	7–330	107±109	2–1702	357±545
Sum of bacteria	1.4×10^8–1.5×10^{12}		1.1×10^9–3.5×10^{12}		2.0×10^7–2.5×10^9		3.0×10^9–2.8×10^{10}	
Sum of E.Coli	1.7×10^7–5.8×10^9		4×10^5–1.7×10^{11}		4.3×10^4–3.5×10^7		1.7×10^7–8.5×10^9	

Note: greywater was a mixture of kitchen water, shower water, laundry and cleaning wastewater.CODt represents total COD, CODs represents soluble COD; TNs represents total soluble Nitrogen; N.D represents undetectable.

blackwater; so additional carbon sources should be considered when developing biological nitrogen and phosphorus removing processes.

Variation of Water Quality at Different Times of Day

According to the results of analyses, shower water and blackwater have less changes of pollutant concentrations in the course of a day; however, combined greywater shows a wide variation in both water quality and quantity; the mean concentrations of the main pollutants of 4-days' analysis at different times through the day were summarized, and are shown in Figure 1 and Figure 2.

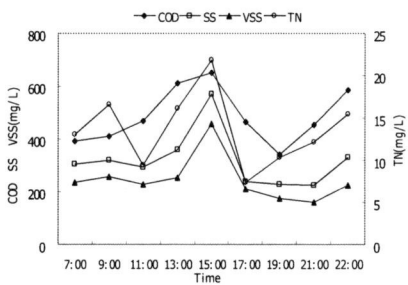
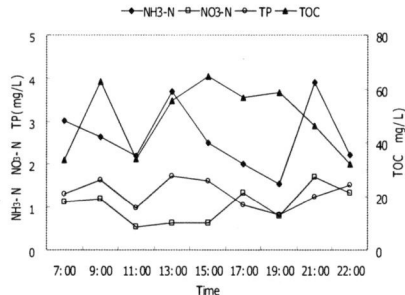

Figure 1 Variation of COD, SS and VSS during different time

Figure 2 Variation of NH$_3$-N, NO$_3$-N, TP and TOC during different time

It is shown that, COD$_{cr}$, SS, VSS and TN have similar variation tendencies at different times of day. Two concentration peaks are seen; and two valleys are found at 11:00 and 17:00. Although the variation tendency for NH$_3$-N, NO$_3$-N, TP and TOC was not obvious, two small peaks and valleys appear at 15:00, 22:00 and 11:00, 17:00 respectively. The high concentration of greywater at 15:00 and 22:00 may have been due to the lessening of activities at home, and low water consumption during those times; during the lunch and dinner preparation times at 11:00 and 17:00, a lot of fresh water consumption caused low pollutant concentration. Variation regularity for greywater quantity through the course of a day has been investigated by Palmquist (2005), in which the times of largest water consumption are corresponded with the climax concentration in this study. In conclusion, the habitants' behaviors affected the components and the concentration of greywater; larger water consumption leads to lower pollutant concentration; and vice versa, lower water consumption leads to higher pollutant concentration in greywater.

Variation Tendencies for Pollutant Concentration During Different Seasons

The long term investigation of greywater, blackwater, kitchen water and shower water lasted for 12 months. Based on analysis of the results over different months, less changing tendencies for shower water and blackwater were found along with the seasons, so they will not be discussed here. The following discussion will focus on the changing tendencies of combined greywater and kitchen water during different seasons.

Through the data, the highest concentration for all kinds of wastewater appears in Winter or in Autumn, while the lower and steady concentrations of main contaminants appear in summer. Take greywater as an example; the variation of different pollutants during months is shown in Figure 3 and Figure 4.

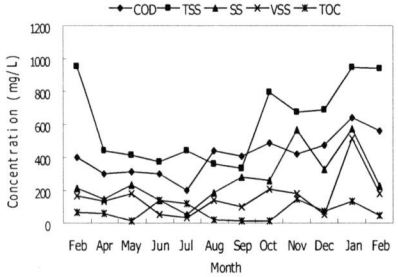

Figure 3 COD$_{Cr}$, SS VSS, BOD$_5$, TN Variation in different month

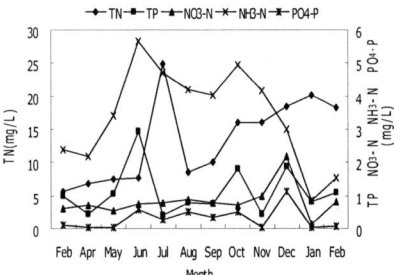

Figure 4 TP, NH$_3$-N, NO$_3$-N, PO$_4$-P, TOC Variation in different month

According to Figure 3 and Figure 4, the concentration of different pollutants in greywater varies with seasonal alternatives, amongst which COD$_{Cr}$, SS, VSS, BOD$_5$ and TN have the similar changing tendencies, except for the big deviation of TN in July, which may have been caused by random circumstances during sampling time. It also can be seen from Table 2 that the average concentrations of pollutants from April to September were much lower than those from October to February in the next year – that is to say, organic pollutants, nitrogen and phosphorus of greywater in Spring and Summer are much lower and only about 60% of those in Autumn and Winter. Not much difference was found for the concentration of TP, NH$_3$-N, NO$_3$-N, PO$_4$-P and TOC during the changing of seasons.

Table 2 The mean concentration of pollutants in greywater and kitchen water in different seasons. (Unit: mg/L)

Categories	COD$_{Cr}$	TN	SS	VSS	TP	NH$_3$-N	TOC
Greywater from April to Sept (Spring/Summer)	310	11.1	151	107	1.1	4.0	71.0
Greywater from Oct to Feb (Autumn/Winter)	493	15.1	352	222	1.2	2.9	62.5
Kitchen water April to Sept (Spring/Summer)	721	45.3	624	383	4.5	3.7	68.4
Kitchen water Oct to Feb (Autumn/Winter)	823	42.7	479	206	3.2	7.3	128.8

DISCUSSION

Feasibility for Separate Discharging and Treating of Domestic Wastewater

The weighted average concentrations of the main parameters of domestic wastewater are calculated by the qualities and the ratio of different wastewaters listed in Table 3.

Table 3 Normal domestic water consumption proportion

Categories	Toilet	Kitchen	Shower	Washing	Total
Percentage (%)	31~32	23~21	31~32	15	100
Volume of consumption (L/EW.d)	40~60	30~40	40~60	20~30	130~190

Note: normal domestic water consumption is supposed to be 130L/EW.d~190l/EW.d.
Data cited from Hellstrom (1997); Shu cun (2002); Palmquist (2005); Maoan (2002); Yue xin (1998).

Figure 5 indicates the comparison of the concentration differences of domestic wastewater and that of combined greywater.

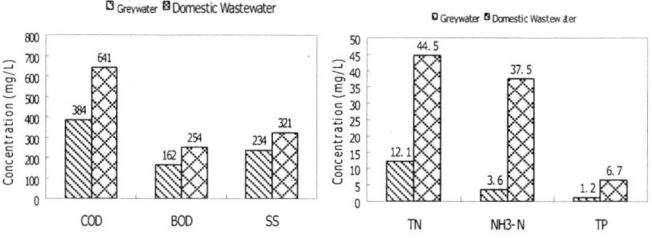

Figure 5 Pollutant concentration comparison between greywater and domestic wastewater

According to Figure 5, one can find that if the greywater is collected and treated together with blackwater, the domestic wastewater will have 250mg/L increase of organic matters, which is 1.7 times as much as that of combined greywater, and the concentrations of TN and TP in domestic wastewater are also 3.7 times and 5.6 times as much as those of greywater. Conversely, if greywater and blackwater are collected and treated separately in communities, it can simplify the greywater treatment procedure and reduce the operating cost significantly; furthermore, the treated water can be reused by nearby inhabitants by using short pipelines to transport reclaimed water. As for blackwater, a flexible treatment process can be selected for removing or recovering nutrients from wastewater, such as sludge or nutrients of black water together with kitchen biological waste, which can be composted to become fertilizer, and utilized in agriculture or landscape work.

Should Kitchen Water be Ranked as Greywater?

Whether kitchen water should be ranked as greywater or not in China needs to be discussed in detail. Considering the high organic or nutrient contents of kitchen water, some scholars have suggested that kitchen water should be discharged separately or ranked as blackwater (Langergraber, 2005). However, three factors will influences the collecting of kitchen water in China.

(1) *The concentration of kitchen water between China and European countries*

The pollutants concentration of kitchen water between China with European countries are listed in Table 4.

It can be seen from the table that the pollutants concentration of kitchen water in China are much lower than those in European countries, especially the concentration for phosphorus, which is only 5% of European countries.

Table 4 Comparison of waste water quality of China and European countries

Parameters	China		European countries
	Range	Average	Range
COD_{Cr}(mg/L)	183~1776	820	936~1380
BOD_5(mg/L)	88~858	386	536~1460
NH_3-N(mg/L)	0.7~62.1	6.4	0.2~23.0
TN(mg/L)	7.4~196.9	47.9	40~74
TP(mg/L)	1.0~10.1	3.9	68~74

Note: The data of China is the results of our investigation, the data for European countries is cited from the statistic results reported by Eriksson (2002).

Other than crashing the solid kitchen waste and discharging it into sewage pipes in most European countries, most solid kitchen waste is dumped into dustbins in China, so many solid-state nutrients, nitrogen and phosphorus are not transferred into kitchen water. In addition, low fatty food consumption in China is another important reason for lower phosphorus content in kitchen water. Considering the components and concentration of kitchen water in China, if kitchen water is ranked as blackwater, it will reduce the concentration of blackwater and influence blackwater treatment and nutrition recovery.

(2) *In the view of grey water treatment*

There are two ways to treat greywater. If the biological way is used to remove the pollutants of greywater, which include kitchen water, the biodegradability of grey water will be improved – that is, it is helpful to treat combined greywater when biological processes are utilized. On the other hand, if greywater is treated by physical or chemical processes, such as membrane separation, the lower the pollutants concentration, the lower consumption of chemical dosage, the less equipment needed and lower cost. Therefore, the absence of kitchen water in grey water will produce better-quality reclaimed water and save on running costs.

(3) *The potential demand of reclaimed water*

After it has been reclaimed, greywater can be used for urban non-potable water usage, including landscape, toilet flushing, road cleaning, car washing, constructing fields and fire-fighting, among which toilet flushing and irrigation for public landscapes are the easiest utilization approaches. Due to the great rainfall difference, the demand for reclaimed water is also different between North and South China. In North China, there is a large potential demand for reclaimed water, which comes from a serious shortage of rainfall and surface source water; while in South China, less irrigation water is required for pubic landscape, and most of the reclaimed greywater can be utilized for toilet flushing.

In conclusion, in South China, toilet flushing is the best motivation for greywater reuse, and kitchen wastewater should be excluded from greywater; in North China, it is suggested that kitchen water should be collected together with other greywater, such as washing water, shower water and laundry water, to satisfy the large demand for reclaimed water.

CONCLUSIONS

Based on a 12-month investigation and discussion of greywater, blackwater, kitchen water and shower water, the following conclusions can be drawn:

- Blackwater has the highest concentration of carbon nutrients, nitrogen, phosphorus, suspended solids and total solids; kitchen water has the second highest concentration of main pollutants; shower water has the fewest contaminants. The COD, nitrogen and phosphorus of blackwater are 2.5–3.5 times, 9.6 times and 15.7 times as much as those of greywater. Excepting shower water, the others have suitable biodegradability.
- According to the investigation of the different times of day, for greywater, the peak and lowest concentration of COD,SS, TN,NH_3-N,TP,TOC appear at 15:00, 22:00 and 11:00, 17:00 respectively, which is in harmony with the water consumption of peaking time and lowest time. In Summer and Autumn, the pollutant concentrations of greywater and kitchen water are lower than those In Winter and Spring, which is about 60% of the later; however, blackwater and shower water had small concentration changes by season.
- When considering the geographical characteristics of China, the potential consumers and the optimization of treating processes, it is suggested that kitchen water can be gathered and treated together with washing water, shower water and cleaning water to reuse in north China; however, kitchen water can be collected with blackwater to recover nutrients in south China.

ACKNOWLEDGEMENT

The authors will be thankful to the financial supported by MOST (The project serial number are NO.2006DFA92690 and 2007DFB90280).

REFERENCES

Ayres R. U. (1996). Statistical measures of unsustainability. *Ecological Economics*, **16**, 239–255.

Eriksson E., Auffarth K., Henze M. and Ledin A. (2002). Characteristics of grey wastewater. *Urban Water*, **4**(1), 85–104.

Friedler E. and Hadari M. (2006). Ecological feasibility of on-site greywater reuse in multistory buildings. *Desalination*, **190**, 221–234.

Hellstrom D. and Karrman E. (1997). Exergy analysis and nutrient flows of various sewerage system. *Water Science and Technology*, **35**(9), 135–144.

Henze (1997). Waste design for households with respect to water, organics and nutrients. *Water Science and Technology*, **35**(9), 113–120.

Jefferson B., Laine A., Parsons S., Stephenson T. and Judd S. (1999). Technologies for domestic wastewater recycling. *Urban Water*, **1**(4), 285–292.

Langergraber G. and Muellegge E. (2005). Ecological sanitation- a way to solve global sanitation problems. *Environment International*, (31), 433–444.

Larsen T. A. and Gujer W. (1996). Separate management of anthropogenic nutrient solutions(Human Urine). *Water Science and Technology*, **34**(3–4), 87–94.

Lazarova V., Hills S. and Birks, R. (2003). Using recycled water for non-potable, urban uses: a review with particular reference to toilet flushing. *Water Science and Technology*, **3**(4), 69–77.

Li F., Wichmann K. and Otterpohl R. (2009). Review of the technological approaches for grey water treatment and reuses. *Science of the Total Environment*, 3439–3449.

Lindon P., Zeman G. and Lintinga G. (2004). Decentralized wastewater treatment and reuse- conception, system and application. Translated by WANG Xiaochang, PENG Dangcong, HUANG Tinglin. Beijing. Chemical industrial publishing house.

Lu W. and Leung A. Y. T. (2003). A preliminary study on potential of developing shower/ laundry wastewater reclamation and reuse system. *Chemosphere*, **52**, 1451–1459.

Maoan, D., Yuhai D., Liping and Q. (2002). Water quality control during treatment of bathing wastewater. *Journal of Harbin University of Civil Engineering and Architecture*, **35**(5), 52–54.

Meisheng N., Hua R. and Jun L. (2004). The water environment plan on ecological community. *Water and Wastewater*, **30**(1), 60–63.

Otterpohl R., Grottker M. and Lange J. (1997). Sustainable water and waste management in urban areas. *Water Science and Technology*, **35**(9), 121–133.

Palmquist H. and Hanaus J. (2005). Hazardous substances in separately collected greyand blackwater from ordinary Swedish households. *Science of the Total Environment*, **348**, 151–163.

Shu cun LI, Yu lin Z. and Wei W. (2002). The application of reclaimed water treatment project in community. The application on utilization project of reclaimed water in the residential quarter. *Journal of Hebei Institute of Architectural Science and Technology*, **19**(2), 19–20.

Water and Wastewater analyses method (the forth version) (2002). Beijing: Chinese Environment Science Publishing House,12.

Wei-hui M., Hong-bin C. and Jin-ning Q. (2008). Progress of domestic wastewater source separation, treatment and resource. *China Biogas*, **26**(4), 15–20.

Yongju Z., Hongbin C. and Qunbiao H. (2008). Review on blackwater treatment and resources. *China Biogas*, **26**(5), 9–14.

Yue xin H. (1998). Primary discuss on the design of reclaimed water project. *Nonferrous Metals Engineering & Research*, **19**(2), 62–65.

Zijie Z. (2003). Water discharge engineering (Fourth Edition). Beijing.

Technical Options for Source-Separated Collection of Municipal Wastewater

Shibao Gao and Jian Zhang*

Wanruo Environmental Engineering & Technology Co., Ltd,
Fuwai Street 28, Beijing 100088
E-mail: j.zhang@envi8.com

Abstract The source separation and reutilisation of municipal wastewater has increasingly gained attention in recent years. The vacuum-based concept for a source-separated collection of municipal wastewater as well as possible application configurations are presented, and some examples of source seperation in practice, from China, are reported.

Keywords Source separation, sanitation, brown water, yellow water, vacuum toilet

INTRODUCTION

The current sanitation system has its origins in early European urbanization, where excreta was flushed by water and mixed with other spent water fractions. This is characterized by gravity sewers with pumping stations, discharging into the surface water bodies with or without cleaning treatment. Until now, all of the practices for wastewater reutilization have been limited to the further treatment of effluents from wastewater treatment plants.

A larger part of organic matter, and most of the nutrients phosphor and nitrogen in municipal wastewater, come from human excreta. Recently the source-separated collection of feces and urine has gained high attention. The core ideology is to keep the balance of the natural material flow, to reduce water consumption and to recover nutrients from wastewater for agricultural purposes. The technical concept is to collect and reutilize feces and urine using dry toilets or micro-flushing toilets, whilst the less polluted water fraction (grey water) will be decentralized, treated, and then reutilised for various purposes.

TOILETS APPLIED FOR SOURCE SEPARATION

The difference between conventional and source separation toilets is that feces (brown water), urine (yellow water), or a mixture of feces and urine (black water) are collected without flushing water, or with very limited amounts of flushing water. Table 1 summarizes the possible source separation toilets with their working principles.

Table 1 Principles of the source separation toilets

Type	Working principle
Composting dry toilet with urine diversion	The urine is collected by a separate drain based on the conventional dry toilet. Composting of feces with structure material such as like sawdust.
Gravity urine diversion toilet	The urine is collected by a separate drain based on the conventional toilet.
(flushing toilet + urine diversion)	water consumption: Urine flushing 0–0.3L Feces flushing <6L
vacuum toilet with micro-flushing	Use of vacuum suction force to reduce water consumption. water consumption: 0.8–1.5 L
vacuum urine diversion toilet	The urine is collected by a separate drain based on the vacuum toilet.
(vacuum toilet with micro-flushing + urine diversion)	Urine flushing 0–0.3L Feces flushing 0.8–1.5 L

The daily water consumption and quantity of effluents of the three kinds of micro-flushing toilets, as well as the dry toilet with urine diversion, are computed for a population equivalent to 1000, with the assumption of toilet visits at the rate of 6 times per day, a total urine amount of 1.2 litres, and a total feces amount of 0.2 litres (Fig. 1). The comparison with amounts from conventional toilets and water saving toilets of the conventional type (3–6 L per flush) shows that water consumption will be reduced by 70–90% by using micro-flushing toilets, and 100% by using dry toilets.

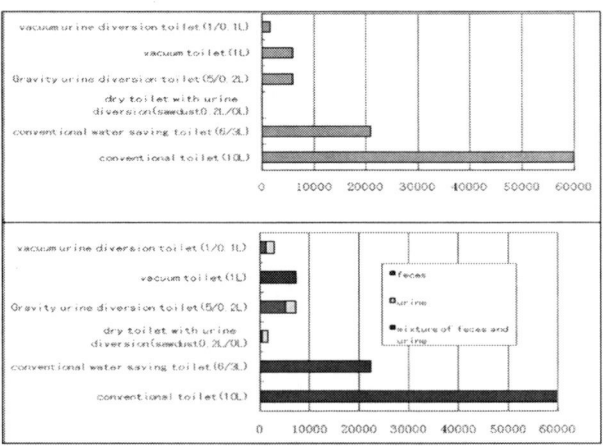

Figure 1 Daily water consumption (in litres) for toilet flushing and effluent quantity for a population of 1000

Meanwhile, all four of these kinds of toilets with a focus of source separation have been applied in demonstration projects in China. Fig. 2 shows the urine diverting micro-flushing toilets applied in the Olympic Forest Park on a large scale.

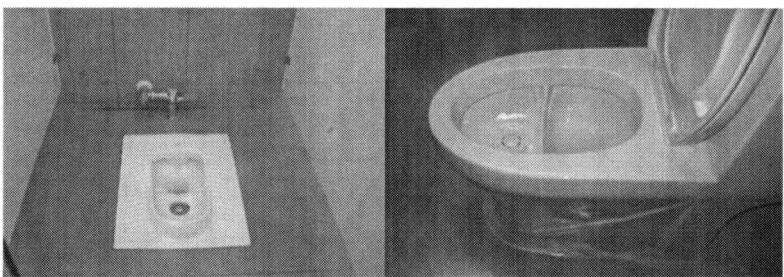

Figure 2 Large scale application of urine separation toilets at the Olympic Forest Park in Beijing

COLLECTION OF FECES AND URINE

Using Vessels and Tank Trucks for Urine and Feces

Vessel and tank truck transportation have typically been used for urine and feces of dry toilets as used, for example, in the eco-settlement demonstration project in Dongsheng, Inner Mongolia, in China.

Though vessel and tank transportation can provide a reliable collection path, this system is unfortunately combined with high manpower demand and low efficiency, and there are various limitations for applying this collection system, particularly for a densely populated area.

Vacuum-Based Collection Systems

In practice, for the implementation and spreading of the source separation strategy, we are faced with various requirements by planners and users, including two essential aspects:

- Semi-centralized collection of brown, yellow or black water for treatment (stabilization or methane generation) or volume reduction with the purpose of allowing further transportation or production of upgraded fertilizer.
- More comfort for transportation conditions, greater acceptance by planners and users, particularly in competition with the conventional sanitation system.

Taking these into account, various vacuum collection systems have been developed and applied in demonstration projects (Zhang *et al.*, 2008a).

The principle of the vacuum collection system is pumping and transporting wastewater using a vacuum as the driving force. In the source-separated collection, the vacuum transportation used for a settlement area of up to 10000 inhabitants as a unit has a good economic efficiency, profiting from the small amount of feces and urine after source separation.

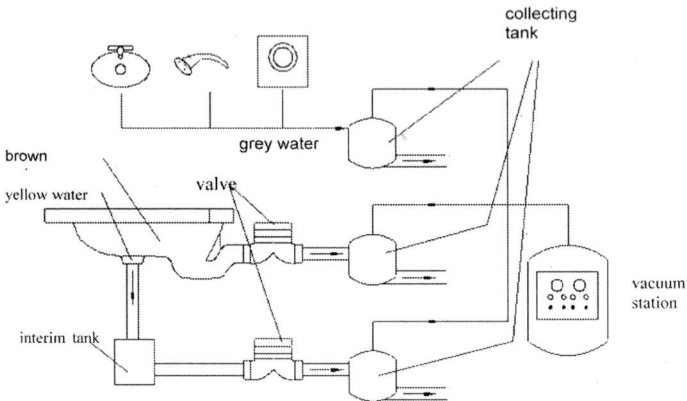

Figure 3 Illustration of possibilities of adopting a vacuum-based source separation strategy

Fig. 3 illustrates the principles of vacuum based source-separated collection of municipal wastewater. In practice, there could be different configurations. For example:

- Black water via vacuum, grey water via a conventional gravity sewer, for cases requiring more centralized collection and treatment of black water, or more decentralized treatment of grey water (for example through constructed wetland).
- Brown and yellow water via vacuum, or grey water via conventional sewers, for situations requiring more centralized collection and treatment of brown and yellow water, and more decentralized treatment of grey water (for example through constructed wetlands.)
- Brown and yellow water via vacuum, grey water via vacuum, for some special local situations – for example, if the settlement is close to water bodies, or in the countryside where dry toilets are not accepted, and construction of conventional sewers is not feasible.

Fig. 4 shows a collection center for feces and urine based on the vacuum system at Qinghua Environment Building, in which brown water is collected by vacuum pipe and yellow water via a gravity pipe.

Figure 4 The collection center for feces and urine, based on the vacuum system at Qinghua University

A vacuum-based source separation system can also be implemented in cases where no vacuum toilets are installed. When conventional gravity toilets or gravity toilets with urine diversion are used, black water, or yellow and brown water can first be collected in wells via gravity pipe(s), and from there the corresponding fractions can be collected and transported via vacuum pipes to a centralized treatment or storage centre. Fig. 5 show such wells in use in a demonstration project for an eco-town in the Town Xiaotangshan, in Beijing.

Figure 5 Collecting wells for yellow, black and grey water

Reutilisation of Feces and Urine

The basic route for a source separation strategy is shown in Fig. 6.

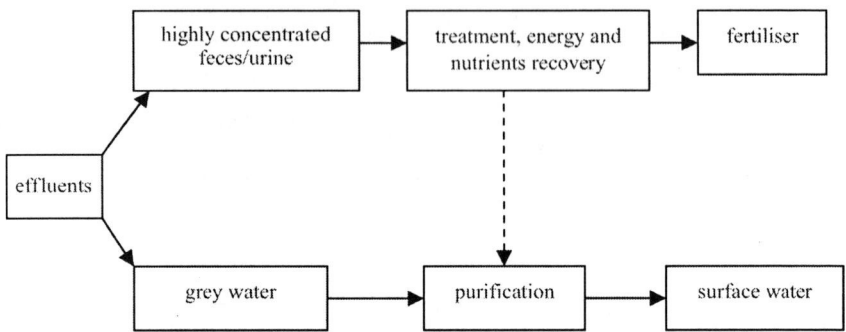

Figure 6 The treatment route for source-separated municipal wastewater.

There are two aims in the treatment process of feces and urine: stabilization and volume reduction, the former being for sanitary requirements, as a pretreatment prior to agricultural use, the latter for the reduction of transportation costs. In countries and towns, the feces and urine can, after stabilisation, be directly used for agricultural purposes if transportation is convenient. In cities, the concentration and enrichment of nutrient salts from feces and urine is more important, in order to reduce transportation costs.

SUMMARY

There are many ways to realize the source separation of municipal wastewater. By using this strategy, feces and urine, the main carriers of the organic load and nutrient salts, will be able to be separately collected.

In the source separation strategy, the grey water can be close to the generation place, de-centralized, treated and reutilised as a resource for outdoor water usage, or to feed a scenic water body with groundwater recharge. The water saving effect, which could be achieved by a conventional wastewater reuse project based on the centralised treatment plant for mixed wastewater, can be achieved by the source separation strategy with much less investment, much less material consumption, and much lower operational expenses. Additionally, there will be the benefit of energy and nutrients being recovered.

There have been many studies reported in the literature which verify the social, environmental and economic benefits of the source separation strategy (Zhang *et al.*, 2008b). The practice of the demonstration projects also demonstrated conclusively that these systems can be realized with wide technical variability without any reduction of the living comfort and hygienic requirement of the users.

The spreading of this new concept requires changing ideas, and its implementation requires that the consensus of the involved circles, and the attention of the urban designers, be obtained.

ACKNOWLEDGEMENTS

The authors thank Zhongguancun Science Park and Ministry of Science and Technology for their Innovation funds 07Z137, and 08C26221100128.

REFERENCES

Zhang J., Zhang J. and Gao S. (2008a). Application of vacuum toilets and vacuum systems or water saving and source separation. *Water & Wastewater Engineering*, **4**(2), 96–99.
Zhang J., Gao S. and Zhang J. (2008b). Concept and application demonstration for ecological sanitation. *China Water & Wastewater*, **24**(2), 10–14.

Performance of Membrane Bioreactor for Onsite Treatment of Higher-Load Graywater

A. Huelgas* and N. Funamizu*

*Department of Environmental Engineering, Hokkaido University, Kita 13, Nishi 8,
Sapporo 060-8628, Japan
E-mail: funamizu@eng.hokudai.ac.jp

Abstract Higher-load graywater (HLGW) is the mixture of kitchen sink wastewater (KSWW) and washing machine wastewater (WMWW) and among the five graywater discharges from the household, these two have high percentages of contribution on the pollution load. The experiment on the effect of OLR was done using 4 lab-scale MBR systems operated at different hydraulic retention time (HRT). The MBR also used to determine the effect of the type of operation modes; continuous and intermittent feeding operations. The experimental results showed that MBR can be operated at HRT of 8 hours and longer for HLGW treatment; and among the two types of membrane used, UF-HF and MF-FP, MF-FP membrane type and configuration gives a better membrane flux performance compared to the former. HLGW treatment using MBR is a promising option if the discharged effluent is reused for non-potable activities like gardening or simply meeting the stringent criteria for effluent discharge to the environment.

Keywords Membrane bio-reactor, Higher loaded gray water

INTRODUCTION

In Onsite Wastewater Differentiable Treatment System (OWDTS), the wastewater discharges from a household is grouped into three: blackwater (feces and urine), higher-load graywater (kitchen sink and washing machine) and lower-load graywater (shower, bath and wash basin) (Lopez Zavala *et al.*, 2002). Volume reduction on water usage and wastewater to be treated is one of the benefits of this concept.

Higher-load graywater (HLGW) is the mixture of kitchen sink wastewater (KSWW) and washing machine wastewater (WMWW) and among the five graywater discharges from the household, these two have high percentages of contribution on the pollution load in terms of chemical oxygen demand (COD), nitrogen (N), and phosphorus (P) components (Almeida *et al.*, 1999; Eriksson *et al.*, 2002). Furthermore, WMWW contains high concentration of the surfactants that is present in the detergent used in washing clothes. Linear alkylbenzene sulphonates (LAS) is the most important surfactant used and considered as a workhorse surfactant in both powder and liquid laundry products (de Guertechin, 1999). Researchers have shown (Patterson *et al.*, 2001; Abu-Hassan *et al.*, 2006) that biodegradation of many surfactants, including LAS, may be restricted between concentrations of 20–50 mgL^{-1} and may be inhibited at higher concentration. Therefore, treatment of HLGW and monitoring its LAS concentration is necessary before it is being discharged to the environment or reused for another purpose. Membrane bioreactor (MBR) has been recently applied to treatment of domestic wastewater and other types of graywater discharges. In this paper, the MBR system is applied for onsite treatment of HLGW and is operated at constant transmembrane pressure (TMP), thus no pump requirement for permeation. Moreover, a simple MBR is used to adapt the fluctuations of the graywater discharges without the need of equalization tank. The following investigations were done: (a) effect of organic loading rate (OLR) on treatment of KSWW only and HLGW mixture using an ultrafiler-hollow fiber (UF-HF) membrane; (b) effect of continuous- and intermittent-feeding operation using a microfilter-flat plate (MF-FP) membrane; (c) comparison with the Johkasou system.

MATERIALS AND METHODS

Experimental Procedure for Determining the Effect of OLR

The experiment on the effect of OLR was done using 4 lab-scale MBR systems operated at different hydraulic retention time (HRT). The first set of these 4 MBRs was used to treat KSWW only; the next set was used to treat the HLGW mixture (1:1 ratio of KSWW and WMWW). The KSWW was obtained at the cafeteria of the Faculty of Engineering at Hokkaido University while the WMWW was obtained from students at the same faculty. A hollow fiber (HF) membrane configuration was used. It was made of poly acryl nitrile with a pore size of 100 kDa (ultrafilter). Other parameters were set constant in all reactors. The TMP and airflow rate were at 5kPa and 2.5 Lmin^{-1}, respectively (Huelgas, 2009).

Samples from the influent, inside the reactor and permeate were obtained for analyses of the following parameters: COD, TKN, NO$_x$-N, TP, PO$_4$-P, and LAS. Membrane flux and MLSS concentrations were also monitored.

Experimental Procedure for Determining the Effect of Continuous and Intermittent-Feeding Operation

Continuous feeding operation

The membrane bioreactor applied in this study employed a microfilter flat-plate membrane system (pore size of 0.4 µm, area = 0.1 m^2, polyolefin) from Kubota Company. The reactor has the following dimensions: 630mm × 90mm × 227mm (H × W × L). The effective volume of 10L was used.

The activated sludge was obtained at Shinkawa Wastewater Treatment Plant in Sapporo, Japan. The 1:1 mixture of the two wastewaters was continuously supplied to the system, an equalization tank is needed in this scenario. A synthetic graywater was used (Huelgas, 2009). Permeate was intermittently (10 min on – 2 min off) withdrawn at a certain TMP. The TMP of 1.5 kPa was set for the first 12 days and increased gradually to 3.0 kPa until day 14th. TMP of 3kPa has been maintained for the remaining of the operation. A water level sensor was used to send signal to the peristaltic pump to supply raw graywater stored in a refrigerator, thereby maintaining the volume of the mixed liquor inside the reactor. Air compressor and diffuser were used to supply air to the system at a flow rate of 10 L min^{-1} and the dissolved oxygen concentration was monitored and maintained above 4 mgL^{-1}. Mixed liquor temperature was maintained at 20°C by a water bath. Samples from the influent, reactor, and permeate were obtained for subsequent analyses.

Intermittent feeding operation

The MBR applied in this study is basically the same as that in continuous-feeding operation. However, the reactor size was modified such that the influx of WW in the morning and the evening will be accommodated without the use of an equalization tank. The MBR has a space to receive this influx of WW during its peak hour discharges (Huelgas 2009). The influent was supplied twice a day. A programmed sequencer is used to automatically supply the raw wastewater into the MBR. Ten liters (10L) of the 1:1 HLGW mixture is discharged evenly from 7:00 to 8:30am in the morning, so 1L of graywater is discharged at 9 minutes interval. In the evening, 1L of KSWW is discharged every 18 minutes from 19:00 to 22:00 that is a total of 10L. Permeate was intermittently (10 min on – 2 min off) withdrawn at a constant minimum TMP induced by a water level difference between the reactor and the permeate. TMP and flux vary throughout the day. Hourly and weekly variations were monitored. Air compressor and diffuser were used to supply air to the system at a flow rate of 10 L min^{-1}.

Samples in the influent, inside the reactor and permeate were obtained for analyses of the following parameters: COD, TKN, NO$_x$-N, TP, PO$_4$-P, and LAS. Membrane flux and MLSS were also monitored.

RESULTS AND DISCUSSION

Effect of Organic Loading Rate

The effect of OLR in the treatment of KSWW only and HLGW mixture was investigated. The influent COD of the KSWW and the HLGW mixture are 1300 mgL^{-1} and 890 mgL^{-1}, respectively. These are higher than the domestic raw wastewater with a range of 250–800 mgl^{-1} (Tchobanoglous *et al.*, 2003), and the total graywater discharge with a range of 495–682 mgl^{-1} (Palmquist and Hanæus, 2005). Table 1 shows the OLR, MLSS and F/M ratio for each corresponding HRT in both series of experiments. The higher value of OLR in the KSWW

$(6.9\ kg_{COD}m^{-3}d^{-1})$ compared to that of the mixture $(5.3\ kg_{COD}m^{-3}d^{-1})$ at HRT = 4 hours is due to the high COD value of the KSWW influent than that of the mixture. The same trend was observed at all other HRTs. Furthermore, the F/M ratio is relatively high for MBR systems treating the KSWW only compared to its corresponding reactors treating the mixture. But in both cases, Reactor 1 has higher F/M ratio compared to Reactors 2, 3 and 4. It has been reported that F/M ratios for MBR systems are $<0.2\ kg_{COD}kg_{MLSS}^{-1}d^{-1}$ and even approaches 0 for a long SRT and high sludge concentration (Stephenson et al., 2000).

It has been observed also that even though Reactor 1 of the KSWW has higher OLR values than that of the mixture, foaming is only observed in the system treating the mixture that is why the reactor was stopped after few days of operation.

Table 1 Operating condition of the reactors in treatment of KSWW and HLGW

	KSWW + WMWW				KSWW			
	R1	R2	R3	R4	R1	R2	R3	R4
HRT	4	8	12	24	4.5	7	12	24
OLR	5.3	2.7	1.8	0.9	6.9	4.5	2.6	1.3
MLSS	9–11	9–11	9–11	6–7	11–13	11–13	11–13	7–9
F/M	0.53	0.27	0.18	0.14	0.58	0.38	0.22	0.16

Table 2 summarizes the characteristics of the permeate. In the treatment of the HLGW mixture, the reactor with longest HRT (24 hours) has the lowest COD range of 12–51 mgL^{-1} (average COD = 29.3 mgL^{-1}). Reactors 2 and 3 have average COD of 82.5 mgL^{-1} and 59.4 mgL^{-1}, respectively. These values are higher compared to the system operated at almost the same condition but treating KSWW only with an average COD of 29, 17 and 17 mgL^{-1} for HRT of 7, 12 and 24 hours, respectively (Huelgas, 2006). The COD is almost double in the system treating the mixture with that system treating KSWW only. This may indicate that some part of the WMWW may not be easily biodegradable.

Table 2 Permeate quality of the reactors in treatment of KSWW and HLGW

mgL^{-1}	KSWW+WMWW			KSWW			
	R2	R3	R4	R1	R2	R3	R4
COD	48–123	28–124	12–51	35–73	11–53	10–26	7–34
	(82.5)	(59.4)	(29.3)	(53)	(29)	(17)	(17)
NH$_4$-N	0.1	0.55	0.06	–	–	–	–
NO$_3$-N	0–0.01	0–0.41	0–3.67	–	–	–	13
PO$_4$-P	–	–	0–0.54	–	0.1	0.8	2.4

Note: values in the parenthesis show the average value.

In Japan, washing machine discharges three sets of wastewater during one cycle of washing clothes. Most of the pollutants including surfactants (like LAS) can be found in the first discharge of the washing machine. The measured total LAS concentration in the first discharge was on the range of $20.59 - 46.24$ mgL^{-1} with an average value of 35.06 mgL^{-1}. Since the influent used in this study was a mixture (1:1) of KSWW and WMWW, the LAS concentration in the influent is around 17.53mgL^{-1}. This value is relatively higher than the reported concentration values of LAS in the domestic wastewater which is on the range of $1 - 15$ mgL^{-1} (Zoller, 2004). This is because the main source of LAS is WMWW and this is more concentrated in the HLGW.

In the treatment of the HLGW mixture, the range of the total LAS concentration in the permeates of reactors 2, 3 and 4 were $37 - 2341$ µgL^{-1}, $11 - 2457$ µgL^{-1}, $8 - 502$ µgL^{-1}, respectively. The lower value in reactor 4 is due to low loading rate and longer sludge retention time which enhances biodegradation of this micro pollutant. Regardless of HRT or loading rate, very high removal rate of LAS was obtained, even up to $> 99\%$. Same removal rate has been observed in other papers with domestic wastewater influent (Temmink, 2004; De Wever, 2004). This high rate of removal can be accounted also to the characteristics of the MBR systems which include complete retention of solids among others. This indicated that there is no inhibition in the biodegradation of LAS at the range of influent concentration around $10.3 - 23.1$ mgL^{-1}.

It was found out that the MBR can be operated at HRT of 8 hours and longer. It corresponds to an OLR of 4.6 kg$_{COD}$m^{-3}d^{-1} and 2.7 kg$_{COD}$m^{-3}d^{-1} for KSWW and HLGW mixture, respectively. Furthermore, it was observed that high COD in the permeate of the system treating the HLGW mixture compared to KSWW only was obtained. It may imply that WMWW has some components that are not easily biodegradable. But $>99\%$ of LAS was removed from the influent LAS concentration of $10.3 - 23.1$ mgL^{-1}, so the remaining organic matter (OM) is not LAS in its original form. Nitrates and phosphates are of low concentrations because these are not abundant in graywater but rather in blackwater. Low flux is observed which opted us to use another membrane, a micro-filter, flat plate (MF-FP) membrane for succeeding experiments.

Effect of Continuous and Intermittent Feeding Operation

Continuous feeding operation

The lab-scale MBR was continuously operated for 87 days. The HRT was at around 10 hours which increased to 16 hours towards the end of the operation. The average HRT during the whole duration is 13.6 hours. This can be attributed to decrease in membrane flux. The MBR has an average flux of 0.22 m^{3}m^{-2}d^{-1}.

The flux decreased from 0.28 $m^3m^{-2}d^{-1}$ to 0.18 $m^3m^{-2}d^{-1}$. MBR system has an advantage of dealing with longer SRT (or even complete retention of sludge) and a high MLSS concentration. The MLSS concentration in this experiment ranges from 10 – 25 gL^{-1}. The average OLR and F/M ratio are 1.21 $kg_{COD}m^{-3}d^{-1}$ and 0.07 $kg_{COD}kg_{MLSS}^{-1}d^{-1}$, respectively. The F/M ratio is lower compared to the previous experiment treating the mixture because no sludge withdrawal was done in this experiment and the MLSS concentration reached as high 25 gL^{-1}.

The qualities of the influent in terms of COD, N and P were measured and the HLGW mixture has a total COD of 675 mgL^{-1}. This is lower compared to that obtained from the real wastewater sample which has a total COD of 890 mgL^{-1}. Although the KSWW was simulated to give the same COD value as that of the real KSWW samples, the WMWW's COD is lower than that of the real samples because the clothes used for washing are basically clean. Therefore, the contribution of pollutants coming from the used clothes is not considered in this experiment. Furthermore, the total LAS concentration in the influent was measured at around 30.8 mgL^{-1}. This concentration is higher than that observed in municipal wastewater plants dealing only with domestic wastewater which has a range of 1 – 15 mgL^{-1} (Zoller, 2004). This is because the main source of LAS is WMWW and this is more concentrated in the HLGW. However, it was furthermore observed that this value is also higher than that obtained from the real WMWW samples which has an average concentration of 17.53 mgL^{-1}.

The average membrane flux of 0.22$m^3m^{-2}d^{-1}$ from day 14 onwards was obtained. Some studies on subMBR treating municipal and domestic wastewater report membrane flux values between 0.12 – 0.96$m^3m^{-2}d^{-1}$ (Stephenson et al., 2000). The low flux observed in the system was due to the operation at constant TMP. The TMP used was also low compared with those observed in the system operated at constant flux. Increasing the constant TMP applied is expected to increase the flux.

Intermittent feeding operation

The reactor was continuously operated for 120 days. The membrane flux was measured throughout the day as the TMP changes due to the intermittent supply of influent. Figure 1a shows the flux against TMP for the morning discharge. The decrease in flux was observed as the weeks proceeded. The flux decline was plotted against time (Figure 1b) at TMP of 2, 3 and 4 kPa. It was found out that it continued to decrease and reached 0.12 md^{-1} for TMP of 3kPa. It was observed that the system continuously treating 1:1 ratio of KSWW and WMWW has a constant flux of 0.2 md^{-1} at TMP = 3kPa (see continuous-feeding operation). The intermittent feeding operation has an advantage of higher flux at TMP = 3kPa), until 60 days of operation.

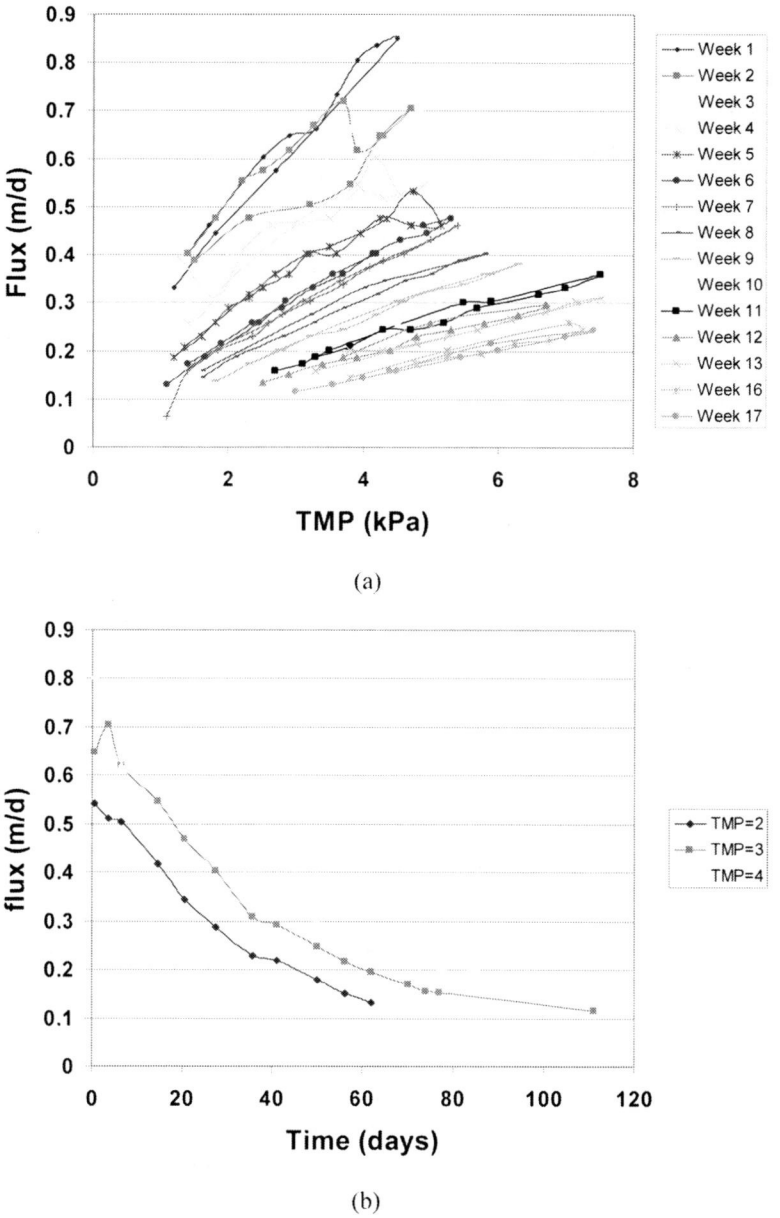

(a)

(b)

Figure 1 (a) Weekly measurement of flux decline against TMP variations (b) Flux decline through time at TMP of 3, 4, and 5 kPa (in the morning discharge)

The COD of the HLGW mixture in the morning and the KSWW only in the evening are 675 mgL⁻¹ and 1050 mgL⁻¹, respectively. The MLSS concentration was maintained at 16 gL⁻¹. Variations on quality of treated wastewater in terms of COD, N and P are measured. Figure 2 shows the COD of the permeate from the composite samples of the morning and evening discharges. It showed that the COD of the permeate from the morning discharge was higher than that of the evening discharge regardless of the fact that the influent COD of the morning discharge is smaller than that of the evening discharge. This has been observed also in previous experiments wherein the permeate obtained from the treatment of the HLGW mixture is higher than that of the permeate obtained from the treatment of KSWW only.

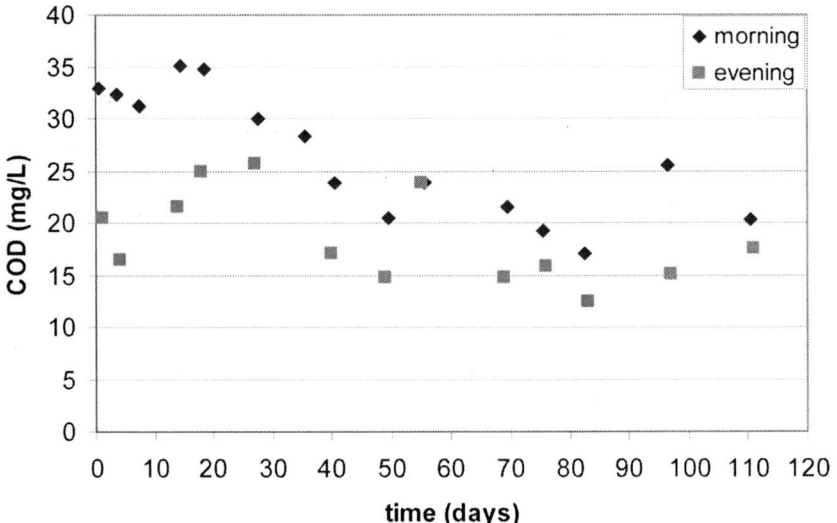

Figure 2 COD of the permeate from the composite samples of the morning and evening discharges

Comparison with Johkasou system

The performance of the three types of Johkasou systems, the anaerobic filter-contact aeration, anaerobic filter – moving bed biofilm and the membrane Johkasou were discussed by Lens et al., 2001. The total tank volume for anaerobic filter-contact aeration, anaerobic filter-moving bed biofilm and membrane Johkasou are 3.55 m³, 3 m³ and 2.488 m³, respectively.

The HLGW treatment using a continuous operation mode requires an equalization tank which volume was set to accommodate a whole day's HLGW discharge. The equalization tank's volume is 0.187 m^3 in addition to the main reactor's volume which is 0.0936 m^3. The total volume is 0.28 m^3. On the other hand, the volume of the subMBR using an intermittent operation mode is 0.187 m^3 which is the sum of the main reactor's effective volume of 0.0936 m^3 and the buffer tank's volume of 0.0936 m^3 (Huelgas, 2009).

The volume of the composting toilet was based on the commercially available "Bio-Lux" (Model: S–15). The main body volume of the composting toilet is 0.38628 m^3 (1m × 0.620m × 0.623m). This was added to the volume of the subMBR for the treatment for HLGW to get the total volume for the treatment using a concept of source separation. Therefore, in the OWDTS system, still the combination in terms of volume is smaller compared to the Johakasou systems.

There is around four times reduction in terms of the total volume needed for the treatment by source separation. However, the maintenance being applied to the Johkasou systems should also be considered with the treatment of the HLGW using subMBR. These can be achieved if longer and application to real household situation is done.

CONCLUSIONS

Onsite treatment of higher-load graywater using subMBR has been investigated. The following has been determined: (a) the effect of organic loading rate on the treatment of the treatment of KSWW only and HLGW using UF-HF membrane; (b) the effect of the continuous and intermittent feeding operation styles on the treatment of HLGW using MF-FP membrane; (c) comparison with Johkasou membrane.

It has been found out that the MBR can be operated at HRT of 8 hours and longer for both the treatment of KSWW only and the HLGW mixture. It was observed that high organic matter in the permeate of the system treating the mixture compared to KSWW only was obtained. It may imply that WMWW has some components that are not easily biodegradable. Furthermore, >99% of LAS was removed. Therefore, the remaining OM is not LAS in its original form. Nitrates and phosphates are of low concentrations because these are not abundant in graywater but rather in blackwater. Low flux is observed which opted us to use another membrane, a micro-filter, flat plate (MF-FP) membrane for succeeding experiments.

Two types of operation style have been investigated, the continuous feeding and the intermittent feeding type of the operation. It has been found out that continuous feeding could give a stable membrane flux throughout the operation but intermittent feeding operation has better membrane flux performance up to a certain time. During the first 60 days, the MBR operated with continuous feeding gave a lower flux at TMP of 3 kPa compared to that operated with intermittent feeding. It has been observed also that the composite sample from the morning discharge has a higher COD than that of the evening discharge regardless of the fact that the influent COD of the latter is higher.

Comparison with the Johkasou system showed that there is around four times reduction in the volume of the treatment facility if source separation is considered.

REFERENCES

Abu-Hassan M. A., Kim J. K., Metcalfe I. S. and Mantzavinos D. (2006). Kinetics of low frequency sonodegradation of linear alkylbenzene sulfonate solutions. *Chemosphere*, **62**, 749–755.

Almeida M. C., Butler D. and Friedler E. (1999). At-source domestic wastewater quality. *Urban Water*, **1**, 49–55.

de Guertechin L. O. (1999). Surfactants: Classification; Handbook of Detergents Part A: Properties. Marcel Dekker, New York, Chapter 2, pp. 7–46.

De Wever H., Van Roy S., Dotremont C., Müller J. and Knepper T. (2004). Comparison of linear alkylbenzene sulfonates removal in conventional activated sludge systems and membrane bioreactors. *Water Science and Technology*, **50**(5), 219–225.

Eriksson E., Auffarth K., Henze, M., and Ledin A. (2002). Characteristics of grey wastewater. *Urban Water*, **4**(1), 85–104.

Huelgas A. (2006). Effect of organic loading rate on onsite treatment of kitchen sink wastewater using subMBR. M. Engineering thesis. Graduate School of Environmental Engineering, Hokkaido University, Sapporo, Japan.

Huelgas A. (2009) Onsite treatment of higher-load graywater by membrane bioreactor. Doctore of Engineering thesis. Graduate School of Environmental Engineering, Hokkaido University, Sapporo, Japan.

Lens P., Lettinga G. and Zeeman G. (2001). Decentralized Sanitation and Reuse. IWA Publishing, pp. 256–280.

Lopez Zavala M. A., Funamizu N. and Takakuwa T. (2002). Onsite wastewater differentiable treatment system:modelling approach. *Water Science and Technology*, **46**(6–7), 317–324.

Palmquist H. and Hanæus J. (2005). Hazardous substances in separately collected greyand blackwater from ordinary Swedish households. *Science of the Total Environment*, **348**, 151–163.

Patterson D. A., Metcalfe I. S., Xiong F. and Livingston A. G. (2001). Wet air oxidation of linear alkylbenzene sulfonate 1. Effect of temperature and pressure. *Industrial & Engineering Chemistry Research*, **40**, 5507–5516.

Stephenson T., Judd S., Jefferson B. and Brindle K. (2000). Membrane Bioreactors for Wastewater Treatment. IWA Publishing, London, UK.

Tchobanoglous G., Burton F.L. and Stensel H.D. (2003). Wastewater Engineering Treatment and Reuse, 4th edn (Int. Ed). Metcalf and Eddy, McGraw-Hill, New York, p. 186, 857 [1819pp].

Temmink H. and Klapwijk B. (2004). Fate of linear alkylbenzene sulfonate (LAS) in activated sludge plants. *Water Research*, **38**, 903–912.

Zoller, U. (2004). Handbook of Detergents: Part B: Environmental Impact. Marcel Dekker, New York, volume **121**, pp. 1–816.

PART FOUR

Ecological/Small Community
Sanitation

Practice of Ecological Sanitation in Beijing: a Demonstration Project

F.-G. Qiu[a,b], X.-D. Hao[a], J.-Q. Li[b], W. Che[b] and D.-Y. Zhang[b]

[a]The R & D Center for Sustainable Environmental Biotechnology, Beijing University of Civil Engineering and Architecture, 1 Zhanlanguan Road, Beijing 100044, P. R. China
[b]Key Laboratory of Urban Stormwater System and Water Environment (Ministry of Education)/ the R & D Center for Sustainable Environmental Biotechnology (Beijing University of Civil Engineering and Architecture), 1 Zhanlanguan Road, Beijing, 100044, China
Fax: +86-10-68322128; E-mail: haoxiaodi@bucea.edu.cn

Abstract The concept of ecological sanitation (ECOSAN) is practiced in a deserted riverside area of 2 ha, which includes urine separation and utilization, biogas production from faeces, animal wastes and biomass, rainwater harvesting, ecological lake, wetland, etc., within the practiced area, water cycle, nutrient cycle and energy cycle are simultaneously formed. In this way, a zero discharge of pollutants is realized. Such an ECOSAN concept is very close to the conventional farming and life style in China, and so the demonstration project is intended as a model for the future of villages in China. The article describes the details of the demonstration project.

Keywords Ecological sanitation (ECOSAN), urine separation, biogas, nutrient, rainwater harvesting, ecological lake, wetland

INTRODUCTION

For a couple of decades, an urbanized level in China has risen to 45.7%, and a total urban population has reached to 607 millions at the end of 2008 (Niu and Pan, 2009). Even in left-over villages, the life of farmers has been totally changed towards a modern style. For examples, farmers' houses have been using flush-toilets; black water is discharged aimlessly; animal wastes are hardly been used as fertilizers; electricity has been even applied for cooking. This trend towards the so-called modern life style has actually deviated from a conventional ecological life style persisting for thousands of years in China, which is definitely not a sustainable way. In this way, resources and energy will be quickly consumed, which will result in deterioration of ecological environment, shortage of freshwater and nutrient, and energy crisis. In practice, this situation has appeared in many areas of China.

Under the circumstance, exploring and demonstrating a future development pattern for villages and even towns has become important and urgent. Nowadays, the future of cities and villages towards sustainability is being emphasized and planned globally; some countries and/or organizations have developed some alternative technologies to achieve the closed loop recycling of water resources, energy, and nutrient in rural and urban communities. Among them, the concept of ecological sanitation (ECOSAN) with urine separation, rainwater harvesting and biogas production has been gradually practiced in some European countries such as Germany and Sweden. (Berndtsson, 2006; Langergraber and Muellegger, 2005). In fact, the conventional farming and life pattern in China is very close to ECOSAN to a large extent. Clearly, changing the conventional pattern towards ECOSAN should be realizable and relatively simple. To convince people (especially for farmers), however, a practical demonstration project of ECOSAN has to be set up, as there is a conventional saying in China: hearing for a hundred times is inferior to seeing it one time.

Although the concept of ECOSAN has been accepted academically and even practiced partially in some areas in China, urine separation toilet is only a main concern in practice, and not a real ECOSAN project has been accomplished yet. With this article, a demonstration project of ECOSAN is introduced, which is located in the suburban area of Beijing and is being constructed under the supervision of Beijing University of Civil Engineering and Architecture (BUCEA) and with the financial support of the Beijing municipal government.

The constructed ECOSAN project is practiced on a deserted riverside with an area of 2 hectares. Besides the concept of ECOSAN, green buildings with energy saving are also reconstructed together.

GENERAL INFORMATION ABOUT THE DEMONSTRATION PROJECT

Contents of the Demonstration Project

Ecological sanitation and green buildings, including urine separation at toilet sources, biogas production with faeces/animal manure and biomass, rainwater harvesting, constructed wetlands for both treating grey water and purifying recycled lake water from collected rainwater, nutrient recycling to crops, heat preservation of buildings, and wind and solar energy utilization.

Project Scale

- Total area of the project: 2 ha.
- Installed toilets: 5 vacuum units for mixed collection of faeces and urine; 9 urine separation units with gravitational flow.

- Vacuum station: equipped with 2 vacuum tanks for greywater and faeces, 2 vacuum pumps and sewage pumps, and automatic control system, each tank's volume is about 350L.
- Anaerobic digester: 2300×2500mm, treatment capacity is about $8m^3/d$.
- Rainwater harvesting: collecting area is about 1565 m^2, of which the road is 1165 m^2, the parking lot is 120 m^2, and the house roof is 280 m^2.
- Wetlands: 120 m^2 for rainwater treatment and 55 m^2 for grey water ($3m^3/d$).
- Urine collecting tank: 700×1800mm, design storage time is about 20–30d.
- Heat preservation: total area of housing insulation is about 530m^2, rooms' outer wall, roofing, and floor insulation properties meet or exceed the current "energy-saving design standards for residential buildings".
- Solar thermal collector: collector area of 20m^2 for the winter heating, 3500 kWh electricity can be saved each year.
- Equivalent population: 30
- Total project budget: $320,000

GEOGRAPHY OF THE PROJECT AREA

The project is practiced on the deserted riverside with the total area of 2 ha, which has been now rented and developed as a resort place. In the resort place, there are the housing rooms of 1200 m^2, a constructed lake of 1800 m^2 (5.5 m deep), a constructed sub-lake of 770 m^2 (2.2 m deep) for purifying lake water, a fish farming pound of 800 m^2 (2.0 m deep), a total farming lands of 200 m^2 for cultivating crops and vegetables and a well for drinking water supply (with a naturally mineral water quality), as shown in Figure 1.

ECOSAN SYSTEM

Description of the ECOSAN Process

The northern and southern housing rooms are installed with 9 urine separation toilets, and urine is collected in a urine tank and then used for crops and vegetables as a fertilizer; faeces is collected in a faeces tank and then transported to the vacuum station and then pumped into the biogas digester for biogas production. The western housing rooms are installed with 5 vacuum toilets; black water (mixture of urine and faeces) is also transported into the biogas digester via the vacuum station. At the same time, grey water from all the housing rooms is transported into the constructed wetland (vertical flow) for grey water treatment also via the vacuum station; treated grey water is directly discharged into the constructed sub-lake for further purification and then recycled gradually into the constructed lake.

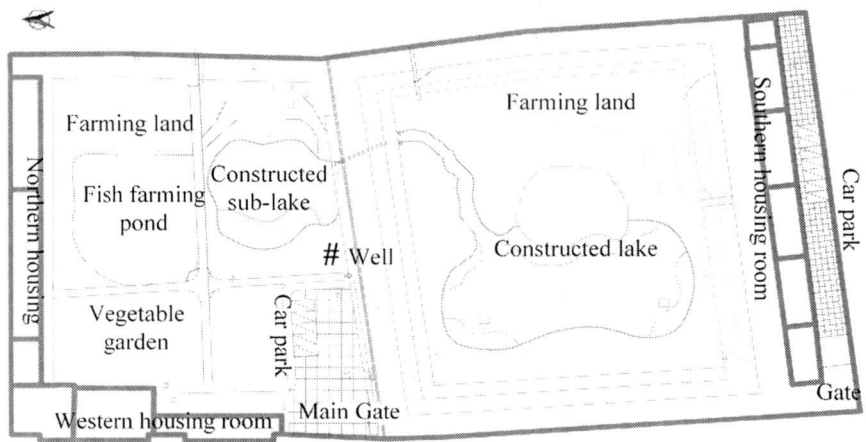

Figure 1 Geography of the project area

The collecting and treating flow sheets of urine, faeces and grey water are shown in Figure 2, and the pipelines' design is shown in Figure 3

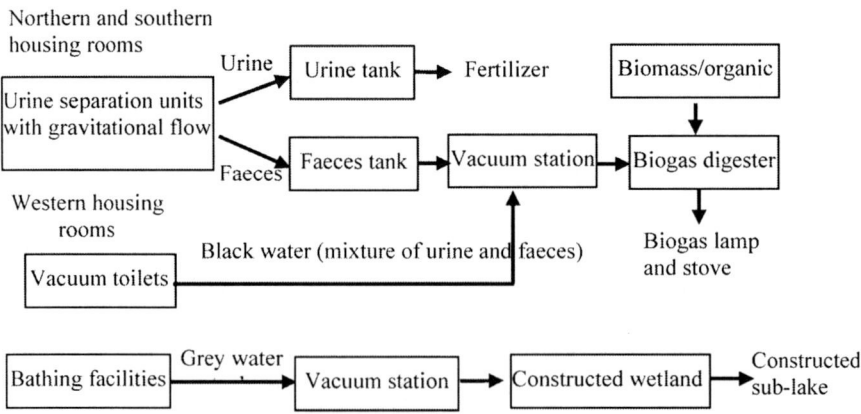

Figure 2 Collecting and treating flow sheets of urine, faeces and grey water

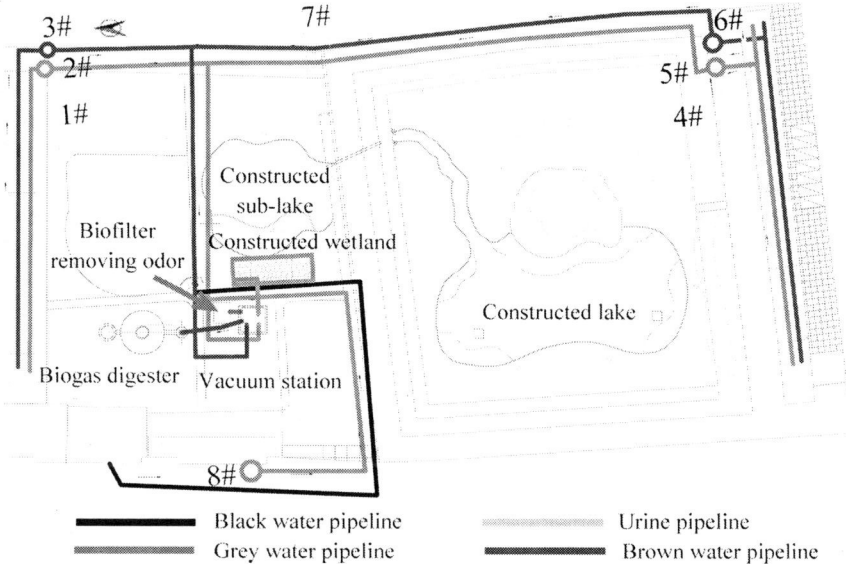

Figure 3   Pipelines' design for the collecting and treating system

Vacuum Station

The vacuum station consists of vacuum tanks, transporting equipment and controlling devices, which is constructed in the form of semi-underground, with the dimension of 3000 × 3000 × 3000 mm. The model of vacuum tanks manufactured by Enviro Systems is WV700. The vacuum station is shown in Figure 4.

Figure 4 Vacuum station and equipment

Biofilter Removing Odour

An off-gas pipeline from the vacuum station is connected with a biofilter (dimension: 2500 × 2000 × 1000 mm, shown in Figure 5) which is constructed to remove odour from faeces and urine. The biofilter is packed from the top to the bottom, with such carriers as in-situ soil (earth layer: 200 mm thick), bark and leaves (covering layer: 50 mm thick), ceramic pellets (transitional layer: 10–20 mm in diameter and 100 mm in thickness), ceramic pellets (holding off-gas layer: 20–40 mm in diameter and 250 mm in thickness), ceramic pellets (absorbing layer: 10–20 mm in diameter and 150 mm in thickness), and sandy soil (bedding layer: 100 mm thick).

Biogas Digester

Faeces is transported from the collecting tank into the biogas digester (ϕ2300 × 2500 mm) via the vacuum station. Biogas (methane: CH_4) produced together with animal wastes and biomass is intended to cook and even illuminate for a family. The biogas digester is equipped with a safe system preventing accidents. The site of the biogas digester is shown in Figure 6.

Figure 5 Biofilter removing odour   **Figure 6** Site of the biogas digester

Tanks Collecting Grey-water, Urine and Faeces

Separated grey-water, urine and faeces from the northern and southern housing rooms flow gravitationally into each collecting tank for storage; urine is then used as the fertilizer after stored for 3 months; faeces is then transported into the biogas digester via the vacuum station; grey-water is then transported on the constructed wetland via the vacuum station.

The collecting tanks of grey-water, urine and faeces are made of brick construction. The tanks' bases are made of concrete (C15) with the thickness of 150 mm; the interior walls (20 mm thick) of the tanks are made of cement/mortar

(1:2.5) mixed with 5% of waterproofing powder. Tanks No. 1 and No. 4 (ϕ700 ×
1800 mm each) shown in Figure 3 are used to collect urine; Tanks No. 2 and No. 5
(ϕ1000 × 1800 mm each) are the tanks collecting grey-water; Tanks No. 3 and No. 6
(ϕ1000 × 1800 mm each) are the tanks collecting faeces; Tank No. 7 (ϕ1000 ×
1800 mm) is a monitoring tank.

Figure 7 Northern housing rooms and a collecting tank under construction

Outdoors Pipelines

Outdoors pipelines are made of polyethylene (PE) material with 75 mm in
diameter. The pipelines transporting faeces and grey-water are buried in an
identical ditch, with a total length of 500 m, which can be seen in Figure 8.

Figure 8 Vacuum outdoors' pipelines and a vacuum lifting tank

Constructed Wetland for Treating Grey-water

The constructed wetland is designed in the type of vertical flow, with a dimension
of 9,000 × 6,400 × 1,200 mm, which is shown in Figure 9. The treating capacity
of the wetland is at a maximum flow rate of 3 m³/d, with an accepting COD

concentration of 360 mg/L. Grey-water is pumped into a distributing system on the wetland, grey-water flows vertically down to the bottom of the wetland and then over-flows to the sub-lake for further purification. Two pumps (QW25–8–22) for transporting grey-water are equipped.

The surface loading of the wetland is designed at 55 mm/d, which needs a total surface area of 55 m². The bed depth of the wetland is designed at 1 m, which results in a hydraulic retention time (HRT) of 5.9 d. The bed of the wetland is packed from the top to the bottom, with such carries as gravels (surface layer: 8–16 mm in diameter and 50 mm in thickness), coarse sands (filtering layer: 0.2–16 mm in diameter and 500 mm in thickness), gravels (transitional layer: 4–8 mm in diameter and 100 mm in thickness), gravels (discharging layer: 8–16 mm in diameter and 200 mm in thickness), and waterproofing cloth with PE membrane (bedding layer: >0.5 mm in thickness).

The distributing pipelines (50 mm in diameter, 50 mm in holes' position and 8 mm in holes' diameter) are made of PE material, which are connected with an influent pipeline (ϕ50 mm). Some valves are installed at the ends of the distributing pipelines and the influent pipelines for emptying. The distributing system is operated in a batch way, working for two times (5 min each time) an hour and stopping for 25 min after each working. The flow rate of the distributing system is designed at Q = 0.0021 m³/s and with a surface loading of μ = 0.55 m/s. The discharging pipelines (ϕ110 mm) with even holes at the bottom are made of PE material as well, which is connected into atmosphere with a tube for the purpose of ventilation/aeration.

On the surface of the wetland, some plants such as reeds, cattails and calami are cultivated at a density of 4 pieces/m². The distributing system is designed with an auto-control program, which can be operated in either a manual or an auto mode.

Figure 9 Constructed wetland and sub-lake

RAINWATER HARVESTING SYSTEM

A rainwater harvesting system is planned and constructed to save water and recharge the constructed lakes. Rainwater from the northern and southern areas are harvested and utilized directly and indirectly. The rainwater harvesting system is shown in Figure 10.

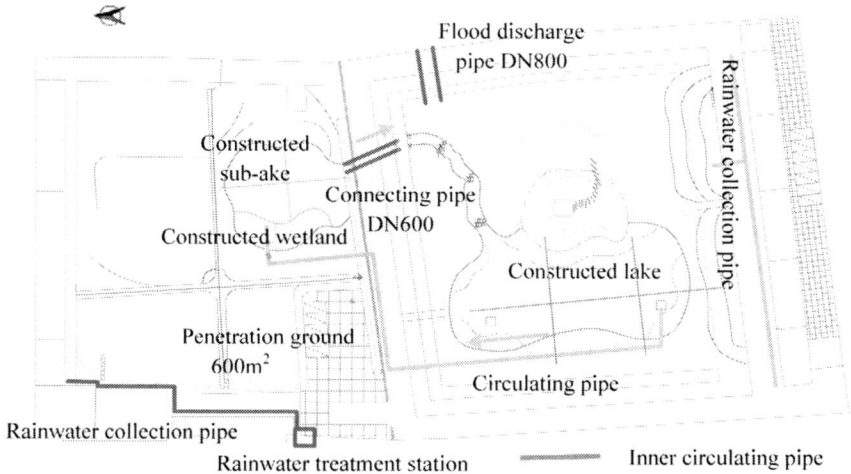

Figure 10 Planned flow-sheet of the rainwater harvesting system

Direct Utilization of Rainwater

Rainwater from the northern and western housing rooms is designed for direct utilization, which mainly consists of recycled filtration, pressurized filtration and advanced filtration by activated carbon. After treated, collected rainwater can be directly used for drinking purpose. The treating process of rainwater is shown in Figure 11.

Rainwater on roofs is collected in a storage tank via a hanged screen intercepting dirt. On the storage tank, an internal recycled filter is combined each other. Filtered rainwater is pumped from the storage tank into a pressurized filter for further purification, with chemicals added for coagulation prior to the pressurized filter. After the pressurized filter, purified rainwater is stored in a tank (Tank 1) for advanced purification. Flowing an advanced filtration with activated carbon and disinfection by UV, purified rainwater is stored in another tank (Tank 2), which is ready for drinking. The filters backwashing water is discharged into the greywater pipe and then treated by constructed wetland.

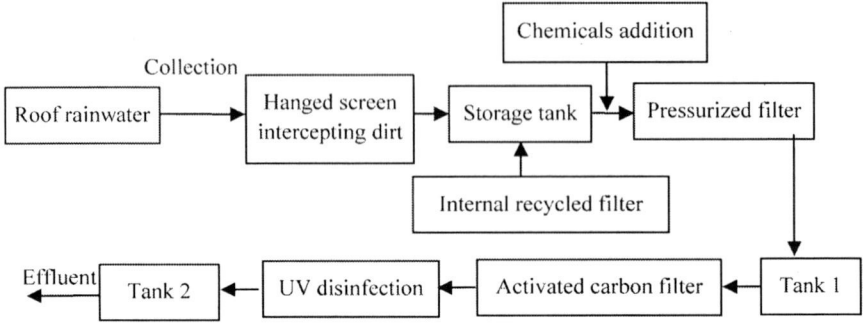

Figure 11 Treating process of rainwater utilized directly for dirking purpose

Main units in treating process shown in Figure 11 are in detail described as bellow: i) storage tank: 8.0 m³; ii) recycled filter: 1,300 mm in length (two sections of 300 mm plus1,000 mm, connected with a flange in between) and 360 mm in diameter, placed horizontally with silica sand (0.8–1.2 mm) as packing carriers and operated at a cross flow rate of 10 m/h and with a backwashing time of 6 min and an washing intensity of 8 L/s · m²; iii) pressurized filter: double packing carries, 1,108 mm in length (anthracite: 400 mm; silica sand: 600 mm; and supporting layer: 100 mm) and 360 mm in diameter, placed vertically, equipped with a dosing unit of chemicals and operated at a flow rate of 12 m/s and with a backwashing time of 8 min and an washing intensity of 16 L/s · m²; iv) Tank 1: 4.0 m³; v) advanced filter with activated carbon: 2 stage filters (alternately used for the first stage) with 1,100 mm in height and 360 mm in diameter each (activate carbon: 1,000 mm; supporting layer: 100 mm), placed vertically and operated at a flow rate of 16 m/h and with a backwashing time of 8 min and an washing intensity of 16 L/s · m²; vi) Tank 2: 4.0 m³.

Indirect Utilization of Rainwater

The purifying system of rainwater for indirect utilization consists of a bioretention areas, a vegetation buffer strips, a constructed wetland and two constructed lakes (an ecological lake and a purifying lake), which collects, treats, stores and utilizes rainwater from the housing rooms' roofs and the ground in the southern area. The purifying system is shown in Figure 12, which has multiple functions of collection, purification, flood control, landscape and entertainment.

Roofs' rainwater from the southern housing rooms is collected in the plants' intercepting device for purification, and both purified and overflowing rainwater is introduced into the constructed lakes. A parking place at the western side is paved with infiltrating bricks, and rainwater nearby either infiltrates into underground or overflows into the ecological lake. Seriously polluted rainwater around the lake flows over the vegetation buffering zone and then enters the ecological lake. Lightly polluted rainwater directly enters the lake.

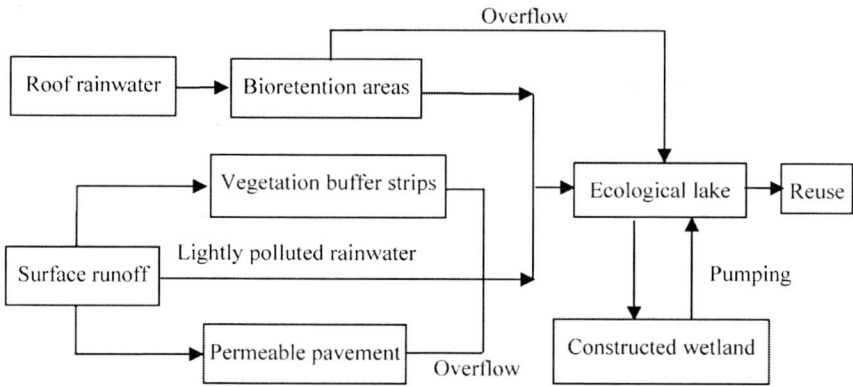

Figure 12 Purifying system of rainwater utilized indirectly for the constructed lake

Some selected plants are cultivated at entrances of rainwater into the lake and at the dead corners for buffering, anti-eroding and plants' purification, with 20% of the total lake surface area. The constructed wetland (Vertical subsurface flow) is constructed at the northern side of the ecological lake, which is used to recycle and purify the lake water. Usually, the lake water is pumped into the wetland for purification and then returns to the ecological lake via the purifying lake.

Some key parameters about the ecological lake are listed as below: i) plants' intercepting area: 14 m²; ii) ecological lake: 2,650 m² (including 530 m² of plants' purifying area), with the banks of grasses, gravels and stones; iii) subsurface flow wetland: 120 m². The rainwater harvesting system of the southern housing rooms and the ecological lake are shown in Figure 13.

Figure 13 Rainwater harvesting system and ecological lake

SUMMARY

The project totally demonstrates the concept of ecological sanitation in the deserted area of 2 ha, which includes urine separation and utilization, biogas production from faeces, animal wastes and biomass, rainwater harvesting, ecological lake, wetland, etc. In this way, three closed cycles for water, nutrient cycle and energy can be realized within the practiced area, which results in a zero discharge of pollutants from the area. The demonstration project is intended as a base of research and teaching for the future of villages.

ACKNOWLEDGEMENTS

The study was financially supported by the Key Project in the National Science & Technology Pillar Program (2006BAJ01B03-02), Funding Project for Academic Human Resources Development in Institutions of Higher Learning Under the Jurisdiction of Beijing Municipality (BJE10016201011), and Scientific Research Common Program of Beijing Municipal Commission of Education (KM200810016006).

REFERENCES

Berndtsson J. C. (2006). Experiences from the implementation of a urine separation system: Goals, planning, reality. *Building and Environment*, **41**(4), 427–437.
Fengrui N. and Jiahua P. (2009). Blue book of cities in China-Annual report on urban development of China (No.1). Social Science Academic Press (China).
Langergraber G. and Muellegger E. (2005). Ecological sanitation—a way to solve global sanitation problems? *Environment International*, **31**(3), 433–444.

Demonstration Project of Ecological Sanitation in Rural Beijing

Wang Huan-sheng[a], Cui Zhi-feng[a] and Wang Kai-jun[b]

[a]Beijing Municipal Research Institute of Environmental Protection, No.59
Beiyingfang Zhongjie, Xicheng District, Beijing, 100037
E-mail: whs1172003@163.com;cuizhifeng@vip.sina.com
[b]Department of Environmental Science and Engineering (DESE) of
Tsinghua University (THU)
E-mail: wkj@tsinghua.edu.cn

Abstract This article introduces a treatment project for rural domestic pollution in suburban areas of Beijing. Through a combination eco-san system, the problems of rural excreta, polluted water and waste pollution are solved based on the concept of separated treatment and source control. Separation of black water and gray water, feces and urine as well as waste is made possible via the eco-san system. Separated urine is used for agricultural purposes after dilution, separated excreta and organic matters are composted together, and low-density small gray water treatment wetland is used as part of the landscape water for the area. With the advantages of lower water consumption, realization of resource recycling, low investment for the whole project, low operation costs and obvious environmental benefits, the system shall be promoted in the rural area.

Keywords Rural domestic pollution, ecological sanitation, small-wetland, waste dividing treatment

In recent years, China has made great progress in the prevention and treatment of environmental pollution, which is transforming from industrial pollution into domestic pollution. With the development of urbanization, domestic pollution in rural areas has become one of the key factors of non-point pollution (Zhang Wei-li *et al.*, 2004). Though great importance has been attached to domestic pollution in rural areas by government departments and environmental protection departments at various levels, there are still limits for the application of different pollution control techniques. It is hard for optimized end treatments to be used in rural areas where the economy is lagging. There have been various problems in the treatment of rural pollution due to lack of appropriate rural domestic pollution control techniques. It costs too much to apply optimized treatments for urban areas in rural areas, but to indulge the pollution will, on the other hand, cause even worse environmental damage. In 2005, the State Development and Reform Commission and the Beijing Municipal Commission of Development and

Reform, cooperating with Sweden and Finland, organized and implemented the model project of eco-san system for solving rural domestic pollution problems, therefore launches the model project research for rural environment pollution control.

STATUS IN QUO OF RURAL DOMESTIC POLLUTION IN SUBURB AREA OF BEIJING

Considering the lack of material for rural domestic pollution situation and integration technique research in China, in the beginning of 2005 the Beijing Scientific Research Institute of Environmental Protection launched a survey in some of the rural areas in suburbs of Beijing on the basic situation of sanitation, waste, water supply and drainage facilities as well as energy facilities (not included in this article).

Environment Facilities in Suburb Rural Area of Beijing

In suburban villages, the type of toilets are mainly traditional squat dry closet (See Table 1 as follows) and excreta are cleared by human force to be used as organic fertilizer for farmland. There is no rain and water pipeline network except for natural channels on main streets for rain. Most domestic wastewater is discharged at source, which causes wastewater flowing in the street. Some residents build negative wells in their yards, and so wastewater filters underground directly. There are small numbers of residents building simple cesspools for flushing toilets and wastewater filters underground with excreta flushing water.

Table 1 Proportion of Different Toilet Types in Suburban Beijing

	Traditional dry closet	Flush toilet	Others
A village in Haidian District	25%	73%	2%
A village in Changping District	95%	5%	0
A village in Daxing District	89%	11%	0
A village in Yanqing District	40%	0	60%

* Flush toilets are mainly cesspools, and other types are mainly double-urn latrines.

Food residue constitutes the major part of domestic waste of residents at a proportion of 50%. Dust and ash constitutes 20% and paper 8% of domestic waste. Collection of waste is mainly mixed collection combined with some human dividing work. In village collection-town transportation-district/city treatment, the collection efficiency is of insufficient quality and there exist serious problems of random dumping incidents.

Major Problems in Villages

The aforesaid villages are all located in suburban areas of Beijing, which makes them appropriate to represent the suburban area of Beijing. Through the survey the following problems came to light:

First, the existing sanitary system poses a great health threat. Concerning flush toilets and squat dry closets, where excreta is used for farmland without being decomposed, most pathogens are still present and will enter the food cycling chain through water cycling and farm crops.

Second, water pollution caused by organic waste composed of cesspool excreta, kitchen waste and other paper, and so on. Untreated excreta that flows with surface flow will cause eutrophication of surface water while leakage of simple cesspools will pollute underground water.

Third, soil pollution caused by solid waste. Domestic waste is made of complicated components which are difficult to degrade, chemical medicines, heavy metals, etc. that will change the physical and chemical nature of soil and therefore affects the growth of crops.

Fourth, replacement of dry closets by flush toilets increases water consumption. With the increase of concentrated water supplies in rural areas, the amount of flush toilets will also increase. During our survey, most residents were found to be planning to build flush toilets. But due to lack of appropriate treatment techniques, flushing water will enter into the water cycling system.

Fifth, there are insufficient implementation criteria for sanitary systems, and comparatively loose management. There is no regular monitoring of water quality and sanitation in place; no criteria for wastewater disposal and discharge, no implementation of sanitary criteria for toilets and excreta disposal, and toilets are built mostly on a random basis. In addition to this, there is no regulation on toilet maintenance, collection and transportation of excreta, no systematic charging standard for drinking water and wastewater disposal with low or no charge, no regulations on charging and waste disposal for solid waste.

IMPLEMENTATION AND RESEARCH OF THE MODEL PROJECT

Considering the aforesaid situation, a village in Changping was chosen for the model project. Eco-san dry closet technology, low-investment and low energy consumption gray water disposal systems and oxygen static composting techniques were used in the project (Wang huan-sheng *et al.*, 2008). The application of eco-san dry closet technology is a treatment for rendering excreta harmless, and makes the use of treated excreta for agricultural purpose possible, and problems of wastewater discharge, transportation and disposal caused by flush toilets also thereby being avoided. The application of dry closet technology can make domestic wastewater into light concentration gray water that is easier to dissolve and which, after the treatment from a low-investment and low energy consumption wetland treatment system can reach B of Level I discharge standard of an urban wastewater treatment plant. Domestic organic waste and excreta can be used safely for agricultural purposes after composting. The remaining waste, with recyclable matters extracted, can be transported to Changping landfill. The major process is shown as follows:

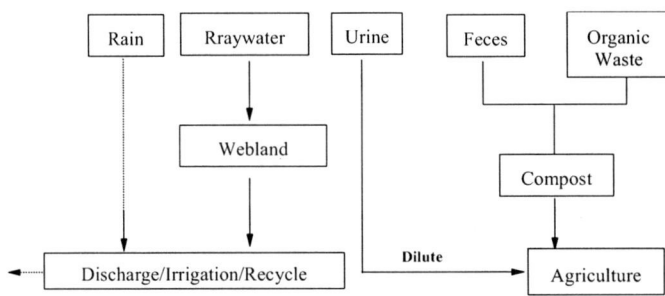

Figure 1 Eco-san System Process

Alteration of Eco-San Toilets

During the survey of housing layout, toilets and domestic wastewater of the village, there were 305 families to be planned for. The amount of families surveyed was 247 and there were 243 families who agreed to have their toilets altered. Of these, 110 families' toilets were suitable for alteration, and the remaining 133 families had new toilets built in their yards. The new type of dry toilets (Figure 2) built in the village reduced the sanitary threat of excreta and the amount of excreta and wastewater to be treated. Through dry toilets, feces are separated from urine and the separated urine is used in farmland or garden, and the separated feces, after composting, can be used as organic fertilizer.

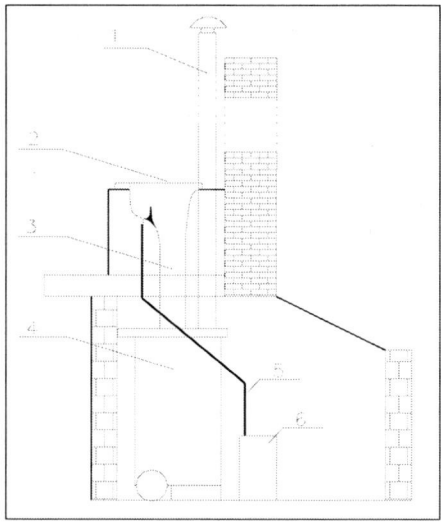

1 Vent-pipe
2 Toilets
3 Pipe
4 Storage tank
5 Urine pipe
6 Barrels of Urine

Figure 2 Outside the toilet and flushing toilet, that separates feces and urine

Domestic Wastewater Treatment System

According to the distribution of residents' houses and the drainage situation in the village, with the concept of combining separated and concentrated treatments, three types were used for wastewater treatment in this project: yard type, street type and concentrated type.

According to our analysis of survey results, concentrated treatment was used for 164 families; comparatively, separated concentrated treatment was used for 74 families divided into 4 sections: street treatment; separated family treatment, i.e. yard treatment was used for the remaining 55 families. The consideration of residents' distribution helped to save investment costs to solve the wastewater treatment problems of Nanzhuang Village.

Yard wetland system

Due to the fact that the houses are not built collectively in rural areas, a collective pipeline and treatment system is not suitable in this situation. The treatment suitable for rural residents is "on the spot" treatment, after the high-density black water has been separated from domestic wastewater, gray water, and after treatment of small wetland treatment system[3], will reach the standard for disposal. The following table is *Effects of small-wetland treatment.*

Table 2 Effect of small-wetland treatment

	Inlet water (mg/L)	Outlet water	Average removal rate	Remarks
COD	140~180	30–40	76%	Hydraulic Load 0.1~0.3m^3/(m^2·d)
SS	25~35	5–8	57%	
Ammonia nitrogen	10~30	0.3–0.8	43%	
TN	10–50	0.5–1	97%	
TP	1–5	0.1–0.35	85%	

Constructed street wetland system

As for residents that live comparatively closely, regional collection and concentrated on-site treatment can be applied to gray water, transported through a collection pipeline, treated at the regional concentrated treatment system. The treatment facility is constructed small-wetland as well. The quality and quantity of wastewater will be tempered at a water collection well to separate suspending matters whose density is higher than water, and then the tempered wastewater will be pumped into the street wetland system with a green landscape formed at the same time among streets and lanes.

Constructed concentrated wetland system

The surface land of the village is comparatively low, the area of the underground constructed wetland is 200m^2 with a hydraulic load of 20cm/d. Water plants planted in the wetland are, for the most part, wild race stem and bulrush. Landform and geographical features were fully taken into consideration in making designs to guarantee an average water distribution and compound flow effect, avoid no-flow areas, and reduce excavation quantity and damage to original water vegetation as much as possible.

An eco landscape system is formed from the original bottomlands and deposed pools and constructed concentrated wetlands. Such a system can be used for wastewater treatment and can also promote regional water circulation. Water filling the system can be use as landscape water or be recycled, making zero waste discharge possible in the region.

Waste Classifying Treatment System

Two kinds of garbage cans were set on each street of the village for inorganic waste (construction waste, dust and ash, metals, abandoned plastic bags and glass products) and organic waste (kitchen waste) respectively. Organic waste (including kitchen waste) took more than 50% of the waste. From then organic waste, food residuals and paper are easy to recycle, and can be used for agricultural purposes after being composted with excreta at an eco station.

The eco station was located in the neighbourhood of the concentrated constructed wetland treatment system, the major effects of which were providing organic accretion (sawdust, smashed leaves) and working as aq feces collection and storage tank. It composted dry feces (0.2kg/d·person) and organic kitchen waste (0.2kg/d·person) with the window composting method. The effective composting volume was 50m^2 and the structural size was $15 \times 4.8 \times 3.6$(h)m.

Building and Operation Cost

The building cost of the whole project was 922,000 Yuan, among which the cost for eco-san toilet alteration was 366,000 Yuan, wastewater treatment system 291,000 Yuan, channel system alteration project 150,000 Yuan, garbage cans 15,000 Yuan and landscape building costing 100,000 Yuan. The operation cost of the whole project was 40 Yuan/day.

PROBE INTO RURAL DOMESTIC POLLUTION TREATMENT MODEL

The above project design for the village was oriented at source reduction, recycling and full process control. The eco-san toilet can save water and reduce excreta pollution, thereby reducing the difficulty of wastewater treatment and guaranteeing the efficiency and economy of light concentration wastewater treatment by small wetland systems. The application of a waste classifying method reduces waste volume as well as transportation costs. The whole project, with the advantages of low investment, low operation costs and effective environmental benefits, is worthwhile promoting in rural areas.

Generally speaking, the traditional concentrated collection and treatment system used for urban pollutants have the disadvantages of high investment, high energy consumption, high maintenance requirements and high dependency on an uninterrupted energy supply, which makes it a weak system and not suitable for rural pollution control. Considering the features of small residing areas, dispersed distribution, poor infrastructure and low economic development, a concentrated collection and treatment method should not be used for control of rural domestic pollution, and a new sustainable model shall be developed to tackle rural domestic pollution.

REFERENCES

Huan-sheng W., Kai-jun, W. and Zhi-feng C. (2008). Investigation of domestic pollution in rural areas in China and discussion on its control mode. *Chinawater & Wastewater*, **24**(20), 20–22.

Wei-li Z., Shuxia W., Hongjie J. and Kolbe H. (2004). Assessment and control strategy for China non-point pollution I. *China Agricultural Science*, **37**(7), 1008–1017.

Current Status of Wastewater Technologies for Small Communities in JAPAN

N. Funamizu*

*Department of Environmental Engineering, Hokkaido University, Kita 13, Nishi 8, Sapporo 060-8628, Japan
E-mail: funamizu@eng.hokudai.ac.jp

Abstract 10% of people in Japan are using the Johkasou decentralized wastewater treatment system. The Johkasou system is a treatment unit for black and gray water. Developed in Japan, it has a compact treatment train with an anaerobic tank, an aerobic reaction basin, a solid liquid separation unit and a disinfection unit. Some Johkasou units are equipped with a membrane unit and can function as a membrane bio-reactor. Other Johkasou systems have an internal recycling system for biological nutrient removal. The maintenance requirements for the Johkasou system are specified by national regulation laws. Japan's night soil treatment plant also plays an essential role in the treatment and disposal of sludge.

Keywords Johkasou, Onsite wastewater management, Night soil treatment system

INTRODUCTION

It is reported that 74.9% of people in Japan are connected to a centralized sewerage system. Of these people, 8.8% are using a decentralized treatment unit (Ministry of Agriculture, Land and Transportation and Environment, 2008). The other 16.3% has its black water collected and treated at the night soil treatment plant. In large cities, where the population can be over half a million, almost everyone can connect their sewer pipe to the centralized sewage system. This is shown in Fig.1. In medium sized cities, with populations from 0.1 million to 0.5 million, the sewage coverage percentage is approximately 80%. Around 10 – 20% of people in Japan are using the so called "Johkasou" decentralized system. For small cities, particularly those with a population of less than 0.05million, the sewerage system is only available to around 35% of people. The main schemes are the decentralized "Johkasou" system, pit latrines and the night soil treatment system. The night soil treatment system is a treatment plant for black water that is collected from the pit latrines at each house. It also treats the collected sludge from the decentralized Johkasou units.

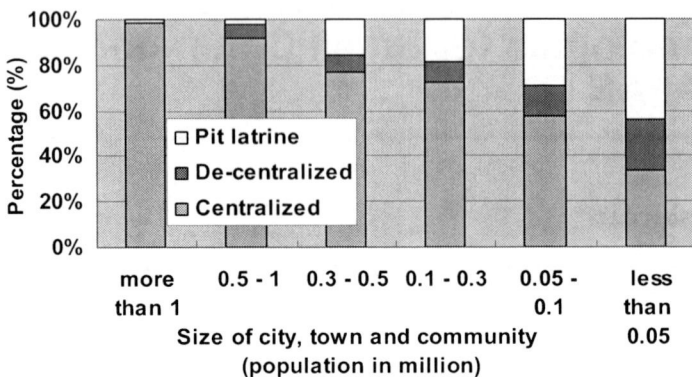

Figure 1 Sanitation systems used in Japan

Used since the 1960's, both the "Johkasou" and the night soil treatment plant are unique Japanese systems. In this paper, the descriptions of these systems are designed to introduce small Japanese communities to the current wastewater technologies that are employed in Japan.

SCHEMES FOR SMALL SCALE WASTEWATER MANAGEMENT IN JAPAN

Decentralized and Centralized Systems for Wastewater Management in Japan

When planning wastewater management, an indicator (or parameter) is essential for selecting which wastewater management scheme is to be applied. Whether decentralized or centralized, the type of system used should be chosen accordingly. Generally speaking, centralized systems are only economically feasible for areas with high population densities. In peri-urban and rural areas, decentralized systems have some distinct advantages. The indicator for the selection of wastewater system should reflect the local population density and geopolitical situation. Construction and subsequent maintenance costs should also be considered.

Japan has introduced an indicator for assessing which wastewater management scheme is to be applied; decentralized or centralized. The indicator is based on the "Critical distance of neighbouring houses". The critical distance is defined by the pipe length and the total costs it requires to construct and maintain (total construction cost as well as one year maintenance costs) of a decentralized system compared with that of a centralized system (Ministry of Environment, Johkasou web site). The local plans for regional wastewater management schemes are usually established by the local government.

Johkasou unit

In Japan, those who are not connected to a sewer pipe service are either using a Johkasou unit or Pit latrine. People who need water for flushing toilets are using a Johkasou. As shown in Fig. 2, there are two types of Johkasou. The standard type, which is very common in Japan, treats gray and black water. The advanced model, a newly developed Johkasou, includes a membrane separation unit for the separation of solids and liquids. It also contains an internal circulation system for nitrogen removal as well as a coagulation unit for phosphorous removal.

The standard Johkasou is a small wastewater treatment unit. It includes the following features: an anaerobic filter tank (flow equalization tank), a bio-filter (aerobic tank) and in some cases a solid/liquid separation unit. Finally, there is a disinfection tank. The anaerobic filter tank is designed for separating solid particles and reducing sludge levels through anaerobic digestion. To promote sedimentation, the anaerobic filter tank is generally divided into two chambers. There are also several types of biological reactors such as suspended growth type and particle bed type with plastic materials. Based on performance results, the standard Johkasou reduces BOD by 90% with less than 20mg/L of BOD in the effluent.

For the removal of BOD, Nitrogen and Phosphorous, an advanced Johkasou system has been developed to include a membrane separation unit. This advanced model has shown to reduce BOD by 97% with less than 5mg/L of BOD in the effluent. In addition to this, there is less than 20mg/L T-N, and less than 1mg/L T-P.

Johkasou systems have been developed for communities such as apartment complexes and clustered residential houses, as well as for just one individual house. The treatment schemes of these medium and large scale Johkasou systems include: a sequencing batch reactor system; a membrane bio-reactor system; and a bio-reactor with contact media.

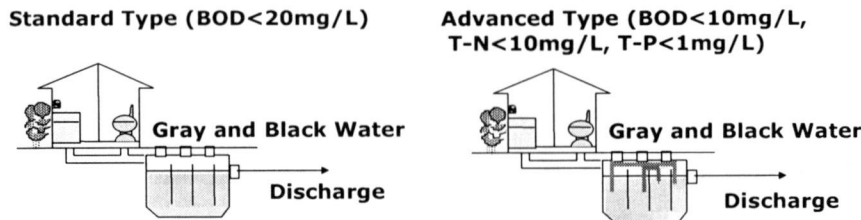

Figure 2 Two types of "Johkasou" in Japan (Adapted form Ebie *et al.*, 2006)

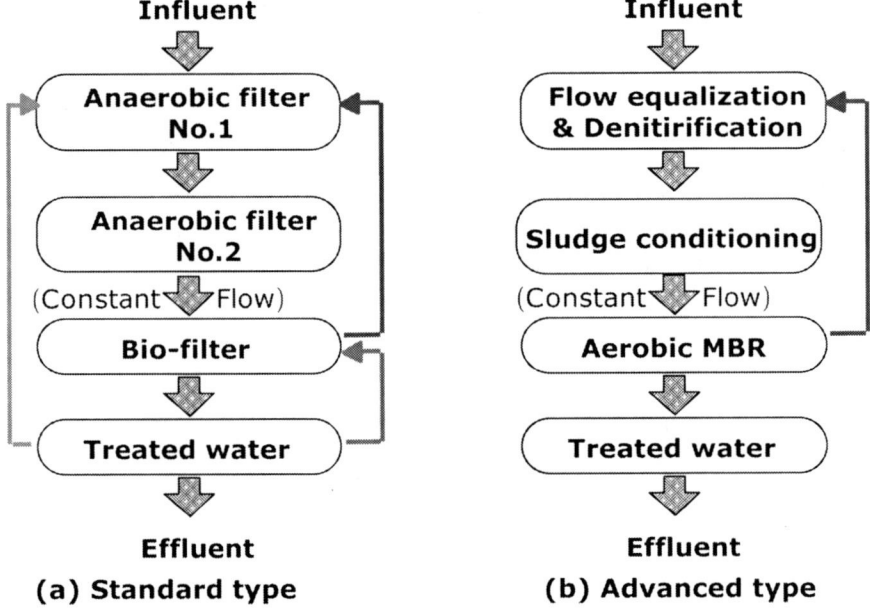

Figure 3 Typical flow chart of a standard type Johkasou (Adapted from Ebie *et al.*, 2006)

Institutional and Finance Framework for Implementation and Maintenance of a Johkasou Unit

With Johkasou units, there are two schemes of ownership: privately owned units and local municipality owned units. In municipality owned systems, instead of installing a communal sewer pipe and asking for a set fee from each household, the municipality install a Johkaso unit in each house. In order to protect the environment and aquatic eco-systems in water systems, wastewater should be managed at a regional and/or watershed scale. As mentioned above, the Johkasou unit is economically more feasible sewage system, particularly in peri-urban and rural areas. Installing a Johkasou at a regional scale is very effective in terms of both its economical and environmental impact. With new systems continuously developing in Japan, this point of view is now rapidly spreading. In order to promote the Johkasou system, it is important to have a good financial support system. National and local government subsidising systems, allow Johakasou users to pay only 10% of the cost. This is shown in Fig. 4. Since the standard cost of one Johkaso for a family (five persons) is about 900,000 JPY (9,000US$), paying 10% of the cost is generally accepted by users (Johkasou System Association, 2009).

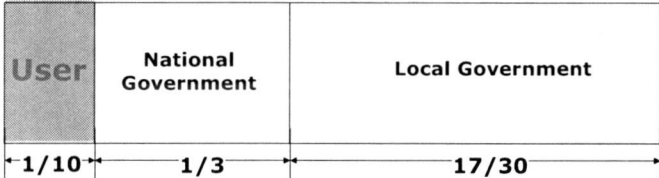

Figure 4 Distribution of cost among user, national and local government

It is well known that the maintenance of the decentralized system is essential. The national regulation law on Johkasou units requires each Johkasou owner to empty the sludge and report its effluent quality at least once a year. Often, the owner employs a maintenance company to empty the sludge and transports it to the night soil treatment plant. The effluent quality is inspected by the authorized organization. In the municipality owned system, the municipality assigns a maintenance company to ensure reliable maintenance. This is an obvious advantage to the municipality owned system.

Night soil treatment plant

The treatment and disposal of sludge withdrawn and collected from Johkasou systems is also important. A night soil treatment plant is used for this. Fig. 5 demonstrates how sludge is treated in Japan. Originally, night soil treatment plants were constructed for treating the black water from pit latrines in houses. However, now they are used for treating, both black water, from pit latrines, and sludge, from the Johkasou.

Figure 6 demonstrates the typical treatment process that occurs in a plant. The plant has a sludge concentration unit, a biological reaction basin, a membrane unit and an advanced treatment unit. The treatment goals of the plant are specified in terms of BOD, suspended solids, total nitrogen content and total phosphorus content. The biological reaction basin is optimized to treat nitrogen components by nitrification and de-nitrification processes. The excess sludge treatment train works as an energy recovery system. It consists of a pre-treatment unit, a methane fermentation unit and an electricity generator. The sludge treatment system can retrieve garbage and manure from livestock, both of which are valuable sources of organic energy. The technology in night soil treatment has continuously been developed and improved to meet the highest environmental requirements in Japan. With technology continuing to progress, the biological membrane component, has now been used in night soil treatment systems for the last 15 years.

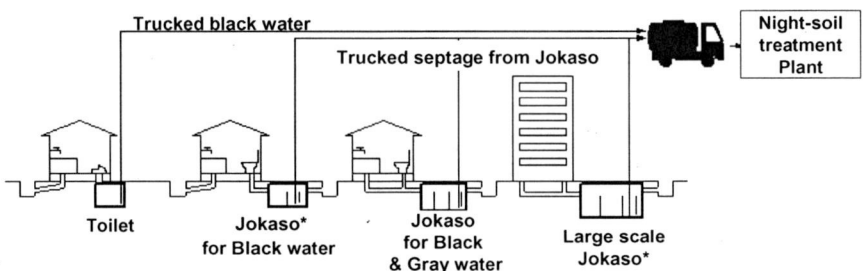

Figure 5 Toilet, Johkasou and Night-soil treatment plant. (Adapted from Ebara corp *et al.*, 2005.)

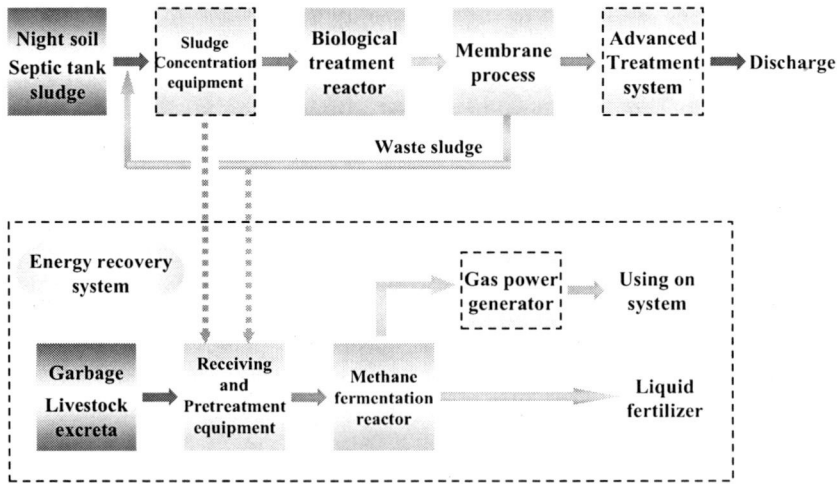

Figure 6 Night soil treatment system as a resource recovery system in Japan (Adapted from Ebara corp *et al.*,2005)

SUMMARY

Although 74.9% of people in Japan are connecting to a centralized sewerage system, about 11 million people (the 8.8% of total population) are using a decentralized treatment unit; the so called "Johkasou". The decentralized Johkasou system is fast becoming common practice for sanitation in small towns and cities in Japan. Sludge is collected and treated at night soil treatment plants and energy is later recovered. Technologies for supporting the Johkasou system, such as membrane technologies and night soil treatment plants, have been continuously developed and advanced to meet the highest environmental requirements in Japan. They are now at their most technologically advanced. These technologies, along with appropriate institutional and financial support, are essential in providing the implementation and maintenance of decentralized wastewater systems in Japan.

REFERENCES

Ebara corporation, KURITA Water Industries Corporation, KUBOTA Corporation, Ex Corporation, Kyoto University (2005). A Proposal for Human and Animal Waste Treatment and Recycling system (HAW-TRECYCLING).

Ebie Y., Itayama T., Inamori Y., Tanaka N. and Kuwabara Y. (2006). Development and Evaluation of Johkasou as a Decentralized Wastewater Management System, Presentation at the CREST Meeting, Sapporo, Japan.

Johkasou System Association (2009). Institutional system related to Johkasou unit, (Johkasou System Association, http://www.jsa02.or.jp/04gyosei/pdf/seido.pdf).

Ministry of Agriculture, Land and Transportation and Environment (2008). Report on coverage population of wastewater treatment in Japan. (Joukasou System Association http://www. jsa02.or.jp/04gyosei/pdf/h19_fukyu2.pdf).

Ministry of Environment. Effective implementation of domestic wastewater treatment facilities, Ministry of Environment, Johkasou web site, http://www.env.go.jp/recycle/jokaso/data/manual02.html.

Development and Practice of Ecological Sanitation System

G.N. Zheng[1] and W. Tu[2]

[1]Municipal Engineering Design Institute, Shanghai Xiandai Architectural Design (Group) Co, Ltd, 258 Road No. 2 Shimen, Shanghai 200041, China
Tel: 86-15821834838; E-mail: jack_53@sohu.com
[2]Huadong Hospital of Fudan University, 221 West Yan'an Road, Shanghai 200040, China

Abstract The basic idea of Ecological Sanitation (Eco San) is classification collection and source treatment of the sewage, so as to achieve a sustainable waste management mode which can accomplish the cycle of nutrients and the reuse of water. In this paper, the basic concept of Ecological Sanitation System was discussed, the water quality and the international up-to-date treatment technologies of the various types of domestic sewage were presented, including the yellowwater, the brownwater, the greywater and others. The article finally put forward suggestions and comments on the global practical application of Ecological Sanitation System.

Keywords Ecological Sanitation, source treatment, yellowwater, greywater, brownwater

INTRODUCTION

The primary mode of sewage disposal in cities today consists of the water-based sewage system whereby flushed water from toilets along with other household-related wastewater is transported for treatment to a WWTP. The fact that increasing water shortage has brought along restriction in the development of society, many scientists have started to ponder seriously on the conventional water discharge system.

Firstly, the conventional water discharge system uses water as a carrier of faecal matter and urine. It is estimated that each person uses up 15000L of water only to flush away 35kg of faecal matter and 500L of urine, thus wasting water which is exhaustible. Secondly, even though the amount of faecal matter and urine is not much, it contains 60% of the total amount of COD, >90% of the total amount of chlorine and phosphorus, >60–70% of the total potassium and most of the E. Coli. Today, purified water from traditional WWTPs still contains 20% nitrogen, >5% phosphorous and >90% potassium and this is the main reason for eutrophication. Finally, due to current limitations in technology, nutrients cannot be recycled and reused, therefore leading to previously agricultural land becoming infertile. The compensation of the lacking nutrients for the agricultural lands depends on mineral fertilizers produced by a large amount of energy and exhaustible resources, thus entailing a considerable amount of waste of natural resources. Above all, the conventional water discharge system is space-occupying, energy-consuming, and lacks the options for water and nutrient reuse.

Ecological Sanitation (EcoSan) System, which represents a holistic concept towards ecologically and economically sound sanitation, aims at that water and nutrient substance are transported in a sustainable, closed-loop cycle, so as to close (local) nutrient and water cycles with as less expenditure on material and energy as possible to contribute to a sustainable development. In this system, human excreta are treated as a resource instead of contaminations and are usually processed on-site and then treated off-site and the nutrients contained in excreta are then recycled by using them. It contributes to the zero-emission of waste and the sustainable development of our society.

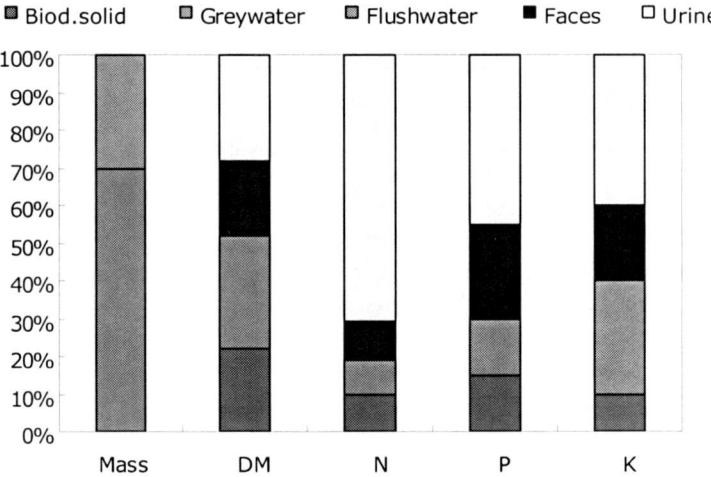

Figure 1 The distribution of mass, dry matter (DM), nitrogen (N), phosphorus (P) and potassium (K) in solid organic household waste and household waste water constituents (Vinnerås, 2001).

For the achievement of optimal recycling utilization of different kinds of materials in the wastewater, each kind of house wastewater, such as yellowwater, brownwater, greywater and blackwater should be collected and treated separately (Yellowwater is separately collected urine or urine containing a small amount of water used for flushing; brownwater is separately collected faecal matter or faecal matter containing a small amount of water used for flushing; greywater is the waste water without excreta, coming mainly from kitchens, bathrooms and laundry; blackwater is the waste water from the toilets, consisting of a mixture of urine and faeces). Approximately 70% of wastewater is greywater, whereby the percentage of nitrogen, phosphorous and potassium are 8%, 15% and 37% respectively (as shown in Figure 1); on the other hand, yellowwater and brownwater contain the highest proportion of nutrients. Therefore, if the amount of water used for flushing was to be reduced, and yellowwater and brownwater were to be collected at their sources or converted to natural resources, not only could water pollution be alleviated but nutrients would also be able to be recycled and reused. At the same time, greywater can be adequately treated using a small-volume processing equipment and discharged or reused.

DEVELOPMENT OF THE TECHNOLOGIES

Technology of Yellowwater

The separate collection and treatment of urine has attracted considerable attention in the engineering community in the last few years and is seen as a viable option for enhancing the flexibility of waste water treatment systems. The composition of stored urine from different collection systems is listed in Table 1. Data from a medical source (Maurer et al., 2006; last column) is given as a comparison for fresh urine.

Table 1 Concentration of urine from different collection systems

Parameter	Concentration					
Source	House-hold	School	Work-place	House-hold	Work-place	Fresh Urine
Dilution	0.33	0.33	0.26	0.75	1	1
PH	9.0	8.9	9.0	9.1	9.1	6.2
N_{tot} (gm^{-3})	1795	2610	1793	3631	9200	8830
$NH_4^+ + NH_3$ (gm^{-3})	1691	2499	1720	3576	8100	463
$NO_3^- + NO_2^-$ (gm^{-3})	0.06	0.07	–	<0.1	0	–
P_{tot} (gm^{-3})	210	200	76	313	540	800–2000
COD (gm^{-3})		1650			1000	–

Storage

Storage offers a possible way of reducing the potential health risks from faecal pathogens. Three storage parameters influence this process: storage time, temperature and pH. Hoglund *et al*. investigated the decay rates of bacterial and viral indicator organisms in stored urine collected from households. The authors concluded that "if stored at 20°C for at least 6 months, urine may be considered safe to use as a fertilizer for any crop" (Hoglund *et al*., 2002). Their experiments showed that temperature was the most crucial parameter influencing the inactivation rates.

Mineo Ikematsu proposed the electrochemical treatment of human urine to enable its storage without the accompanying unpleasant odor. In laboratory-scale experiments, the time-dependent variation in the pH of human urine, after the addition of urease, could be suppressed by chlorine produced via the electrochemical treatment of diluted human urine. Ureolysis was quantified by pH increase within 100h. This suppression occurred as a result of an irreversible change in the conformation of urease that resulted in its inactivation at an oxidation-reduction potential (ORP) of ca.240mV or above. Due to the electrochemical inactivation of urease during the entire storage period of urine, the hydrolysis of urea in urine, which results in the production of the unpleasant odor due to ammonia formation, can be avoided (Mineo *et al*., 2007).

Reverse osmosis

In reverse osmosis membranes, the retention of ammonium is better than its uncharged form (ammonia) and therefore the retention performance depends strongly on the pH. Researchers acidified stored urine to pH 7.1 in order to prevent permeation of ammonia. At a pressure of 50bar, a maximum concentration factor of 5 could be achieved, resulting in the following recoveries of nutrients in the retentate: ammonium: 70%; phosphate: 73%; potassium: 71% (Maurer *et al*., 2006).

Anammox process

Under anaerobic conditions, ammonium and nitrite are converted mainly to nitrogen gas. The formation of nitrite stops half way through the process, producing a 1:1 ammonium/nitrite solution. Udert *et al*. (2003) added this solution to anammox sludge from a pilot plant treating digester supernatant. At 30°C they measured a denitrification rate of $1000g_N$ $m^{-3}d^{-1}$ and the ratio of total ammonia to nitrite elimination was $1:1.18 \pm 0.07$. The results of these experiments show that nitrogen can be removed from source-separated urine with anammox. A combination of nitrification and anammox reactors could eliminate 75–85% of the nitrogen, leaving an ammonium nitrate solution (Udert *et al*., 2003).

Struvite(MgNH₄PO₄)

Magnesium ammonium phosphate ($MgNH_4PO_4H_2O$), also known as struvite, MAP or AMP, is an attractive precipitate because it conveys two dominant waste water nutrients in solid form. Ronteltap *et al.* investigated the conditional solubility product and the equilibrium reactions for urine (Mariska *et al.*, 2007). The simplified solubility product was determined with $[Mg] \cdot [NH_4^+ + NH_3] \cdot [Portho] = 10^{-7.6}M^3$ (pH=9 and a ionic strength=0.68), where [Mg] is the concentration of dissolved magnesium, $[NH_4^+ + NH_3]$ the measured ammonium+ammonia, and [Portho] the dissolved ortho-phosphate.

Technology of Brownwater

From the hygienic point of view, faeces should be considered to contain pathogens and should therefore be treated before their nutrient content is utilized. The most common faecal treatment to achieve this sanitization is composting.

Composting is a self-heating microbial aerobic degradation of organic wastes. During the process, carbon dioxide, water and heat are produced and oxygen is consumed. Microbial respiration during composting produces heat (Haug, 1993). The temperature varies within the composting mass, with higher temperatures in the central parts of the mass. To expose the material in the outer low temperature zones to higher temperatures, the material should be mixed (Haug, 1993;Epstein, 1997). To increase temperature, heat loss can be reduced by insulation (Haug, 1993; Epstein, 1997). With appropriate insulation, compost temperatures increase. When these higher temperatures (>50°C) are maintained for a sufficient period (1week), the compost is sanitized (Schnning and Stenstrm, 2004; WHO, 2006).

C. Niwagaba researched on composting of human faeces with food waste. Temperatures were monitored in three 78-L wooden compost reactors fed with faeces-to-food waste substrates (F:FW) in wet weight ratios of 1:0, 3:1 and 1:1, which were observed for approximately 20 days. To achieve temperatures higher than 15°C above ambient, insulation was required for the reactors. Use of 25 mm thick styrofoam insulation around the entire exterior of the compost reactors and turning of the compost twice a week resulted in sanitising temperatures (50°C) to be maintained for 8 days in the F:FW=1:1 compost and for 4 days in the F:FW=3:1 compost. In these composts, a reduction of >3log10 for E.coli and >4log10 for Enterococcus spp. was achieved. The F:FW=1:0 compost, which did not maintain 50°C for a sufficiently long period, was not sanitised, as the counts of E.coli and Enterococcus spp. increased between days 11 and 15 (Niwagaba *et al.*, 2009).

Ash and saw dust were also applied separately to source-separated faeces during the collection phase. The die-off of E. coli in the faeces/ash mixture was faster initially (first 7 days) compared to that achieved in the faeces/saw dust mixture even though the die-off achieved after 30°C 50 days was nearly similar for both mixtures. E. coli was not detected in faeces/ash after about 2 months, but was detected after 2 months in the faeces/saw dust mixture. Enterococcus spp. did not decrease below detection in faeces/ash or faeces/saw dust mixture but high numbers (difference of about 2logs) were detected at all times in faeces/saw dust than in faeces/ash mixture. The difference in the die-off in the mixtures of faeces/ash and faeces/saw dust was attributed to the differences in the characteristics of the additives, namely, high alkaline mineral content (giving high pH) and lower moisture content of ash compared to saw dust. It is recommended to increase use of ash as additive over saw dust in urine diversion dry toilets (Niwagaba et al., 2009).

Technology of Greywater

Greywater is generated as a result of the living habits of the people involved, the products used and the nature of the installation and, therefore, its characteristics are highly variable (Table 2). The kitchen greywater and the laundry greywater are higher in both organics and physical pollutants compared to the bathroom and the mixed greywater. All types of greywaters show good biodegradability in terms of the COD:BOD ratios (Li, 2009).

Table 2 The characteristics of greywater by different categories (Li et al., 2009)

	Bathroom	Laundry	Kitchen	Mixed
PH	6.4–8.1	7.1–10	5.9–7.4	6.3–8.1
TSS (mg/l)	7–505	68–465	134–1300	25–183
Turbidity (NTU)	44–375	50–444	298.0	29–375
COD(mg/l)	100–633	231–2950	26–2050	100–700
BOD (mg/l)	50–300	48–472	536–1460	47–466
TN (mg/l)	3.6–19.4	1.1–40.3	11.4–74	1.7–34.3
TP (mg/l)	0.11– >48.8	ND– >171	2.9– >74	0.11–22.8
Total coliforms (CFU/100ml)	$10–2.4\times10^7$	$200.5–7\times10^5$	$>2.4\times10^8$	$56\text{-}8.03\times10^7$
Faecal coliforms (CFU/100ml)	$0–3.4\times10^5$	$50–1.4\times10^3$	–	$0.1–1.5\times10^8$

Membrane technology

Cornelia Merz treated greywater with membrane technology. A 3L-lab-scale membrane bioreactor (MBR) treating the shower effluent from a sports club in Rabat, Morocco, was operated with a hollow fibre membrane for 137 consecutive days. Treatment in the MBR reduced the COD-load on average by 85% and a mean effluent concentration of 15mgL was obtained. Temperature and biomass concentration did not have a marked influence on removal performance. BOD_5, TKN, NH_4^+, Total phosphorus, Surfactants, Faecal coliforms concentration were reduced by 94%, 63%, 72%, 19%, 97% and 99% respectively (Merz *et al.*, 2007).

Constructed wetland

A. Gross (Gross *et al.*, 2007) conducted a research on treatment of greywater by constructed wetland. The system was based on a combination of vertical flow constructed wetland with water recycling and trickling filter, and was termed recycled vertical flow constructed wetland (RVFCW). The RVFCW was efficient at removing virtually all of the suspended solid sand biological oxygen demand, and about 80% of the chemical oxygen demand after 8h. Fecal coliforms dropped by three to four orders of magnitude from the irinitial concentration after 8h, but this was not always enough to meet current regulations forum limited irrigation. The treated greywater had no significant negative impact on plants or soil during their study period.

Coagulation processes

In a study lead by Pidou *et al.*, the coagulation processes and the magnetic ion exchange resin process were applied for shower greywater treatment (Pidou *et al.*, 2008). At optimal conditions, coagulation with aluminium salt reduced COD, BOD, turbidity, TN and PO_4^{3-} from 791 mg/L, 205 mg/L, 46.6NTU, 18 mg/L and 1.66 mg/L in the influent to 287 mg/L, 23 mg/L, 4.28NTU, 15.7 mg/L and 0.09 mg/L respectively. The total coliforms, the E. coli and the faecal enterococci in the reclaimed greywater are all less than 1/100 mL. Coagulation with ferric salt achieved similar treatment efficiencies as that obtained with aluminium salt. The coagulation processes in Pidou's study in 2008 were able to reduce the BOD concentration to less than 30 mg/L but fail to decrease the turbidity to less than 5NTU. The COD, BOD, turbidity, TN and PO_4^{3-} were decreased by the magnetic ion exchange resin to 272 mg/L, 33 mg/L, 8.14NTU, 15.3 mg/L and 0.91 mg/L respectively. The total coliforms, the E.coli and the faecal enterococci in the reclaimed greywater are 59/100 mL, 8/100 mL and less than 1/100 mL. The magnetic ion exchange resin process failed to reduce the turbidity and the BOD to the levels required for both unrestricted and restricted reuses. The coagulation process and the magneticion exchange resin process have minor effects on the removals of both TN and PO_4^{3-}.

Up flow anaerobic sludge blanket

Tarek A. Elmitwalli conducted a research on greywater treatment by an up flow anaerobic sludge blanket (UASB) reactor (Elmitwalli *et al.*, 2007). High values of maximum anaerobic biodegradability (76%) and maximum COD removal in the UASB reactor (84%) were achieved. The results showed that the colloidal COD had the highest maximum anaerobic biodegradability (86%) and the suspended and dissolved COD had similar maximum anaerobic biodegradability of 70%. Furthermore, the results of the UASB reactor demonstrated that a total COD removal of 52–64% was obtained at HRT between 6 and 16h. The UASB reactor removed 22–30% and 15–21% of total nitrogen and total phosphorous the greywater, respectively, mainly due to the removal of particulate nutrients. The characteristics of the sludge in the UASB reactor confirmed that the reactor had a stable performance. The minimum sludge residence time and the maximum specific methanogenic activity of the sludge ranged between 27 and 93 days and 0.18 and 0.28kgCOD/(kgVSd).

CASE STUDY

Naoko proposed one of the new integrated waste treatment systems: an "sustainable sanitation system" that includes separation of the black water from water system by a non-flushing toilet (bio-toilet), and a gray water treatment based on a biological and ecological concept (Naoko et al., 2006). Sustainable sanitation system also converts the domestic waste to soil conditioners and fertilizers, for farmland use. The availability of this system was investigated by analyzing the sawdust used in the bio-toilet and the quality of the effluent in the household wastewater treatment facility. As the result, the water content of the sawdust did not exceed 60% in any of the sampling points and the BOD and COD of the effluent of the household wastewater treatment facility were below 10 and 20 mg/L respectively, due to the low loading. Compared to the pollution load on the water environment created by the conventional system, it was found that the effluent of the house has a lower load than the tertiary treatment and the volume of the water consumption is 75% of the conventional system.

Source separation in a housing estate is realized firstly in Germany at "Lubeck-Flintenbreite", for 350 inhabitants in a densely populated rural area (Günter et al., 2005). The installed system comprises a strict separation of blackwater, greywater and stormwater. The treatment of stormwater and greywater takes place in swales respectively in constructed wetlands. It was planned but up to now not implemented that blackwater together with organic waste should be treated an aerobically (producing biogas for energy and heat production).

The vacuum toilet system has been running for 2 years with only minor technical problems. The flushing system which has been optimised during operation needs only about 0.7L per flush. The daily mean drinking water consumption of 77L per person therefore is significantly low compared to the German average 129L. Peaks in spring and summer time are caused by garden irrigation. After a time of accustoming the vacuum toilets are accepted and are seen more hygienic than conventional flushing toilets.

SUGGESTIONS AND COMMENTS

- Following the present water resources crisis in our country, Ecological Sanitation (EcoSan) System can simultaneously achieve the recycling utilization of the water resources and nutrients, which is of strategic significance on the exchange and recycling of the water resources and materials between urban and rural areas.
- The effective source separation and treatment technology is the basis for the implementation of Ecological Sanitation (EcoSan) System. The progressive development of wastes recycling technology provides technical supports for controlling health risks and reducing treatment costs.
- The practice of Ecological Sanitation (EcoSan) System should be adapted to local conditions. It should be carried out as demonstration projects in water-scarce urban regions where sewage treatment systems are inadequate, and then gradually expanded.
- The management, maintenance and oversight mechanisms of Ecological Sanitation (EcoSan) System should be explored. We should implement multi-faceted, flexible management approach from different aspects, such as public education and participation, availability of funds, processing technology etc.

REFERENCES

Elmitwalli T. A. and Otterpohl R. (2007). Anaerobic biodegradability and treatment of greywater in upflow anaerobic sludge blanket (UASB) reactor. *Water Research*, **41**(6), 1379–1387.

Gross A., Shmueli O., Ronen Z. and Raveh E. (2007). Recycled vertical flow constructed wetland (RVFCW)-a novel method of recycling greywater for irrigation in small communities and households. *Chemosphere*, **66**(5), 916–923.

Hoglund C., Stenstrom T. A. and Ashbolt N. (2002). Microbial risk assessment of sourceseparated urine used in agriculture. *Waste Management & Research*, **20**(2), 150–161.

Ikematsu M., Kaneda K., Iseki M. and Yasuda M. (2007). Electrochemical treatment of human urine for its storage and reuse as flush water. *Science of theTotal Environment*, **382**(1), 159–164.

Langergraber G. and Muellegger E. (2005). Ecological sanitation—a way to solve global sanitation problems? *Environment International*, **31**(3), 433–444.

Li F. Y., Wichmann K. and Otterpohl R. (2009). Review of the technological approaches for greywater treatment and reuses. *Science of the Total Environment*, **407**(11), 3439–3449.

Mariska, R., Max, M. and Willi G. (2007). Struvite precipitation thermodynamics in source-separated urine. *Water Research*, **41**(5), 977–984.

Maurer M., Pronk W. and Larsen T.A. (2006). Treatment processes for source-separated urine. *Water Reseach*, **40**(17), 3151–3166.

Merz C., Scheumann R., Hamouri B. E. and Kraume M. (2007). Membrane bioreactor technology for the treatment of greywater from a sports and leisure club. *Desalination*, **215**(1–3), 37–43.

Nakagawa N., Otaki M., Miura S., Hamasuna H. and Ishizaki K. (2006). Field survey of a sustainable sanitation system in a residential house. *Journal of Environmental Sciences*, **18**(6), 1088–1093.

Niwagaba C., Kulabako R. N., Mugala P. and Jönsson H. (2009). Comparing microbial die-off in separately collected faeces with ash and sawdust additives. *Waste Management*, **29**(7), 2214–2219.

Niwagaba C., Nalubega M., Vinnerås B., Sundberg C. and Jönsson H. (2009). Benchscale composting of source-separated human faeces for sanitation. *Waste Management*, **29**(2), 585–589.

Pidou M., Avery L., Stephenson T., Jerey P., Parsons S. A., Liu S. M., Memon F. A. and Jeerson B. (2008). Chemical solutions for greywater recycling. *Chemosphere*, **71**(1), 147–155.

Udert K. M., Fux C., Münster M., Larsen T. A., Siegrist H. and Guje W. (2003). Nitrification and autotrophic denitrification of source-separated urine. *Water Science and Technology*, **48**(1), 119–130.

Vinnerås B. (2001). Faecal separation and urine diversion for nutrient management of household biodegradable waste and waste water. *Department of Agricultural Engineering*, Report 245 SLU, Uppsala.

Assessing the Sustainability of Innovations in Urban Ecological Sanitation: *Erdos Eco-town Project*

J. McConville[a], A. Rosemarin[a], Z. Li[b] and A. Flores[c]

[a]Stockholm Environment Institute, Kräftriket 2B, SE-10691 Stockholm, Sweden
Tel: +46 73 707 8605, Fax: +46 8 674 7020
E-mail: jennifer.mcconville@sei.se, arno.rosemarin@sei.se
[b]University of Science and Technology, Haidian District, Beijing 100083, China
E-mail: zifulee@yahoo.com.cn
[c]Centre for Sustainable Development, Department of Engineering, University of Cambridge,
St. John's College, Cambridge CB2 1TP, United Kingdom
E-mail: aeflores@alum.mit.edu

Abstract The China-Sweden Erdos Eco-town Project was a first bold attempt to install dry urine-diverting sanitation systems in new multi-story buildings at a relatively large-scale level of implementation. The project has been in some ways a "living laboratory" for the ca 3000 inhabitants housed in the 832 apartments in 42 buildings. The design of the system was part of a vision for a sustainable urban future with improved efficiency in water consumption and reuse of nutrient resources. However, to achieve sustainability the various system components including technical, institutional and social aspects, need further improvements. This paper critically analyses the sustainability of the project and offers lessons for future implementation of more sustainable sanitation systems.

Keywords Eco-Sanitation, Multi-Story, Sustainability, Urban

INTRODUCTION

The concept of eco-towns is becoming increasingly popular as the world struggles to meet environmental challenges and reduce the human footprint on the Earth. Although there are many new designs and pilot projects for sanitation that are testing ways to save water, recycle nutrients and/or save energy few attempts have been made to introduce these in the urban setting The technical, institution and social challenges are large and there is much resistance to innovation among both users and sector professionals. The Sustainable Sanitation Alliance is trying to improve the situation and has worked out a definition for a sustainable sanitation system as one that is *"economically viable, socially acceptable, and technically and institutionally appropriate, and it protects the environment and natural resources"* (SuSanA, 2008). The different criteria outlined in this definition can be a starting point for assessing specific cases and furthering learning for improving the sustainability of urban sanitation.

This paper presents one such case from northern China where a dry toilet solution was eventually changed to waterborne sewerage after three years of operation. It takes a critical look at the original design and subsequent operations, particularly from how the different social, institutional, economical, technical, health and environment perspectives can affect the objectives for an ecologically friendly and sustainable sanitation system.

ERDOS ECO-TOWN PROJECT

The China-Sweden Erdos Eco-Town Project (EETP) was initiated in 2003 by the Dongsheng District Government and Stockholm Environment Institute (SEI) as part of the EcoSanRes Program. The project was meant to provide opportunities for meeting high environmental ambitions and conserving water through implementing a dry ecological sanitation system in this semi-arid region. Dongsheng City is located in the south-western part of Inner Mongolia on the central part of the Erdos plateau. This part of China receives rainfall of 300–400 mm per year. At the time the project was initiated, about one-third of the Dongsheng City was forced to ration water.

The EETP is an ambitious attempt to provide innovation to the urban sanitation sector. Specific objectives for the project were to provide dry, ecological sanitation solutions for human excreta and greywater management in multi-story building using onsite facilities to increase recycling of waste products and reduce water consumption. Completed in 2006, the eco-town consisted of 832 apartments in 42 buildings as a new housing development with onsite greywater treatment in addition to the collection of urine and feces fractions and onsite composting.

System Design

The design of the Erdos Eco-Town Project (EETP) was meant to showcase sophisticated dry, urine diverting sanitation, a separate greywater system and organic and solid waste facilities in an urban environment. The working concept behind the system design is that the separation of waste streams will increase the efficiency of treatment and facilitate the recycling process. For example, the majority of harmful bacteria and viruses in human excreta are found in feces, whereas urine can be safely handled after a relatively short storage period (WHO, 2006). Thus, separation of urine can significantly reduce the volume and nutrient load on the treatment facilities, as well as facilitate the recycling of nutrients through use of urine as a liquid fertilizer. This system design focused on the separation of three main waste streams: feces, urine, and greywater (Figure 1).

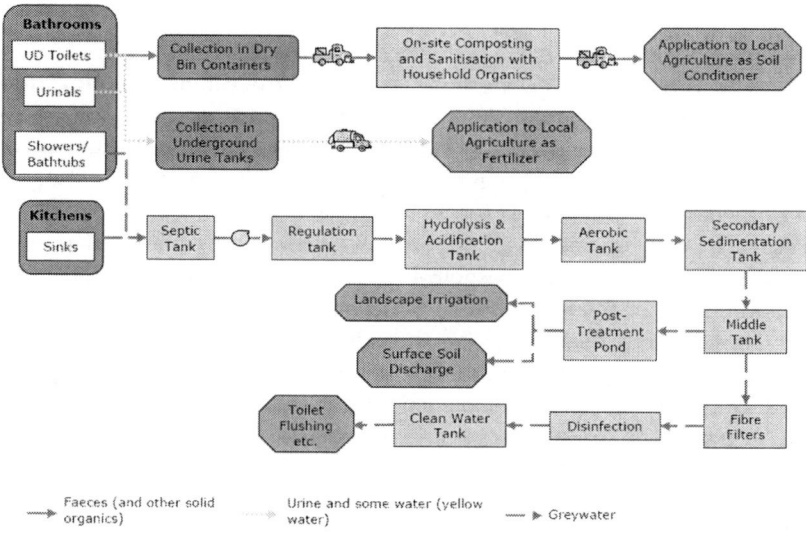

Faeces (and other solid organics) ·····> Urine and some water (yellow water) ——▸ Greywater

Figure 1 Diagram showing flow streams from urine, faces and greywater in the EETP (diagram by Flores)

The urine and feces flow streams start at the urine-diverting dry toilets and urinal which divert the liquid and solid excreta from each other. A turning bowl feces receptacle in the dry toilet is connected to a chute so that after use the feces drop vertically down into the basement where it is collected in bins. Sawdust is manually added to the feces after each toilet use in order to keep the contents of the bins dry and reduce odor. The feces collection bins are connected to a ventilation system that is meant to vent odor from the bins to vent pipes on the roof. The bins are collected in a truck by the maintenance workers and taken to the on-site indoor thermal compost plant where the feces are processed into an organic fertilizer product for agricultural application. The pipes from the urine hole in the toilet and the urinal run to underground urine tanks where the urine is collected and stored. A pump-truck is used to empty the urine tanks and the stored urine can be applied as a fertilizer to local agriculture.

The greywater in the system comes from the kitchen sinks, showers and bathtubs. It flows into a separate pipe system to an on-site greywater treatment plant with a capacity of 80L/person/day. The plant uses primary sedimentation, anaerobic treatment, activated sludge and aerobic bio-film treatment, secondary sedimentation and a holding pond (3700 m³) to treat the water to the National Code for Grade II effluent standards.

System Operation

This project was managed and implemented by a unique private-public partnership between the Government of Dongsheng, and a private developer. The Government managed the project with collaborative supervisory and advisory support from SEI, while the developer agreed to implement an innovative sanitation approach in a real-estate business venture in return for access to the land area for the project and a promised government tax rebate incentive.

The day-to-day operations of the eco-town were run by a maintenance team located on-site. The team was responsible for cleaning the basements, emptying/ transporting the fecal bins, operating the compost plant and emptying the urine tanks. The feces were treated at the on-site indoor thermal ($50°C$) batch composting plant. However, many households threw solid waste into the toilets so that it was necessary to install an additional manual separation of solids from organics in the compost plant. The resulting organic fertilizer was distributed to local organic farmers who used the product in their production. The local government decided against using the urine on urban green areas due to potential odor problems near inhabited buildings, so most of the urine was trucked to the city landfill where it helped fuel the breakdown of organic waste.

The project office operated a telephone hot-line so that households could call if there were problems with the dry toilet. The maintenance team was responsible for responding to these calls and repairing the toilets at the household level. In general, the majority of the complaints from households had to do with failure of mechanical parts in the toilet, odor from the ventilation system, broken fans in the ventilation system, and crystallization of urine in odor locks.

Despite a competent servicing system for the toilet, the majority of households remained dissatisfied with having to use these toilets. During household surveys, most people responded that the dry system was inconvenient and that they preferred flush toilets (Flores *et al.*, 2009). They complained about odor problems and that it was complicated to clean. In 2008, households began to boycott paying the service fees for the maintenance and operation of the buildings. In December of 2008, a household committee was formed to defend the legal rights of the households and to improve management of the eco-town. The committee eventually decided that the dry toilets should be changed to water flush models. They complained to the project office government representatives for the project, and even the mayor. In July of 2009, with the termination of the SEI project contract in sight, the government decided to switch the system to flush toilets. The system has retained the greywater pipe system and onsite treatment plant. The urine tanks are being used as sedimentation tanks and the capacity of the treatment plant is being increased. Sludge collected is being composted onsite.

A number of dry toilets have been retained in order to further develop the toilet design and ventilation system.

A SUSTAINABLE SYSTEM?

Achieving a sustainable sanitation system in the urban setting depends on successfully addressing several key aspects. According to the Sustainable Sanitation Alliance (SuSanA, 2008), a sustainable sanitation system must satisfy the following criteria:

- **Health and hygiene:** reducing the risk of exposure to pathogens and hazardous substances that could affect public health.
- **Environment and natural resources:** involves the required energy, water and other natural resources for construction, operation and maintenance of the system, potential emissions to the environment and the degree of recycling and reuse.
- **Technology and operation:** incorporates the functionality, robustness, flexibility to change, and ease of use of the system.
- **Financial and economic issues:** relate to the capacity of households and communities to pay for the installation and operation of the system, as well as a cost-benefit assessment.
- **Socio-cultural and institutional aspects:** these criteria ensure that the system is socially acceptable, as well as the existence of a supporting institutional framework, including legal regulations and distributions of responsibilities.

Extensive monitoring data have been collected during operation of the EETP regarding user satisfaction, system performance, material flows, nutrient production and operation costs. The following sections take a critical look at how the project met these criteria for sustainability.

Health and Environmental Impacts

The primary objective of a sanitation system is to protect and promote human health by preventing the cycle of disease. Assessment of the sustainability of a sanitation system from the health perspective means determining its ability to reduce the risk of exposure to pathogens and hazardous substances that could affect public health (SuSanA, 2008). The treatment measures within the system; greywater treatment with activated sludge, urine storage and thermal composting at 50–60°C, all result in elimination of bacterial pathogens at the end of the pipe. However, from a risk perspective, one particular area of concern for the EETP dry system is the handling of fecal matter by the maintenance personnel during collection and transporting of the waste. Although they are covered in protective clothing and masks there is ample opportunity for contaminating surfaces with raw fecal material. Furthermore, improper use of the bins by residents pouring water down the fecal chutes results in heavy water-filled bins that need to be manually cleaned by personnel. The fecal handling procedure needs to be improved, possibly with more mechanization, to further reduce risk exposure.

The other main objective with a sanitation system is to protect the environment. Environmental protection demands consideration of the emissions discharged from the sanitation system during use, but also the amount of energy, water and other natural resources required for construction, operation and maintenance of the system, The degree of recycling and reuse practiced can be seen as positive environmental impacts. The results of a comprehensive life-cycle assessment (LCA) of the system sustainability comparing it to a conventional sewerage system indicates the potential of dry systems to contribute to reduced water consumption, the recovery of valuable resources from domestic wastewater, and reduced eutrophication (Flores *et al.*, 2009). However, there is some critique that the dry system consumes greater amounts of natural resources compared to the conventional waterborne sanitation system. This is in large part due to the conventional system's centralized and larger-scale design, allowing it to benefit from economies of scale in both construction and operation. In addition, the current design of the dry system is quite material-intensive because of the basements required for fecal collection and storage. The dry system's transport requirements are also quite energy-intensive as the receiving farms for compost and treated urine are several kilometers away. Further efforts are needed to improve efficiency in material consumption and reduce transportation so that the system can be more competitive and sustainable.

Socio-Cultural and Institutional Aspects

The socio-cultural and institutional aspects cover a broad range of factors that can have significant impacts on sustainability, yet at the same time are harder to quantify and account for in planning. This study looks at these aspects from two angles; institutional management and social acceptance.

Institutional Management

The public-private partnership set up in the beginning of this project was a new approach to institutional cooperation between the municipality, a commercial developer and a research institute. It was recognized that the management of an onsite system such as the EETP, required different model from the prevailing centralized, government-managed model. However, there are few precedents or existing legislation to control such agreements and there were many conflicts over the quality of the workmanship, lack of inspection, responsibilities for retrofitting, and compensation for the developer (e.g. they had been promised tax rebates that the government actually was not in a position to give). In reality, the institutional model was still strongly government controlled.

The original idea was to set up a non-profit holding company in order to provide a long-term operations and maintenance facility. The idea was that income could be derived from sales of water, and urine and compost as fertilizers. However, after construction this option was not deemed feasible and instead the government maintains responsibility for operations and maintenance of the sewage treatment and compost plants. They justify the system based on the reduced use of drinking water, zero sewage loading to the centralized municipal plant, no need to install large collector pipes, and reduced solid waste load. However, the government did not show interest in exploiting potential markets for the reuse products, mainly because the volumes have been limited. A major bone of contention in the dry system was the cost of collection and maintenance which the local government did not want to take on itself. Further work is needed to develop new regulations and management models that could promote the end-use of sanitation products, improve quality control, and improve linkages between the sanitation and agriculture sectors.

Social Acceptance

The weakest link in this project by far was household acceptance and as a result the sustainability of the solutions was always in question because of user resistance. A common comment from households was that the toilets were awkward to use and explaining their function to visiting family relatives and friends was considered an embarrassment and unnecessary burden. Other complaints centered around the use of sawdust as a desiccant while defecating which caused the upwelling of dust while sitting, as well as the difficulty for children and elderly to operate the system unassisted on their own. However, by far the most important problem in this project is the presence of odor. For several reasons relating to both poor building quality but also inadequate design there was often persistent occurrence of unacceptable odor. After 2–3 years of use, most of the households were of the opinion that the dry toilets as installed were not acceptable in the long term and further improvements would be necessary if the dry system was to be retained.

Even though the project has been continually retrofitted and upgraded to fix technical problems, many households are dissatisfied. After a few years there was a strong voice from the users committee to change the entire system to conventional flush toilets. They were frustrated with continued malfunctions of the system, but it also negatively affected their social life as they are ashamed to have guests using the "strange" toilet.

Although when interviewed, households acknowledged that the dry system could save valuable water, and reduce pollution load, they also retained the attitude that dry toilets were something for rural communities and not for modern urban multi-story apartments. In a survey conducted by Flores in spring of 2009, 30% of households appreciated that the system saves water but only 5% knew that it reduced pollution and was environmentally friendly. This indicates that they did not value the system or understand the reasons that it was installed. They merely saw it as an inconvenience and desired change. Only 8% of those interviewed said they would recommend the system.

Of course, training and management of households have been a central concern since household level maintenance and proper use are critical for the overall functioning of the system. For example, many users tended to use the dry toilet as a solid waste bin for disposal of household items and kitchen wastes. However, even with targeted information campaigns and demonstration, there was still a need to install a pre-composting rotating cage filter device to remove non-degradable items. In general, the behavioral change needed to shift to dry sanitation was a slow process for a segment of the population while others adapted quickly. In part the problem was a lack of understanding by the design team of who would be using the dry system (many had flush systems prior to moving to the eco-town), and in part there was a lack of education on the need for such a system. The urbanites that purchased these apartments were not prepared to spend time on these unfamiliar household tasks and would prefer to retain a "flush and forget" mentality without thinking of environmental consequences.

Financial Aspects

Financial and economic issues relate to the capacity of households and communities to pay for sanitation, including the construction, operation, maintenance and necessary reinvestments in the system. It also includes costs such as environmental pollution and health hazards, and benefits such as increased agricultural productivity and subsistence economy, employment creation, improved health and reduced environmental risks.

A cost-benefit study was carried out by Tsinghua University (Zhou et al., 2007) comparing the dry system with a waterborne conventional system of similar size in Dongsheng. According to the results of Zhou et al. (2007), the capital cost of the dry system is 2.2 times greater than that of the wet system: US$480 versus US$217 per capita. The difference in capital cost is mainly driven by the high cost of construction of the basement to house the dry system's fecal collection system. The results from Tsinghua University also show that the O&M cost of the dry system is 2.5 times greater than that of the wet system: US$26 versus US$11 per capita per year. These values do not include actual and potential incomes generated from the sales of recovered resources such as urine, compost, and water from the dry system and water from the wet system. If an economic value could be given to organic fertilizer produced in the dry system, then these O&M costs could be reversed. The main economic savings under present cost conditions are the water savings which amounted to about 30% less water used compared to the conventional system. Assuming the saved water is used for industrial purposes instead, the value of the saved water is relatively high.

The benefits to environmental protection, human waste recycling and resource conservation over the long-term would be much more significant than the added initial cost today. The future of urbanization especially in arid zones of the world depends on how efficient the limited water resources are used. Also innovative sanitation systems can be integrated with the solid waste and agricultural sectors in order to produce positive synergies. This project has made a major contribution towards these directions.

Technical Aspects

Technology and operation aspects incorporates the functionality (ability to meet standards) and the ease with which the entire system (the collection, transport, treatment and reuse and/or final disposal) can be constructed, operated and monitored by the local community and/or the technical teams of the local utilities. Furthermore, the robustness of the system; its vulnerability (towards power cuts, water shortages, floods, earthquakes etc.); and the flexibility and adaptability of its technical elements to the existing infrastructure and to demographic and socio-economic developments are important aspects.

The ability of the EETP dry system to meet treatment standards was rated as neutral in an LCA study (Flores *et al.*, 2009). This rating is due to the EETP being unable to meet the greywater discharge standards consistently. Testing of the EETP greywater treatment system in 2007 indicated that it generally met Total-P and NH_3-N requirements, but could not meet COD consistently (Zhu, 2008). Much of this can be explained by the exceptionally low use of water by the households and therefore low dilution of the greywater (Chinese discharge standards are based on concentrations).

For the ease of use perspective, the dry sanitation system faced a number of operation and maintenance challenges. In general, it is a more complex system to operate and maintain because of the separate waste streams involved. A survey of 100 households in April and May 2009 (Flores, 2009) found that the average user response to the question "How convenient/easy is the dry toilet to use?" was 2, on a scale of 1 to 9, with 1 being "very inconvenient/difficult". The inconvenience of the dry toilets was primarily associated with the need to separate streams of urine and faces, the use of sawdust, and the difficulty of maintaining them clean. From the maintenance team perspective, the urine collection and disposal system and the fecal collection, treatment, and disposal system are fairly easy to operate and maintain. The procedures are not technically complex, although they are somewhat labor-intensive and unpleasant in the case of the fecal management system. The operation and maintenance of the greywater treatment plant requires a skilled worker, but operation of the entire system is well within the capacity of the local utilities to provide.

The dry sanitation technology for multi-story buildings is not standard and off-the-shelf, nor are there building norms for ventilation of such systems. A major challenge in this project was the lack of interest from the developer and the local government to fully test and approve the technology before scaling up. The market pressure, speed of urbanization, poor workmanship and lack of inspection all led to a weakening of the final product provided to the consumers. The technology was not designed to be able to cope with inadequate plumbing and building skills and poor building materials. The toilet prototypes performed well in the lab since they were built using stainless steel parts and proper pipe fittings. But when it came to scaling up, many hidden flaws were only detected after a few years of operation. The design of the system could not take into account the unanticipated occurrence of such poor workmanship and lack of building inspection.

The project has been a source of much learning both in the development and testing of prototypes and also full-scale implementation and maintenance. In particular, modifications in the urine-diverting toilet and odor traps have been tested. Ventilation of the feces bins has also undergone development including placing the bins in sealed cabinets and improved basement pipe and fan alignments venting to the roof level. Many corrections were added when it was discovered that the original installations were faulty and hadn't followed the blueprint designs. Considerable experience has been gained in the area of indoor air pressure management since toilet ventilation was directly affected by kitchen and bathroom fans and opening of windows as well as changes in outdoor air pressure. Successful ventilation improvements were introduced by using a urine-diverting dry toilet manufactured by Separett with its own small fan (9 W) unit venting upwards to the roof and not down to the basement. These toilets were also modified to be connected to the basement bins. They have been running for several months, are well accepted by the households and represent the present state of the art for dry toilets in multi-story buildings.

The dry system does appear to be adaptable since there have been successful up-grades of toilet models such as the Separett, and the shift to flush toilets was not overly complicated and took about 3–4 months for the ca 800 flats. The current decentralized sewage system at EETP is in itself a novelty since it uses the small pipes and has onsite treatment. However, there may be questions regarding its vulnerability to water shortages. There may come a day when the vulnerability of the waterborne system will push decision-makers to again consider dry sanitation options in Dongsheng City.

CONCLUSIONS

Site-specific factors such as local culture and policies, proximity to agriculture, existing sanitation infrastructure, and water supply availability play a major role in the success of innovative sanitation projects. The future of urbanization and urban agriculture especially in arid and semi-arid zones of the world depends on how efficient the limited water and nutrient resources are used. This project has made a major contribution in exploring what levels of innovation are possible.

The results of the sustainability discussion above show the potential of dry systems to contribute to reducing water consumption, the recovery of valuable resources from domestic wastewater, and reduced eutrophication. The first two benefits are particularly relevant in the context of the Dongsheng District, which suffers from water shortage and poor-quality soils. However, the dry system proved to be unsustainable in this context for a number of reasons. Technological problems with odor control were a major factor in the removal of the dry toilet system. Improvements of the ventilation system, toilet design, and fecal management will also contribute to improved sustainability from both a technical and a societal perspective. The dry sanitation technology is less mature than the waterborne system and needs further development of management models, quality control, supporting regulation, and standards for the plumbing and building sectors before the institutional aspects are fully sustainable. In addition, this project highlights the need for education and that developing a local interest in the project is critical. Successful multi-story dry sanitation systems in Germany and Sweden have shown that the families in these buildings tend to be environmentally interested and therefore are willing to take on more responsibility for their system (Berger, 2003; Kvarnstrom et al., 2006). Innovative sanitation solutions need a better marketing strategy to sell the advantages of these systems to the households as well as the authorities. All stakeholders should see how these systems contribute to reduced water consumption and the recovery of valuable resources, as well as improving their lifestyles. Making urban sanitation more sustainable remains a major challenge of balancing several complex and critical criteria, but by learning lessons from each innovation progress is being made.

REFERENCES

Berger W. (2003). *Results in the Use and Practice of Composting Toilets in Multi Story Houses in Bielefeld and Rostock, Germany*, 1st International Dry Toilet Conference, University of Tampere, Finland, 20th–23rd August 2003.

Flores A. (2009). Unpublished data – household survey results from April and May 2009.

Flores A., Rosemarin A. and Fenner R. (2009). Evaluating the sustainability of an innovative dry sanitation (Ecosan) system in China as compared to a conventional waterborne sanitation system. In: *Proc. WEFTEC Oct 2009 Florida* (in press).

Government of Dong Sheng District, Erdos Municipality (2003). Introduction to Urban Development of the Dong Sheng District.

Kvarnström E., Emilsson K., Richert Stintzing A., Johansson M., Jönsson H., af Petersens E., Schönning C., Christensen J., Hellström D., Qvarnström L., Ridderstolpe P. and Drangert J. O. (2006). *Urine Diversion – One Step Towards Sustainable Sanitation*, EcoSanRes Publication Series, Report 2006–1.

Sustainable Sanitation Alliance (Susana) (2008). Sustainable Sanitation for Cities, Thematic Paper version 1.2, Sustainable Sanitation Alliance, Cities Working Group.

World Health Organization (WHO) (2006). WHO guidelines for the safe use of wastewater, excreta and greywater, volume IV. *Excreta and Greywater Use in Agriculture*, WHO Press, Geneva, Switzerland.

Zhou L., Li J., Yu C. and Wang Y. (2007). *Erdos Eco-Town Project of Inner Mongolia: Economic Evaluation of the Ecosan System*. Available from EcoSanRes, Stockholm Environment Institute, Stockholm, Sweden.

Zhu Q. (2008). Report on Technical Summary of the China-Sweden Erdos Eco-Town Project – From February 2003 to July 2007. Available from EcoSanRes, Stockholm Environment Institute, Stockholm, Sweden.

PART FIVE
Nutrient Management and Recovery

Brief Overview and Assessment of Potential Technologies for Beneficial Recovery of Ammonia and Phosphate from Various Types of Wastewater

W. Rulkens

Prof. Dr. Ir. Wim Rulkens
Wageningen University, Sub-Department of Environmental Technology
P.O. Box 8129, 6700 EV Wageningen, Netherlands
E-mail: wim.rulkens@wur.nl

Abstract Currently the focus in treatment of wastewater is increasingly on beneficial reuse of the effluent and beneficial recovery of energy and valuable compounds. In that respect nitrogen and phosphorous containing components can be considered as valuable products. In this paper a brief and indicative overview and a brief evaluation of the most important potential technologies for recovery of nitrogen and phosphorous containing components such as ammonia, nitrate, phosphorous and phosphate from wastewater streams are given.

Technically there are many feasible recovery scenarios for these components. Treatment scenarios in general consist of one or more of the following treatment steps/ treatment technologies: precipitation, crystallization, stripping, acid vapor scrubbing, transmembrane chemo-sorption, reverse osmosis, nanofiltration, electrodialysis, ion-exchange, wet oxidation, hydrothermolysis, anaerobic digestion and nitrification.

Beneficial recovery of nitrogen and phosphorous containing compounds is in general economically only feasible if the recovery process is part of a complete treatment process of the wastewater in which removal of nitrogen and phosphorous containing components and prevention that these components are emitted to the environment is an essential part of the treatment process.

The economic feasibility of the recovery processes for nitrogen and phosphorous containing compounds is also strongly governed by the market value of the products. Most products from recovery processes for nitrogen and phosphorous containing compounds have the potential that they can be used as a fertilizer in agriculture or in aquaculture. Phosphorous containing products can also be used as a substitute for phosphate rock.

Keywords Wastewater, recovery, ammonia, nitrate, phosphorous, phosphate, technologies, overview

INTRODUCTION

Nitrogen and phosphorous containing compounds are abundantly available in many types of wastewater. During the past ten years increasing attention has been paid to the removal of nitrogen and phosphorous containing components from wastewater and slurries aimed to achieve concentration levels in the effluent which satisfy the standards for discharge onto surface water. To that aim treatment processes have been developed and applied to convert the nitrogen present in nitrogen containing compounds into nitrogen gas and to remove phosphorous containing compounds chemically by precipitation or biologically by concentration into the surplus sludge.

However, currently the focus in treatment of wastewater is not only on the production of an effluent of high quality that can be discharged onto surface water, but also more and more on beneficial reuse of the effluent and beneficial recovery of (sustainable) energy and valuable compounds from the wastewater. In that respect phosphorous (as phosphate) and nitrogen (as ammonia) can be considered as valuable compounds. The production process of phosphate (and phosphorous) from phosphate rock requires a lot of energy and results in emissions of pollutants and large amounts of wastes. Furthermore there is a scarcity of phosphate rock of relatively high quality. In recent years this has caused an increase in the price of phosphate rock. Ammonia is produced from an energy source, such as natural gas, by means of the Haber-Bosch process. To produce one kg ammonia an energy amount of 37 MJ (natural gas) is necessary (Hellstrom 2003). Also the production of ammonia results in emissions of pollutants. Furthermore it can be expected that in short term the demand for ammonia and phosphate will strongly increase due to the increasing demand for meat (animal production) and the increasing demand for fertilizer that is used for the production of food and biomass. Therefore the challenge is currently to recover nitrogen and phosphorous containing compounds from the wastewater and slurries for reuse (Wilsenbach *et al.* 2003). This challenge is strengthened by the increasing energy costs and by the increasing interest in decentralized sanitation systems where a strong focus exists on recovery and reuse.

There are many processes to recover ammonia and phosphate from wastewater. However, the development stage and the specific products of these processes vary strongly. This is due to the variation in volume of the various wastewater streams and the composition of nitrogen and phosphorous containing compounds in these wastewater streams: municipal wastewater, source separated urine and faeces, rejectwater from sewage sludge digestion processes, industrial wastewater from the food and fertilizer industry, concentrates and slurries (such as liquid manure and sewage sludge). Practical application currently often fails because of the high costs and the low or uncertain benefits of the obtained products.

The aim of this paper is to give a brief and indicative overview and a brief evaluation of the most important potential technologies for recovery of ammonia, nitrate, phosphorous and phosphate from wastewater streams. The overview might be useful in practice in the selection of the most appropriate technology and in the identification of future technological challenges.

RECOVERY PROCESSES FOR AMMONIA AND NITRATE

Introduction

Nitrogen compounds in wastewater can in general be present as ammonia, nitrate or organic nitrogen containing compounds. Organic nitrogen containing compounds can be converted into ammonia or nitrate. Beneficial recovery of ammonia or nitrate is in general part of an integral, complete treatment system for a wastewater stream. Various types of physical/chemical and biological treatment processes for recovery of these components are available.

Physical/Chemical Processes to Recover Ammonia

Several physical/chemical treatment steps are available to recover directly ammonia from wastewater and slurries (Rulkens *et al.* 1998; Altinbas *et al.* 2002). Some of these processes are based on processes applied in the chemical process industry. Most of these processes have been extensively elaborated in literature and handbooks.

- Stripping of ammonia with air in combination with absorption of ammonia from the stripping gas in an inorganic acid (Kabdasli *et al.* 2000; Wager *et al.* 2009). In air stripping, the specific property is utilized that ammonia has a high relative volatility at high pH values. The first step in this process consists of the addition of lime or sodium hydroxide to the wastewater or slurry in order to increase the pH. After this step the wastewater is stripped with air at ambient or elevated temperature. Ammonia present in the stripping gas is absorbed into a concentrated acid solution of HCl, H_2SO_4, H_3PO_4, or HNO_3. In this way a concentrated aqueous solution (30–50%) of an inorganic ammonium salt is obtained. In general these concentrates can be slightly polluted with other components. In case the ammonia containing wastewater contains also CO_2 it might be useful to remove this CO_2 by stripping the wastewater prior to the addition of chemicals for increase of the pH. This can save chemicals necessary for increase of the pH for the ammonia stripping process. The removal efficiency of ammonia is high. The final concentration in the treated wastewater or slurry can be low.
- Steam stripping of ammonia in combination with absorption of ammonia in water. In this modification of the stripping process steam is used as stripping

gas instead of air. This stripping process is applied at elevated temperatures. The ammonia containing water phase is concentrated by means of a distillation process at elevated pressure. In this way a concentrated solution of ammonia in water is obtained (Drese 1988).

- Acid vapour scrubbing of ammonia from the gas phase during evaporation. In this process the ammonia containing wastewater stream is evaporated and the ammonia is recovered from the vapour by contacting this vapour with a concentrated ammonium sulphate solution, which is dispersed in the vapour phase in small droplets from a spraying tower. The ammonium sulphate solution is recycled, during which precipitated ammonium sulphate is recovered from the solution. To prevent condensation of water from the vapor, the scrubbing solution has to be heated. At high concentrations of ammonia the process is expected to be economically feasible. In general this process is applicable in case of evaporation or drying of slurries and concentrates that contain high concentrations of ammonia (Van Voorneburg *et al.* 1995).

- Transmembrane chemo-sorption of ammonia (Klaassen and Van Voorneburg 1995; Klaasen *et al.* 1996). In this process, hydrophobic, microporous, hollow-fiber membranes with gas-filled pores are used. The gas phase enables transport of gaseous ammonia through the membrane, while it forms a barrier for other components. The first step in this process is removal of colloidal and suspended particles from the wastewater. Lime or sodium hydroxide is added to raise the pH of this liquid in order to increase the concentration of volatile gaseous NH_3. The wastewater is then fed to the hollow-fibre membrane module. An aqueous solution of an inorganic acid, for example H_2SO_4, is led along the external area of the membrane. Ammonia diffuses from the slurry through the gas-filled pores of the membrane to this acid solution. Finally a concentrated solution of an ammonium salt is obtained that might be used as a fertilizer. In principle a high recovery efficiency and a low final concentration of ammonia in the effluent can be obtained. Based on current experience with the process it is doubtful whether the process is economically feasible.

- Precipitation of ammonia as struvite (Lehmkuhl 1990; Celen *et al.* 2001; Altinbas *et al.* 2002). At high pH and high concentrations ammonia can be precipitated as magnesium ammonium phosphate hexahydrate, $MgNH_4PO_4 \cdot 6H_2O$ (MAP or struvite). The solubility of this mineral in water is about 98.7 mg/l. Struvite can be used as a fertilizer. The first step in the precipitation process is to increase the pH of the wastewater to just above 9 by adding lime or sodium hydroxide. To reduce the amount of lime or sodium hydroxide it might be useful to strip CO_2 from the solution before the precipitation process, if helpful after a decrease of pH by adding an inorganic acid. After this step a stoichiometric amount of soluble magnesium and phosphate ions (magnesium hydrophosphate, magnesium oxide, magnesium chloride, soluble phosphate) is added to the wastewater. The obtained precipitate, struvite, is separated from the wastewater by settling or filtration. With struvite precipitation it is not possible to get a final low ammonia concentration in the effluent.

- Release of ammonia from struvite. It is possible to evaporate ammonia from the struvite precipitate by heating it and to recover magnesium hydrophosphate for reuse. This saves phosphate and magnesium salts. The evaporated ammonia can be absorbed in water or in a concentrated aqueous solution of a mineral acid. Another possibility to release ammonia from struvite is by acid treatment of the wet or dry precipitate. Also in this case magnesium hydrophosphate can be reused in the precipitation process (Zhang $et\ al.$ 2004).

- Phosam process. In this process the ammonia present in the vapour of a stripping process or a drying process is absorbed into a concentrated solution of $(NH_4)H_2PO_4$ where ammonia is bound to $(NH_4)_2HPO_4$. The absorbing liquid is heated and ammonia desorbs as a concentrated gas flow. From this gas phase the ammonia can be recovered as a concentrated ammonia solution in water or as a concentrated aqueous solution of an ammonia salt. After the desorption process the absorbing liquid, containing $(NH_4)H_2PO_4$, is recirculated. The process is primarily aimed for application in the chemical industry.

- Ion-exchangers (Mercer $et\ al.$ 1970; Koon and Kaufman 1975; Liberti and Passino 1981; Bolto and Pawlowski 1987; Harland 1994). In this process the wastewater is treated with an appropriate selective cation exchanger which absorbs ammonia in the form of ammonium ions from the wastewater. Prior to this absorption step it is necessary to remove colloidal and suspended particles which can disturb the ion exchange process and to decrease the pH of the wastewater to a level that all ammonia is present as ammonium ions. After the absorption step a regeneration step of the ion-exchanger with a concentrated inorganic acid such as H_2SO_4 is applied. This results in a concentrated solution of an inorganic ammonium salt. Application of concentrated nitric acid is not possible because there is a serious risk on explosions, depended on the type of ion exchanger and the operating conditions.

- Electrodialysis. In electrodialysis membranes are used which are either selectively permeable for cations or selectively permeable for anions. These types of membranes are arranged alternately in series in an electrical field between two electrodes. With this system the ions present in the wastewater that is fed to this system can be separated in a flow in which the ions are concentrated and a flow which contains only a low residual concentration of ions. In case of ammonium ions the recovery of ammonia from the concentrate can occur according to the methods already discussed for concentrated wastewater streams. It is not possible with this process to obtain very low residual ammonia concentrations in the wastewater.

- Reverse osmosis. In this process a semi-permeable membrane is used that is permeable for water and that has a low permeability for all types of soluble compounds. The water is forced through the membrane by application of a pressure. The treatment process results in a relatively large permeate flow consisting of water with a low concentration of soluble compounds and a relatively small concentrate flow in which most of the soluble compounds, including ammonia, are concentrated. The concentration level of ammonia in the permeate strongly depends on the separation efficiency of the membrane. In general the separation efficiency of a membrane for ammonia is lower than for ammonium ions. Recovery of ammonia from the concentrate can occur according to the methods already discussed for concentrated wastewater streams. Low final concentrations of ammonia (in the permeate) are possible.

Physical-Chemical Treatment Processes to Recover Nitrate

The most important technologies to recover nitrate from wastewater streams are selective ion-exchange (Grinstead and Jones 1971), electrodialysis, and reverse osmosis. With selective ion exchange it is possible to obtain a high concentration of a nitrate salt and a low final nitrate concentration in the treated water. This is also the case with reverse osmosis. Electrodialysis and reverse osmosis are, however, not selective for nitrate. Electrodialysis and reverse osmosis are aimed to concentrate the nitrate to a higher concentration level that might be attractive for a further concentration by means of evaporation.

Physical-Chemical and Biological Conversion of Organic Nitrogen Containing Compounds into Ammonia or Nitrate

Organic nitrogen can be converted into ammonia or nitrate by physical-chemical or biological treatment. Examples of this kind of treatment are wet oxidation, wet hydrolysis at high temperature and anaerobic digestion.

- Wet oxidation at subcritical conditions. During wet oxidation at subcritical conditions (temperature between 150 and 374°C, high pressure), organic matter present in the wastewater is oxidised by adding oxygen. At these conditions organic nitrogen containing compounds present in the wastewater are converted to ammonia. The high temperature of the medium offers an ideal opportunity for stripping or evaporating ammonia from the liquid and to recover it from the resultant vapour as ammonium sulphate, ammonium phosphate or ammonium nitrate.

- Hydrothermolysis (Heesink *et al.* 1995). In this process, the wastewater is heated in a closed vessel to supercritical conditions at a temperature of 500–600 °C. As a result, organic substances are hydrolysed. This process produces a combustible gas phase containing H_2, CH_4 and CO_2 as the main compounds, a solid fraction consisting of ash and insoluble salts, and water containing some dissolved compounds. During this process, 80–100% of the nitrogen compounds are converted to soluble salts. Ammonia can be stripped from the treated water. This process is still in the development phase.
- Anaerobic digestion. During anaerobic digestion of wastewater or sludges part of the organic compounds in the wastewater, especially those that can be easily biodegraded, are converted into biogas. The latter may provide energy for other treatment steps. Nitrate and organic bound nitrogen are for the most part converted to ammonia. Experience with this process has been gained on a large scale.
- Nitrification. Organic nitrogen containing compounds and ammonia can be converted biologically by nitrification into nitrate. After removal of suspended and colloidal particles from the nitrified wastewater nitrate can be recovered by ion exchange or can be concentrated by electrodialysis or reverse osmosis. Nitrification can be useful if in the further processing of the wastewater or concentrates the valuable nitrogen containing compounds should be non-volatile.

Potential Markets for the Products of the Recovery Processes for Ammonia and Nitrate

Potential markets for the products of the recovery processes for ammonia and nitrate are:

- Struvite and concentrated solutions of mono-ammonia phosphate, diammonia phosphate, ammonia-nitrate, ammonia-sulphate and ammonia in water can basically be used in the agricultural sector as a fertilizer. The applicability, however, strongly depends on the quality of the products, costs and acceptance by the farmers.
- Concentrated solutions of ammonia in water or ammonia-nitrate might basically be used as a raw material in the fertilizer production industry. (Roland 1982). However, the acceptance is in general very low because of the small amounts compared with the production capacity of the fertiliser

industry and the possible presence of small amounts of other compounds than nitrogen containing compounds.

- Concentrated solutions of ammonia in water can be used in detoxification processes of NO_x containing exhaust gases
- Aqueous solutions of ammonia, ammonia-nitrate, and ammonia-phosphate can be used for the production of algae and duckweed. (Oron *et al.* 1988). These products can be used for the production of animal or fish feed. A requirement is that the solutions are free from toxic pollutants. Algae can also be used for the production of diesel fuel or for the production of pharmaceuticals. Duckweed can also be used as a biomass for the production of sustainable energy.

RECOVERY PROCESSES FOR PHOSPHATE AND PHOSPHOROUS

Introduction

The physical state of phosphorous containing compounds in wastewater such as municipal wastewater, industrial wastewater from the fertilizer industry or food industry and in slurries such as pig manure, can vary strongly. Phosphorous containing compounds can be present as soluble phosphate, as inorganic precipitates, or bound to colloidal and suspended particles. Organic phosphorous containing components in the wastewater can be converted chemically or microbiologically into (ortho)phosphate.

Recovery of phosphate from wastewater and slurries, aimed at beneficial reuse, can occur by precipitation or crystallization, or by processing phosphorous containing ashes, obtained in incineration processes of phosphorous containing wastes or concentrates, into phosphorous. Beneficial recovery of phosphate or phosphorous requires appropriate pretreatment steps of the wastewater or slurry. Because of the strong variation in physical state of the phosphorous containing compounds in wastewater or in slurries and, as a consequence, the different ways phosphate or phosphorous can be recovered from wastewater streams or slurries, three characteristic wastewater streams will be discussed here. These wastewater streams are industrial wastewater, municipal wastewater and pig manure.

Recovery of Phosphate from Industrial Wastewater

Phosphorous containing compounds in industrial wastewater, such as wastewater from the fertilizer producing industry or the food industry are in general present as phosphate. Several (standard) methods are available for the recovery of phosphate from these types of wastewater streams. The most important methods will briefly be discussed in the next.

- Precipitation as calciumphosphate at elevated pH by adding calcium hydroxide, often after stripping of CO_2, to save chemicals for pH increase. Several modifications of this process exist. One modification is crystallization by using calcite as seeding material and calciumhydroxide to increase the pH (Donnert and Salecker 1999).
- Precipitation as magnesiumphosphate at elevated pH by adding magnesium-hydroxide or magnesium oxide.
- Precipitation as aluminum phosphate or iron phosphate by adding Al salts respectively Fe salts.
- Precipitation or crystallization as struvite ($MgNH_4PO_4.6H_2O$), similar to the process of struvite formation for removal of ammonia from wastewater (Schultze-Rettmer 1991; Moerman et al. 2009; Sanchez et al. 2009).
- Concentration of phosphate by means of reverse osmosis, nanofiltration or electrodialysis in case of low phosphate concentrations, followed by a precipitation or crystallization process applied to the concentrate.
- Concentration of the phosphorous containing compounds in sludge in case that a biological wastewater treatment system, also including enhanced biological phosphorous removal, is applied (Stratful et al. 1999). By anaerobic treatment the phosphorous present in this sludge can be released as phosphate and can be recovered by precipitation or crystallisation from the retentate obtained after mechanical dewatering of the digested sludge.

To achieve an appropriate technical performance of the precipitation or crystallization process and to obtain products of high quality careful management of the process conditions is required. Compared with precipitation processes crystallization processes result in a product of high density and low water content. Often also the presence of other components in the product of a crystallization process is lower. With struvite precipitation/crystallization it is not possible to obtain a low final phosphate concentration in the liquid phase. With the other precipitation/crystallization processes in general a low final concentration of phosphate can be achieved. However, it has to be noted that in most struvite precipitation or crystallization processes, the final product is never a pure product, but contains always a certain fraction of other compounds, depended on the applied process conditions (Hao et al. 2009).

Recovery of Phosphate or Phosphorous from Municipal Wastewater

The concentration of phosphorous containing compounds in municipal wastewater is in the order of 10 mg/l. Beneficial recovery of phosphate at this concentration level is only feasible if a phosphate concentration step is included in the treatment process for the municipal wastewater. There are two processes to remove and concentrate phosphorous containing compounds. One group of methods is to concentrate phosphorous containing compounds in the (surplus) sewage sludge by chemical precipitation with calcium hydroxide, iron salts or aluminum salts. The other group of methods is enhanced biological concentration of phosphorous containing compounds in the (surplus) sewage sludge (Stratful et al. 1999). Several methods are available to recover phosphate or phosphorous from this sludge (Booker et al. 1999; Woods et al. 1999). Detailed information is also given by Balmer (2004), Roeleveld et al. (2004), Cornel et al. (2009), Montag et al. (2009) and Schaum et al. (200(9). Ek et al. (2006) discuss the possibilities to recover nutrients from urine and rejectwater from anaerobically digested sludge. Most important methods are briefly mentioned.

- Extraction of sludge with inorganic acids, such as sulphuric acid at ambient or elevated temperature. In the extraction process phosphorous compounds are dissolved as phosphates. After separation of the liquid phase phosphate can be selectively removed by precipitation, ion exchange or nanofiltration. Heavy metals will also dissolve in the extraction process. It is possible to remove these heavy metals prior to the precipitation of phosphate by selective precipitation with sodium sulfide.
- Anaerobic treatment of the sewage sludge to release phosphorous containing compounds as soluble phosphate (Stratful et al. 1999). After separation of the sludge particles phosphate can be recovered as calcium phosphate or struvite from the liquid phase (retentate/filtrate) by precipitation or crystallization processes (Doyle 2002; Britton et al. 2009).
- In case of biological phosphorous removal it is possible to produce a phosphate rich supernatant liquor in a side stream of the wastewater treatment process. By means of a crystallization process it is then possible to produce calcium phosphate pellets that can be used as a substitute for phosphate rock.
- Incineration of phosphorous containing sewage sludge followed by liquid extraction of the phosphate from the ash of the incineration process with sulphuric acid. After extraction the pH of the extracting agent is stepwise increased. The first precipitate is aluminum phosphate. Heavy metals present in the extracting agent precipitate at higher pH. In this way a heavy metal free aluminum phosphate can be obtained.
- Incineration of phosphorous containing sewage sludge and use of the ash as a substitute for phosphate rock in the industrial production of phosphorous (Schipper and Korving 2009).
- Incineration of phosphorous containing sewage sludge and use of the ash directly as fertilizer or via the fertilizer production industry. If the

concentration of toxic heavy metals in the ash such as Cd, Cu and Zn is too high these heavy metals can be removed by thermo-chemical treatment. (Hermann, 2009). To that aim chlorine containing compounds such as magnesium chloride or potassium chloride or added to the ash. By heating this mixture to temperatures above 1000 °C the heavy metals evaporate as heavy metal chlorides and are removed as heavy metal waste from the gas phase. In case the ash is used by the fertilizer industry there are also specific requirements regarding the maximum amount of heavy metals in the ash. However, there or no set requirements for the maximum amount of iron.

Recovery of Phosphate or Phosphorous from Pig Manure

Pig manure (slurry manure) is characterized by a high concentration of dry solids (100 g/l, 80% is organic), a high concentration of Kjeldahl-N (8 g/l, mainly ammonia), a high concentration of phosphorous (5 g/l as P_2O_5), and a high concentration of potassium (8 g/l as K2O). Pig manure can therefore be considered as a valuable resource for the production of energy and the recovery of P, N, K for beneficial reuse. In the development of treatment processes for pig manure a strong focus is given on the recovery of these valuable compounds. Phosphorous is for the major part bound in insoluble form in the suspended and colloidal manure particles. Part of the phosphorous is present as phosphate in the liquid fraction of the manure. By means of an anaerobic treatment process of the manure the concentration of phosphorous in the liquid fraction can be increased. The distribution of phosphorous between liquid fraction and solid fraction is also influenced by the pH. Greaves *et al.* (1999) gives a review about the various modifications phosphorous that might be present in manure. Several processes are available to recover phosphate from manure. Some processes have already been discussed in more detail in the foregoing sections. These will be mentioned very briefly.

- Precipitation or crystallisation of phosphate from the liquid fraction as calciumphosphate or struvite (Daumer *et al.* 2009).
- Precipitation of phosphate at high pH from the liquid fraction (of calf manure) as potassium struvite. Whether ammonium struvite or potassium struvite is produced in the precipitation process depends on the composition of the liquid phase. Potassium struvite is produced at low ammonium concentrations and high potassium concentrations (Schuiling and Andrade 1999).
- Mechanical separation of the solid fraction from the manure followed by drying and incineration or by direct incineration. In the incineration process an ash is produced with a relatively high P_2O_5 content. In principle this ash can be used directly as fertilizer or can be used as a substitute for phosphate rock.

Potential Markets for the Products of the Recovery Processes for Phosphate or Phosphorous

Potential markets for the products of the recovery processes for phosphate are:

- Caliumphosphate, struvite and potassium struvite can directly be used as fertilizer. A requirement is a sufficiently low concentration of toxic heavy metals and toxic organics. There is some discussion about the acceptable amount of iron.
- Calciumphosphate and struvite can also be used for the industrial production of fertilizer, provided that the composition satisfies the requirements (Durrant *et al.* 1999).
- Ash from incineration processes of phosphorous containing sludge can be used as a fertilizer. If necessary, toxic heavy metals can be removed by wet chemical processes or thermal processes.
- Ash from incineration processes of phosphorous containing sludge can also be used as a substitute for phosphate rock. Depended on the composition this substitute for phosphate rock can be used in the phosphorous producing industry and/or in the fertilizer producing industry.

ASSESSMENT OF THE VARIOUS TECHNOLOGIES AND PRODUCTS

In the assessment of the various recovery scenarios for nitrogen and phosphorous containing compounds many aspects have to be considered and elaborated such as pretreatment and posttreatment processes, amounts of chemicals that are necessary, energy requirements and quality of the energy, quality and related market value of the products, scale and character of the experience with the treatment systems, environmental impact, social acceptance of the treatment process and the products, investment and operating costs, national and local environmental policy regarding wastewater treatment and manure treatment. Looking to the literature only a few cases have been elaborated in detail and provide results that might directly be useful for other cases. However, these results are always very specific and translation to other potential cases is often not possible without further research and development. Nevertheless, from the previous some general statements can be derived which might be useful for the development of specific recovery scenarios at other conditions and situations:

- Technically there are many feasible recovery scenarios. However, special attention has to be given to prevent the presence of small amounts of toxic heavy metals or toxic organic pollutants such as endocrine disruptors in the products. The presence of small amounts of toxic components can strongly reduce the market value of the products.

- The economical feasibility of the recovery processes for nitrogen and phosphorous containing compounds is strongly governed by the market value of the products. Only in case that industrial producers of fertilizers or producers of phosphate can use the product as a substitute for phosphate rock it is more or less clear what the market value might be of the product.

- Most products from recovery processes for nitrogen and phosphorous containing compounds have the potential that they can be used as a fertilizer in agriculture or in aquaculture. However, the development of a sustainable market in this area requires a constant product, a certification of the product quality with respect to the fertilizer value, a permanent distribution structure, a guarantee of regular supply, a competitive price compared with artificial fertilizers, and an acceptance by the users.

- Economically the beneficial recovery of nitrogen and phosphorous containing compounds is strongly governed by the energy prices, especially in case of recovery of ammonia or nitrate.

- Beneficial recovery of nitrogen and phosphorous containing compounds is in general economically only feasible if the recovery process is part of a complete treatment process of the wastewater or slurry in which removal of nitrogen and phosphorous containing components and prevention that these components are emitted to the environment is an essential part of the treatment process.

- Removal of nitrogen and phosphorous containing compounds from a wastewater or slurry and prevention of emission of these compounds to the environment are currently essential parts of a complete treatment process for a wastewater or slurry. In that respect methods for beneficial recovery of nitrogen and phosphorous containing components are in technical and economic respect in competition with the standard methods of removal of these components. These standard methods are in general very cost efficient.

- The aim to recover nitrogen and phosphorous containing compounds is the reduction of the industrial production of ammonia and phosphorous because this production requires a substantial amount of energy and causes also environmental pollution. In the assessment of the environmental sustainability of the various recovery options also the energy necessary for the recovery process and the environmental pollution caused by the recovery process has to be taken into account (Mulder 2003).

FINAL CONCLUSIONS

Many technical options exist for the beneficial recovery of ammonia, nitrate, phosphate or phosphorous containing components from wastewater. Most of these options are focused on the production of a fertilizer component. The economic feasibility of these options is still a matter of uncertainty. Key aspects are the market value of the product and the costs of recovery. To stimulate the recovery of nitrogen and phosphorous containing components for beneficial reuse cooperation with fertilizer producing firms and the development of markets for application of the obtained products in agriculture or aquaculture are required. Helpful in this development might also be a clear development of environmental policy that stimulates reuse of these products. Because the development of technologies for recovery of nitrogen and phosphorous containing compounds is in fact in the beginning phase there are still opportunities to improve the recovery technology, to reduce its costs and to improve the quality and thus also the market value of the products.

REFERENCES

Altinbas M., Ozturk I. and Aydin A. F. (2002). Ammonia recovery from high strength agro industry effluents. *Water Science & Technology*, **45**(12), 189–196.

Balmér P. (2004). Phosphorus recovery – an overview of potentials and possibilities. *Water Science & Technology*, **49**(10), 185–190.

Bolto B. A. and Pawlowski L. (1987). *Wastewater Treatment by Ion-Exchange*. J. W. Arrowsmith Ltd. Bristol, UK. ISBN: 0419133208.

Booker N. A., Priestley A. J. and Fraser I. H. (1999). Struvite formation in wastewater treatment plants: opportunities for nutrient recovery. *Environmental Technology*, **20**, 777–782.

Britton A., Prasad R., Balzer B. and Cubbage L. (2009). Pilot testing and economic evaluation of struvite recovery from dewatering centrate at HRSD's Nansemond WWTP. *International Conference on Nutrient Recovery from Wastewater Streams. Vancouver, British Columbia, Canada*, K. Ashley, D. Mavinic and F. Koch (eds.), IWA Publishing, London, pp. 193–202. ISBN: 9781843392323.

Celen I. and Turker M. (2001). Recovery of ammonia as struvite from anaerobic digester effluents. *Second International Conference on Recovery of Phosphates from Sewage and Animal Wastes*, March, 12–13, Noordwijkerhout, Holland.

Cornel P. and Schaum C. (2009). Phosphorus recovery from wastewater: needs, technologies and costs. *Water Science & Technology*, **59**(6), 1069–1076.

Daumer M. L., Béline F. and Parsons S. A. (2009). Chemical recycling of phosphorus from piggery wastewater. *International Conference on Nutrient Recovery from Wastewater Streams. Vancouver, British Columbia, Canada*, K. Ashley, D. Mavinic and F. Koch (eds.), IWA Publishing, London, pp. 339–350. ISBN: 9781843392323.

Donnert D. and Salecker M. (1999). Elimination of phosphorus from waste water by crystallization. *Environmental Technology*, **20**, 735–742.

Doyle J. D. and Parsons S. A. (2002). Struvite formation, control and recovery. *Water Research*, **36**, 3924–3940.

Drese J. (1988). Stripping as a cheap solution for the livestock waste problem. *Pt/Procestechniek*, **43**(4), 43–45. (in Dutch).

Driver J., Lijmbach D. and Steen I. (1999). Why recover phosphorus for recycling, and how? *Environmental Technology*, **20**, 651–662.

Durrant A. E., Scrimshaw M. D., Stratful I. and Lester J. N. (1999). Review of the feasibility of recovering phosphate from wastewater for use as a raw material by the phosphate industry. *Environmental Technology*, **20**, 749–758.

Ek M., Bergström R., Bjurhem J. E., Björlenius B. and Hellström D. (2006). Concentration of nutrients from urine and reject water from anaerobically digested sludge. *Water Science & Technology*, **54**(11–12), 437–444.

Greaves J., Hobbs P., Chadwick D. and Haygarth P. (1999). Prospects for the recovery of phosphorous from animal manures: a review. *Environmental Technology*, **20**, 697–708.

Grinstead R. R. and Jones K. C. (1971). Nitrate removal from waste waters by ion exchange. *Journal of Water Pollution Control Research Series*, Report No. 17010, Environmental Protection Agency, Washington D.C., January 1971.

Hao X. D., Wang C. C., Lan L. and Van Loosdrecht M. C. M. (2009). A quantitative method analyzing the content of struvite in phosphate-based precipitates. *International Conference on Nutrient Recovery from Wastewater Streams .Vancouver, British Columbia, Canada*, K. Ashley, D. Mavinic and F. Koch (eds.), IWA Publishing, London, pp. 79–88. ISBN: 9781843392323.

Harland C. E. (1994). *Ion Exchange: Theory and Practice*, The Royal Society of Chemistry, UK. ISBN: 0851864848.

Heesink A. B. M., Versteeg G. F., Osse S. L. J. and Ter Maat H. F. (1995). Hydrothermolysis of pig slurry: An evaluation. *Technical Report*. Procédé Twente BV, Enschede, the Netherlands. (in Dutch).

Hellstrom D. (2003). Exergy analysis of nutrient recovery processes. *Water Science & Technology*, **48**(1), 27–36.

Hermann L. P-recovery from sewage sludge ash – technology transfer from prototype to industrial manufacturing facilities. *International Conference on Nutrient Recovery from Wastewater Streams. Vancouver, British Columbia, Canada*, K. Ashley, D. Mavinic and F. Koch (eds.), IWA Publishing, London, pp. 405–415. ISBN: 9781843392323.

Kabdasli I., Tunay O., Ozturk I., Yilmaz S. and Arikan O. (2000). Ammonia removal from young landfill leachate by magnesium ammonium phosphate precipitation and air stripping. *Water Science & Technology.* **41**(1), 237–240.

Klaassen R., Jansen A. E., Van Voorneburg F., and De Dooij A. C. P. (1996). *Ammonia Removal from Aqueous Streams with Transmembrane Chemo-Sorption (TMCS)*, TNO report, TNO, Apeldoorn, the Netherlands. (in Dutch).

Klaassen R. and Van Voorneburg F. (1995). *Removal of Ammonia from Liquid Manure and Waste Waters with Membrane Chemo-Sorption*, TNO report, TNO, Apeldoorn, the Netherlands. (in Dutch).

Koon J. H. and Kaufman W. J. (1975). Ammonia removal from municipal wastewaters by ion exchange. *Journal WPCF*, **47**(3), 448–465.

Lehmkuhl J. (1990). Verfahren für die Ammonium-Elimination. WLB Wasser, Luft und Boden, 11–12, 46–48. (in German).

Liberti L. and Passino R. (1981). An ion exchange process to recover nutrients from sewage. *Resources and Conservation*, **6**, 263–273.

Mercer B. W., Ames L. L., Touhill C. J., Van Slyke W. J. and Dean R. B. (1970). Ammonia removal from secondary effluents by selective ion exchange. *Journal WPCF*, **42**(2), 95–107.

Moerman W., Carballa M., Vandekerckhove A., Derycke D. and Verstraete W. (2009). Phosphate removal in agro-industry: pilot and full-scale operational considerations of struvite crystallisation. *International Conference on Nutrient Recovery from Wastewater Streams, Vancouver, British Columbia, Canada*, K. Ashley, D. Mavinic and F. Koch (eds.), WA Publishing, London pp. 245–255. ISBN: 9781843392323.

Montag D., Gethke K. and Pinnekamp J. (2009). Different strategies for recovering phosphorus: technologies and costs. *International Conference on Nutrient Recovery from Wastewater Streams. Vancouver, British Columbia, Canada*, K. Ashley, D. Mavinic and F. Koch (eds.), IWA Publishing, London, pp. 159–167. ISBN: 9781843392323.

Mulder A. (2003). The quest for sustainable nitrogen removal technologies. *Water Science & Technology*, **48**(1), 67–75.

Oron G., De-Vegt A. and Porath D. (1988). Nitrogen removal and conversion by duckweed grown on wastewater. *Water Research*, **22**(2), 179–184.

Roeleveld P., Loeffen P., Temmink H. and Klapwijk B. (2004). Dutch analysis for P-recovery from municipal wastewater. *Water Science & Technology*, **49**(10), 191–199.

Roland L. D. (1982). Some options for treating liquid wastes from nitrogen fertilizer factories. *Effluent and Water Treatment Journal*, 216–220.

Rulkens W. H., Klapwijk A. and Willers H. C. (1998). Recovery of valuable nitrogen compounds from agricultural liquid wastes: potential possibilities, bottlenecks and future technological challenges. *Environmental Pollution*, **102**(S1), 727–735.

Sanchez A., Barros S., Mendez R. and Garrido J. M. (2009). Phosphorus removal from an industrial wastewater by struvite crystallization into an airlift reactor. *International Conference on Nutrient Recovery from Wastewater Streams. Vancouver, British Columbia, Canada*, K. Ashley, D. Mavinic and F. Koch (eds.). IWA Publishing, London, pp. 89–97. ISBN: 9781843392323.

Schaum C., Cornel P. and Jardin N. (2009). Phosphorus recovery from sewage sludge ash: possibilities and limitations of wet chemical technologies. *International Conference on Nutrient Recovery from Wastewater Streams. Vancouver, British Columbia, Canada*, K. Ashley, D. Mavinic and F. Koch (eds.). IWA Publishing, London pp. 659–670. ISBN: 9781843392323.

Schipper W. and Korving L. Full-scale plant test using sewage sludge ash as raw material for phosphorus production. *International Conference on Nutrient Recovery from Wastewater Streams. Vancouver, British Columbia, Canada*, K. Ashley, D. Mavinic and F. Koch (eds.), IWA Publishing, London, pp. 591–598. ISBN: 9781843392323.

Schuiling R. D. and Andrade A. (1999). Recovery of struvite from calf manure. *Environmental Technology*, **20**, 765–768.

Schulze-Rettmer R. (1991). The simultaneous chemical precipitation of ammonium and phosphate in the form of magnesium-ammonium phosphate. *Water Science & Technology*, **23**(4–6), 659–667.

Stratful I., Brett S., Scrimshaw M. B. and Lester J. N. (1999). Biological phosphorus removal, its role in phosphorus recycling. *Environmental Technology*, **20**, 681–695.

Van Voorneburg F., Ten Have P. J. W., Sneijders J. H. and Schneiders L. H. J. M. (1995). *Acid Vapour Scrubbing of Ammonia in a Pig Slurry Evaporator*. TNO-rapport. Ref. no. R95-218, TNO, Apeldoorn, the Netherlands. (in Dutch).

Wager F., Wirthensohn T., Corcoba A. and Fuchs W. (2009). Air stripping of ammonia from anaerobic digestate. *International Conference on Nutrient Recovery from Wastewater Streams. Vancouver, British Columbia ,Canada,* K. Ashley, D. Mavinic and F. Koch (eds.), IWA Publishing, London, pp. 719–735. ISBN: 9781843392323.

Wilsenbach J. A., Maurer M., Larsen T. A. and Van Loosdrecht M. C. M. (2003). From waste treatment to integrated resource management. *Water Science & Technology,* **48**(1), 1–9.

Woods N. C., Sock S. M. and Daigger G. T. (1999). Phosphorus recovery technology modeling and feasibility evaluation for municipal wastewater treatment plants. *Environmental Technology,* **20**, 663–679.

Zhang S., Yao C., Feng X. and Yang M. (2004). Repeated use of MgNH4PO4.6H2O residues for ammonium removal by acid dipping. *Desalination,* **170**, 27–32.

Recovering Pure Struvite from Wastewater Near the Neutral pH

X. D. Hao[a], C. C. Wang[a] and M. C. M. van Loosdrecht[b]

[a] The R & D Centre of Sustainable Environmental Biotechnology, Beijing University of Civil Engineering and Architecture, Beijing 100044. P. R. of China
[b] Dept. of Biochemical Engineering, Delft University of Technology, Julianalaan 67, 2628 BC Delft, the Netherlands
E-mail: xdhao@homail.com

Abstract Precipitate harvested from conventional chemical precipitation of struvite under different pH values and from a novel electrochemical deposition method was characterized and analyzed by XRD, IR and element analyses. The experimental results revealed that the optimal pH ranges for having high struvite content (>90%) were respectively at 7.5–9.0 in pure water solution and 7.0–7.5 in tap water solution. The experiments also demonstrated the feasibility of electrochemical deposition for recovering pure struvite from the tested solutions at a neutral pH value with a high removal efficiency, which could be easily realized at a lower voltage. The study aimed at engineering application recovering purer struvite from wastewater at the neutral pH value.

Keywords struvite, chemical precipitation, electrochemical deposition, low pH value, wastewater

INTRODUCTION

Chemical precipitation of struvite ($MgNH_4PO_4 \cdot 6H_2O$) from phosphate-rich wastewater (anaerobic supernatant, urine and/or animal wastes) is increasingly getting global attention for resource recovery and closing nutrient cycles. Recovered struvite can be potentially used as a direct or an indirect valuable fertilizer.

Some processes recovering struvite have been theoretically and experimentally investigated by chemists, biochemists and civil engineers (Lee *et al.*, 2003; Hao and van Loosdrecht, 2006; Yigit *et al.*, 2007; Kim *et al.*, 2007; Ronteltap *et al.*, 2007; Wilsenach *et al.*, 2007; Pastor *et al.*, 2008). Precipitation of struvite is mainly controlled by such factors as $Mg^{2+}/NH_4^+/H_nPO_4^{n-3}$ concentrations, pH and temperature. The influences of Ca^{2+} and other factors on precipitation of struvite have been also investigated (Stratful *et al.*, 2001; Le Corre *et al.*, 2005). Among others, pH seems an extensively emphasized factor controlling formation of struvite. According to the available references (Battistoni *et al.*, 1997; Miles and Ellis, 2001; Stratful *et al.*, 2001; Altinbas *et al.*, 2002; Jaffer *et al.*, 2002; Suzuki *et al.*, 2002; Nelson *et al.*, 2003), alkaline pH values in the range of 9.0–10.7

are generally considered to be a favourite condition forming pure struvite. Furthermore, X-ray diffraction technology (XRD) is the most popular analyzing method used to determine the presence of struvite in precipitates.

However, it was observed that a remaining ammonium content in an anaerobic supernatant was much higher than a theoretical molar ratio (Mg:N:P=1:1:1) when a series of experiments recovering struvite from wastewater were conducted at pH>9 (Hao et al., 2008b). Although the recovered precipitate samples resulting from the supernatant were analyzed by XRD, an accurate content of struvite could not be easily determined by the positions and intensities of XRD peaks. In other words, XRD is only a qualitative method determining the existence of struvite in phosphate-based precipitate. The high remaining ammonium in the supernatant revealed that at pH>9 more phosphate-based compounds rather than struvite might be formed in the recovered precipitate.

Based on the above experimental observation, the alkaline pH range forming pure struvite was strongly doubted. This initiated an experimental plan to ascertain the relationship between optimal pH range and pure struvite formation. Above all, an appropriate method determining the struvite content in recoved precipitate became a crux. For this reason, element analyses preceded by dissolution was developed and applied in the experiments to indirectly calculate the struvite content based on the NH_4^+-N in recovered precipitate (Hao et al., 2008a). According to the element analyses, the optimal pH ranges were determined near a neutral pH range for froming purer struvite (90%) in the recovered precipitate (Hao et al., 2008a), which are totally different from the reported pH range (>9.0) in literature.

Besides pH, moreover, Ca^{2+} is also an important factor affecting formation of struvite. Experiments demonstrated that pH>8.0 in real wastewater containing a high concentration of Ca^{2+} resulted in more impurities in recovered precipitate due to the existence of Ca^{2+} (Hao et al., 2008a). It was therefore concluded that purer struvite precipitate from real wastewater should be controlled at the neutral pH range (<8.0) (Hao et al., 2008a). However, the reaction rate of struvite-based precipitate at the neutral pH must be quite slow, which implied that engineering application with conventional precipitation of struvite was hadly possible at the neutral pH. Under the circumstance, a further study had to followed to speed up formation of struvite at the neutral pH. The following experiments demonstrated that some technical measures could be employed to such an aim, and so increasing thermodynamic driving force, electrochemical deposition method and strengthening by carrier precipitation like micro-sand and PAM (Poly(acrylamide)) were respectively tested.

With this article, formation of pure struvite was tested by conventional chemical precipitation and electrochemical deposition at the neautral pH. X-ray diffraction (XRD), infrared spectrum (IR), and element analyses were employed to analyze the struvite content in recovered precipitate.

EXPERIMENTS

Analytical grade chemicals such as $NaH_2PO_4 \cdot 2H_2O$, $MgSO_4 \cdot 7H_2O$, NH_4Cl, $NH_4H_2PO_4 \cdot 2H_2O$ and NaOH were used as the reagents forming struvite. A relatively pure struvite (the labelled purity: 99.0%) was used as a reference compound, which was commercially purchased from Alfa-Aesar (US).

Chemical Precipitation of Struvite

Under an identical method described by Hao *et al.* (2008a), two series of experiments were conducted with pure water and tap water as solvents respectively. The tap water in Beijing mainly consists of ground water, with a high mineral content: $c(Ca^{2+}) = 2.17$ mM and $c(Mg^{2+}) = 1.34$ mM.

Electrochemical Deposition of Struvite

Setup of electrochemical deposition

The setup of electrochemical deposition is shown in Figure 1. It is actually an electrolytic cell composed of platinum pole electrode as an anode (Lanlike Co., Ltd), with working area $\Phi 0.1$ cm $\times 0.5$ cm and a nickel pole electrode (Lanlike Co., Ltd) as a cathode, with working area $\Phi 0.1$ cm $\times 0.5$ cm. The effective volume of the setup was 250 ml, and the space between the anode and the cathode was about 5.0 cm.

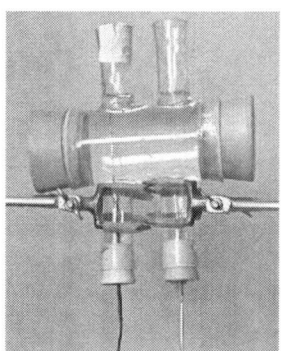 

Figure 1 Setup (left) for the electrochemical deposition of precipitate and the obtained precipitate (right)

Preparation of electrolyte

I: 1.5 mmol $NH_4H_2PO_4 \cdot 2H_2O$ was dissolved in 0.15 L ultra pure water; II: 1.5 mmol $MgSO_4 \cdot 7H_2O$ was dissolved in 0.15 L ultra pure water. The equal volumes of solution I and II were mixed together, and then moved into the setup, in which the concentrations of three main ions (Mg^{2+}, NH_4^+ and $H_nPO_4^{n-3}$) were in the ranges of normal values found in waste streams such as the supernatant from anaerobic digesters. The pH value was adjusted to the desired point (7.0–7.5) by adding the NaOH solution of $1 \ mol \cdot L^{-1}$.

Formation of struvite by electrochemical deposition

As soon as electricity was exerted in the system, some bubbles were released both from the anode and the cathode. At the same time, pH in solution near the cathode increased slowly to 7.0–7.5, and then precipitate appeared and layered on the electrodes, as can seen in Figure 1 (right).

Characterization of Struvite

Crystal characterization of the dried precipitate was performed by Rigaku D/max IIIA X-ray diffractometer with CuKα radiation ($\lambda = 1.5406$Å), and the intensity data were collected in the range of $10° \leq \theta \leq 80°$ at 298(2) K. Infrared (IR) spectra was recorded using KBr pellet on PerkinElmer spectrum 100 Fourier Transform infrared spectrophotometer in the range of 400–4000 cm^{-1}. The SEM images were observed with TM-1000 (Hitachi, Japan). The concentration of phosphorus was analyzed using IRIS Advantage Inductively Coupled Plasma Atomic Emission Spectrometer, and the concentration of Mg^{2+}, Ca^{2+} and NH_4^+-N were determined using Dionex DX-120 Ion Chromatograph.

RESULTS AND DISCUSSION

Formation of Struvite

In principle, struvite formation should be accompanied with the presence of Mg^{2+}, NH_4^+ and $H_nPO_4^{n-3}$. The experiments indicated that precipitation of struvite resulted in a rapid decrease on pH, as also observed by other researchers (Boistelle *et al.*, 1983), which is due to the fact that a low pH makes HPO_4^{2-} become the major form of phosphate in liquid. As a result, struvite formation can be expressed as Equation 1.

$$Mg^{2+} + NH_4^+ + H_nPO_4^{n-3} + 6H_2O \rightarrow MgNH_4PO_4 \cdot 6H_2O + (n+1)H^+ \qquad (1)$$

In fact, struvite formation should be mainly controlled by pH, initial relative ionic (Mg^{2+}, NH_4^+ and PO_4^{3-}/HPO_4^{2-}/ $H_2PO_4^-$) concentrations and other ions like

Ca^{2+}. Among them, pH could be a key factor controlling struvite formation (shape, morphology and purity).

Electrochemical deposition is a new method to trap struvite from phosphate-rich wastewater. As electricity was exerted, some bubbles were released both from the anode and cathode of the setup. The generation of OH^- increased the interfacial pH (up to 7.5) near the cathode (Gabrielli *et al.*, 2006), as illustrated by Equation 2 and Equation 3, which was subjected to the diffusion rate of oxygen towards the cathode and the consumption rate of OH^- by the chemical reactions forming precipitate. Along with the increase of OH^- near the cathode, Mg^{2+}, NH_4^+ and $H_nPO_4^{n-3}$ ions might react with each other to initiate the nucleation and growth of struvite as shown in Equation 1, then struvite precipitate were formed and layered on the cathode electrode.

$$O_2 + 2H_2O + 4e^- \rightarrow 4OH^- \tag{2}$$

$$H_2O + e^- \rightarrow \frac{1}{2}H_2 + OH^- \tag{3}$$

XRD Analysis of Precipitates

XRD was used to qualitatively determine the major compositions in the harvested precipitate, as shown in Figures 2 and 3.

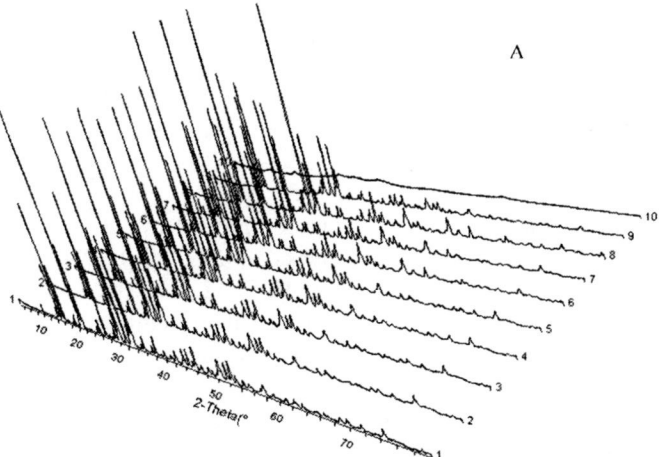

Figure 2 XRD patterns of precipitates obtained by traditional struvite precipitation under different pHs in pure water (A) and tap water (B) [1-purchased struvite, 2-pH= 7.5, 3-pH=8.0, 4-pH=8.5, 5-pH=9.0, 6-pH=9.5, 7-pH=10.0, 8-pH=10.5, 9-pH=11.0, 10-pH=11.5.]

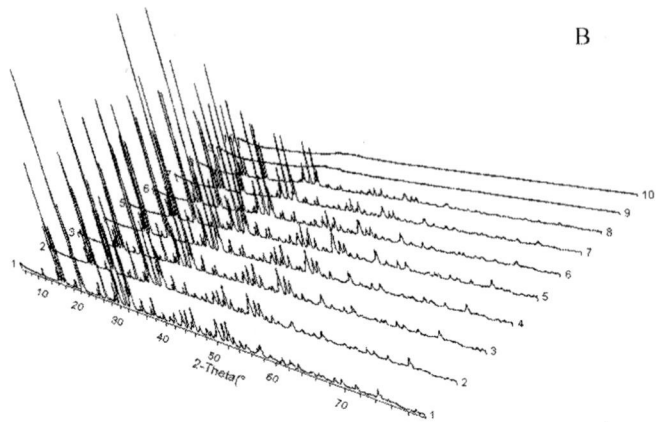

Figure 2 *(Continued)*

The 2-theta angles of (001) and (002) diffraction characteristic peaks (20.85° and 33.27°) of the samples obtained from the pure water solution (pH 7.5–9.0) and the tap water solution (pH 7.0–8.5) match well with the XRD patterns of the data base (PDF#15-0762) and AR grade $MgNH_4PO_4 \cdot 6H_2O$, which indicates that the harvested precipitate and the pure struvite are very much alike. However, many noise peaks appear in XRD patterns of the precipitate obtained from the pure water solution pH>9.5, while the characteristic peaks of struvite become weaker, and some (001) and (002) diffraction characteristic peaks of $Mg_3(PO_4)_2 \cdot 4H_2O$ and even $Mg(OH)_2$ appear, which indicates that the harvested precipitate contained not only the desired struvite but also some impurities like $Mg_3(PO_4)_2 \cdot 4H_2O$ and even $Mg(OH)_2$.

Due to the negative impact of Ca^{2+} (Le Corre *et al.*, 2005), some impurities were precipitated at pH>9.0, which can be proved by the appearance of (001) and (002) diffraction characteristic peaks of $Mg_3(PO_4)_2$, $Mg(OH)_2$, $Ca_3(PO_4)_2$ ($K_{sp}=2.1\times10^{-33}$) and $CaHPO_4$ ($K_{sp}=1.8\times10^{-7}$). Figure 2 indicates that (001) and (002) diffraction characteristic peaks of struvite completely disappeared at a high pH (>10.5), which reveals almost no struvite existence but only some other precipitated compounds.

The XRD patterns of struvite obtained from electrochemical deposition are shown in Figure 3, in which the purchased pure struvite was used to compare the XRD patterns between the purchased struvite and harvested precipitate. As shown in Figure 3, no clear differences on both position and intensity of the peaks can be observed by comparing the XRD patterns from the pure struvite and harvested precipitate.

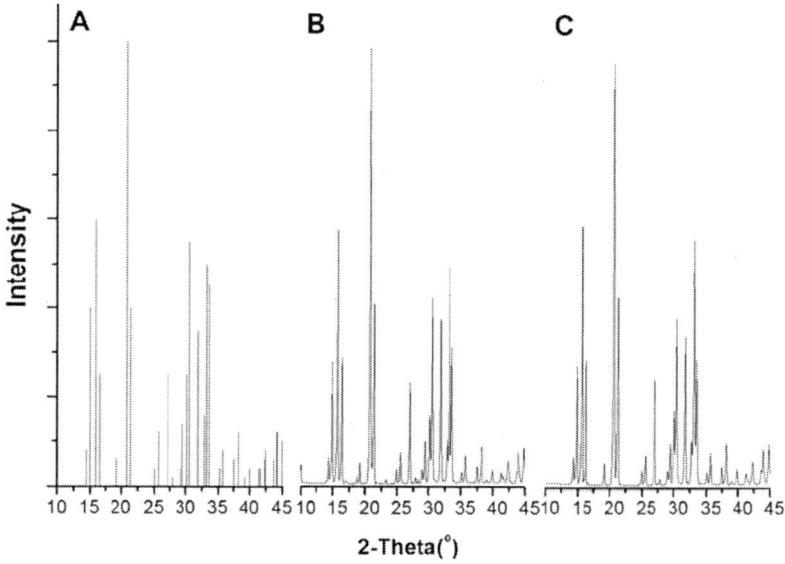

Figure 3 XRD patterns of the purchased pure struvite and harvested precipitate on the Pt sheet by electrochemical deposition (A: standard pattern PDF#15-0762; B: purchased struvite (labelled purity: 99.0%); C: harvested precipitates)

Infrared (IR) Spectra

As shown in Figure 4, the infrared spectra of the precipitate obtained from the pure water solution (pH = 7.5–11.0) and tap water solution (pH = 7.5–10.0) reveal strong absorption bands at 455 cm^{-1}, 568 cm^{-1}, 1,000cm^{-1} and 1,430 cm^{-1}, in which the bands at 455 cm^{-1}, 568 cm^{-1}, 1,000 cm^{-1} could be assigned as $v_2(PO_4^{3-})$, $v_4(PO_4^{3-})$ P-O bending and $v_3(PO_4^{3-})$ antisym stretching respectively, while the 1,430 cm^{-1} could be assigned as NH_4^+ v_4 asym bending (Banks *et al.*, 1975). But the positions and shapes of these characteristic peaks/bands of the precipitates obtained at a higher pH changed seriously, which indicates the content of PO_4^{3-} and NH_4^+ decreased, even decreased nearly to zero.

The precipitate harvested from electrochemical deposition was also identified by infrared spectra. The assignments of the main peaks (462.07 cm^{-1}, 571.49 cm^{-1}, 1,007.05 cm^{-1} and 1,435.24 cm^{-1}) for the harvested precipitate match well with the reported values of pure struvite

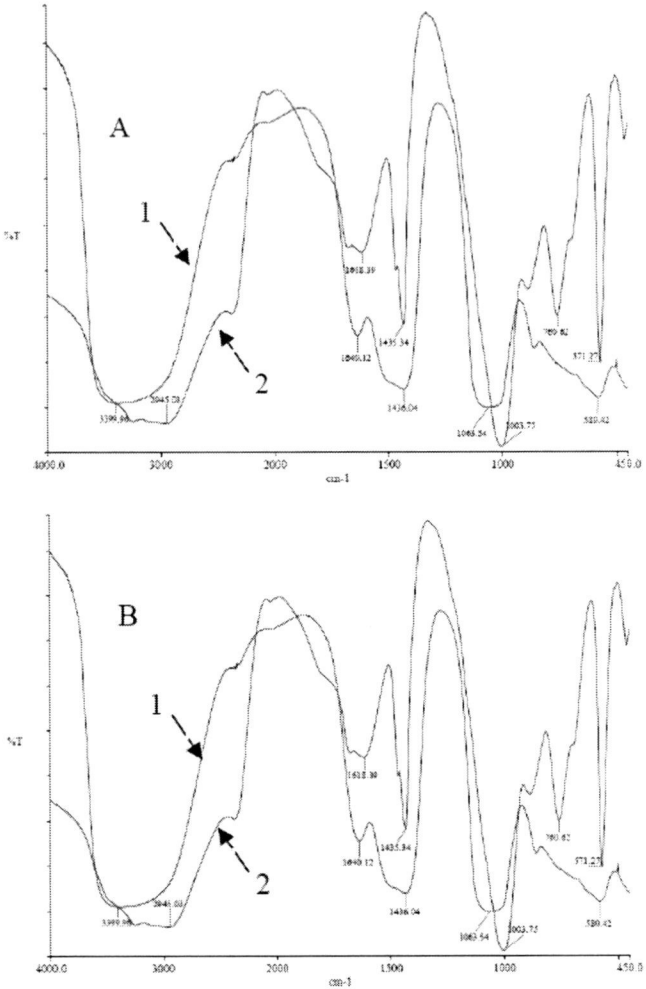

Figure 4 IR spectrum of the precipitate obtained by chemical precipitation under different pHs in the pure water system (A) and tap water system (B) (1-pH = 7.5, 2-pH = 11.5)

The Struvite Content in Precipitate

The element analyses preceded by a dissolution method developed by Hao *et al.* (2008a) could be a quantitative and efficient method to determine the exact struvite content in the precipitate. The struvite contents in the precipitate obtained from conventional chemical deposition from the ultra pure water solutions decreased gradually in the pH range of 7.5–10.5, and then decreased sharply at pH > 10.5. The struvite contents in the precipitate from the tap water solutions were controlled not only by pH but also by the Ca^{2+} content in the original solutions. The struvite contents in the precipitate were 96.8% and 95.7% at pH = 7.0 and 7.5 respectively. At pH > 7.5, the struvite content decreased sharply, down to 15.5% at pH = 10.5. An even higher pH (> 10.5) resulted in the complete disappearance of struvite in the precipitate (Hao et al., 2008a).

The dissolution method developed by Hao et al. (2008a) was also used to perform the element analyses of the harvested precipitate layered on the Pt sheet by electrochemical deposition. The results reveal that the concentrations of NH_4^+-N, TP and Mg in the harvested precipitate were at 0.6329 mM, 0.6401 mM and 0.6530 mM respectively. As a result, the content of struvite in the precipitate is thus calculated at 97.1%, which means that the harvested precipitate contained the very high content of pure struvite. The phosphate removal efficiencies in the original solutions were also analyzed; the results reveal that up to 94.5–96.1% of the phosphate removal effciency was achieved, which demponstrates that electrochemical deposition could satisfy both a high purity of struvite and a high removal efficiency of phosphate.

CONCLUSIONS

All the results and analyses demonstrated that the struvite content in the precipitate was controlled by both pH and Ca^{2+} in the original solutions. The optimal pH ranges for the purer struvite content (> 90%) were found respectively at 7.5– 9.0 for the ultra pure water solution and at 7.0–7.5 for the tap water (mainly consisting of ground water) solution. A high Ca^{2+} content in tap water resulted in more calcium compounds, which will inhibit the formation of struvite in the precipitate. It is therefore proposed to perform struvite precipitation at a neutral pH (<8.0) in real wastewater.

The experiments also indicated the feasibility of electrochemical deposition for recovering purer struvite from wastewater at a neutral pH, which could be easily realized at a low voltage (DC = 3–12 V). Furthermore, this method could realize both the high phosphate removal effciency in solution and the high content of struvite in recovered precipitate. This approach would make the engineering application of recovering pure struvite possible at the neutral pH. Further large-scale experiments have to be followed to ascertain key controlling factors affecting formation of struvite, and to develop some new setups with cheaper and stable anodes and cathodes.

ACKNOWLEDGEMENTS

The study was financially supported by National Natural Science Foundation of China (50978013), the China's High-tech R & D (863) Program (2006AA06Z320), Funding Project for Academic Human Resources Development in Institutions of Higher Learning Under the Jurisdiction of Beijing Municipality (PHR20100508 and PHR 201008372) and the Scientific Research Common Program of Beijing Municipal Commission of Education (KM200910016009).

REFERENCES

Altinbas M., Yangin C. and Ozturk I. (2002). Struvite precipitation from anaerobically treated municipal and landfill wastewaters. *Water Science and Technology*, **46**(9), 271–278.

Banks E., Chianelli R. and Korenstein R. (1975). Crystal chemistry of struvite analogs of the type $MgMPO_4.6H_2O$ (M = potassium(1+), rubidium(1+), cesium (1+), thallium(1+), ammonium(1+). *Inorganic Chemistry*, **14**, 1634–1639.

Battistoni P., Fava G., Pavan P., Musacco A. and Cecchi F. (1997). Phosphate removal in anaerobic liquors by struvite crystallisation without addition of chemicals: preliminary results. *Water Resources*, **31**, 2925–2929.

Boistelle R., Abbona F. and Madsen H. E. (1983). On the transformation of struvite into newberyite in aqueous systems. *Physics and Chemistry of Minerals*, **9**(5), 216–222.

Gabrielli C., Maurin G., Francy-Chausson H., Thery P., Tran T. T. and Tlili M. (2006). Electrochemical water softening - principle and application. *Desalination*, **201**, 150–163.

Hao X. D., Dai J., Cao Y. and Hu Y. S. (2008). Experimental study on the effect of COD/P ratios and phosphate recovery on a BNR system. *Environmental Science*, **30**(11), 3098–3103. (in Chinese).

Hao X. D. and van Loosdrecht M. C. M. (2006). Model-based evaluation of struvite recovery from P-released supernatant in a BNR process. *Water Science and Technology*, **53**(3), 191–198.

Hao X. D., Wang C. C., Lan L. and van Loosdrecht M. (2008). Struvite formation, analytical methods and effects of pH and $Ca2+$. *Water Science and Technology*, **58**(8), 1687–1692.

Jaffer Y., Clark T., Pearce P. and Parsons S. (2002). Potential phosphorus recovery by struvite formation. *Water Resources*, **36**, 1834–1842.

Kim D., Ryu H., Kim M., Kim J. and Lee S. (2007). Enhancing struvite precipitation potential for ammonia nitrogen removal in municipal landfill leachate. *Journal of Hazardous Materials*, **146**(1–2), 81–85.

Le Corre K., Valsami-Jones E., Hobbs P. and Parsons S. (2005). Impact of calcium on struvite crystal size, shape and purity. *Journal of Crystal Growth*, **283**(3–4), 514–522.

Lee S., Weon S., Lee C. and Koopman B. (2003). Removal of nitrogen and phosphate from wastewater by addition of bittern. *Chemosphere*, **51**, 265–271.

Miles A. and Ellis T. (2001). Struvite precipitation potential for nutrient recovery from anaerobically treated wastes. *Water Science and Technology*, **43**(11), 259–266.

Nelson N., Mikkelsen R. and Hesterberg D. (2003). Struvite precipitation in anaerobic swine lagoon liquid: effect of pH and Mg:P ratio and determination of rate constant. *Bioresource Technology*, **89**(3), 229–236.

Pastor L., Mangin D., Barat R. and Seco A. (2008). A pilot-scale study of struvite precipitation in a stirred tank reactor: conditions influencing the process. *Bioresources Technology*, **99**(14), 6285–6291.

Ronteltap M., Maurer M. and Gujer W. (2007). Struvite precipitation thermodynamics in source-separated urine. *Water Resources*, **41**, 977–984.

Stratful I., Scrimshaw M. and Lester J. (2001). Conditions influencing the precipitation of magnesium ammonium phosphate. *Water Resources*, **35**(17), 4191–4199.

Suzuki K., Tanaka Y., Osada T. and Wike M. (2002). Removal of phosphate, magnesium and calcium from swine wastewater through crystallization enhanced by aeration. *Water Resources*, **36**, 2991–2998.

Wilsenach J., Schuurbiers C. and van Loosdrecht M. (2007). Phosphate and potassium recovery from source separated urine through struvite precipitation. *Water Resources*, **41**, 458–466.

Yigit N. and Mazlum S. (2007). Phosphate recovery potential from wastewater by chemical precipitation at batch conditions. *Environmental Technology*, **28**, 83–93.

Comparison of Activated Alumina and Coal Sand Filter Media for Phosphorus Removal

Junling Wang[a,b], Yajun Zhang[a] and Junqi Li[b]

[a]School of Environment and Energy Engineering, Beijing University of Civil Engineering and Architecture, Beijing 100044, China
[b]Key Laboratory of Urban Stormwater System and Water Environment, Ministry of Education, Beijing, 100044, China
E-mail: wangjunling@bucea.edu.cn

Abstract The performance of activated alumina and coal sand was tested for removing phosphorus in water filtration process. Experiments demonstrated that the turbidity removal efficiency of activated alumina was somehow lower than coal sand. The removal efficiency (60–80%) of activated alumina on TP was higher than that (40–60%) of coal sand. The effect of pH on removing phosphorus was minor with coal sand. The STP (soluble total phosphorus) removal efficiencies of activated alumina and coal sand fell in the ranges of 50–90%, 20–50% respectively, on which an optimal pH value was controlled at 5–6. The SRP (soluble reactive phosphorus) removal efficiencies of activated alumina and coal sand ranged from 50% to 90% and from 20% to 50% respectively when the pH value was controlled at its optimum. The PP (particulate phosphorus) removal efficiencies of activated alumina and coal sand were in the range of 50–65%, 65–85% respectively, and pH was not a major controlling factor. The different performances of the different media on phosphorus removal were mainly caused their different adsorptive capacities, on which activated alumina was stronger than coal sand on removing STP.

Keywords Filter media, activated alumina, coal sand, phosphorus removal, drinking water

INTRODUCTION

Generally the proportion of organic carbon, nitrogen and phosphorus is 100:10:1 for microorganism regrowth in pipe water, therefore, decreasing phosphorus content in drinking water can restrain the regrowth of microorganism. It was reported that phosphorus would replace the organic matter as the limitation factor of microbiological regrowth in drinking water with high organic matter concentration (Lehtola *et al.* 2001, Sang *et al.* 2003). Most surface water sources have been polluted by organic matter severely in China, thus the phosphorus will be one limitation factor in Chinese drinking and raw water.

Many researchers have studied technology for removing phosphorus from drinking water. It was reported that membrane treatment technology (Dietze *et al.* 2003) and the ozonation process (Nishijima *et al.* 1997) were two techniques to remove phosphorus in drinking water, but these methods were not efficient enough. Some researchers concluded that traditional water treatment processes could not achieve efficient performance for phosphorus removal (Jiang *et al.* 2004a, 2004b), so new filter media to improve the removal of phosphorus were explored for the purposes of keeping drinking water biologically stable. Activated alumina was a kind of media with high adsorptive capacity for many pollutants, but its ability to remove phosphorus from drinking water has not been revealed; moreover, application feasibility of activated alumina as filter media should be verified.

Experimental Material

Filter media: granular activated alumina, silica sand and anthracite, characteristics of these shown in Tab. 1.

Filter apparatus: Two filter columns made of polymethyl methacrylate. The height of each column was 2 meters, and the inner radius 35 cm. Piezometer tubes, flow meters and backwashing pipes were installed. In one column 700 mm activated alumina filter media were filled, and in the other column 350 mm anthracite and 350 mm silica sand were filled in the upper and lower halves.

Table 1 Sorts and physical characteristics of filter media

Item	Activated alumina	Silica sand	Anthracite
Granular size/mm	1.2	0.5–1.25	1–1.8
Density/(g·cm-3)	1.04	2.64	1.60
Porosity/%	44.9	32.1	52.7

Experimental Methods

The velocity of filtration is 5m/h under experimental condition, controlled by a valve. The pH value of raw water was adjusted by adding NaOH or HCl with concentration 5% of each. The experimental environment's temperature was 15°C.

The backwashing flow for the coal sand filter was 4–6L/(s·m²) for 10 minutes. Activated alumina was dipped in 5% sulfuric acid solution for 24 hours, and washed by distilled water before being used. Activated alumina was regenerated by being back washed for 30 minutes using 10% alumina sulfate solution, subsequently washed with distilled water.

RESULTS

TP and Turbidity Removal

Raw water was prepared with the addition of KH_2PO_4 and a little clay into tap water, and turbidity was 2.55. TP (Total Phosphorus) was 39.19 μg/L, STP (Soluble Total Phosphorus) 31.34 μg/L.

From the start of filtration, effluent was sampled at 10 minutes, 0.5, 1, 3, 5, 7, 10, 15 and 17 hours; then TP and turbidity of samples were tested. The results are shown in Fig. 1.

According to Fig. 1, TP removal efficiency by activated alumina was more effective than that of coal sand, but their turbidity removal efficiency levels were the opposite. If 10 µg/L TP in effluent was the limited line, the average TP removal rate was in the range of 70–80% by activated alumina, and 45%–55% by coal sand. The breakthrough time of activated alumina filtration was 9 hours, and 12 hours for coal sand.

In raw water main phosphorus form was SRP (Soluble Reactive Phosphorus). The adsorptive capacity of activated alumina for SRP was 3000 µg/L (Wang et al. 2007, 2008), but coal sand filter media did not have adsorptive ability for phosphorus, so TP removal rate by activated alumina was higher than coal sand.

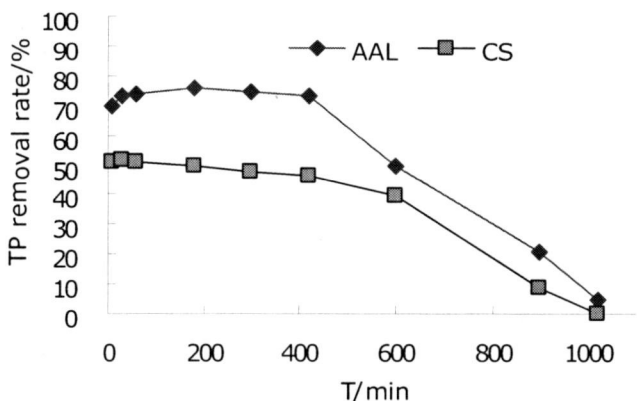

Figure 1 Comparison of TP removal rate by activated alumina and coal sand filter media (AAL—activated alumina, CS—coal sand)

The average activated alumina granular size was 1.2 mm in these experiments, but the coal sand filter medium granular size was 0.5–1.8 mm. The gradation of coal sand was proper and the scale of size was wider than activated alumina, so the holding performance for particles was better than activated alumina. It was reported that anthracite performed well for the removal of small sized particles with a diameter from 2 µm to 3 µm (Zhang et al. 2007). In addition, the shape of activated alumina granular was spherical. But the shapes of both sand and anthracite granules were irregular. They could hold suspended solids easily. The turbidity of water was influenced by the quantity of suspended solid particles in most cases; and so coal sand was superior to activated alumina for turbidity removal. Moreover, the physical holding ability is the dominant method for removing particles and turbidity of water.

Effect of pH

The concentration of TP, STP, PP (Particulate Phosphorus) and SRP was 61.14 µg/L, 39.7 µg/L, 21.4 µg/L and 32.9 µg/L respectively in raw water. The influent pH value as adjusted at 4, 5, 6, 7, 8, 9 and 10. Water samples were obtained from column effluent at 10, 15, 30, 60, and 120 minutes after filtration started.

TP removal

TP removal rate would be the highest with pH value 4–6. Under suitable pH value conditions the removal rate could rise from 60% to 80%, and acid environment condition was benefit for TP removal by activated alumina. For coal sand media, the effect of pH was indistinct, because the main way to remove TP in filtration was to hold the PP by physical methods. The function of holding was hardly affected by pH value. The result shown was that the TP removal rate by coal sand media was at the range of 40%–60%, lower than 60%–80% by activated alumina.

STP removal

With different pH values, the removal efficiency for STP was remarkably different, especially the STP concentration in effluent reached the lowest level when the pH value was 4, and after 20 minutes the effluent STP concentration was lower than 10 µg/L. As for coal sand filter, the effect of pH was not obvious. As a result, effluent STP concentration by coal sand ranged from 20 µg/L to 30 µg/L, but that by activated alumina was at the range of 10–20µg/L. With different pH values, the removal efficiency for STP by activated alumina was much better than coal sand.

It was reported that adsorptive capacity of activated alumina for phosphorus was higher than coal sand (Ding *et al*. 2002), and the optimum pH value was at the range of 5–6; undoubtedly it was close to the result of this experiment.

SRP removal

The SRP removal rates vary with pH value. Under acid conditions (pH = 4–5) the effluent SRP concentration was the lowest, especially when pH value was 4 and the effluent SRP concentration was lower than 10 µg/L. As for the use of coal sand filters, the impact of raw water pH was unapparent on effluent SRP concentration levels. It was concluded that SRP removal rate by activated alumina ranged between 50% and 90%, but 20%–50% by coal sand. The SRP could not be removed by physical holding behavior in filtration, so the higher removal rate by activated alumina could only be explained that adsorption was the main method for SRP removal. The removal rate for SRP by coal sand was very low because it didn't have the adsorptive ability for phosphorus.

PP removal

With different pH value, the removal rate for PP by both activated alumina and coal sand varied irregularly; however, the effluent PP concentration was different at range of 0–14 µg/L by activated alumina filter and 0–7 µg/L by coal sand. It was concluded that the PP removal rate by coal sand was higher than that by activated alumina whatever the pH was, and removal rate of PP by activated alumina and coal sand were at the range of 50–65% and 65–85% respectively.

Comparison

STP removal efficiency by activated alumina was better than coal sand; and for PP removal it was the converse, as is shown in Fig. 2 and Fig. 3. The removal rate by activated alumina was higher than that by coal sand, because the proportion of STP to TP was very high in raw water, the removal rate of TP was mainly affected by STP content.

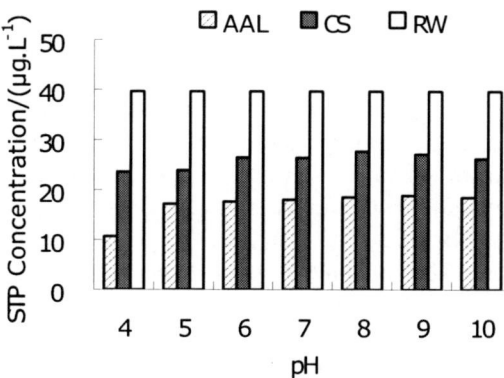

Figure 2 Comparison of STP concentration by activated alumina, coal sand and raw water with different pH value (AAL—activated alumina, CS—coal sand , RW—raw water)

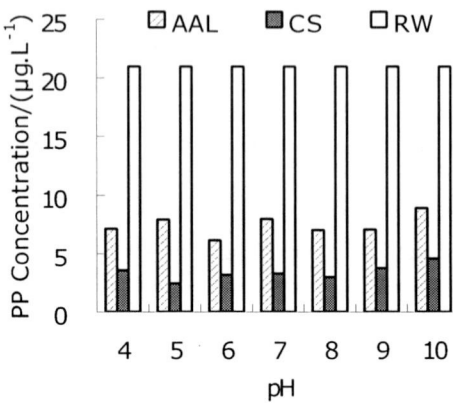

Figure 3 Comparison of PP concentration by activated alumina, coal sand and raw water with different pH value (AAL—activated alumina, CS—coal sand, RW—raw water)

Discussion

In fact, activated alumina is a type of porous material. Particulate matter in water can be removed by settling, inertia, holding and water kinetic mechanism, finally adhered on the surface of activated alumina granular. Additionally, activated alumina itself coagulates by numerous microparticles. There are a great deal of inner micropores in three types. One is in the midst of the crystal pore, the second is in the midst of the subcrystal pore, and the third is the lacuna pore. The average pore size is at range of 3–5 mm (Dong et $al.$ 1995). All of these pores form the large inner surface. The biggest specific surface area of some kinds of activated alumina is more than $300\,m^2/g$. The adsorptive force for anion by activated alumina is followed as (Ding et $al.$ 2002): $OH^->PO_4^{3-}>F^->SO_4^{2-}>I^->Br^->Cl^->NO_3^-$, so the adsorptive force for phosphorus is at the top of this range. This explains the main cause of strong adsorptive capacity for phosphorus in filtration.

In addition, the surface isoelectric point pH value of activated alumina is 9.5 more than quartz sand 4–5 and anthracite 0.7–2.2, so in water there is a great deal of OH^- on the surface of the activated alumina, ion exchange occurs easily. The surface isoelectric point pH value means that both coal and sand do not have adsorption ability even if they had large surface area, therefore, removal rate by them only depend on the holding capacity for PP removal.

CONCLUSION

Experimental study showed that activated alumina can remove TP with higher efficiency and turbidity, and with lower efficiency than coal sand.

Under different pH condition, TP removal rate by activated alumina was higher than coal sand, the average removal rate for phosphorus by activated alumina being in the range of 60%–80%, but 40%–60% by coal sand. STP removal rate by activated alumina and coal sand were 50%–90%, 20%–50% respectively, and SRP removal rate by activated alumina ranged from 50% to 90%, which was higher than 20%–50% by coal sand. The PP removal rate by activated alumina ranged between 50% and 65% that was lower than 65%–85% by coal sand. Activated alumina could remove TP efficiently because its high adsorptive capacity for SRP, and SRP was the main component of TP in raw water. The turbidity and PP removal rate depended on the physical method, so removal rate of turbidity and PP by coal sand was higher than activated alumina.

In order to improve the phosphorus removal in drinking water to meet biological stability, it is suggested that activated alumina filter media should be used in filtration as STP is in higher proportion to TP in raw water.

ACKNOWLEDGEMENTS

This work was financed by Major Projects on Control and Rectification of Water Body Pollution (2009ZX07317–005, 2009ZX07419-006) and National Key Technology R&D Program (2008 BAJ08B13-04). We thank the laboratory staffs of the laboratories of School of Environment and Energy of BUCEA.

REFERENCES

Dietze A., Wiesmann U. and Gnirss R. (2003). Phosphorus removal with membrane filtration for surface water treatment. *Water Science and Technology: Water Supply*, **3**(5), 23–30.

Ding W. M. and Huang X. (2002). Progress of studies on phosphorus removal from wastewater by adsorbents. *Chinese Journal of Environmental Engineering*, **3**(10), 23–27.

Dong W. Y., Su Y. L. and Wang W. X. (1995). Preparation and characterization of spherical activated alumina. *Industrial Catalysis*, (2), 36–41.

Jiang D. L. and Zhang X. J. (2004a). Relationship between phosphorus and bacterial regrowth in drinking water. *Chinese Journal of Environmental Science*, **25**(5), 57–60.

Jiang D. L. and Zhang X. J. (2004b). Phosphorus in drinking water and it's removal in conventional treatment process. *Chinese Journal of Environmental Science*, **24**(5), 796–801.

Letola M. J., Miettinen I. T., Vartiainen T., Myllykangas T. and Martikainen P. J. (2001). Microbially available organic carbon, phosphorus and microbial growth in ozonated drinking water. *Water Research*, **35**(7), 1635–1640.

Nishijima W., Shoto E. and Okada M. (1997). Improvement of biodegradation of organic substance by addition of phosphorus in biological activated carbon. *Water Science and Technology*, **36**(12), 251–257.

Sang J. Q., Yu G. Z., Zhang X. H. and Wang Z. S. (2003). Relation between phosphorus and bacterial regrowth in drinking water. *Chinese Journal of Environmental Science*, **24**(4), 81–84.

Wang J. L., Feng C. M., Yang Y. L. and Li G. B. (2008). Experiment on phosphorus removal by activated alumina. *Journal of Beijing University of Technology*, **34**(6), 621–625.

Wang J. L., Wu J.Q., Long Y. J. and Li G. B. (2007). Comparison on phosphorus and turbidity removal by activated alumina and other filter materials. *Chinese Journal of Environmental Engineering*, **1**(10), 18–21.

Zhang X. L., Tang S. Q. and Li T. (2007). Comparison of enhanced filtration effect between bio-ceramic filter and anthracite filter. *China Water and Waste Water*, **23**(1), 72–76.

The Application of the Probiotics Principle to Convert Biomass into Organic Fertilizer

Saburo Matsui

Emeritus Professor Kyoto University, Matsui Consulting Firm of the Environment 10–45 Uchihata-cho, Hanazono, Ukyo-ku, Kyoto City, Japan 616-8045
Fax 81.75.464.5860
E-mail: m36@3kankyo.co.jp

Abstract Japan imports 100% of its phosphate for agriculture and industry. We know that phosphorus is a more rapidly depleting resource than oil. In addition to this, Japan faces severe problems with eutrophication in both fresh and marine waters. Nitrate contamination in ground water, is also an impending problem. In order to alleviate both of these problems, less chemical fertilizer must be used. However, when reducing the usage of chemical fertilizer, agricultural production must be sustained in order to avoid famine. This can be done by enhancing the recycling of nutrients. Sludge from sewage contains a high nutrient content. It also contains some hazardous materials such as heavy metals, drugs, industrial chemicals, etc.

Organic farming has advantages in three folds: the recycling of nutrients; the usage of humus as soil amendment; and the application of probiotics. Organic farming does not utilize insecticides, herbicides or fungicides. Avoiding these chemicals, it utilizes beneficial microbes in its composting methods. In order to develop beneficial microbes in composting, probiotics are applied. This type of agriculture practices three types of composting methods used both separately and in combination with following bacteria species: Bacillus; Lactic acid; and Actinomycetous. In general, these bacteria types are harmless to human health. In fact, they can enhance natural human immunity. When applying probiotics to agricultural fields, the microbes excrete auxin and/ or cytokinin which are major plant hormones for growth and fruit bearing. These types of bacteria are also beneficial to humans, as well as to agricultural fields. Supporting the need to reduce the usage of chemical fertilizer in agriculture, the application of probiotics has great potential for further development.

Solving the problem of sludge contamination with its hazardous materials such as heavy metals, drugs, industrial chemicals, etc is a major issue. I discovered that sub-critical water oxidation of sludge can destroy most hazardous organic chemicals by strong hydrolysis. There is also the possibility of fixing heavy metals in silica soil so that they are not dissolved in water.

Keywords Organic Farming, Biomass, Sewage Sludge, Probiotics, Bacillus bacteria type, Lactic acid bacteria type, Actinomycetous type

INTRODUCTION

The installation of water infrastructure has greatly contributed to the production of clean water. In developed countries, such as Japan, more than 80% of its sewage sludge and industrial sludge are incinerated. Sludge incineration needs kerosene to ensure cleaner gas emissions. However, this practice faces another problem. In order to meet the Kyoto Protocol and beyond, there needs to be a reduction of GHG emissions from all public sectors. Water environmental engineers need to seek for solutions other than incineration.

Japan imports 100% phosphate for agriculture and industry. We know that phosphorus is a more rapidly depleting resource than petroleum. Another point to consider is that Japan faces severe eutrophication problems in both fresh and marine waters. In addition to this, there are problems with nitrate contamination in ground water. Both problems necessitate less use of chemical fertilizers such as nitrogen and phosphate. However, in doing this, we must keep agricultural production levels high in order to avoid famine. This can be done by recycling the nutrients from urban waste and then using them for agricultural fertilization. (Matsui, S, *et. al.* 2001, 2002).

Sewage sludge contains high nutrient content. It also contains hazardous materials such as heavy metals, drugs, industrial chemicals, etc. The use of innovative technology is necessary to overcome those problems. However, the reuse of organic waste as an agricultural fertilizer still provides a fundamental solution. In this study, it was found that there are many farmers in Japan who successfully practice organic farming. When visiting these sites, it seemed that the principle of composting was generally successful.

Organic fertilizer has advantages in four folds; the recycling of nutrients; the use of humus as a soil amendment; the application of probiotics; and the taste and flavor of products. Organic farming does not utilize insecticides, herbicides or fungicides. Avoiding those chemicals, it utilizes beneficial microbes in its compositing methods. To develop beneficial microbes in composting and to supply them into soil, the use of probiotics is applied.

CHEMICAL FERTILIZER, PESTICIDE AND HERBICIDE

Based on a massive use of chemical fertilizers, pesticides and herbicides, a large scale monoculture of major crops can be produced to provide enough food for human kind. The major food supplying countries, however, practice modern agricultural methods that are not sustainable. Problems arise such as soil deterioration, land and water ecosystem destruction, ground water contamination, and the depletion of economical phosphorus, etc. In spite of efforts to alleviate matters, through the use of eco-toxicity and agrochemicals, there are still reports of ecological deterioration in agriculture fields. In a recent report, it was ironically discovered that Japanese farmers who apply pesticide and herbicide to their fields cultivated for commercial use, would never apply them to their fields, cultivated for their own consumption. Pesticides and herbicides are used in order to control continuous crop hazards. However, in the past, farmers did not use them. What technology did they apply to overcome the problem?

The topic of Chemical fertilizer has been brought into question further. The massive use of nitrogen fertilizer brings a whole new dimension to the problems relating to the global water contamination. Lakes and reservoirs are continuously being deteriorated by eutrophication, changing ecosystems, food webs and the extinction of certain species. The contamination of water by Nitrogen is the biggest global issue next to global warming.

THE EU ORGANIC FARMING APPROACH

According to the EU regulations on organic farming, "organic farming is an agricultural system that seeks to provide you, the consumer, with fresh, tasty and authentic food while respecting natural life-cycle systems." To achieve this, organic farming relies on a number of objectives and principles, as well as common practices designed to minimize the human impact on the environment, while ensuring the agricultural system operates as naturally as possible (EU organic farming 2010)

Typical organic farming practices must include:

* Wide crop rotation as a prerequisite for an efficient use of on-site resources
* Very strict limits on chemical synthetic pesticide and synthetic fertilizer use, livestock antibiotics, food additives and processing aids and other inputs
* Absolute prohibition of the use of genetically modified organisms
* Taking advantage of on-site resources, such as livestock manure for fertilizer or feed produced on the farm
* Choosing plant and animal species that are resistant to disease and adapted to local conditions
* Raising livestock in free-range, open-air systems and providing them with organic feed
* Using animal husbandry practices appropriate to different livestock species

When farmers practice animal husbandry following these principles, they must not apply antibiotics to their animals. Farmers must obtain feed that is cultivated without chemical synthetic pesticides or synthetic fertilizers. Animal dung is an excellent fertilizer for such feed that does not include antibiotics. It is important to establish a cycle of food, organic farming and organic waste. It would be very difficult to practice these methods in large scale agriculture, but in principle, these practices should be applied. By following these principles, Scientists and engineers can help develop true organic farming.

In order to develop these practices, technologies on how to prepare and supply good/effective compost for agricultural fields under variety conditions of farming must be established. There are many types of organic waste that can be used for producing compost. This can include left over harvest crops, vegetable and fruits, unused parts of plants, livestock manure and raw garbage. It is still important, at this stage, for farmers of organic farming to produce good/effective compost for organic farming.

Organic farming has advantages in at least four areas; recycling nutrients; utilizing humus as soil amendment; limiting the use of chemical synthetic pesticides; and producing tasty and authentic food. In order to achieve success in these four areas, the application of probiotics to compost can be used. Selecting a good species of bacteria for organic waste is crucial in producing effective compost that sustains good bacteria in the end product. When compost is applied to soil in fields, the bacteria starts to proliferate and work as a microbial chemical against bad microbes. This is one of the methods of probiotics agriculture.

THREE GROUPS OF GOOD/EFFECTIVE BACTERIA FOR PROBIOTICS AGRICULTURE

In traditional methods of Japanese composting, at least three groups of compositing bacteria were used individually and, or in combination. The following species were used: *Bacillus bacteria groups*, *Lactic acid bacteria groups* and *Actinomycetous groups*. These bacteria species are in the group of Gram positive and protect agro-products from cropping hazards. They do this by expelling against various bad worms and insects, such as nematodes with potatoes and some types of insects with soybeans and maize. They are also effective in controlling fungi such as powdery mildew, downy mildew, *phythium* (damping off with many plants), *plasmodipophora brasscae* (club-root with the cabbage family); *Cruciferae* (plants, and fusarium of wilt with tomato and banana) (Matsui, S, 2009).

Important points to consider when using these three bacteria with different types of organic waste are: oxygen supply, temperature control and moisture control.

Bacillus Species Composting

Bacillus species includes *Bacillus subtilis, Bacillus subtilis* var. *natto, Bacillus thuringiensis,(Bt),* etc. They are strict aerobic bacteria which rapidly decompose any type of organic waste such as animal dung, raw garbage, sewage sludge and many others. Strict aerobic conditions need to be provided from the start to the end of composting. There needs to be an optimum supply of air coming from the bottom of the composting yards and a roof protecting the mound of compost from rainwater. If the mound is big enough, the temperature in the mound becomes very high, exceeding 85°C or more. This is called hyper thermophilic composting. Such high temperatures are enough to kill viruses, Gram negative bacteria, and the eggs of worms and insects etc. Since most of the bacterial pathogens, dangerous to humans and animals are in this group, this type compost products are safe. Due to aerobic conditions, anaerobic bacteria cannot grow. Spore forming bacteria in the Gram positive group may survive, but these are generally not harmful to humans or crops. There is an exception, however, *Bacillus anthracis,* which is the most dangerous type of bacteria. During composting of organic waste that contains protein, producing high ammonia is unavoidable. Decomposing amino acids creates the odor problem. This requires good control of ventilation in composting yards. It takes at least 6–7 weeks to stabilize the compost in the middle-latitude zone. It also requires 6 to 7 compartments in order operate the shifting and mixing of the mound. When brought into agricultural fields, a lot of spores of the *Bacillus species* remain. They then start to germinate, working as a microbial pesticide or herbicide or fungicide.

Lactic Acid Bacteria Composting

Lactic acid bacteria species include *Lactobacillus sp, Bifidobacterium sp, Enterococcus sp, Lactococcus sp, Pediococcus sp, Leuconostoc sp.,*etc. They can grow in either aerobic or anaerobic condition, rapidly decomposing any type of organic waste. They some times work together with *Yeast species* such as *Saccharomyces sp. Schizosaccharomyces sp.* When decomposing organic waste that is rich in carbohydrate, they prefer more anaerobic conditions. It is not necessary to provide aerobic conditions from start to finish in composting. However, it is still necessary to provide a roof over the mound of compost for rain water protection and good drainage from the bottom of the mound. When the mound is big enough, temperatures inside the mound become high, exceeding 65 to 85°C. This is thermophilic composting. If the temperature is not high enough, some gram negative bacteria may survive. However, if the pH of the compositing mound drops below 5, many Gram negative bacteria may die off in the acidic conditions. Pathogenic viruses cannot survive. Spore producing Gram positive bacteria may survive. It is possible that some of the eggs of worms and insects may survive under the acidic conditions. It takes 2–3 months, depending on the composting temperature in the middle-latitude zone, to get full matured compost. When the organic waste in composting contains carbohydrate, fat, oil, and protein, the bacteria can produce lactic and amino acids. This drops the pH of the compost, and can help growth of plants in fields. Odor problems, due to the emission of ammonia, are not as severe as Bacillus species composting. It is important to ensure avoiding the survival of pathogenic yeast, such as *Candida Pneumocystis jiroveci* and pathogenic lactic acid like *Clostridium perfringens.*

Actinomycetous Species Composting

Actinomycetous species include *Actinoplanes sp, Ampullariella, Dactylosporangium sp, Streptomyces sp*, etc. If grown in strict aerobic conditions, they can decompose any type of organic waste. They are particularly good at decomposing oil, fat, and plant cell walls with lignin, latex and chitin. In composting, it is necessary to provide aerobic conditions from start to finish. Protection from rain water must be provided over the mound as well as good drainage from the bottom. When the mound is big enough, the inside temperature becomes relatively high, exceeding 35 to 65°C. This is mesophilic composting. As the temperature is not high enough, some gram negative bacteria may survive. Other Gram negative bacteria may die off under the long starving conditions. Pathogenic viruses cannot survive in prolonged composting periods. Spore producing Gram positive bacteria may survive as some of the eggs of worms and insects may survive. It takes 3–5 months, depending on the composting temperatures and air supply in the middle-latitude zone, to obtain matured compost. When composting with organic waste that contains carbohydrate, fat, oil, and protein, the bacteria produces low molecule organic and amino acids. This drops the pH of the compost, and can help growth of plants in agricultural fields. The matured compost has a unique smell of earth or fungus as they form a fine, whitish, net of bacteria covering the surface of the compost mound. Bark compost is commercially available and is produced by Actinomycetous species compsoting. Since many types of antibiotics are produced by the fermentation of *Streptomyces*, *Actinomycetous species* in the compost, they are effective in controlling some of the pathogens to crops and vegetables. There are several important pathogenic *Actinomycetous species* including: *Mycobacterium tuberculosis, Mycobacterium leprae, Corynebacterium diphtheriae, and Mycobacterium bovis.*

AUXIN AND CYTOKININ

It is understood that the *Bacillus species* produces auxin *that is a key plant hormone for growth.* In this study, it was found that the *Actinomycetous species* produces cytokinin which is another key plant hormone for extending leaves and bearing fruit. It was also discovered that the some *Lactobacillus species* produces cytokinin. When good compost is added to soil, those types of composting bacteria continue excreting auxin and/or cytokinin. Both stimulate plant growth and enhance bearing fruit. The sweetness in fruit, comes from its sugar content. The application of bark compost, greatly improves the sugar content of strawberry whilst suppressing the problem of continuous cropping. Bacillus species composting greatly helps to stop nematodes of nemtodosis with potatoes and enhances sweetness of the potatoes. Lactic acid bacteria composting stimulate growth of plants and accelerates bearing in fruit. The flavor and taste of leaf vegetables are very distinct. The sweetness of fruits such as mandarin orange, melon, dragon fruit, etc. is effectively enhanced. At the same time, crop hazards are controlled.

It was found that Japanese consumers recognized the *good taste* in organic farming product. Due to its quality, in terms of safety as well as taste, the Japanese accept high prices. In this study, the application of good bacteria to organic farming is called probiotics.

auxin

cytokinin

COMMONALITY OF GOOD BACTERIA AMONG HUMANS, ANIMALS AND PLANTS

Good bacteria are often used to process food such as alcohol, vinegar, yogurt, cheese, etc. They are also used as industrial applications, such as *Clostridium sp* (the production of acetone, methanol and butanol), *Clostridium acetobutylicum* (the production of acetic acid from alcohol), *Acetobacter aceti* (the production of sugar solution to acetic acid), *Mycoderma aceti* (the production of lactose to lactic acid, etc). Retting jute for fibers employs activities such as: *Bacillus subtilis, B. polymyxa; Clostridium tertium and C. felsimium*. Silage preparation using *Streptococcus sp* and *Lactobacillus sp* is key in supplying good feed to cattle. The Curing of tea and tobacco leaves is processed by activities of *Bacillus megatherium* (tea and tobacco). *Mycococcus candisans* (tea). *Bacillus subtilis natto* is utilized to produce the popular food in Japan, Natto (fermented soybeans).

Many good and bad bacteria species are used in the human immune system, such as *Escherichia.coli, Lactobacillu sps, Streptocopccu sp, Staphylococcus sp* located in human intestine. Some synthesize Vitamin K and B and help with food fermentation. Both within and without the body, friendly bacteria are always competing with infectious bacteria for space. We inhale and ingest virulent bacteria on a daily basis. Without the normal oral and intestinal flora, disease would be much more common, and far more severe. Intestinal bacteria are expelled and renewed daily.

Amongst the bacteria, *Lactobacillus acidophilus* and other lactic acid bacteria aid the digestion of lactose. This is done through the production of niacin and folic acid, discouraging the colonization of other bad bacteria. They also assist in the recycling of amino acids from bile to digestion. Another area they are found is the vagina. Protecting the womb by producing lactic acid, it discourages fungal growth. It is the reason why women are susceptible to infections of *Candida* when taking some types of antibiotics. Large groups of lactic acid bacteria that are utilized for preparation of yogurts, cheese, pickled vegetables, etc can aid in dairy consumption. They are practical in terms of immune responses and digestion, as well as for alleviating lactose intolerance. More than half of the body's immune tissue is located in the lining of the small intestine. When taking antibiotics, a person is much more susceptible to infection. This is because the good bacteria, having been depleted, leaves space for other bacteria to grow. This also occurs when antibiotic lotions or creams are used in excess upon the skin.

Other bacteria examples are *Staphylococcus epidermidis* and *Proprionibacterium acnes which* are less virulent flesh-eating bacteria. *Staphylococcus aureus* could take their places if given the opportunity. *P. acnes* is often the cause of skin acne. However, it is usually commensal, meaning that exists on the skin without doing any harm. They are in the opportunistic pathogen category. The bacterium lives off of fatty acids and sebaceous fluid secreted by the pores of the skin. *Streptococcus mutans* is present in the mouth and converts sucrose, sugar, to lactic acid. If not kept in check, this species can cause dental plaque and tooth decay.

Good bacteria, beneficial to the human body, can also be beneficial to other mammals such as cattle. Three groups of bacteria namely ***Bacillus bacteria groups***, ***Lactic acid bacteria groups*** and ***Actinomycetous groups*** are also useful for cropping. This provides a solid foundation for the development of probiotics in agriculture, namely organic farming. The use of probiotics in agriculture has great potential for further development; it supports the need to reduce the application of chemical fertilizer and agrochemicals.

THE APPLICATION OF SUBCRITICAL WATER REACTION TO SEWAGE SLUDGE AND ORGANIC WASTE

Composting of any organic waste can be done, so long as the pretreatment of organic waste has been properly processed. Sludge from large cities contains many synthetic chemicals including heavy metals, industrial, medical and natural hazardous substances, which calls into question the quality of the compost. To resolve this issue, the application of a sub-critical water reaction to sewage sludge pretreatment can be implemented. The water critical point is found at 374°C(647°K) and has a pressure of 218atm(22.1MPA). Beyond this point, the water status takes on supercritical conditions where strong oxidation and reduction reactions can take place. At this stage, it is possible to decompose any organic and inorganic compounds into an atomic status. This reaction takes place under high pressures and temperatures that require a robust metallic reactor. Economically, it is very difficult to operate. However, subcritical water reactions can be established and operated in more economical ways. Technology for this is developing rapidly in Japan and in other parts of the world. According to theoretical analysis and bench scale experimental results, this could be a promising application to use organic waste.

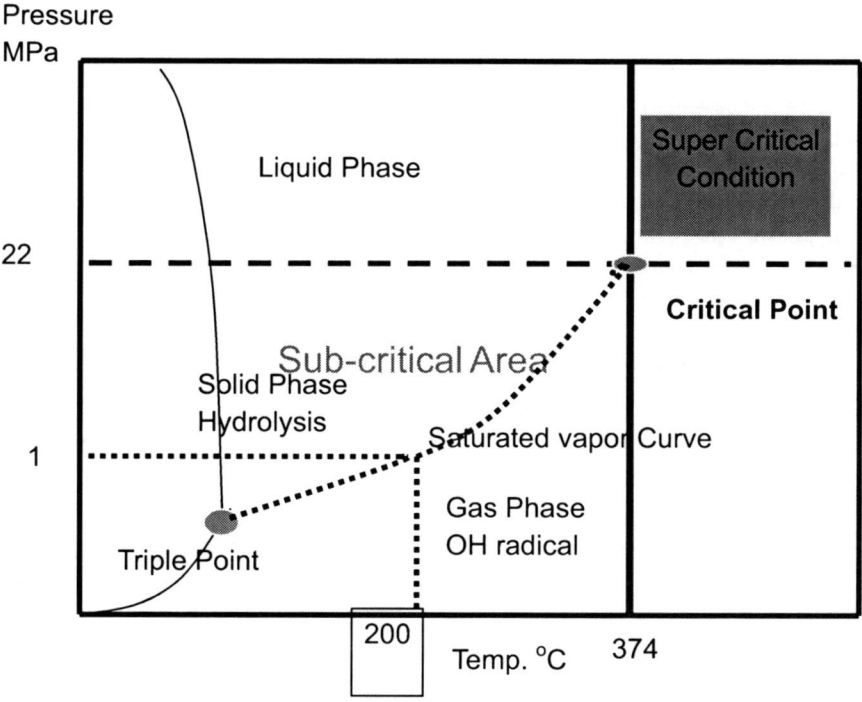

Figure 1 Water State at different Pressures and Temps

The subcritical water reaction follows the saturated vapor curve in Fig. 1. The area below the curve is in the gas phase where a strong OH radical reaction takes place. The area above the curve is in the solid phase where a strong hydrolysis reaction occurs. The ionizing power of water is strongest at a temperature around 250°C, so hydrolysis takes place the most around this point. Almost no oxidation occurs, and no production of CO_2 occurs. As the temperature increases, the dielectric constant of water decreases to those of methanol, ethanol and benzene, so that the extraction power of such subcritical water becomes very strong and can extract oil out of organic waste. Depending on the content of organic waste and the operational conditions of pressure and temperature, the reaction product might vary enormously. It can lead to the extraction of oil or other valuable substances such as the production of organic acids, or plant fibers, or plastic monomers, etc. Hazardous organic chemicals can be decomposed by strong hydrolysis, undergoing partial OH radical reaction. After the process of careful subcritical water reaction, most of the hazardous chemicals change into other, more harmless substances. Treated under different operational conditions, raw garbage is used to produce a product which can feed animals.

Sewage sludge can be processed into low molecule organic acids such as acetic acid. This is a source for methane fermentation. In traditional sludge digestion, two stages of anaerobic digestion are required: the digestion of acido-genesis and the digestion of methano-genesis. The subcritical water reaction completes the acido-genesis stage within one hour, so that methano-genesis immediately follows, giving fast energy production. As for heavy metals, research shows that it is possible to fix metals in the form of silicate amorphous by adding silicate material. If the treated sludge contains heavy metals that are not soluble under strong acid solutions, this guarantees the use of treated sludge in agriculture.

REFERENCES

Matsui S, Hevze M, Ho G. and Otterpohl R. (2001). *Emerging Paradigms in Water and ASnitation in Frontiers in Urban Water Management-Dead Lock or Hope.* IWA publishing, UK.

Matsui S. (2002). *The Potential of Ecological Sanitation, Japan Review of International Affairs*, No. 4. Vol. 16. The Japan Institute of International Affairs, Tokyo EU organic farming (2010). http://www.organic-europe.net/europe_eu/default. asp.

Matsui S. (2009). Probiotics principle that can help organic farming. *Journal of Environmental & Sanitary Engineering Research*, **31**. The Association of Environmental & Sanitary Engineering Research, Kyoto University.

Nitrification of Source Separated Urine in a Sequencing Batch Reactor

H. R. Mackey[a], Y. Chen[a], G. H. Chen [a]* and Mark M. C. van Loosdrecht[b]

[a]Department of Civil and Environmental Engineering, The Hong Kong University of Science and Technology, Clear Water Bay, Kowloon, Hong Kong, China
[b]Department of Biotechnology, Delft University of Technology, Julianalaan 67, NL-2628 BC Delft, The Netherlands.
E-mail: ceghchen@ust.hk

Abstract This paper summarizes recent work on nitrification of human urine separated at source and characteristics of human urine are reported. Full nitrification of diluted urine was achieved without inhibition in a sequencing batch reactor (SBR) under a loading rate of 1.1 kg-N/m^3.d, when necessary amount of base was added for maintaining pH at around 7. Granulation of the nitrifying biomass was observed with a long settling time (90 min).

Keywords Human urine, nitrification, granular sludge, SBR.

INTRODUCTION

Urine is a critical component of municipal wastewater. Although comprising less than 1% of the bulk volume of municipal wastewater (Maurer *et al.*, 2006) it contributes around 80% and 50% of the total nitrogen and phosphorous loads respectively to a municipal wastewater treatment plant (Sperandio *et al.*, 2008). This means separation and treatment of urine at source can provide many benefits such as a significant reduction of nutrient load and hence footprint at centralized treatment plants, opportunity for efficient recovery of nutrients and the ability to dose nitrate into the sewer to reduce sulfide production.

There are many available options for the treatment or pre-treatment of source separated urine depending on the final goal of effluent reuse or disposal. These include but are not limited to ammonia stripping, acidification for stabilisation, ion exchange or electrodialysis for nutrient recovery, membrane filtration for further concentration of pollutants, anaerobic oxidation such as Anammox to produce nitrogen gas or aerobic oxidation such as in an aerobic Sequencing Batch Reactor (SBR) to produce nitrite and/or nitrate.

The effectiveness of the chosen treatment process relies heavily on the level of hydrolysis of the urine. Urine initially comprises predominantly organic nitrogen in the form of urea but quickly hydrolizes to ammonia. This generates a significant rise in pH and precipitates containing phosphorous, nitrogen, magnesium and calcium such as struvite and apatite as shown in the data of Table 1. Therefore depending on whether conversion of chemicals or recovery is desired hydrolysis should be either promoted or prevented. This chapter mainly focuses on aerobic conversion of the organic nitrogen into nitrate using an SBR.

Table 1 Composition of fresh and hydrolyzed urine

Parameter	Unit	Fresh urine			Hydrolized urine		
		1	2	3	4	2	3
pH		6.2	6.2	6.4	9.0	9.0	9.2
TN	mg-N/L	8830	9200	5980	7909	9200	5900
TKN	mg-N/L	–	–	–	–	–	5600
NH_4^++NH_3	mg-N/L	463	480	374	7572	8100	5380
COD	mg-O_2/L	–	10000	6700	–	10000	6700
K	mg/L	2737	2200	1200	3485	2200	1200
Na	mg/L	3450	2600	3067	2842	2600	3067
Ca	mg/L	233	190	134	40.4	0	13.4
Mg	mg/L	119	100	55	4.5	0	1.01
Cl	mg/L	4970	3800	5077	6772	3800	5077
TP	mg-P/L	800–2000	740	735	606	540	550

1. Ciba-Geigy (1977); 2 Udert *et al.* (2006); 3 Chen (2009); 4 Kirchmann and Pettersson (1995).

NITRIFICATION USING SBR

Background

The SBR is a batch process reactor that operates on a four-stage cycle: namely fill, aerate, settle and decant. It is commonly used in small to moderate scale operations for both industrial and municipal wastewater treatment (Irvine and Ketchum, 1988; Artan and Orhon, 2005). It includes benefits such as small footprint by combining the reaction and settling tanks into a single entity, ideal plug-flow kinetics (Silverstein and Schroeder, 1983), easy control of feeding and cycle parameters such as hydraulic retention time (HRT) and sludge retention time (SRT) by altering feeding and decanting periods, and operation versatility to meet different treatment requirements by adding anoxic or anaerobic periods into the cycle (Arora *et al.*, 1985) and/or changing the settling and decant parameters (Liu *et al.*, 2005).

Treatment of high strength wastewaters is not uncommon with SBR technology. SBRs have been used successfully for treatment of landfill leachates such as at the Chandler Landfill, Brisbane, Australia (Doyle *et al.*, 2001). The preconstruction bench-scale testing achieved full nitrification with nitrogen loading rates of up to 5.91 gN/L.d and influent ammonia concentrations of 300–800mg–N/L. The full-scale plant is successfully in operation, initially achieving only partial nitrification under initial high leachate loads but has since managed full nitrification with a decrease in landfill leachate supply. The ammonia and chemical oxygen demand (COD) are similar to that which would be expected from diluted urine, hence, SBR is regarded as a suitable reactor for urine nitrification.

Urine Nitrification Chemistry

Currently, little research has focused on full nitrification of human urine, although there have been studies into partial nitrification of human urine (Udert *et al.*, 2003; Wilsenach and van Loosdrecht, 2006). The two significant issues with full nitrification of urine are inhibition from free ammonia (FA) and free nitrous acid (FNA), and inhibition by pH, caused by consumption of all available alkalinity. Nitrification of urine is a three-step processes. The first step is hydrolysis and often takes place outside of the reactor. This is undertaken by the enzyme urease (urea amidohydrolase) produced by a large number of bacteria where the overall reaction can be described as:

$$NH_2(CO)NH_2 + 2H_2O \rightarrow NH_3 + NH_4^+ + HCO_3^-$$

At room temperature with the appropriate urease-positive bacteria present in the collection systems complete or nearly complete hydrolysis of urea to ammonia could occur (Udert *et al.*, 2003). The following step is the conversion of ammonia to nitrite, carried out by ammonia-oxidising bacteria (AOBs). The combined reaction including cell growth requirements is given by (Wiesmann and Libra, 1999):

$$NH_4^+ + 1.98HCO_3^- + 1.38O_2 \rightarrow 0.0182C_5H_7O_2N + 0.98NO_2^- + 1.04H_2O + 1.89H_2CO_3$$

This reaction consumes nearly two moles of bicarbonate alkalinity per mole of ammonia. The next step is carried out by nitrite-oxidising bacteria (NOBs).

$$NO_2^- + 0.02H_2CO_3 + 0.49O_2 + 0.005NH_4^+ + 0.005HCO_3^- \rightarrow 0.005C_5H_7O_2N + NO_3^- + 0.015H_2O$$

Theoretically, the complete nitrification of ammonia to nitrate requires 7.07 mg alkalinity as $CaCO_3$ and 4.07 mg O_2 for every mg of NH_4^+-N where the urine provides approximately half of the required alkalinity. Both AOBs and NOBs are autotrophic bacteria and are therefore slow growers compared to their heterotrophic counterparts. The predominant AOB and NOB species are *nitrosomnas* and *nitrobacter* respectively. Both are sensitive to environmental conditions such as pH, temperature, FA and FNA conditions. A summary of the optimum conditions are given in Table 2. At low pH conditions FNA inhibition occurs, while at high pH FA inhibition can occur.

Table 2 Optimum conditions for urine nitrification

Parameter	Suitable Nitrosomonas Range[1]	Suitable Nitrobacter Range[1]	Optimum Nitrification Range[2]
Temperature (°C)	5–30	5–40	28–32
pH	5.8–8.5	6.5–8.5	7.2–8.0
DO (mg/L)			2–3
Generation time (hr)	8–36	12–60	
Sludge yield (g VSS/g substrate)	0.04–0.13	0.02–0.07	
NH_3-N(mg/L)	<10	<0.1	<0.1
NO_2-N(mg/L)	<1	<1	<1
SRT (d)			4–6

[1]Chen (2009)
[2]Wiessmann *et al.* (2007)

Critical Considerations

Hydrolysis of the influent ureum is an important factor in successful nitrification of human urine. In a study of full nitrification of human urine in an SBR by Chen (2009) it was found that in the initial month of operation high levels of total Kjeldahl nitrogen (TKN) remained in the effluent, indicating that the urea in the urine was not hydrolysed within the reactor. However, after the first month the reactor seemed to acclimatise to the influent and complete hydrolysis occurred. The reactor was started-up with biomass from a reactor fed with synthetic urine influent deficient in COD to promote autotrophic biomass. This may provide a possible solution to minimize struvite formation in the collection systems which ia mainly caused by an increase in pH during urea hydrolysis.

Due to the slow growth rate of autotrophic bacteria great care is required at start-up to prevent wash-out of the autotrophic biomass. In the initial weeks the effluent solids may well be in the same order of magnitude as the autotrophic biomass production. This can be overcome by quickly increasing the loading rate, starting with extended settling periods and implementing a temporary secondary settling system to recycle effluent solids during the first few weeks.

Due to the high N-to-COD ratio a higher proportion of nitrifying bacteria is present within the biomass (Chen, 2009; Liu *et al.*, 2007). This has been found to be beneficial to settleability of the biomass providing SVI values significantly below 100 even in the absence of granulation (Doyle *et al.*, 2001). Biomass high in autotrophic bacteria has been found to be more compact and dense aiding in settleability (Doyle *et al.*, 2001; Liu *et al.*, 2004; Liu *et al.*, 2007), possibly due to a strong correlation between growth rate and granule density (de Kreuk and van Loosdrecht, 2004).

Inhibition from FA can be managed by controlling the feeding pattern of the sludge. By splitting the feeding into three evenly spaced periods over a 12-hour cycle, Chen (2009) was able to achieve nitrification with no inhibition even with influent ammonia concentrations of up to 1590 mg-N/L corresponding to a high TKN loading rate of 1.1 kg-N/m^3.d. By using such a feeding pattern the bulk liquid FA concentration was kept below 7.25 mg-N/L which meant that for a short period in the cycle *nitrobacter* may have been inhibited but not in such a way that affected the overall nitrification. The reactor was dosed with Na_2CO_3 whenever the pH fell below 7.3. In the study the urine accounted for 41% of the total alkalinity required to maintain the desired pH. The actual addition of alkalinity required will depend on pH setpoints used. The pH setpoint determines the trade off between alkalinity dosage required and allowing the system to maintain the maximum nitrification rate.

Granulation

Our recent research focus has been placed on urine nitrifying biomass granulation. Granulation is the self-immobilization of cells into clusters. Granule biomass exhibits very good settling properties enabling a high biomass concentration and a small reactor footprint. Extensive researches have been conducted on heterotrophic granulation in SBRs with low N-to-COD ratios. Chen (2009) reported substantial granules formed during the urine nitrification in SBR. This finding was unexpected because the SBR reactor was operated with a long settling time and hence slow critical settling velocity, which should not promote the granulation. It has been well documented that one of the critical selection pressures for an aerobic SBR to achieve granulation is a short settling period (in the order of 10 minutes) or a high critical settling velocity (Buen *et al.*, 2000; Liu *et al.*, 2005, 2007; McSwain *et al.*, 2004). This is determined by the settling time and withdrawal period, the principle being that slowly settling bacteria are washed out, naturally selecting faster settling granules. Liu *et al.* (2008) found that the settling time required for granulation of nitrifying granules is less than that for heterotrophic bacteria. However, the mean diameter of nitrifying granules has generally been less than 1 mm (Liu *et al.*, 2008; Shi *et al.* 2008) compared with heterotrophic granules ranging from 1–6 mm (de Kreuk and van Loosdrecht, 2006; Li, 2009; Liu *et al.*, 2009). Owing to much slower growth rate of autotrophic biomass than heterotrophic biomass, nitrifying granules could become less sensitive to some operational conditions such as settling time but granulation may become more sensitive to loading rate and mixing condition which will be further studied.

Summary

Full nitrification of human urine in SBR is feasible and autotrophic granules occurred. Granulation of urine nitrifying biomass is crucial for providing a cost-effective and space-saving means of treating source-separated human urine at site. This enables full nitrification of urine for discharge of nitrified urine into sewers to replace current nitrate chemical dosing for controlling H_2S production in rising mains as well as inducing in-sewer denitrification and soluble COD removal, resulting in downsizing of centralised sewage treatment works (Jiang *et al.*, 2010).

ACKNOWLEDGMENT

The authors wish to thank Hong Kong Research Grants Council for their financial support for part of this work (grant 611607).

REFERENCES

Arora M. L., Barth E. F. and Umphres M. B. (1985). Technology evaluation of sequencing batch reactors. *Journal Water Pollution Control Federation*, **57**(8), 867–875.

Artan N. and Orhon D. (2005). Mechanism and design of sequencing batch reactors for nutrient removal, TR 19. IWA Publishing, London.

Beun J. J., Van Loosdrecht M. C. M. and Heijnen J. J. (2000). Aerobic granulation. *Water Science and Technology*, **41**(4–5), 41–48.

Chen Y. (2009). *Full Nitrification of Human Urine in a Sequencing Batch Reactor*. M.Phil thesis, Dept. Civil and Environmental Engineering, The Hong Kong University of Science and Technology, Hong Kong.

Ciba-Geigy (1977). *Wissenschaftliche Tabellen Geigy, Teilband Korperflussigkeiten* (Scientific Tables Geigy. Volume body fluids), 8th edn, Basel, German.

Doyle J., Watts S., Solley, D. and Keller J. (2001). Exceptionally high-rate nitrification in sequencing batch reactors treating high ammonia landfill leachate. *Water Science and Technology*, **43**(3), 315–322.

Irvine R. L. and Ketchum Jr. L. H. (1988). Sequencing batch reactors for biological wastewater treatment. *Critical Reviews in Environmental Control*, **18**(4), 255–294.

Kirchmann H. and Pettersson S. (1995). Human urine - Chemical composition and fertilizer use efficiency. *Fertilizer Research*, **40**, 149–154.

de kruek M. K. and van Loosdrecht M. C. M (2004). Selection of slow growing organisms as a means for improving aerobic granular sludge stability. *Water Science and Technology*, **49**(11–12), 9–17.

Jiang F., Chen G. H. and van Loosdrecht M. C. M (2010). Urine Nitrification and sewer discharge to realize in-sewer denitrification to simplify sewage treatment in Hong Kong, 7th IWA LET Conference in Arizona, June 2–4, 2010 (poster paper).

Li A. J. (2009). *Determining Factors for Aerobic Sludge Granulation in Bioreactors: Mechanism Analysis, Mathematical Modeling and Experimental Verification.* Ph.D thesis, Dept of Civil Engineering, The University of Hong Kong, Hong Kong.

Liu X. W., Sheng G. P. and Yu H. Q. (2009). Physicochemical characteristics of microbial granules. *Biotechnology Advances*, **27**, 1061–1070.

Liu Y., Wang Z. W., Qin L., Liu Y. Q. and Tay J. H. (2005). Selection pressure-driven aerobic granulation in a sequencing batch reactor. *Applied Microbiology and Biotechnology*, **67**, 26–32.

Liu Y., Qin L. and Yang S. F. (2007). *Microbial Granulation Technology for Nutrient Removal from Wastewater.* Nova Science Publishers, New York.

Liu Y. Q., Wu W. W., Tay J. H. and Wang J. L. (2008). Formation and long-term stability of nitrifying granules in a sequencing batch reactor. *Bioresource Technology*, **99**, 3919–3922.

Mackey H. R. (*in progress*). *Granular Nitrification of Urine in a Sequencing Batch Reactor.* M.Phil thesis, Dept. of Civil and Environmental Engineering, The Hong Kong University of Science and Technology, Hong Kong.

Maurer M., Pronk W. and Larsen T. A. (2006). Treatment processes for source-separated urine. *Water Resources*, **40**(17), 3151–3166.

McSwain B. S., Irvine R. L. and Wilderer P. A. (2004). The influence of settling time on aerobic granules. *Water Science and Technology*, **50**(10), 195–202.

Silverstein, J. and Schroeder E. D. (1983). Performance of SBR activated sludge processes with nitrification/denitrification. *Journal Water Pollution Control Federation*, **55**(4), 377–384.

Shi X. Y., Yu H. Q., Sun Y. J. and Huang X. (2009). Characteristics of aerobic granules rich in autotrophic ammonium-oxidizing bacteria in a sequencing batch reactor. *Chemical Engineering Journal*, **147**, 102–109.

Sperandio M., Pambrun V. and Paul E. (2008). Simultaneous removal of N and P in a SBR with production of valuable compounds: application to concentrated wastewaters. *Water Science and Technology*, **58**(4), 859–864.

Tsuneda S., Park S., Hayashi H. and Hirata A. (2001). Enhancement of nitrifying biofilm formation using selected EPS produced by heterotrophic bacteria. *Water Science and Technology*, **43**(6), 197–204.

Udert K. M., Larsen T. A. and Gujer W. (2006). Fate of major compounds in sourceseparated urine. *Water Science and Technology*, **54**(11–12), 413–420.

Wiesmann U., Chio I. S. and Dombrowski E. (2007). *Fundamentals of Biological Wastewater Treatment.* Wiley-VCH, Wenheim.

Wiesmann U., and Libra J. (1999). Special aerobic wastewater and sludge treatment processes and process combinations. *Biotechnology: Environmental Processes*, **11a**, 373–415.

Wilsenach J. A. and van Loosdrecht M. C. M. (2006). Integration of processes to treat wastewater and source-separated urine. *Journal of Environmental Engineering*, **132**(3), 331–341.

PART SIX

Treatment of Separated and Combined Used Water and Solids

Compositing Toilet: Its Functions and Design Procedure

N. Funamizu, M.A. Lopez Zavala, R. Itoh, S. Hotta and T. Kakimoto

Department of Environmental Engineering, Hokkaido University. Kita 13, Nishi 8, Sapporo 060-8628, Japan
E-mail: funamizu@eng.hokudai.ac.jp

Abstract Since the Onsite Wastewater Differentiable Treatment System (OWDTS) was introduced, several research initiatives have been conducted to support this approach. This paper summarizes some of the research findings of a study conducted on a laboratory-scale, to assess the decomposition process of faeces and urine in the composting toilet. The main aim was to establish appropriate criteria for the proper design and operation of the system. In the study, the following aspects were discussed: the characterization of faeces; the biological activity in the composting reactor; the modelling of the aerobic biodegradation process; the effects of temperature and moisture content on the decomposition process; the nitrogen process; the kinetics of water in the composting matrix; the decline in pathogens and the fate of pharmaceuticals; the toxicity and characterization of the organic matter produced; and the criteria for designing the bio-toilet.

Keywords Composting toilet, design, organic matter, pharmaceuticals, toxicity, water

INTRODUCTION

The Onsite Wastewater Differentiable Treatment System is an ecological sanitation approach with a high potential to achieve sustainable treatment and management of domestic wastewater (Lopez Zavala *et al.*, 2002). In this system, the treatment of toilet waste, using the composting toilet, is an essential and key process. There are several benefits that can be expected, for example: *i)* conservation of approximately 30% of fresh water; *ii)* elimination of sources of pathogens from the domestic wastewater stream; *iii)* enhanced stabilization of organic matter contained in toilet wastes; *iv)* recovery and recycle of nutrients contained in excreta; *v)* reduction of wastewater flow and volume of materials; and *vi)* inhibition of pathogens and reduction of the risk they represent for the public health; *vii)* control of micro-pollutants such as pharmaceuticals.

The composting toilet considered here (Fig. 1) differs from conventional composting systems in several ways: 1) Saw dust is used as an artificial soil matrix for biological reactions; 2) the composting reactor is provided with heating and mixing systems that ensure a continuous thermophilic-aerobic biodegradation process and a uniform temperature distribution; 3) the moisture content in the reactor is kept in the range 50–60% by heating and ventilation; 4) traditional composting systems have a batch configuration, whereas the composting toilet here is a continuous feed system with constant reaction conditions in terms of temperature and the moisture content of the saw dust (Lopez and Funamizu, 2005a).

Since black water contains: 1) organic matter; 2) water; 3) nutrients; 4) pathogenic micro-organisms; 5) micro-pollutants such as hormones and pharmaceuticals, several research activities have been conducted to evaluate and an analyse the functions of the composting toilet. In this paper, a summary of these results are presented.

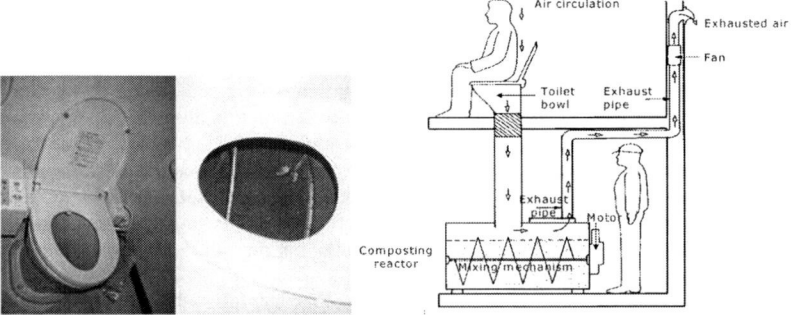

Figure 1 Composting toilet

EVALUATION OF THE COMPOSTING TOILET

Characteristics of faeces (Lopez Zavala *et al.* 2002b)

The characterization of faeces using sawdust as a matrix was conducted, concentrating on the biodegradability of its organic matter. Two methods were used for this purpose. In the first method, expressed in terms of COD, COD measurements, experimental oxygen utilization rate profiles (OUR) and simulation were used. In the second method, expressed in terms of mass units, measurements of physical parameters such as total solid (TS) and volatile solid (VS) were employed.

In terms of COD, faeces were divided into two fractions: biologically inert material 20% and slowly biodegradable organic matter 80%. Characterization in these terms is useful for modelling purposes and is essential for establishing the criteria for the proper design and operation of bio-toilet systems. As for mass units, faeces were characterized into three fractions, fixed solids 16%, non-biodegradable volatile solids 21%, and biodegradable volatile solids 63%. Characterizing faeces in these terms is an affordable way of evaluating the quantity and composition of the residue that accumulates in the composting reactor. This interferes with the design and operation of the composting toilet system.

The Biodegradation of organic matter and its model

Effect of organic loading (Lopez and Funamizu, 2005a)

Organic load was not a limiting factor for the aerobic biodegradation of feces. TS, VS, and COD reductions in the order of 56%, 70%, and 75%, respectively, were observed irrespective of the organic load.

Mathematical model (Lopez and Funamizu, 2004a, Hotta and Fuanmizu, 2009)

A bio-kinetic model was introduced to describe the aerobic biodegradation of faeces during the composting process of faeces. This model includes three processes for carbonaceous material degradation: 1) the hydrolysis of organic matter; 2) the growth of heterotrophic bacteria; 3) the decay of bacteria. The model was prepared using activated sludge modelling techniques and formulations. Stoichiometric parameters were adopted from literature on activated sludge systems. Kinetic parameters were estimated by conducting batch tests for several organic loadings and by using respirometry, curve-fitting techniques, and sensitivity analysis. Feasibility and applicability of these parameters were assessed by conducting faeces intermittent-feeding tests and by simulating the experimental OUR profiles.

Effect of temperature (Lopez and Funamizu, 2004b)

Temperature is one of the most important factors affecting microbial growth and biological reactions. In this study, the effect of temperature on the aerobic biodegradation of faeces is described. This is done through the comparison and analysis of experimental oxygen utilization rates (OUR). Profiles were obtained from batch tests which were conducted at several temperatures covering mainly mesophilic and thermophilic ranges. Additionally, the effect of temperature was incorporated into the bio-kinetic model and the simulation of experimental OUR profiles was conducted. Results show that mesophilic and thermophilic microorganisms behave differently according to temperature; additionally, results suggest that the optimum temperature in terms of faeces biodegradability is within the thermophilic range (nearly 60°C). The enzymatic activity of microorganisms at 70°C was remarkably diminished. For better predictions in the mesophilic range, two fractions of slowly biodegradable organic matter were identified: easily hydrolyzable organic matter (XSe) and slowly hydrolyzable organic matter (XSs).

Effect of moisture content (Lopez and Funamizu, 2005b)

The composting process is determined by the biological response of different microorganisms kept under different conditions of moisture content. Low moisture content (< 65%) ensures the aerobic degradation of faeces, whilst high moisture levels (≥ 65%) causes both aerobic and anaerobic decomposition. Because anaerobic conditions occur at levels of high moisture content, (≥ 65%), the process generates odour and VFA emissions. In addition, simultaneous aerobic and anaerobic processes at high moisture content levels causes the increase of sulphate concentrations and the formation of nitrites in the sawdust matrix. This occurs, despite the fact that the composting process was conducted at thermophilic temperatures. At low levels of moisture content, anaerobic emissions, nitrification products and increased sulphate concentrations were not detected. Critical moisture was the moisture content which caused the highest degradation rates, the beginning of odour, anaerobic emissions, and an increase in sulphate concentrations. Thus, critical moisture may be adopted as a simple physical operation parameter. It was found to be approximately 65%.

Fate of nitrogen (Hotta and Funamizu, 2007a,b)

The operation test of the composting toilet showed that most of the nitrogen volatizes as NH_3 gas during the composting process. By separating faeces using urine diverting toilet bowls, and supplying only faeces to the composting toilet, it is possible to achieve nitrogen recovery. This is because most of the nitrogen is contained in the urine. The experimental results also showed that 1) the time lag was between the peek time of the CO_2 production rate and the peek time of the ammonium nitrogen volatilization rate; 2) composting only faeces led to high nitrogen recovery levels, so approximately 10% to 20% of nitrogen in the faeces supplied was volatized from the system as an ammonia gas.

Drying kinetics of water from the compost matrix (Tanaka *et al.* 2009)

Drying the mixture of human waste and sawdust is an essential process for the dry composting toilet system. That said, the moisture content is an important factor for the biodegradation process and it should be kept around 65%. A stable drying rate can be obtained by applying appropriate planning and operation policy measures. In this study, the effects of operational conditions on the drying rate were examined, using both lab-scale equipment, and a full-scale toilet system. The drying rate is shown by the mass transport model. It demonstrates the adequacy calibration of water-vapour pressure at each drying surface for each device. The mass transfer coefficient of a full-scale toilet can be estimated by using the results from the lab scale equipment and by considering the circulation air flow by the non-dimensional expression. The drying rate can then be estimated under the operational conditions. This can be applied to the design and planning of the full scale composting toilet system.

Pathogen decline (Nakata *et al.*, 2003)

Temperatures over 45°C (thermophilic range) were more effective in inhibiting the coliforms (colony-formation) and bacteriophages (plaque-formation). This consequently results in the inactivation of pathogens. At 60°C, the formation of coliform bacteria colony was inhibited in just 1.17 hours. However, at 45°C the inactivation occurred after 8 and 24 hours, respectively. At temperatures lower than 45°C, it took a much longer period to obtain a 6-log reduction of CFU. The decaying process of bacteriophage showed the same profile. The temperature distribution in the bio-toilet system was not uniform. The non-uniformity of temperature distribution meant that it was not so effective in inhibiting pathogens. Results of risk assessment calculations showed that a reduction of compost withdrawal brings infection risk to an acceptable level (1×10^{-4} per year). This is achieved by mixing the sawdust *i)* 20 times per day during 2 days or *ii)* by mixing it 15 times per day, for 3 days, after the operating bio-toilet was last used. A more effective way to control infection risk is to eliminate the low temperature zone which will reduce the reaction time. This is especially effective, when using a low mixing frequency, such as 2 times per day, thus increasing the volume temperature.

Fate of pharmaceuticals (Kakimoto and Funamizu, 2007a,b)

A test, revealing the effect of amoxicillin on the composting process of faeces was conducted. In addition, a test, looking at the possibility of bacterial reactivation by intermittent faeces feeding was conducted. This study established several facts: 1) a significant reduction of amoxicillin occurred within thirty minutes. This is because this antibiotic has a very sensitive ring structure (beta-lactam ring); 2) The amoxicillin dose system delayed the onset time of the biodegradation of faeces. In doing so, it reduced the maximum biodegradation rate and lengthened the time for the degradation of faeces; 3) In the case of 100 µg/g-dry, (the estimated amount of amoxicillin excreted by one person, in a toilet, in one day), 40% less faeces were treated than in the control; 4) A simulated change in the initial microbial count was not able to explain the experimental result, but a simulation which changed the initial bacterial count and its activity coincided with the experimental result. We, therefore, inferred that the effect of amoxicillin was not only to reduce the initial microbial count but also to weaken the bacterial activity; 5) the intermittent feeding test showed that, with 10 and 100 µg/g-dry, the respiratory activity recovered in a few days, but that no such reactivation could be observed in the 1000 µg/g-dry case. The result of the intermittent feeding test implies that the initial quantity of the surviving bacteria had had a very strong effect on their reactivation.

Characterization of organic matter in compost (Narita *et al.*, 2005)

In order to assess the stability of the compost generated by the composting toilet, the characteristics of the organic matter (DOM) was examined. This was done using the matter extracted from the compost, obtained through the thermophilic aerobic biodegradation of the faeces. The conclusions resulting from this study are that: 1) The main component of DOM from the bio-toilet are solutes over 30,000 Da of molecular weight (40%); on the other hand, micro-molecules (MW< 1,000 Da) constituted more than 60% of the DOM from other samples, (90% in the liquid samples); 2) The DOM stabilization level reached in the composting reactor of the bio-toilet system was greater than that shown by the DOM from the other sources; 3) Stabilization of DOM in the bio-toilet system was characterized during by, *i,)* an increase and then a decrease of DOM with MW < 1,000 Da; *ii)*, a decrease and then an increase mainly of DOM with MW > 30,000; and *iii)*, a decrease, then an increase, and finally a decrease of the TOC/E260 for almost all size cut-off. To complete the stability of OM in the bio-toilet, it will take over 10 days; 4) the net production of LPS in wastewater treatment plants is greater than that of the bio-toilet. Regarding the biodegradability of soluble microbial products, a small KDO generation in the bio-toilet can be a sign of more complete DOM stabilization.

DESIGN OF COMPOSTING TOILET

The size of the composting reactor is determined by conditions such as: 1) the water loading rate due to daily contributions of urine, water contained in faeces, and water for cleaning the toilet bowl; 2) Drying rate, i.e., evaporation rate of water contained in urine and faeces and cleaning water; 3) Organic loading rate due to daily feeding of faeces and toilet paper; i.e., faeces-sawdust ratio (F/S); 4) Biodegradability and biodegradation rate of toilet wastes which are affected by several factors, such as environmental conditions like temperature, moisture content, pH, oxygen availability, etc; 5) Mixing frequency (Lopez and Funamizu, 2006).

The average daily faeces and urine production rates per capita per day found in the literature are approximately 130 g (wet basis) for faeces and 1200 ml for urine (Almeida *et al.*, 1999; Del Porto and Steinfeld, 2000). The water content of faeces is approximately 82%. Thus, the daily water-loading rate of human excreta totalized 1307 ml per capita per day (or 1325 g per capita per day, if density of urine is 1.015 g/cm^3). No studies have been conducted to evaluate the amount of water needed for cleaning the bio-toilet, but assuming the quantity used for that purpose is relatively small, cleaning water may be considered negligible for design purposes. The design of the composting reactor must ensure that the water loaded in one day is evaporated so that the bio-toilet is able to receive water loads the next day. If the water content of faeces is 82%, the daily organic loading in the dry form of faeces, is 23.5 g per capita per day. Compared with the water-loading rate, the organic loading rate is very low. Therefore, as mentioned previously, the size of the composting reactor will be mainly determined by the water-loading rate.

The required drying surface area is estimated using the drying theory. This takes into account the average drying rate, within the range 50 to 60% and the data of the critical hourly water loading (from contributions of urine and water contained in faeces). The volume of the sawdust matrix must ensure that 1) moisture content becomes approximately 60% during the critical water loading to the system, and 2) critical water load is totally evaporated in the critical time (3 hours), i.e., moisture content of the sawdust matrix becomes approximately 50% after the critical drying period.

SUMMARY

Based on the results of experimental research, criteria for the proper design and operation of the composting toilet system were established. The results and discussions of each research activity were vast and cannot all be included in this paper. However, if readers are interested in more details, they are advised to consult the references included in previous sections. Amongst the different approaches of reducing the load of micro-pollutants to the aquatic environment, the composting toilet might provide a sustainable. The pilot plant study began in 2005. Now, the performance of composting toilets and gray water treatment systems are being evaluated (Itoh *et al.*, 2005).

REFERENCES

Hotta S. and Funamizu N. (2007a). Biodegradability of fecal nitrogen in composting process. *Bioresource Technology*, **98**(17), 3412–3414.

Hotta S. and Funamizu N. (2009). Simulation of accumulated matter from human feces in the sawdust matrix of the composting toilet. *Bioresource Technology*, **100**(3), 1310–1314.

Hotta S., Noguchi T. and Funamizu N. (2007b). Experimental study on nitrogen compo-
    nents during composting process of feces. *Water Science and Technology*, **55**(7),
    181–186.
Itoh R., Fukuda M., Itayama M., Kiji M., Yokota M. and Funamizu N. (2005). A pilot
    study of the onsite wastewater differentiable treatment system in Chichibu sity,
    Japan. *Proceedings of 9th International Conference ECOSAN India*, Mumbai, India,
    pp. 214–220.
Kakimoto T. and Funamizu N. (2007b). Factors affecting the degradation of amoxicillin in
    composting toilet. *Chemosphere*, **66**, 2219–2224.
Kakimoto T., Osawa T. and Funamizu N. (2007a). Antibiotic effect of amoxicillin in
    human excrement on the feces composting process, and reactivation of bacteria by
    intermittent feeding of feces. *Bioresource Technology*, **98**, 3555–3560.
Lopez Zavala M. A., Funamizu N. and Takakuwa T. (2002a). Onsite wastewater
    differentiable treatment system: modeling approach. *Water Science and Technology*,
    46(6–7), 317–324.
Lopez Zavala M. A., Funamizu N. and Takakuwa T. (2002b). Characterization of feces for
    describing the aerobic biodegradation of feces. *Journal of Environmental Systems*,
    720, 99–105.
Lopez Zavala M. A., Funamizu N. and Takakuwa T. (2004a). Modeling of aerobic bio-
    degradation of feces using sawdust as a matrix. *Water Research*, **38**(5), 1327–1339.
Lopez Zavala M. A., Funamizu N. and Takakuwa T. (2004b). Temperature effect on
    aerobic biodegradation of feces using sawdust as a matrix. *Water Research*, **38**(9),
    2406–2416.
Lopez Zavala M. A., Funamizu N. and Takakuwa T. (2005a). Biological activity in the
    composting reactor of the bio-toilet system. *Bioresource Technology*, **96**(7), 805–812.
Lopez Zavala M. A. and Funamizu N. (2005b). Effect of moisture content on composting
    process in a biotoilet systes. *Compost Science & Utilization*, **13**(3) 208–216.
Lopez Zavala M. A. and Funamizu N. (2006). Design and operation of the bio-toilet
    system. *Water Science and Technology*, **53**(9), 55–61.
Nakata S., Lopez Zavala M. A., Funamizu N., Otaki M. and Takakuwa T. (2003).
    Temperature effect on pathogens decline in the bio-toilet system. *Proceedings of Dry
    Toilet 2003, 1st International Dry-Toilet Conference*, Tampere, Finland, pp. 131–139.
Narita H., Lope Zavala M. A., Iwai K., Itoh R. and Funamizu N. (2005). Transformation
    and characterization of dissolved organic matter during the thermophilic aerobic
    biodegradation of faeces. *Water Research*, **39**(19), 4693–4704.
Tanaka A., Funamizu N., Ito R. and Masoom P. M. (2009). Estimation of water evapora-
    tion rate from composting toilet. *Presented at the Third International Dry Toilet
    Conference*, August 2009, Tampere, Finland.

Treatment of Brownwater
Results of Mesophilic Tests in
Stahnsdorf/Germany

A. Wriege-Bechtold[a], M. Barjenbruch[a], A. Peter-Fröhlich[b] and B. Heinzmann[b]

[a] Technische Universität Berlin, TIB 1-B16, Gustav-Meyer-Allee 25, 13355 Berlin, Germany
[b] Berliner Wasserbetriebe, Cicerostraße 24, 10719 Berlin, Germany
E-mail: alexander.wriege-bechtold@tu-berlin.de

Abstract All flows of domestic wastewater currently discharge into one pipeline. This includes water from all domestic outlets such as flushing toilets, dish washers, washing machines, showers etc. In the so called end-of-pipe-system, fresh water is used to transport wastewater to the sewer system and then on to the wastewater treatment plant (WWTP). At this point, the nutrients from the energy-rich wastewater are separated from the pollutants. The nutrients and trace elements contained in excrement can then be used for fertiliser, or can be recycled for other uses. A relevant percentage, however, is still in the effluent which can cause the eutrophication in rivers and lakes. A major problem in conventional waste water systems is their high consumption of fresh water. These systems are also energy-intensive and the purification of wastewater causes carbon dioxide emissions. Tests in Stahnsdorf/ Germany have shown that the digestion of brownwater and bio-waste as a co-substrate, can lead to compromising results.

Keywords Alternative sanitary systems, source separation, close the loop, no-mix-toilets, biogas, fertiliser, save water

INTRODUCTION

In Germany, there are about 0.5 million kilometres of sewers, 20 % of which are old and need restoring (ATT *et al.* 2008). At the time of their construction, both population growth and water use was increasing. However, today there is now decreasing water consumption and a declining population. Many people have started to save water due to rising water prices. The need for sewers to be flushed causes costs to rise. The fact that wastewater treatment plants (WWTPs) have become so under loaded, means that the treatment process is suboptimal.

The construction of conventional sanitary systems (end-of-pipe-systems), especially in arid areas, lies at odds with the ubiquitous water shortage. Novel sanitary systems provide an opportunity for sustainable water use. Nutrients, in wastewater, can be used as a resource through water reuse, energy production and fertiliser substitution.

With the existence of current sewer systems, as they are, in Central Europe, it is difficult to establish change immediately. Novel sanitary systems are considered, only as an export business for developing countries and sometimes for sparsely populated areas in developed countries. However, these new sanitary systems can be implemented in peri-urban and urban areas with high density populations. Rapidly growing mega-cities, with little or no waste water infrastructure, provide particularly interesting examples of this.

NOVEL SANITARY SYSTEMS

Alternative Sanitary Systems are resource-orientated systems, linking water and nutrients in wastewater treatment.

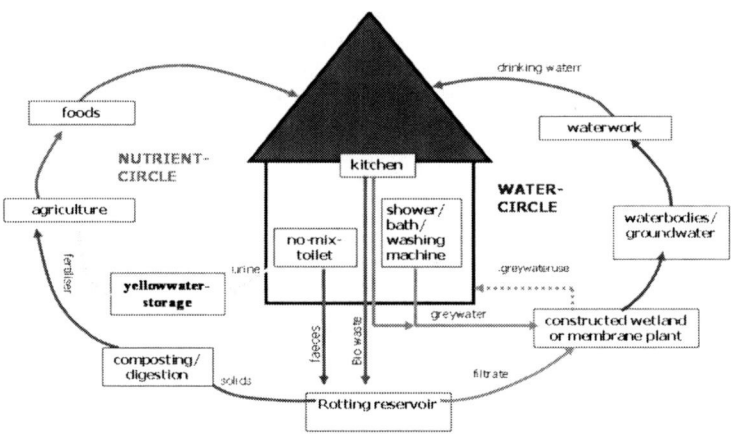

Figure 1 Different flows of household wastewater

Wastewater from households is parted in 3 different flows (Figure 1). Yellow water refers to urine with, or without, flush water. Yellowwater is collected in urinals and toilets. Brownwater is the mixture of faeces and flush water from toilets. Blackwater is a combination of brownwater plus yellowwater. Greywater is the effluent from showers, dish washers, washing machines and hand wash basins. Figure 2 shows the different loads produced in a day, per capita (in percent of total load).

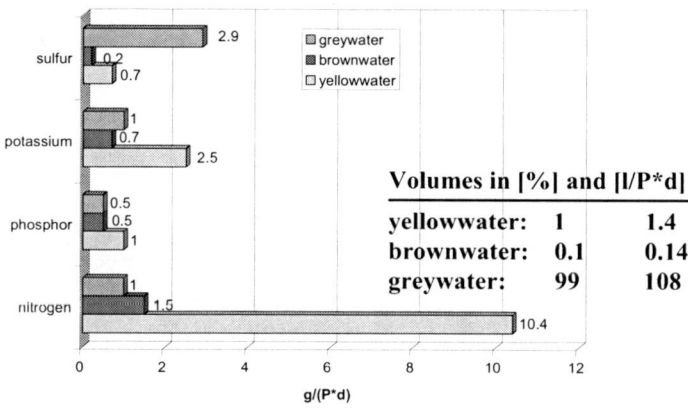

Figure 2 Amount of different substances in waste water [3]

Greywater is, volumetrically, the largest component of wastewater. However, it is no major hygienic concern. It contains very little nutrients and only some chemical residues from washing machines, dishwashers and showers. The treatment of greywater is simpler than the treatment of yellow or brownwater. The COD load is 40 % of the total load, but the concentration is low because of the high volume. Greywater can be treated by constructed wetlands, membrane reactors, moving bed reactors etc.

The largest amount of nutrients is contained in yellow water. For instance, 81 % of the total nitrogen load is in the yellow water. Almost comparable, is the load of potassium. When used for fertiliser, it is better to collect the urine with only a low volume of flush water, or even without any at all. Although yellow water has less hygienically critical components than brownwater, it might contain the most hormones or medical residues. Various treatment processes have been successfully tested for this, such as vacuum evaporisation, steam stripping, precipitation, ozonisation and so on. Most yellowwater treatment facilities require a lot of energy and are tested on a pilot plant scale. It is generally possible to produce concentrated fertiliser from urine.

Because of its high pathogen content, brownwater is hygienically critical. It is energy-rich and consists of organics, nutrients and trace elements.

TREATMENT OF BROWNWATER

Models and Measures for Collection of Brownwater

In general, there are two main types of treatment for brownwater: aerobic (composting or vermi-composting) and anaerobic (digestion). The best available treatment technique depends on the dilution of flush water. Different kinds of toilets use different varieties of flush water (see Table 1).

Table 1 Kinds of toilets for brownwater collection and corresponding volumes of flow (Oldenburg, 2005), modified

Kind of toilet		Advantages +/Disadvantages −
no-mix-toilet		+ low consumption of flush water
▨	6l	+ small-scale dilution
▨	1,2l	+ reuse of nutrients possible
Vacuum no-mix-toilet		+ low consumption of flush water
▨	3l	− high complexity
▨	1,2l	+ reuse of nutrients possible
		− test stage
no-mix-composting-toilet and		+ low tech, no need for flush water
no-mix-dry-toilet		− increasing maintenance effort
▨	0,15l	− high space requirement
▨	1,2l	

Pre-treatment, it is useful to separate the solids from the liquids in the brownwater. Pre-treatment technologies are:

- sedimentation
- centrifugation
- flotation
- filtration
- screening

To generate brownwater the use of no-mix-toilets is essential.

Applied composting material should have a water content of between 40 and 60 %. It should also have a neutral pH-value (Naudascher 2000). Due to its structure and organic load, compost applied to farming or gardening, is more of a soil conditioner, than a fertiliser. In agriculture, it helps to increase the available water capacity of soils. Composting is a low tech process and is often used for small scale applications such as a one-family-house.

The digestion of brownwater has a higher technical complexity than composting. Here, it is possible to obtain a better degree of stabilisation and an increased production of energy by biogas. In practise two different temperature ranges can be used: mesophilic (35 °C) and thermophilic (55 °C) digestion. The differences are shown in Table 2.

Table 2 Advantages/disadvantages of thermophilic/mesophilic operating mode of the biogas plant

	Thermophile (55°C)	Mesophilic (35°C)
disinfection	possible	not possible
HRT	about 10 days	about 20 days
biogas production	higher	lower
process stability	lower	higher
energy demand	higher	lower
experience	less	more
complexity	higher	lower

The digestion process, lasting 20 days, is even shorter than composting (3–4 months, by using worms for vermi-composting it decreases to 1–2 months). The time requirement for composting, therefore, is higher. Stabilised sludge can also be used as soil conditioner in agriculture, like compost.

Test with brownwater digestion in Stahnsdorf Germany

In a project near Berlin some vacuum no-mix-toilets were installed in an office building. At the same time, some gravity no-mix-toilets were installed in an apartment building.

The results of the tests, using a biogas plant, show a good production of biogas. Results also show that there is a possibility to use bio-waste from households as a co-substrate. The measured values of treated sludge show that it is possible to use it as fertiliser in general agriculture (Peter-Fröhlich et al., 2007).

The biogas plant in Stahnsdorf is a two stage fixed bed digester with a volume of 300 L. The biogas plant consists of a balancing tank with a stirrer and the biogas reactor, itself. At first, the biogas plant was operated in thermophilic mode, the first stage being the acidification reactor, and the second one as methanogenic. For mesophilic mode the fixed band was removed. The brownwater from the office building ran down to the balancing tank. Sedimentation was used as the method of pre-treatment for the brownwater. Settled material was then loaded into the biogas plant.

Due to some technical difficulties in the thermophilic mode, there was a change to mesophilic digestion in the process. A test was carried out, which considered the digestion of brownwater, taken from the same source, but mixed with bio waste from the kitchens as a co substrate.

After settlement, 62.5 % of the brownwater (with low solids) overflowed to a buffer. In first test period 37.5 % (with high solids) was fed to the biogas plant. In the second test period (with bio waste) it changed to 53.5 % overflow and 46.5 % feed because of the increasing total influent to the settlement tank of nearly 30 % (from 33 to 42 litres per day on average). The calculated specific total of COD loading for the 1st test period (without bio waste) was 0.5 kg COD/(m³*d). The mean hydraulic retention time (HRT) was 27 days. In the second test period using bio waste as a co-substrate it decreased to 18 days, but the specific total COD loading increased up to 1.3 kg COD/(m³*d).

In the period using the bio waste, the biogas plant produced 377 L biogas/kg vDR_{input}. In the first period it was 438 L biogas/kg vDR_{input} or rather 285 L CH_4/kg vDR_{input}. 71 % of the volatile dry matter (vDR) and 75 % of the chemical oxygen demand (COD) were degraded in the digestion process without bio waste. In conventional digesters in wastewater treatment plants, the degradation of vDR is usually around 50 %. The digestion in Stahnsdorf is the first treatment step in the cleaning process. However, in conventional wastewater treatment plants, it is the last step, after the mechanical and biological cleaning processes. COD is already degraded in the first two processes. With bio waste used as a co-substrate, the degradation ratio decreases to 67 % for vDR and 68 % for COD. This is caused by the higher load and the shorter HRT. The HRT in the test period with bio waste was 2 days shorter than the optima in Table 2 (Billmaire *et al.*, 2001). This provides little loading space for digesters at 1.5 kg vDR/(m³*d). The loading specifics are, therefore, in a minimal to optimal range. The CH_4-content decreased from 65 % in the first test period to 57 % in the second one. The decreasing biogas production in the second test period may be resultant of the particle size of the bio waste. The vDR was not degraded as well as in first test period. A better milling of bio waste will be tested on a laboratory scale to check the influence of this parameter.

The additional inflow of bio waste, in the test period had a co-substrate of about 1,044 g/d. Calculated via COD load app, 5 persons are connected to the biogas plant. In Germany, the average quantity of bio waste accumulated per person each day is nearly 200 g. DR in bio-waste was 26 % and vDR was 24 %. In Table 3 all results of the two test periods are summarised.

Table 3 Results of mesophilic tests without and with bio waste as co-substrate

			Mesophilic without bio-waste 7 weeks	Mesophilic with bio-waste 6 weeks
quality	HRT	[d]	27	18
	reactor influent	[L/d]	12,3	19,7
DR	conc. influent	[mg/L]	0,9	1,8
	conc. effluent	[mg/L]	0,3	0,7
	degrated	[%]	67	61
vDR	conc. influent	[mg/L]	0,7	1,5
	conc. effluent	[mg/L]	0,2	0,5
	degrated	[%]	71	67
	sp. loading tot	[kg/(m³*d)]	0,3	0,9
COD	conc. influent	[mg/L]	12,364	22,186
	conc. effluent	[mg/L]	3,063	7,014
	degrated	[%]	75	68
	sp. loading tot	[kg/(m³*d)]	0,5	1,3
biogas	quality	[L/d]	31,2	88,7
	CH_4-content	[%]	65	57
	CH_4-content	[L/d]	20	51
	gas yield	[L biogas/ (kgvDR*d)]	438	377

CONCLUSIONS AND FUTURE ASPECTS

Conventional wastewater treatment is state of the art in Middle Europe. Most of the nutrients in wastewater are removed in wastewater treatment plants. However, a part of them is discharged out to rivers and lakes which can effect eutrophication. Rapidly growing mega-cities have real problems with the collection and treatment of incoming wastewater. Sewage systems, if existent, are overloaded. Insufficient sewage purification causes disease and environmental pollution. Furthermore, there is a scarcity of fresh water. There is not enough fertiliser for agriculture anywhere. Alternative sanitary concepts are now available to citizens. Applying various methods and techniques, the new concepts accommodate the different flows of wastewater, like greywater, brownwater and yellowwater. In this study, the treatment of brownwater was investigated. For brownwater there are two main treatment possibilities: composting and digestion. Digestion produces usable biogas e.g. for heating, cooking and power supply. The implementation of such novel sanitary concepts is easier in urban areas than in low density areas. It is, however, still possible.

Mesophilic digestion tests show good results in terms of biogas production and degradation. The biogas production seems to be gradable. Additional tests are already running.

CO_2-emissions will be reduced by the use of biogas. It will also be reduced by the substitution of high energy produced mineral fertiliser. In theory, the implementation of alternative sanitary systems in multi-storey buildings of urban areas is possible. However, more research into the adaptation of SBR and membrane plants is required.

Alternative sanitary systems are a good alternative to the conventional urban sewer systems.

REFERENCES

ATT et al. (2008). Branchenbild der deutschen Wasserwirtschaft 2008, Arbeitsgemeinschaft Trinkwassertalsperren e. V. und andere, wvgw Wirtschafts- und Verlagsgesellschaft Gas und Wasser mbH, Bonn 2008.

Billmaier K. et al. (2001). Co-Fermentation von biogenen Abfällen in Faulbehältern von Kläranlagen, Merkblatt. Berichte zur Umwelt, Bd. 22, Ministerium für Umwelt und Naturschutz, Landwirtschaft und Verbraucherschutz NRW, Düsseldorf.

DWA (2008). Neuartige Sanitärsysteme, Deutsche Vereinigung für Wasserwirtschaft, Abwasser und Abfall; 1. Auflage, Hennef.

Naudascher I. (2000). Kompostierung menschlicher Ausscheidungen durch Verwendung biologischer Trockentoiletten, Uni Karlsruhe - Institutsverlag Siedlungswasserwirtschaft 2001 , 222 S.; Univ., Diss., Karlsruhe.

Oldenburg M. (2005), Speech at Conference of Department of Urban Water management, University of Rostock.

Peter-Fröhlich A. et al. (2007). Sanitation Concepts for Seperate Treatment of Urine, Fae ces and Greywater (SCST) – Kompetenzzentrum Wasser Berlin. http://www. kompetenz-wasser.de/ SCST22.0.html.

U.N. Population Division (2009). http://www.unpopulation.org.

Treatment of Domestic Sewage in an Anaerobic Baffled Reactor at Ambient Temperature

Li Qingxue, Wu Ping and Xiao Wei

College of Urban Construction, Hebei University of Engineering, Handan, 056038, P.R. China
E-mail: liqingxue_610@126.com

Abstract An Anaerobic Baffled Reactor was operated for 345 days under different conditions to assess the performance of treating raw domestic sewage. The reactor was operated at temperatures in the range of 9°C~29.2°C and at hydraulic retention times (HRTs) in the range of 6~24h. To assess the self-inoculation potential of the ABR, the start-up was carried out, without seed sludge, at a HRT of 24h. The start-up ABR was achieved in 60days. The results showed that the ABR was very effective in the treatment of domestic sewage and was capable of sustaining hydraulic shock loads. The averaged removal efficiencies of total chemical oxygen demand (CODt) and suspended solids (SS) of ABR were more than 75.3% and 92.5% at an HRT between 12 and 24h, respectively. Compartment-wise profiles indicated that most of the CODt removal occurred in the first compartment. While the HRT was decreased from 12h to 6h, CODt and SS removal efficiencies of ABR were found to be reduced to 55% and 70%, respectively. At this stage, the ambient temperature was only 15°C. Therefore, long HRT should be adopted at a lower temperature.

Keywords Domestic wastewater, anaerobic baffled reactor (ABR), hydraulic retention time (HRT), hydraulic shock loads, ambient temperature

INTRODUCTION

With the rapid economical development and improvement of living standards in Chinese rural areas, the discharge capacity of domestic sewage wastewater is increasing every year. It is estimated that more than 25 millions m^3 of domestic sewage wastewater is discharged into nearby soil and water bodies without being subjected to any kind of treatment. This is due to inadequate sewage networks and lack of funds. Decentralised and low-cost processes are considered to be a better choice for rural areas (Lens *et al.*, 2001). Anaerobic technologies are the core of sustainable, decentralised treatment (Lettinga, 1996; Hammes *et al.*, 2000). The success of anaerobic systems is due to process simplicity, low operation costs and the independency of electricity (Al-Jamal, 2009).

Currently, intensive research works have been conducted on the treatment of dilute sewage wastewater at ambient temperatures using anaerobic reactors, such as the upflow anaerobic sludge blanket (UASB) and the expanded granular sludge bed (Elmitwalli *et al*., 2002; Halalsheh *et al*., 2005). However, these technologies are not suitable for the decentralized treatment of rural sewage in most developing countries as they need complex maintenance and control, skilled operators and manufacturers (Boller, 1997). One of the innovative reactor designs, developed to implement this technology, is an anaerobic baffled reactor (ABR) developed by McCarty and co. (McCarty, 1981). It is known as a high-rate anaerobic reactor. It has a number of advantages over other reactors such as: good solid retention, low bed bypass, high to tolerance to hydraulic and organic shock loads, and stable reactor performance (Barber and Stuckey, 1999). In addition, it has many potential advantages, i. e. simple design, low maintenance requirements and low operating and capital costs.

Recent publications have revealed the potential of ABR for effective treatment of domestic and municipal wastewater (Foxon *et al*., 2004; Feng *et al*., 2008; Bodkhe, 2009). Bodkhe (2009) used a modified ABR to treat municipal wastewater at an HRT of 6h and at a temperature of 35°C. The removal efficiencies in SS, BOD, and COD were found to be 86%, 87% and 84% respectively. Feng (2008) developed a bamboo carrier ABR to treat domestic sewage at 28°C. The total COD removal efficiency of 69% has been achieved at a hydraulic retention time (HRT) of 18h. The successful treatment of low strength wastewaters in an ABR at mesophilic temperatures has been reported. However, the temperature of domestic sewage is in the range of 10–30°C and lower than mesophilic temperatures. Several studies reported that reduced temperatures resulted in deteriorations of anaerobic process performance. Therefore, anaerobic treatment systems able to operate at a low temperature offer substantially lower treatment costs for low temperature wastewaters, and could potentially increase the range of anaerobic treatment to domestic sewage.

The main objectives of this research were to asses the process performance of on-site ABR for the treatment of domestic sewage under ambient conditions and to increase the knowledge on the system design. To achieve these objectives, ABR was operated under ambient conditions at HRT of 6–24h for thirteen months and a temperature of 9–29°C.

MATERIALS AND METHODS

Experimental set-up

A laboratory scale ABR was fabricated using 8mm thick transparent plexi glass sheets. It measured 524mm long, 170mm width with a height of 460mm. The ABR reactor, with a total volume of 40.98L and an effective volume of 30.44L, was divided into four compartments by vertical baffles. The undivided compartments were divided by a hanging baffle into two chambers: the upflow chamber and downflow chamber. The ratio of the upflow chamber width and the downflow chamber width was 3:1. The lower portions of the hanging baffles were bent at 55° to direct the flow to the centre of the upflow chamber. This was to achieve better contact of feed and bio-solids. Wastewater sampling ports were located 100mm from the top of the reactor in the centre of upflow chambers of each compartment, while sludge sampling ports were located 50mm above the bottom of the reactor. The influent feed was pumped using a variable speed peristaltic pump (BT01-100, Longer, China).

Wastewater characteristics

The domestic sewage was taken from the Campus of the Hebei University of Engineering. The main characteristics of the sewage used in this study are presented in Table 1.

Table 1 Characteristics of the influent domestic sewage

Parameter	Concentration
Temperature (°C)	9–29.2
pH	7.12–8.46
Alkalinity(mg/L as $CaCO_3$)	170–600
SS(mg/L)	44.7–1381.2
CODt (mg/L)	225–1772
CODdis (mg/L)	71.4–588
Total Kjeldhal nitrogen (mg/L)	21.5–90.3
Ammonia nitrogen (mg/L)	18.2–65.6

Analytical methods

The samples for the influent and effluent were collected twice every week. They analyzed the total chemical oxygen demand (CODt), total suspended solids (TSS), ammonium nitrogen (NH_4^+-N) as described by the standard methods (State Environment Protection Administration of P.R. China, 2002). Raw samples were used for measuring total COD(CODt), 0.45μm membrane filtered samples for dissolved COD(CODdis). The suspended COD (CODss) was calculated as the difference between CODt and CODdis. The tests for volatile fatty acids (VFA) and alkalinity were carried out as described by He Yanling(1999). The pH was measured with a pHS-3C pH/mv meter. TOC concentration and total nitrogen concentration (TN) was evaluated by a TOC analyzer (TOC-V$_{CPH}$, Shimadzu).

Experimental procedure

The ABR reactor that treated high-strength organic wastewater eight years ago was not seeded with anaerobic digested sludge in this experimentation. The reactor was operated at an HRT of 24 h and at the ambient temperature of 20.0°C~24.0°C during start-up. The CODt removal was only 10% in the beginning of the start-up period (3th day). As the start-up time went on, the CODt removal efficiency increased and was up to 40% on the 28th day. High-quality CODt removal was noticed on the 60 th day. The removal efficiencies of CODt were more than 80%. Hence, the reactor performance in terms of COD removal indicated the success of start-up.

A series of experiments were conducted after the start-up. The HRT of the bioreactor was gradually reduced from 24h to 18h, to 12h, then to 6h. The performance of the reactor was studied at various HRTs. Once a steady state of operation was achieved at that particular HRT, the HRT was shifted to the next lower value and so on by increasing the volumetric flow rate of influent wastewater. A pseudo-steady state of operation was believed to have been achieved when the variation in effluent COD and SS values were found to be insignificant. Performance of the reactor studied for 343d at different HRTs is shown in various figures and discuss in the following paragraphs.

RESULTS AND DISCUSSION

ABR performance

COD removal

The daily variations in the COD concentration of the ABR influent and effluent, and COD removal are shown in table 2. The influent total COD concentration was in the range of 255 to 1772 mg/L. Soluble COD ranged between 71.8 and 588 mg/L. In the domestic wastewater, the major part of CODt is CODss followed by CODdis. The particulate (suspended+colloidal) COD represented 68.2%~75.5% of domestic wastewater CODt. The results demonstrated that the higher CODt removal was achieved. The CODt average removal efficiencies were found to be 91.7%, 88.7%, 75.3%, 55.4% and 55.2% at HRT 24, 18, 12, 8 and 6h, resulting in the effluent CODt concentrations of 31.7~125.4mg/L, 31.3~108mg/L, 74.3~259mg/L, 287~361mg/L and 283~428mg/L, respectively. The removal efficiencies of CODt in the ABR were found with decreasing the HRT. Decreasing the HRT to 6~8h reduced the CODt removal only to around 55%. The ABR had the lower CODt removal at HRT of 6~8h because of the lower temperatures of 15.1~19.2°C and lower HRTs. The low temperature was expected to affect the CODss removal negatively (Mahmoud et al., 2003). Previous research has demonstrated that the performance of single stage UASB systems at low temperatures is limited by the slow hydrolysis of entrapped solids that accumulate in the sludge bed (Zeeman and Letting, 1999).

The average effluent CODt concentrations were observed to be lower than 100mg/L at HRTs higher than 18h, which was below the discharging standard limit (National Environmental Protection Agency of PR China, 1996). Most of the effluent CODt was in dissolved form. The CODdis in the effluent of ABR represented 46%~100% of CODt. Removal of particulate fraction of organics was found to be greater than soluble fraction.

Table 2 The performance of ABR at ambient temperature

HRT (h)	Temperature (°C)	pH	VFA (mg/L)	Alkalinity (mg/L)	CODt removal (%)	SS removal (%)
24	19.0–28.5	7.70–8.86	55–130	220–550	88.6–94.6 (91.7)	88.5–100 (94.5)
18	24.1–29.4	7.48–8.39	70–110	170–410	80.9–94.6 (88.7)	76.9–94.7 (93.8)
12	20.7–25.3	7.83–8.07	110–170	450–570	63.7–86.1 (75.3)	76.6–93.7 (92.6)
8	15.1–19.2	7.36–8.09	145–200	470–550	39.5–74.7 (55.4)	58.6–79.4 (69.7)
6	15.7–15.8	7.44–7.66	125–170	420–510	40–71.3 (55.2)	61.3–80.2 (70.9)

Suspended solids removal

A superior performance of the reactor in terms of SS removal was observed as shown in Table 2. The effluent SS concentration generally increased with a decrease in HRT. The average SS removal efficiencies were 92.9%, 93.8%, 92.6%, 69.7% and 70.9%, resulting in the effluent SS concentrations of 11–76.5mg/L, 10.1–52mg/L, 20–66mg/L, 68.4–120mg/L and 89–191mg/L at HRTs of 24, 18, 12, 8 and 6h, respectively. SS removal efficiencies seemed to be affected slightly by variations in the HRT range between 12 and 24h. The SS removal efficiency was above 92.6% at HRT of 12 h. However, SS removal efficiency decreased significantly as HRT was decreased from 12 to 8 h.

Nitrogen removal

The daily variations in the TN and NH_4^+ concentration of the ABR influent and effluent are shown in Fig. 1 and Fig. 2. The difference in NH_4^+ and TN concentration between the influent and effluent of ABR was insignificant. The achieved NH_4^+ and TN removal efficiencies were rather low. Even negative NH_4^+ removal efficiencies were observed occasionally. Foxon et al. (2005) reported an increase in the levels of ammonical nitrogen by about 10–15% along the reactor.

This was due to the conversion of organic nitrogen in the influent under anaerobic conditions. Because of the increasing concern over eutrophication of surface waters and strict regulations on nitrogen discharges, direct anaerobic treatment of wastewater would necessitate aerobic or physical-chemical post-treatment.

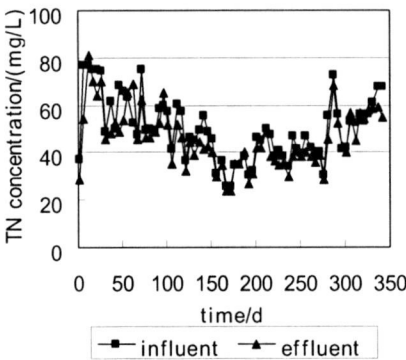

Figure 1 The variations of TN concentration of the ABR influent and effluent

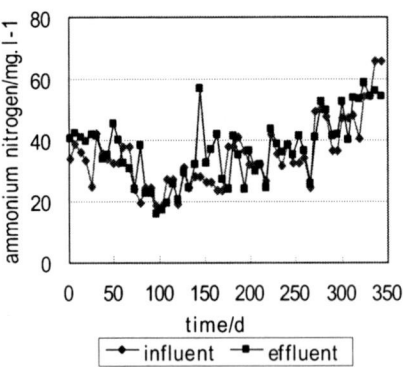

Figure 2 The variations of ammonium nitrogen concentration of the ABR influent and effluent

Compartment-wise profiles

The average characteristics of CODt along the length of the ABR at different HRTs are given in Fig. 3. Maximum CODt removal took place in the first compartment of the ABR. The results showed that the average CODt removals in the four compartments were 74.9%, 13.4%, 1.3% and 0% at HRT of 24h. The significant decreases of CODt concentrations were observed within the first one compartment. In Compartment 1, CODt removal decreased from 74.9% at HRT of 24 h to 69.3% at HRT of 12h. With every increase in HRT, more and more COD was transferred to later Compartments (i.e. 2, 3, 4) and anaerobic micro-organisms housed in these compartments played a greater and greater role in the biodegradation of organics. Manariotis and Grigoropoulos (2002) also found that while treating low strength wastewater using ABR at an HRT of 12 h, most of the organic matter was removed in the first two compartments. They observed that CODt removals in the first three compartments were 56.1%, 22.4% and 5.3%, respectively.

The pH along the length of the ABR is given in Fig. 4. The pH in the first compartment was lowest. It gradually increased as wastewater moved through the later compartments. The pH in the effluent of ABR was found to be above 7 at HRTs from 6h to 24h. Dama et al. (2002) have also found a lower pH in earlier compartments as the acidogenesis and acetogenesis predominate in these compartments. The pH values increase from the first compartment to the last compartment down the reactor due to the degradation of VFA in the later compartments.

Response to hydraulic over-loadings

Domestic wastewater is characterized by strong fluctuations in organic matter, particularly in the flow rate. In order to test the reactor capability of withstanding the flow variations, the effect of transient hydraulic shock loads on reactor performance in terms of chemical oxygen demand (COD) removal was examined. The reactors were operated at 24h HRT and 27.3°C as a base-line condition. Hydraulic shocks with an HRT of 6 h (an increase of 4 times in the influent flow rate) were applied to the reactors for 24 h. The baseline conditions resulted in 93.2% CODt removal; however, when the HRT decreased to 6h, the reactor performance deteriorated and the COD removal efficiency dropped to 58.6%. It was found that the ABR was very stable to large transient shocks, and while biomass loss was substantial, it recovered back to its baseline performance only 9 h after the shock ceased.

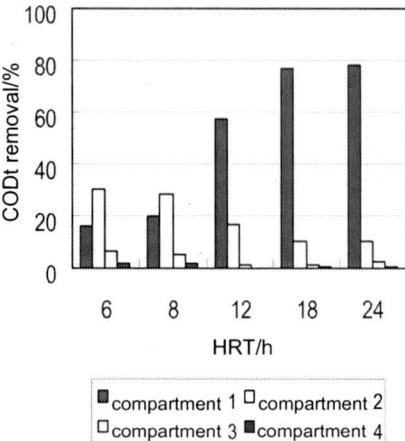

Figure 3 CODt removal efficiencies in the compartments of ABR

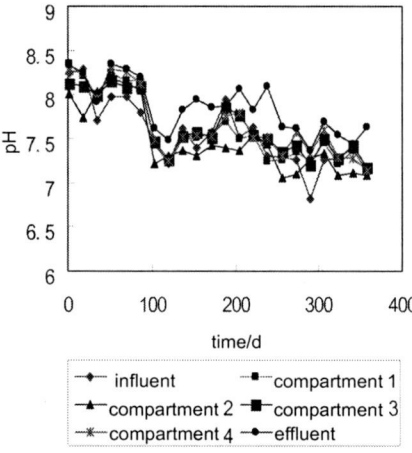

Figure 4 pH profile in the compartments of ABR

CONCLUSIONS

A long-term operation (345days) of ABR indicated that the ABR reactor was suitable for the decentralized treatment of rural sewage at an ambient temperature. Despite the strong fluctuations in influent wastewater characteristics, the treated effluent quality was consistent at a particular HRT. The CODt average removal efficiencies were found to be 91.7%, 88.7%, 75.3%, 55.4% and 55.2% at HRT

24, 18, 12, 8 and 6h, respectively. The most CODt had been removed at the first and the second compartments, while other compartments played little role in removing CODt. Temperature has a significant effect on CODt removal in ABR treating domestic wastewater. Therefore, long HRT should be adopted at a lower temperature. ABR alone may not meet the treatment requirements for domestic wastewaters in rural areas without post-treatment to remove residual COD and to remove or to recover nutrients.

REFERENCES

Al-Jamal W. and Mahmoud N. (2009). Community onsite treatment of cold strong sewage in a UASB-septic tank. *Bioresource Technology*, **100**, 1061–1068.

Barber W. P. and Stuckey D. C. (1999). The use of the anaerobic baffled reactor (ABR) for wastewater treatment: a review. *Water Research*, **33**(7), 1559–1578.

Bodkhe S. Y. (2009). A modified anaerobic baffled reactor for municipal wastewater treatment. *Journal of Environmental Management*, **90**(8), 1–6.

Boller M. (1997). Small wastewater treatment plants – a challenge to wastewater engineers. *Water Science and Technology*, **35**(6), 1–12.

Dama P., Bell J., Foxon K. M., Brouckaerr C. J., Huang T., Buckley C. A., Naidoo V. and Stuckey D. C. (2002). Pilot-scale study of an anaerobic baffled reactor for the treatment of domestic wastewater. *Water Science and Technology*, **46**(9), 263–270.

Elmitwalli T. A., Sklyar V., Zeeman G. and Lettinga G. (2002). Low temperature pretreatment of domestic sewage in an anaerobic hybrid or an anaerobic filter reactor. *Bioresource Technology*, **82**, 233–239.

Feng H. J., Hu L. F., Mahmood Q., Qiu C. D., Fang C. G. and Shen D. S. (2008). Anaerobic domestic wastewater treatment with bamboo carrier anaerobic baffled reactor. *International Biodeterioration Biodegradation*, **62**, 232–238.

Foxon K. M., Pillay S., Lalbahadur T., Rodda N., Holder F. and Buckley C. A. (2004). The anaerobic baffled reactor (ABR): An appropriate technology for on-site sanitation. *Water Institute of South Africa (WISA) Biennial Conference*, Cape Town, South Africa, 2–6 May 2004.

Halalsheh M., Sawajneh Z., Zu'bi M., Zeeman G., Lier J., Fayyad M. and Lettinga G. (2005). Treatment of strong domestic sewage in a 96 m3 UASB reactor operated at ambient temperatures: two-stage versus single-stage reactor. *Bioresource Technology*, **96**, 577–585.

Hammes F., Kalogo Y. and Verstraete W. (2000). Anaerobic digestion technologies for closing the domestic water, carbon and nutrient cycles. *Water Science and Technology*, **41**(3), 203–211.

Lens P., Zeeman G. and Lettinga G. (2001). *Decentralised Sanitation and Reuse; Concepts, Systems and Implementation*. IWA Publishing, London, UK.

Lettinga G. (1996). Sustainable integrated biological wastewater treatment. *Water Science and Technology*, **33**(3), 85–98.

Mahmoud N., Amarneh M. N., Al-Sa' ed R., Zeeman G., Gijzen H. and Lettinga G. (2003). Sewage characteristics as a tool for the application of anaerobic treatment in Palestine. *Environmental Pollution*, **126**, 115–122.

Manariotis I. D. and Grigoropoulos S. G. (2002). Low-strength wastewater treatment using an anaerobic baffled reactor. *Water Environment Research*, **74**(2), 170–176.

McCarty P. L. (1981). One hundred years of anaerobic treatment digestion 1981. In: *Anaerobic Digestion*, vol. 1., Hughes (ed.), Elsevier Biomedical Press, pp. 3–21.

State Environment Protection Administration of PR China (2002). *Methods for Monitor and Analysis of Water and Wastewater*, fouth edn, China Environmental Science Press, Beijing (in Chinese).

Van Lier J. B. (2008). Current and future trends in anaerobic digestion: diversifying from waste (water) treatment to resource oriented conversion techiques. *Water Science and Technology*, **58**(3), 85–98.

Zeeman G. and Lettinga G. (1999). The role of anaerobic digestion in closing the water and nutrient cycle at community level. *Water Science and Technology*, **39**(5), 187–194.

An Innovative Integrated Reactor System for Simultaneous Removal of Carbon, Sulfur and Nitrogen Based on Biological Niches

A.J. Wang, C. Chen, C.S. Liu, N.Q. Ren and D.J. Lee

State Key Laboratory of Urban Water Resource and Environment, Harbin Institute of Technology (SKLUWRE, HIT), Harbin 150090, China
E-mail: waj0578@hit.edu.cn

Abstract An innovative biological wastewater treatment system for the removal of organic carbon, sulfur and nitrogen was developed to consist of three reactors, i.e. sulfate reduction and organic matter removal (SR-CR), autotrophic and heterotrophic denitrifying sulfide removal (A&H-DSR) and nitrification (AN). The essential operational parameters for SR-CR reactor are sulfate and organic loading rates, hydraulic retention time (HRT), COD/SO_4^{2-} ratio and pH, and for A&H-DSR reactor are sulfide and nitrate loading rates, HRT, pH, S^{2-}/NO_3^- ratio and COD/NO_3^- ratio. Laboratory tests identified the optimal conditions for SR-CR reactor include HRT≥24h, sulfate loading rate ≤7.5 kg SO_4^{2-}/m^3•d and organic loading rate ≤10 kgCOD/m^3•d, COD/SO_4^{2-}≥2 and pH≥6.5. For the A&H-DSR process, the optimal conditions were found to be sulfide loading rate ≤6.0kg S^{2-}/m^3•d, nitrate loading rate ≤3.5 kg NO_3^-/m^3•d, S^{2-}/NO_3^-≥1, COD/NO_3^-≥1.25:1 and pH≥7.5. When operating at so-determined optimal conditions sulfate, ammonia and organic matter removals of 99%, 90% and 99%, respectively, were obtained, so to recover ≥90% elemental sulfur (S^0) from the sulfide-laden wastewater streams. The predominant functional organisms include *Clostridiaceae* sp., *Desulfomicrobium* sp., *Methanosaeta* sp. in the SR-CR reactor, and *Sulfurovum* sp., *Pseudomonas aeruginosa* and *Denitratisoma* sp. in the A&H-DSR reactor.

Keywords Desulfurization, denitrifying sulfide removal, denitrification, bio-phase separation, elemental sulfur, control strategies

INTRODUCTION

Sulfate and ammonia-laden wastewaters from pharmaceutical industry, food fermentation and paper pulp need treatment prior to their safe disposal. Sulfate and ammonia removal are conventionally achieved using two separate systems: sulfate reduction and sulfide oxidation and nitrification and denitrification. Reyes-Avila *et al.* (2004), Manconi *et al.* (2006) and Chen *et al.* (2008a) developed denitrifying sulfide removal (DSR) process which could simultaneously remove sulfide, nitrate and organic carbon from high strength sulfide and nitrate containing wastewater in a single reactor. Sulfate-laden waters or wastewaters are commonly observed with sulfide frequently being yielded in the action of biological sulfate reduction. Incorporation of sulfate reduction process with the DSR process provides an efficient way to deal with sulfate and nitrate-laden wastewaters. Taking into account possible degradation reactions of organic carbon matters by acidogens and methanogens (MPB) (Mizuno et al., 1994), the authors proposed an innovative integrate system that can simultaneously remove sulfate, nitrate and organic carbons (Figure 1).

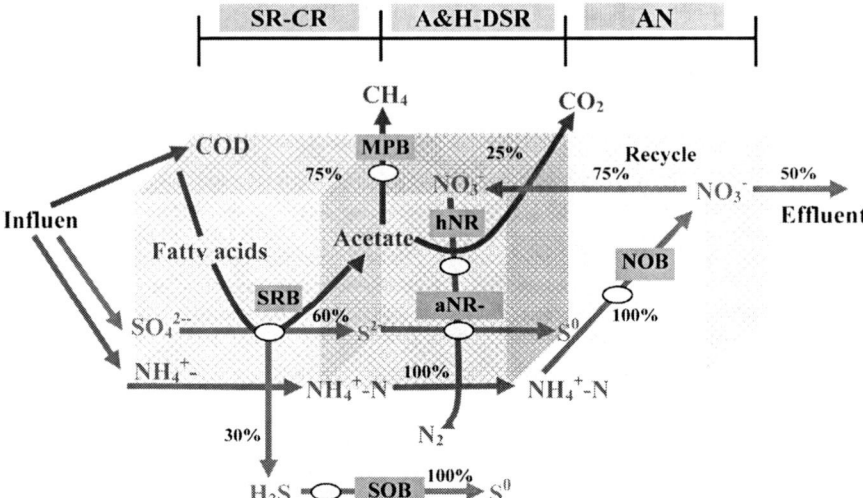

Figure 1 The proposed simultaneous C-S-N removal process system based on bio-phase separation principle. Note: SR-CR: sulfate and organic substance oxidation; A&H-DSR: autotrophic and heterotrophic combination denitrifying sulfide removal; AN: nitrification; SRB-sulfate reduction bacteria; hNRB-heterotrophic denitrifiers; aNR-SOB-autotrophic denitrifiers; MPB-methane production bacteria; SOB-sulfide oxidation bacteria; NOB- ammonia oxidation bacteria; Recycle rate −200%

This system consists of three units: a) sulfate reduction and organic matter removal (SR-CR), b) autotrophic and heterotrophic denitrifying sulfide removal (A&H-DSR), and (c) nitrification (AN). In the SR-CR process, organic matters are first degraded to fatty acids (e.g. lactate, butyrate, acetate) and/or solvents (e.g. ethanol) and then to CH_4. Sulfate reducing bacteria (SRB) easily consumes fatty acids. The autotrophic denitrifiers (aNR-SOB) converts sulfide to elemental sulfur using nitrate as electron acceptor (NO_3^- to NO_2^-); while the heterotrophic denitrifiers (hNRB) utilized converts the yielded sulfide and residual fatty acids (mainly acetate) from

SR-CR for denitrification (NO_2^- to N_2). The AN stage transformed all ammonium in streams to nitrate using ammonium-oxidizing bacteria (AOB), with produced nitrate being carried over to A&H-DSR with recycle. Alkalinity produced in denitrification can keep suspension pH high so to minimize release of hydrogen sulfide (H_2S) from the SR-CR and A&H-DSR reactors. Ecological niches are achieved to maximize performance of this proposed integrated SR-CR/A&H-DSR/AN system.

Scope and Objectives

In this paper, an innovative wastewater treatment system for the simultaneous removal of organic carbon, sulfur and nitrogen are detailed. The expanded granular sludge bed (EGSB) reactors were fabricated for the SR-CR and A&H-DSR stages, and a cross-flow aerobic reactor is used for the AN stage. Functional microorganisms are cultivated in each unit so the biological niches can be individually achieved. Optimal operational parameters for maximizing system performance are searched for.

CURRENT KNOWLEDGE

Traditional nitrogen removal processes

Numerous nitrogen compounds such as ammonium and nitrate can induce europhication of water bodies (Baker, 1998; Oenema and Roest, 1998). The traditional microbial processes of autotrophic nitrification combined with heterotrophic denitrification (such as A/O or A²O) are conventionally applied for nitrogen removal in wastewater treatment. In brief, ammonium is first oxidized to nitrate or nitrite by autotrophs under aerobic condition, and then the yielded nitrate or nitrite is denitrified to di-nitrogen gas by heterotrophic denitrifers under anaerobic or anoxic conditions.

Traditional sulfur removal process

Combined anaerobic sulfate reduction and aerobic sulfide oxidation or photosynthetic sulfur oxidation are conventionally used for sulfur removal in wastewater treatment (Buisman et al., 1990). Sulfate in water is first reduced to sulfide by SRB under anaerobic condition, and then the produced sulfide is oxidized by sulfide oxidation bacteria or photosynthetic sulfur bacteria to sulfur under appropriate operational conditions.

Simultaneous sulfur and nitrogen removal process

Some chemolithoautotrophic microbes can use inorganic sulfur compounds such as sulfide, sulfur, or thiolsulfate as electronic donors, and nitrate or nitrite as electronic acceptors to drive denitrification (Kuenen et al., 1992; Wang et al., 2005).

Wang *et al.* (2005) developed a simultaneous desulfurization and denitrification (SDD) process utilizing a single autotrophic strain, *Thiobacillus denitrificans*, in a CSTR. Approximately 75% of sulfide was converted to S^0 at loading rate of 0.6 kg-S $m^{-3}d^{-1}$. Manconi *et al.* (2006) determined that activated sludge in a CSTR can perform autotrophic denitrification to completely remove sulfide and nitrate at loading rates respectively of 0.11 kg-NO_3^--N $m^{-3}d^{-1}$ and 0.28 kg-S^{2-}-S $m^{-3}d^{-1}$.

Krishnakumar *et al.* (1999) utilized an isolated chemolithoautotrophic strain, *Thiobacillus denitrificans*, to removal sulfide at oxidation rate of approximately 0.9g S^{2-} $gVSS^{-1}$ h^{-1} under denitrifying conditions. Nemati *et al.* (2006) cultivated a chemolithoautotrophic denitrifer, *Thiomicrospira* sp. CVO, in a continuous bioreactor to achieve complete removal of sulfide at volumetric loading rate of 1.6 mmol/h.

ASSESSMENT

Optimization of operating parameters in the SR-CR reactor

Based on the scheme of Figure 1, the SR-CR stage should efficiently degrade organic matter and sulfate following Eq. (1):

$$NH_4^+ + \text{organic matter} + SO_4^{2-} \rightarrow NH_4^+ + HCO_3^- + CH_4 + HS^- + \text{VFAs} \tag{1}$$

The operational conditions of SR-CR process should favor the growth of SRB and MPB. It is well accepted that COD/SO_4^{2-}, sulfate loading rate, pH, HRT and organic loading rate are the key factors influencing the performance of sulfate reduction and removal of organic matter to methane (Parkin, 1991; Mizuno *et al.*, 1998). Table 1 lists the tested parameters and the optimal conditions so determined for maximizing COD removal are: COD/SO_4^{2-} >2.0, sulfate loading rate ≤7.5 $kgSO_4^2$-S $m^{-3}d^{-1}$, pH ≥6.5, HRT >24h and organic loading ≤10kg COD $m^{-3}d^{-1}$. Under this condition, removal of both sulfate and COD exceeded 80% at 2000 mgL^{-1} sulfate and 20000 mgL^{-1} COD.

Optimization of operating parameters in the A&H-DSR reactor

The A&H-DSR reactor is to simultaneously remove sulfide produced in the SR-CR reactor and nitrate recycled from the AN reactor. This unit is performed by synergistic action of both autotrophic and heterotrophic denitrifiers. Restated, sulfide is used as electron donors by autotrophic denitrifers with nitrate as electron acceptors. With limited nitrate, sulfide is principally converted to elemental sulfur according Eq. (2). The nitrate is removed by hetertrophic denitrifiers using organic carbon such as VFAs as electron donors (Eq. (3)):

$$NO_3^- + HS^- \rightarrow N_2 + S^0 \tag{2}$$

$$VFA + NO_3^- \rightarrow N_2 + H_2O + CO_2 \tag{3}$$

The known factors affecting DSR process include S^{2-}/NO_3^-, sulfide loading rate, pH, COD/NO_3^-, and nitrate loading rate (Chen *et al.* 2008a, 2008b). Table 2 lists the tested results. The optimal conditions are determined to be: sulfide loading rate$\leq$6.0kg S^{2-} $m^{-3}d^{-1}$, nitrate loading rate$\leq$3.5kg NO_3^--N $m^{-3}d^{-1}$, pH~7.5, $S^{2-}/NO_3^- \geq 1:1$ and $COD/NO_3^- \geq 1.25:1$. Under the so-determined operational conditions, the removal of sulfide, nitrate and organic matter can reach 100%, 90% and 95%, respectively. Chen *et al.* (2009) experimentally confirmed that most sulfide consumed is converted to S^0 and nitrate to N_2, consistent with the findings by Chen *et al.* (2008a). N_2O and CO_2 are negligibly yielded throughout the test. Additionally, the activity of SRB is constrained by the limited quantities of residual sulfate and fatty acids from the SR-CR reactor.

Table 1 Operating condition tests for optimization of SR-CR process

Parameters	The tested condition	Optimal conditions and removal efficiencies
COD/SO_4^{2-}	1:1→2:1→3:1→4:1→ 5:1→10:1	COD/SO_4^{2-}>2:1 SO_4^{2-} removal > 80%, COD removal > 90%
Sulfate loading (kg SO_4^{2-}-S $m^{-3}d^{-1}$)	1.0→2.0→3.0→5.0→ 7.5→10	Sulfate loading <7.5 kg, COD/SO_4^{2-}>2:1 SO_4^{2-} removal >80%, COD removal > 80% pH≥6.5
pH	5.0→6.5→7.0→7.5	SO_4^{2-} removal >80%, COD removal> 80%
HRT (h)	10h→15h→24h→30h	HRT>24h,COD/SO_4^{2-}>2:1 SO_4^{2-} removal >80%, COD removal> 80%
organic loading (kg COD $m^{-3}d^{-1}$)	2.0→4.0→6.0→10→ 15→20	Organic loading <10 kg,COD/SO_4^{2-}>2:1 SO_4^{2-} removal >80%, COD removal> 85%

Overall efficiency of the integrated C-S-N removal system

Local biological niches are established in the integrated C-S-N removal system. Restated, the sulfate and organic matters in wastewater firstly enter the SR-CR reactor where the effluent (rich in sulfide) is produced (Eq. (1)). Then the sulfide-laden effluent is fed to the A&H-DSR reactor coupled with the recycled flow from the AN reactor, providing substrates for reactions in Eqs. (2) and (3). The effluent with ammonia is introduced to the AN reactor to generate nitrate via nitrification (Eq. (4)). A fraction of the treated wastewater in AN reactor is recycled back to the A&H-DSR reactor with the rest being discharged.

$$O_2 + NH_4^+ \rightarrow NO_3^- + H_2O \tag{4}$$

Table 2 Operating condition tests for optimization of A&H-DSR process

Parameter	The tested condition	Optimal conditions and removal efficiencies
pH	$7.5 \rightarrow 8.5 \rightarrow 10$	pH~7.5 S^{2-} removal >95%, NO_3^- removal≥95%
S^{2-}/NO_3^-	$5:2 \rightarrow 5:3 \rightarrow 5:4 \rightarrow 5:5$ $\rightarrow 5:6$	$S^{2-}/NO_3^- \geq 5:5$, S^{2-} removal >95%, NO_3^- removal≥95%
COD/NO_3^-	$0.75:1 \rightarrow 1:1 \rightarrow 1.25:1$ $\rightarrow 2:1$	$COD/NO_3^- \geq 1.25:1$ S^{2-} removal >95%, NO_3^- removal≥95%
Sulfide loading (kg S^{2-} m^{-3}d^{-1})	1.0 kg$\rightarrow$2.0$\rightarrow$3.0$\rightarrow$4.0$\rightarrow$ 5.0$\rightarrow$6.0$\rightarrow$8.0	6.0 kg S^{2-} removal >95%, NO_3^- removal≥95%
Nitrate loading (kg NO_3^- m^{-3}d^{-1})	0.5 kg$\rightarrow$1.0$\rightarrow$2.0$\rightarrow$2.5$\rightarrow$ 3.0$\rightarrow$3.5$\rightarrow$4.0	Nitrate loading≤3.5kg, S^{2-} removal >95%, NO_3^- removal≥90%

Table 3 The overall efficiencies of the simultaneous C-S-N removal system

	Operations and removal efficiencies	Effluent
Sulfate loading	7.5 kgSO$_4^{2-}$-Sm^{-3}d^{-1}, removal rate>99%	SO$_4^{2-}$≤180 mgl^{-1}
Ammonia loading	3.5 kg NH$_4^+$-N m^{-3}d^{-1}, removal rate> 90%	NH$_4^+$≤20 mgl^{-1}
Organic loading	10 kg COD m^{-3}d^{-1}, removal rate > 99%	COD≤100 mgl^{-1}
S^0 reclamation	6.0 kg S^0-m^{-3}d^{-1}, S^0 recovery ~90%	

Table 3 lists the integrated test with operational parameters listed in Tables 1 and 2. The overall removal efficiencies of sulfate, ammonia and organic matter in the integrated SR-CR/A&H-DSR/AN system reach 99%, 90% and 99%, respectively, under the loading rates of 7.5 $kgSO_4^{2-}$-S $m^{-3}d^{-1}$, 3.5 kg NH_4^+-N $m^{-3}d^{-1}$ and 10kg COD $m^{-3}d^{-1}$. The effluent concentrations of sulfate, ammonia and COD are $\leq$180mgL^{-1}, $\leq$20 mgL^{-1}, and $\leq$180 mgL^{-1}. The production rate of S^0 reaches 6.0kg $S^0 m^{-3}d^{-1}$, about 20 times higher than that reported by Reyes-Avila's (2004).

Characterization of microbial community in the C-S-N removal system

Figure 2 shows the DGGE profiles of microbial communities in the SR-CR reactor under sulfate loading rate of 3.75 and 7.5 kg SO_4^{2-}-S $m^{-3}d^{-1}$, and in the A&H-DSR reactor under sulfide loading rate of 3.0 and 6.0 kg S^{2-}-S $m^{-3}d^{-1}$, respectively.

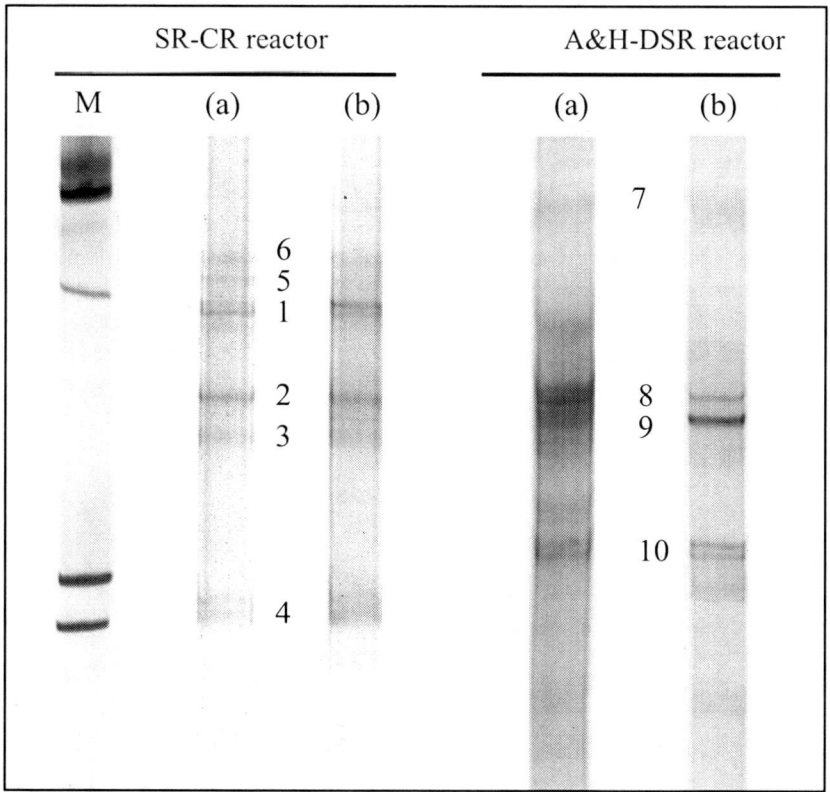

Figure 2 DGGE analysis of microbial communities for SR-CR reactor at sulfate loading rates of (a) 3.75 kgm^{-3}d^{-1} and (b) 7.5 kgm^{-3}d^{-1} and A&H-DSR reactor at sulfide loading rates of (a) 3.0 kgm^{-3}d^{-1} and (b) 6.0 kgm^{-3}d^{-1}

The sequences with similarity exceeding 99% are considered as the same operational taxonomic unit (OTU). In the SR-CR reactor samples, bands 1, 2 and 4 corresponding to *Clostridiaceae* sp., *Methanosaeta* sp. and *Desulfomicrobium* sp. are present. *Clostridiaceae* sp., heterotrophic species, can utilize certain sugars and various single amino acids for fermentation and produce acetate as major fatty acid product (Brisbarre *et al.*, 2003). *Desulfomicrobium* sp. is an incomplete oxidizing SRB (Rozanova *et al.*, 1988) and reduces sulfate using various fatty acids, ethanol or H_2 as electron donors. *Methanosaeta* sp. is an obligate-aceticlastic methane-producing bacteria which is likely responsible for the methane production using acetate in the SR-CR reactor (Sumiko *et al.*, 2008).

In the samples of A&H-DSR reactor, bands 8, 9 and 10 corresponding to *Sulfurovum* sp., *Pseudomonas aeruginosa* and *Denitratisoma* sp. are present. The *Sulfurovum* sp. is a chemolithotrophic bacteria using sulfur, sulfide or thiosulfate as electron donors and nitrate as electron acceptors (Inagaki *et al.*, 2004), which may be responsible for the sulfide removal. *Pseudomonas aeruginosa* and *Denitratisoma* sp. are the heterotrophic denitrifiers and are likely responsible for organic carbon and nitrate removal (Fahrbach *et al.*, 2006). The *Pseudomonas aeruginosa* can utilize various organic substrates as electron donors and carbon source. The *Denitratisoma* sp. can utilize fatty acids (C_2 to C_6) including isobutyrate, crotonate, DL-lactate, pyruvate, fumarate and succinate as electron donors and nitrate as electron acceptors. The microbial community in the AN unit comprises *Nitrobacter* sp. and *Nitrospina* sp. which perform nitrification and oxidize ammonia to nitrite or nitrate (data not shown), consistent with the results reported by Zhang *et al.* (2008). Based on the DGGE analysis, the functional strains identified in the integrated C-S-N removal system are summarized in Figure 3. Also, an ecological model of the interactions among functional organisms in this process system is shown in Figure 4.

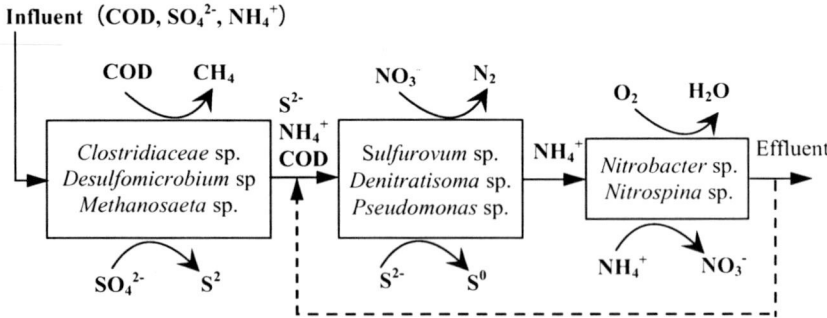

Figure 3 A simplified bio-phase separation pattern identified in the simultaneous C-N-S removal process system

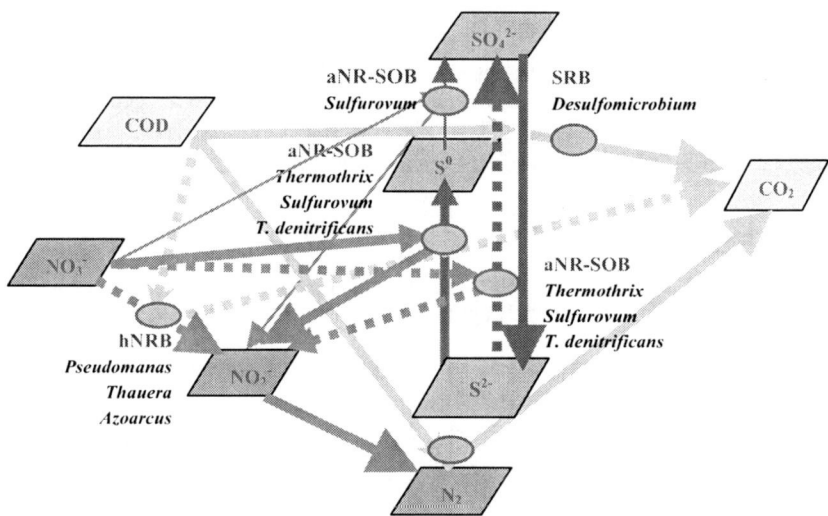

Figure 4 Ecological model of functional organisms in the simultaneous C-N-S removal process system

RECOMMENDATIONS AND NEEDS

Effective control of operational parameters such as HRT and pH is required to maintain local biological niches in each units subjected to change in wastewater quantity and quality. We propose that the HRT in SR-CR, A&H-DSR and AN reactor are set at 24h, 8h and 4h, respectively, so as to lead to a total HRT of at 36h. The volumes of the SR-CR, A&H-DSR and AN bioreactors can be designed as 1:1:2.5 to achieve individual optimal loading rates. Moreover, internal recycling rates and recycling stream from AN are adjusted to reach needed upflow velocities. Other important operational parameters discussed in **ASSESSMENT** should also be emphasized, which include a) COD/SO_4^{2-} and pH in the SR-CR reactor; b) pH, S^{2-}/NO_3^- and COD/NO_3^- in the A&H-DSR reactor; as well as c) pH, ammonia concentration and DO in the AN reactor.

This study reveals the feasibility of applying an integrated SR-CR/A&H-DSR/ AN system can effectively remove nutritious compounds together with organic matter for sulfate and ammonia containing wastewater. The major advantages are summarized as follows: (a) simultaneous removal of organic carbon, nitrogen and sulfur in one integrated system; (b) efficient elemental sulfur production rate is obtained; and (c) less sludge production due to anaerobic degradation, thereby reducing costs of sludge treating.

ACKNOWLEDGEMENTS

The authors would like to thank the Financial supports by National Natural Science Foundation of China (No.50878062), by The National Science & Technology Pillar Program During The Eleventh Five-Year Plan Period (2008BADC4B01), and by The National Scientific and Technological Project (2008ZX07207-005-002).

REFERENCES

Anderson G. K. and Sau C. B. (1984). State of Art of Anaerobic Digestion for Industrial Applications in the United Kingdom. *Proc. 39th Ind. Waste Conf.*, Purdue Univ., pp. 783–793.

Anderson G. K. (1986). Fate of COD in an Anaerobic System Treating High Sulfate Bearing Wastewater. *Paper Presented at Int. Conf. on Toxic Waste Treatment*, Washington D. C. USA.

Brisbarre N., Fardeau M. L. and Cueff V. (2003). *Clostridium caminithermale* sp. nov., a slightly halophilic and moderately thermophilic bacterium isolated from an Atlantic deep-sea hydrothermal chimney. *International Journal of Systematic Evolutionary Microbiology*, **53**, 1043–1049.

Buisman C. J. N., Bert G. and Ijspeert P. (1990). Optimization of sulfur production in a biotechnological sulfide-removing reactior. *Biotechnology and Bioengineering*, **38**, 50–56.

Baker L. A. (1998). Design considerations and application for wetland treatment of highnitrate waters. *Water Science and Technology*, **38**(1), 389–395.

Chen C., Ren N., Wang A., Yu Z. G. and Lee D. J. (2008a). Simultaneous biological removal of sulfur, nitrogen and carbon using EGSB reactor. *Applied Microbiology and Biotechnology*, **78**, 1057–1063.

Chen C., Wang A. J. and Ren N. Q. (2008b). Biological breakdown of denitrifying sulfide removal process in high-rate expanded granular bed reactor. *Applied Microbiology and Biotechnology*, **78**, 1057–1063.

Chen C., Wang A. J., Ren N. Q., Lee D. J. and Lai J. Y. (2009). High-rate denitrifying sulfide removal process in expanded granular sludge bed reator. *Bioresource Technology*, **100**, 2316–2319.

Fahrbach M., Kuever J. and Meinke R. (2006). *Denitratisoma oestradiolicum* gen. nov., sp. nov., a 17β-oestradiol-degrading, denitrifying betaproteobacterium. *International Journal of Systematic Evolutionary Microbiology*, **56**, 1547–1552.

Fu Z., Yang F., An Y. and Xue Y. (2009). Simultaneous nitrification and denitrification coupled with phosphorus removal in an modified anoxic/oxic-membrane bioreactor (A/OMBR). *Biochemical Engineering Journal*, **43**, 191–196.

Kuenen J. G., Robertson L. A. and Tuovinen O. H. (1992). The genera *Thiobacillus, Thiomicrospira, and Thiosphaera*. In: *The Prokaryotes*, A. Balows, H. G. Truper, M. Dworkin, Harder W and K.-H. Schleifer (eds.), vol. 3, Springer, New York, pp. 2638–2657.

Gadekar S., Nemati M. and Hill G. A. (2006). Batch and continuous biooxidation of sulfide by Thiomicrospira sp. CVO: Reaction kinetics and stoichiometry. *Water Research*, **40**, 2436–2446.

Inagaki F., Takai K., Nealson K. H. and Horikoshi K. (2004). Sulfurovum lithotrophicum gen. nov., sp nov., a novel sulfur-oxidizing chemolithoautotroph within the epsilon-Proteobacteria isolated from Okinawa Torugh hydrothermal sediments. *International Journal of Systematic Evolutionary Microbiology*, **54**, 1477–1482.

Kim J., Guo X., Behera S. K. and Park H. (2009). A unified model of ammonium oxidation rate at various initial ammonium strength and active ammonium oxidizer concentrations. *Bioresource Technology*, **100**, 2118–2123.

Manconi I., Carucci A., Lens P. and Rossetti S. (2006). Simultaneous biological of sulphide and nitrate by autotrophic denitrification in an activated sludge system. *Water Science and Technology*, **53**, 91–99.

Mizuno O., Li Y. Y. and Noike T. (1994). Effects of sulfate concentration and sludge retention time on the interaction between methane production and sulfate reduction for butyrate. *Water Science and Technology*, **30**, 45–54.

Mizuno O., Li Y. and Noike T. (1998). The Behavior of sulfate-reducing bacteria in acidogenic phase of anaerobic digestion. *Water Resources*, **32**, 1626–1634.

McCartney D. M. and Oleszkiewicz J. A. (1993). Competition between methanogens and sulfate reducers: Effect of COD:sulfate ratio and acclimation. *Water Environment Research*, **65**, 655–658.

Oenema O. and Roest C. W. (1998). Nitrogen and phosphorus losses from agiculture into surfacewaters: the effect of policies and measures in The Netherlands. *Water Science and Technology*, **37**, 19–30.

Parkin G. F. (1991). Anaerobic filter treatment of sulfate-containing wastewaters. *Water Science and Technology*, **23**, 1283–1291.

Ren N. Q., Wang A. J. and Han H. J. (2006). Development of combined biological technology for treatment of high-strength organic wastewater and results of case studies. *Journal of Ocean University of China*, **15**, 311–316.

Reyes-Avila J., Razo F. and Gomez J. (2004). Simultaneous biological removal of nitrogen, carbon and sulfur by denitrification. *Water Resources*, **38**, 3313–3321.

Rozanova E. P., Nazina T. N. and Galushko A. S. (1988). Isolation of a new genus of sulfate-reducing bacteria and description of a new species of this genus, *Desulfomicrobium apsheronum* gen. nov., sp. nov. *Mikrobiologiya*, **57**, 634–641.

Sumiko J., Hirasawa J. S. and Sarti A. (2008). Application of molecular techniques to evaluate the methanogenic archaea and anaerobic bacteria in the presence of oxygen with different COD: Sulfate ratios in a UASB reactor. *Anaerobe*, **14**, 209–218.

Tajima K., Aminov R. I., Nagamine T., Ogata K., Nakamura M., Matsui H. and Benno Y. (1999). Rumen bacterial diversity as determined by sequence analysis of 16SrDNA libraries. *FEMS Microbiology Ecology*, **29**, 159–169.

Wang A., Du D., Ren N. Q. and Groenestijn van J. (2005). An innovative process of simultaneous desulfurization and denitrification by Thiobacillus denitrification. *Journal of Environmental Science and Health*, **A40**, 1939–1949.

Watanabe Kodama K. Y. and Harayama S. (2001). Design and evalution of PCR primers to amplify bacterial 16S nbosomal DNA fragments used for community fingerprinting. *Journal of Microbiological Methods*, **44**, 253–262.

Wang A. J., Du D. Z., Ren N. Q. and van Groenestijn J. W. (2005). An innovative process of simultaneous desulfurization and denitrification by Thiobacillus denitrification. *Journal of Environmental Science and Health Part A-Toxic/Hazardous Substances & Environmental Engineering*, **40**(10), 1939–1949.

Zhang Y., Xin M. and Gao W. (2007). Advances on nitrifying bacteria and its application in wastewater nitrification. *Enviromental Pollution & Control*, **6**, 1–7.

Characterization of Polyhydroxybutyrate-rich Aerobic Granules in an SBR under Nitrogen Deficient Conditions

Jin Wang, Wen-Wei Li, Xian-Wei Liu, Zheng-Bo Yue and Han-Qing Yu*

Department of Chemistry, University of Science & Technology of China, Hefei, 230026 China
E-mail: hqyu@ustc.edu.cn

Abstract Polyhydroxybutyrate (PHB)-rich aerobic granules were cultivated in a sequencing batch reactor (SBR) under nitrogen deficient conditions through a two-step strategy. In the first step the PHB-storage ability of activated sludge were enhanced by limiting both oxygen and ammonia initially, and PHB content of sludge increased to $43.1 \pm 2.0\%$ at day 70; In the second step granular sludge was cultivated to get a high PHB volumetric productivity. During granulation, the settling ability of the PHB-rich sludge continuously improved. The mature PHB-rich granular sludge, with a PHB content of $40 \pm 4.6\%$, presented a buff color and regular morphology of elliptical and flat shape regardless of the diameter difference. In addition, the surface properties of activated sludge changed significantly during the cultivation. The surface charge, extracellular polymeric substances (EPS) content and the sludge hydrophobicity all increased during granulation, while the surface energy of sludge decreased to a relatively steady state accompanied with the growth of granular sludge. This study demonstrates that the metabolism of intracellular storages induced microbial to produce EPS, which favored the formation of aerobic granules.

Keywords Activated Sludge, Granulation, Polyhydroxybutyrate (PHB), Sequencing Batch Reactor (SBR), Surface Properties, Wastewater

INTRODUCTION

Polyhydroxyalkanoates (PHAs) are constituted of a large and versatile family of polyesters produced by various bacteria. PHAs are receiving considerable attention because of their potential use as renewable and biodegradable plastics and as a source of chiral synthons (Kessler *et al.*, 2001). The first microbial polyhydroxyalkanoate (PHA) discovered is polyhydroxybutyrate (PHB). PHB has been found to accumulate in many microorganisms, typically Gram-negative and -positive species and *archaebacteria* (Kessler *et al.*, 2001). However, the high production costs of PHB, associated with the expensive raw materials, low productivity and complex equipment, e.g., aseptic operation, have hampered their large-scale production and wide use (Choi and Lee, 1997; Wendlandt *et al.*, 2001).

In the last decade, mixed cultures started to be adopted for PHB production (Choi and Lee, 1997; Kessler *et al.*, 2001). Activated sludge is found to be able to accumulate PHB as carbon and energy storage materials under unsteady conditions arising from an intermittent feeding regime and variation in the presence of an electron acceptor (van Loosdrecht *et al.*, 1997). The PHA production in mixed cultures like activated sludge offers many advantages over pure culture, such as low cost, simple process control, no requirement of monoseptic processing, and an improved utilization of wastes (Satoh *et al.*, 1998). Thus, a considerable effort has made to produce PHA using activated sludge (Majone *et al.*, 1996; Beccari *et al.*, 1998; Chau and Yu, 1999; Beun *et al.*, 2000, 2002; Serafim *et al.*, 2004; Dias *et al.*, 2005; Dionisi *et al.*, 2006). However, in the previous work the PHB-producing reactors, mainly dominated by flocs with a high sludge volume index (SVI) value, generally have a low biomass retention. This leads to a low volumetric productivity and thus restricted PHB productivity.

Compared with sludge flocs, granules have dense and strong structures, which lead to fast settling and solid-liquid separation. Moreover, they enable high biomass retention and withstand high-strength wastewater and shock loadings. Sequencing batch reactor (SBR) has been found favorable for formation of aerobic granules (Su and Yu, 2005; Liu *et al.*, 2006). The formation of different granular sludge under various operational conditions has been extensively investigated. PHB has also been found to commonly present in granules. So far, however, very little information on the formation and characterization of PHB-rich granular sludge is available.

Therefore, the main objective of this work was to improve the PHB productivity of activated sludge by cultivating PHB-rich aerobic granules under ammonium-deficient conditions through two steps. In the first step, the PHB-storage ability of activated sludge was enhanced, and the granulation of the sludge was subsequently facilitated. In the second step, activated sludge was subjected to aerobic dynamic substrate feeding for a long period under ammonia-limiting conditions. The sludge properties during the cultivation were evaluated to explore the formation of the specific PHB-rich aerobic granules. The results obtained from this work may provide useful information for improving PHB production by using activated sludge.

MATERIALS AND METHODS

Reactor and Operations

A laboratory-scale SBR with a working volume of 2 L was utilized for culturing the activated sludge. This SBR was operated sequentially as 4 min of influent filling, 334–344 min of aeration, 5–15 min of settling and 2 min of effluent withdrawal. The hydraulic retention time (HRT) was 12 h and the sludge retention time (SRT) was not more than 10 days. The airflow was maintained constant at a required airflow rate by a needle-valve mass flow controller mounted on the main airline. The reactor was operated without pH control; the temperature was kept at $25 \pm 1°C$ using a belt heater and a temperature controller.

The seed sludge used for the SBR was taken from an aeration tank in Wangxiaoying Municipal Wastewater Treatment Plant, Hefei, China. The seed sludge had a mixed liquor suspended solids (MLSS) concentration of 9.2 g/L and a sludge volume index (SVI) of 54.2 mL/g. Its specific gravity and the settling velocity were 1.006 and 7.0 m/h, respectively. Sludge of 0.9 L was inoculated into the SBR, resulting in an initial MLSS concentration of 4.2 g/L in the reactor.

The synthetic wastewater used in the SBR was composed of (in mg/L): $CH_3COONa·3H_2O$, 800-1600; $MgSO_4·7H_2O$, 600; NH_4Cl, 80-120; EDTA, 100; K_2HPO_4, 92; KH_2PO_4, 45; $CaCl_2·2H_2O$, 70 and 1 mL of trace elements solution. The trace element solution contained (in mg/L): H_3BO_3, 50; $ZnCl_2$, 50; $CuCl_2$, 30; $MnSO_4·H_2O$, 50; $(NH_4)_6Mo_7O_{24}·4H_2O$, 50; $AlCl_3$, 50; $CoC_{12}·6H_2O$, 50 and $NiCl_2$, 50. The pH of the salt solution was adjusted to 7.0 by the addition of NaOH solution.

Analytical Methods

Microbial morphological observation was conducted by using an optical microscope (Olympus CX41). The granule size was measured using an image analysis system (Image-pro Express 4.0, Media Cybernetics) with an Olympus CX31 microscope and a digital camera (Olympus C5050). From the imaging results, four parameters were calculated. Aspect ratio is defined as the ratio between the length of major axis and that of minor axis in an ellipse equivalent to the object (i.e., an ellipse with the same area, first and second degree moments). Form factor (FF) is defined as the ratio of the granule's area to the area of a disc with the same perimeter as the granule [Eq. (1)].

$$FF = 4\pi \frac{area}{perimeter^2} \qquad (1)$$

The roundness (R) of a granule is defined as the ratio of the granule's area to the area of a disc with a diameter equal to the length of the granule [Eq. (2)]. The roundness looks like the form factor but instead of the perimeter, it uses the length of the granule, what makes it more sensitive to how elongated the granule is, rather than how irregular its outline may be (Russ, 1990).

$$R = \frac{4}{\pi} \frac{\text{area}}{\text{length}^2} \qquad (2)$$

Minimum Feret diameter (minFD) is the minimum distance between parallel tangentstouching opposite sides of an object.

The inner structure of granule was observed using both scanning electron microscope (SEM) and transmission electron microscope (TEM). The extracellular polymeric substances (EPS) of sludge were extracted and determined according to Frolund et al. (1996). The contents of extracellular carbohydrates and proteins were determined as described previously (Wang et al., 2007).

The ζ potentials of EPS extractions were measured using a ζ potential analyzer (Zetasizer, Maivern Inc. Germany). The hydrophobic nature of the activated sludge particles was determined by contact angle, which was measured by axisymmetric drop shape analysis following the method proposed by Duncan-Hewitt et al. (1989). Water and 1-bromonaphthalene were used as the probing liquids.

Measurement of chemical oxygen demand (COD), MLSS, mixed liquor volatile suspended solids (MLVSS), SVI and specific oxygen uptake rate (SOUR) was performed using the Standard Methods (APHA, 1995). Dissolved oxygen (DO) was determined using an InPro 6000 polarographic oxygen sensor (Mettler Toledo Co., Switzerland). The settling velocity of granule was measured by recording the time taken for individual granule to fall from a certain height in a measuring cylinder.

The determination of PHB concentration was carried out using a gas chromatograph (GC9790, Wenlin Co., China) equipped with a 30 m × 0.32 mm × 0.25 μm HP-5 column. The temperatures of both the injector and detector were kept at 250°C. The initial temperature of column was set at 120°C for 2 min, followed with a ramp of 10°C/min to 240°C and kept for 2 min. PHAs purchased from Aldrich Inc. (403113, Milwaukee, USA) were used as the standard. The sludge PHB content (% PHB, in g PHB/g MLVSS) was expressed as:

$$\%PHB = \frac{PHB}{MLVSS} \times 100 \qquad (3)$$

where MLVSS includes active biomass (X) and PHB.

The surface free energy of activated sludge was evaluated with the data of contact angle measurement (JC2000A, Powereach Co., Shanghai) (Sheng and Yu, 2006). According to the geometric-mean equation, the surface free energies can be separated into two components, i.e., a non-polar or van der Waals, γ^{LW}, and a polar or acid-base, γ^{AB} (Bos *et al.*, 1999). The pure liquid (L) contact angles (θ) can be expressed as:

$$\cos(\theta) = -1 + 2(\gamma_B^{LW} \cdot \gamma_L^{LW})^{1/2} \cdot \gamma_L^{-1} + 2(\gamma_B^{AB} \cdot \gamma_L^{AB})^{1/2} \cdot \gamma_L^{-1} \qquad (4)$$

where γ_B and γ_L are the bacterial and the liquid surface free energies. The values of γ_B^{LW} and γ_B^{AB} could be estimated from Eq. (4) with the contact angle data. The surface energy of bacterium is expressed as:

$$\gamma_B = \gamma_B^{LW} + \gamma_B^{AB} \qquad (5)$$

RESULTS

Enrichment of Sludge with Intermittent Feeding

In pure cultures, PHB is synthesized as a consequence of an external nutrient limitation in the presence of excess carbon source (Wendlandt *et al.*, 2001; Wang *et al.*, 2007; Wang and Yu, 2007). Previous studies have demonstrated that the specific PHB productivity for mixed cultures could be improved under N- or O-limiting conditions (Third *et al*, 2003; Dias *et al.*, 2005; Dionisi *et al.*, 2006). In order to maximize the intracellular PHB content, both oxygen and ammonia limiting strategies were adopted in the initial enrichment for the PHB-producing activated sludge.

Figure 1 shows the variations of sludge PHB content as well as MLSS, SVI and COD removal of the SBR during sludge enrichment and granulation under aerobic dynamic substrate feeding operation. In Runs 1 and 2 (the initial 37 d after the inoculation), the DO in the reactor was limited by decreasing the air velocity from 0.2 to 0.04 m³/h, and gradual increase of PHB content was observed in this period (Figure 1B). Meanwhile, the low DO resulted in poor sludge settling ability, with the SVI increased rapidly to 300 mL/g (Figure 1B). This caused severe wash out of sludge. The MLSS declined sharply to below 3000 mg/L during this period (Figure 1A). Thus, the oxygen limitation strategy was altered in the middle of Run 2 by re-increasing the air velocity to 0.2 m³/h. As a result, the SVI drop rapidly to around 130, showing a remarkable improvement in the sludge settling ability under increased hydraulic shear force. However, the PHB content kept rising at almost the same rate as in oxygen limitation condition (Figure 1B). This implies that oxygen limitation is not an indispensible factor for PHB accumulation in sludge. Similar results were also obtained by Serafim *et al.* (2004), who found that the PHB content of sludge increased from 30% to 40% when the oxygen flow rate increased from 0.013 to 0.033 m³/h.

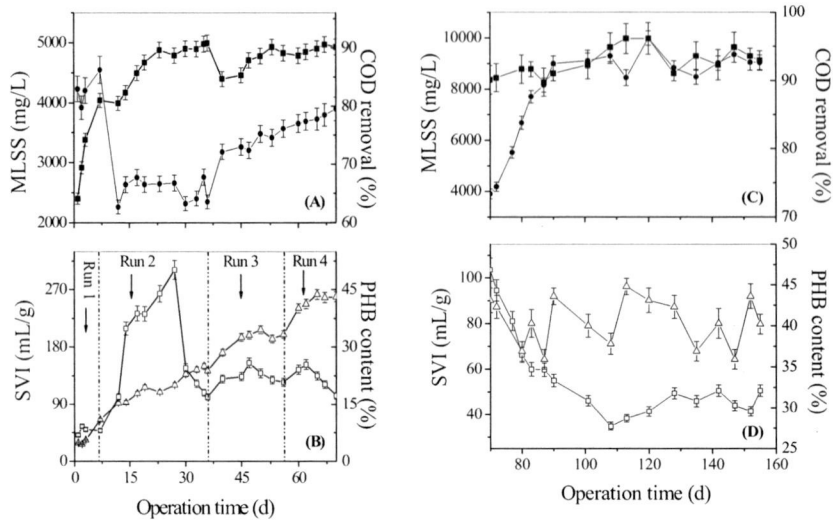

Figure 1 Performance of the SBR with dynamic substrate feeding operation: (●) MLSS; (■) COD removal efficiency; (□) SVI; and (□) PHB content

A stable state was achieved for SVI and PHB content in Run 3, whereas the MLSS and COD removal kept growing gradually during this period. An ammonia-limitation strategy was applied in Run 4, where the ammonia concentration was reduced and the COD/N ratio was increased from 48 to 72. This operation strategy was based on the assumption that the fraction of substrate derived for cell growth could be controlled by limiting the nitrogen availability in the medium, thus the PHB storage would increase. As shown in Figure 1B, the PHB content in sludge increased to 43.1 ± 2.0% under nitrogen limitation condition (Figure 1B). It is clearly evidenced that limiting the ammonia availability in medium in the periodic operation favors the enrichment of microorganisms with good PHB storage capability and competitiveness with other microorganisms (Salehizadeh and van Loosdrecht, 2004).

Granulation of PHB-Rich Activated Sludge

The PHB-storage ability of sludge had been improving during the two-month cultivation, as shown in Figs.1A and B. However, the SVI of the floc sludge remained at a relatively high level of around 100 mL/g, which resulted in a lower volumetric productivity. In this case, although the intracellular PHB content in sludge increased, the entire PHB productivity of the SBR was not high because of the low biomass retention. The granulation of PHB-rich activated sludge started to occur after cultivation for about 80 d. Figures 1C and 1D show that the reactor performance improved continuously in terms of COD removal efficiency and SVI. After operation of 80 days, tiny granules were observed in the reactor. In order to accelerate granulation, the settling time was further reduced to 5 min at this point. After 100-day operation, the MLSS concentration stabilized at 7.5–10.0 g/L and the reactor was dominated by mature granular sludge (Figure 2) with little flocs. The PHB content was kept at 40 ± 4.6% (Figure 1D). Aerobic granules could be cultivated at low substrate COD/N ratio of 20:1 (Zheng *et al.*, 2005). However, in this work the granules were cultivated at a higher COD/N ratio (about 70:1). This suggests that COD/N ratio is not the key factor governing the formation of granular sludge.

The mature PHB-rich granules show a regular morphology, as shown in Figure 2. The granules had a diameter of 1.0–3.8 mm and a buff color. A close examination by SEM revealed that the PHB-rich granules have porous inner structure. Such structure is likely to facilitate the transport of nutrients and substrate. It can also been seen that there were mainly three bacterial morphologies in the granules, namely: rod, coccus and filament. TEM image shows that PHB presented in the inner of the granules.

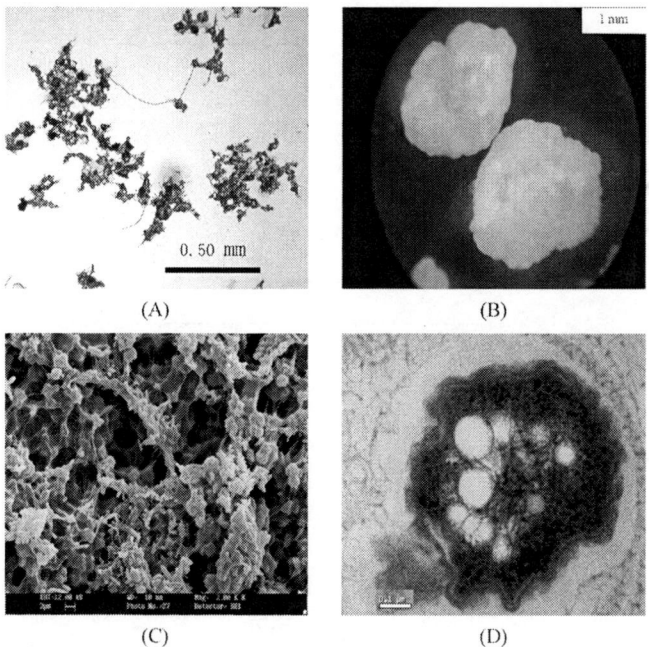

(A)   (B)

(C)   (D)

Figure 2 Microscopic images of (A) seed sludge; (B) PHB-rich granular sludge; (C) SEM photos and (D) TEM photos of PHB-rich granular sludge

PHB Production of Granular Sludge

Figure 3 shows the performance of a typical granule-based SBR in one operation cycle. The transition points of DO increase and OUR decrease were used to differentiate the "feast" and "famine" periods (Figure 3A). The DO concentration decreased immediately after substrate addition, but remained almost constant during the "feast" period, and then rose again after the carbon source became exhausted. The OUR profiles agree well with the variation of DO concentration. In the "famine" phase the OUR kept decreasing until the end of the cycle, reaching the endogenous respiration state. After the depletion of acetic acid, PHB was slowly consumed. At the end of the cycle the PHB concentration almost recovered the original level. On the other hand, as shown in Figure 3B, the concentration of NH_4Cl declined more rapidly in the feast phase than the famine phase, indicating that the growth rate of biomass slowed down in the famine period.

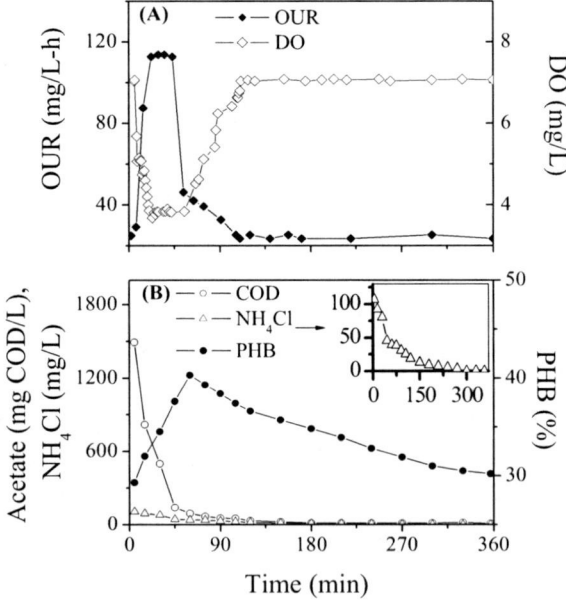

Figure 3 Typical profiles of COD, PHB, ammonia, DO and OUR during a cycle

Changes of Sludge Properties

As shown in Figure 1, the SVI value, which reflecting the flocculation ability of sludge, varied with the reactor operation. On the other hand, the flocculation of microorganisms is related to their surface properties, such as surface charge, hydrophobicity and EPS content (Maier *et al.*, 2000).

The evolution of the contact angle is illustrated in Figure 4. Since water is a polar liquid, while 1-bromonaphthalene is a non-polar liquid, the different contact angle data could represent the polar and non-polar components of the bacterial surface energy. The water contact angles decreased rapidly within the initial three weeks, but then increased significantly and remained at a relatively high level. In contrast, the change of 1-bromonaphthalene contact angles was insignificant, which fluctuated around 48.9°.

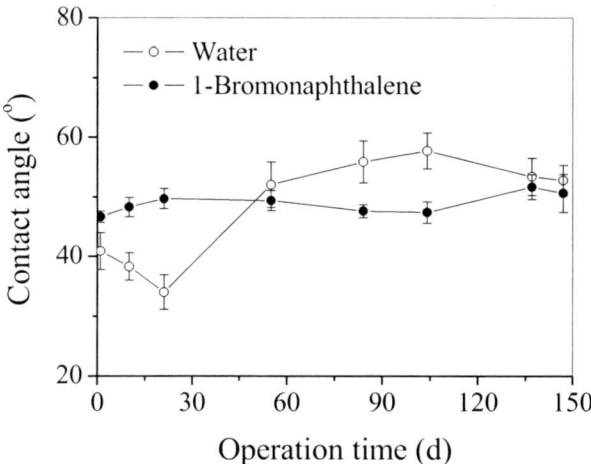

Figure 4 Contact angles of sludge during cultivation

Figure 5 shows the variation of EPS components and charge density of EPS extracted in different cultivation periods. Extracellular carbohydrates were the major component of EPS in the sludge, which varied considerably during cultivation. After an increase in the initial three weeks, the content of extracellular carbohydrates underwent a rapid decreased rapidly and then a gradual rise. The sludge zeta potential, total EPS concentration and extracellular carbohydrates/proteins ratio show a similar variation tendency (Figure 5).

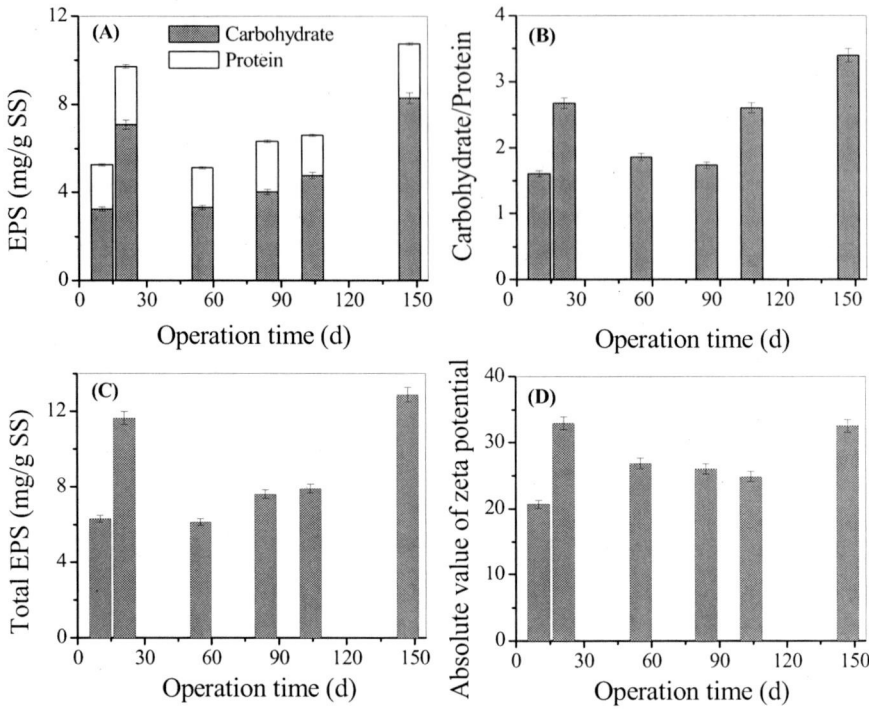

Figure 5 Changes of EPS content in sludge and z during cultivation

DISCUSSION

Morphological Observation of Granular Sludge

The dimensionless parameters are very sensitive to the examined variations in shape and irregular boundary of the object. In addition, the form factor is particularly sensitive to the roughness of the object boundary (Russ, 1990). The aspect ratio and roundness can reflect the circular degree of an object, that is, objects with shape close to circle have roundness and aspect ratio with values close to one. To investigate the influences of diameter variations on the morphology of granules, a total of 100 mature PHB-rich granules taken from the SBR were examined. As shown in Figure 6A, the aspect ratio changed little and the correlation was as low as 0.03. This indicates that the diameter has little influence on the morphology of granules. This was further confirmed by the variation of the roundness and form factor along with the minimal Feret diameter, as shown in Figs. 6B and 6C. Nearly all of the granules exhibited flat-elliptical morphologies regardless of the diameter difference.

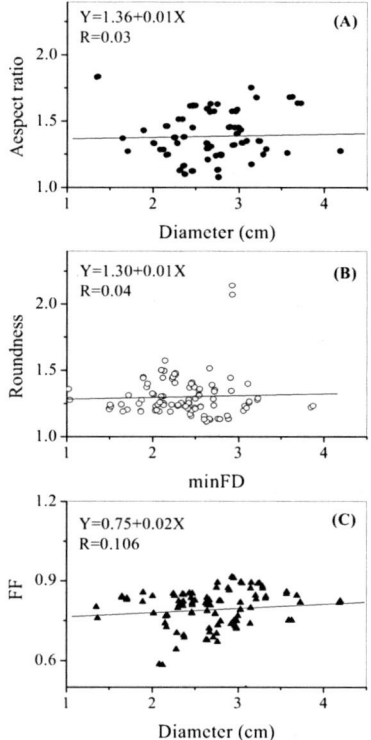

Figure 6 Scatter plots of (A) aspect ratio; (B) roundness; and (C) form factor with diameters (minimal Feret diameters) of granular sludge

In addition to the flat-elliptical morphology, the granules also exhibited a lighter color compared with other reported aerobic granules. This difference might be partially attributed to the nitrogen limitation cultivation condition. It is recognized that microorganisms can cope with stressful environmental conditions by developing sophisticated cooperative behaviors and intricate communication capabilities. With such capabilities, microbial community may develop complex spatiotemporal patterns in response to adverse growth conditions (Ben-Jacob *et al.*, 2000). Thus, it can be inferred that the morphology of the PHB-rich granules might be due to the self-modulation of microorganisms with the cultivation at high COD/N ratios.

Granule EPS

A high carbohydrates content could facilitate cell-to-cell interaction and further strengthen microbial structure through formation of a polymeric matrix. However, in recognition of the contribution of extracellular carbohydrates to biogranulation, the importance of hydrophobicity and charge also need to be accounted for in the biogranulation process (Liu *et al.*, 2006). As shown in Figures 4 and 5, the sludge surface properties varied considerably in the cultivation, especially before and after the oxygen limitation abandonment. The water contact angle decreased, while the carbohydrates in EPS increased initially and then declined after the oxygen limitation release. The shear force has a positive effect on the production of extracellular carbohydrates and hydrophobicity of cell surface (Liu *et al.*, 2006). In this work, the granular sludge showed much higher hydrophobicity than the bioflocs. Thus, a high hydrophobicity is favorable for strengthening cell-cell interaction and might be the main force for the initiation of granulation.

The cell surface hydrophobicity and charge are related to the sludge EPS (Sheng and Yu, 2006). The 1-bromonaphthalene contact angle presented a positive relationship with the EPS content (Figure 7). Since the 1-bromonaphthalene and water contact angles reflect the hydrophilic and hydrophobic abilities of sludge respectively, the results were consistent with the previous reports, where EPS content of sludge exhibited a negative effect on the water contact angle (Sheng and Yu, 2006).

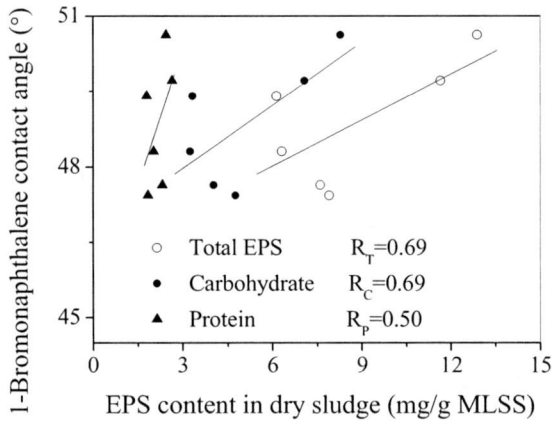

Figure 7 Correlation between 1-bromonaphthalene contact angle of sludge and EPS content

Furthermore, a close correlation between the zeta potential and EPS was observed (Figure 8). Similar relationship between total EPS content and surface charge can be found in literature (Mikkelsen and Keiding, 2001). The surface charge may be resulted from the production of extracellular carbohydrates which boasts a amount of carboxylic and hydroxy groups. On the other hand, according to the DLVO theory, increased surface charge would lead to increased repulsive electrostatic interactions, which are detrimental to the agglomeration. However, in this study the granulation of sludge seemed scarcely influenced by the increased surface charge. In addition, only those cells with a low EPS content would be dominated by the electrostatic interaction whereas cells with rich EPS would be dominated by the polymeric interaction (Tsuneda et al., 2003). Thus, although the surface charge increased with the EPS content, the repulsive electrostatic interactions are too weak to resist sludge from flocculation. Meanwhile, the special role of polymeric entanglement was enhanced by the increasing EPS content, facilitating agglomeration of flocs to compact granules.

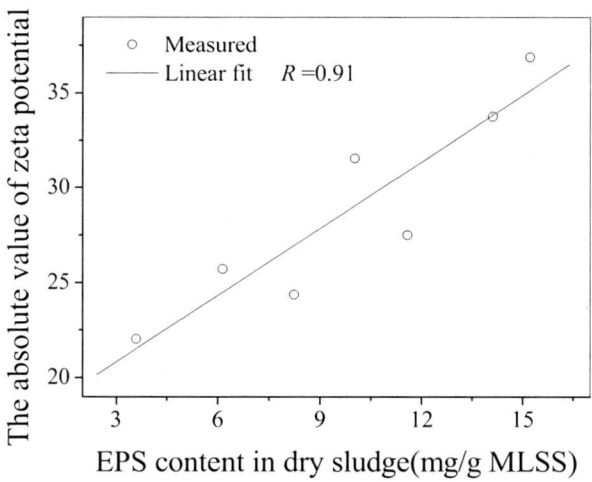

Figure 8 Correlation between zeta potential and EPS content

The asymmetric surface forces on the particles results in the surface energy, which determines the critical (i.e., stable) particle size and may therefore be decisive with respect to the thermodynamic barrier to the formation of new particles. Thus, it is an important parameter in determining particle formation and growth (Artelt et al., 2005). However, little is known about this micro-agglomeration so far.

The change of surface energy during sludge cultivation was investigated in the present study. In a thermodynamic sense, reducing cell hydrophobicity would lead to an increase in the excess Gibbs energy of the surface. However, in an attempt to reduce the high surface energy, the liquid surrounding the microorganisms tend to maintain the spheric face while the microorganisms would try to self-aggregate from liquid phase to form a new larger solid phase, namely microbial aggregates. As shown in Figure 9, the reduced hydrophobicity resulted in an increased surface energy of sludge, which contributed to the later flocculation of flocs. Besides, the surface energy of sludge decreased to a relatively steady state accompanied with the growth of granular sludge. As for a special substance, the higher granularity, the smaller surface energy it has. Thus, when the growth of granules tended to stabilization, the surface energy would stop decreasing.

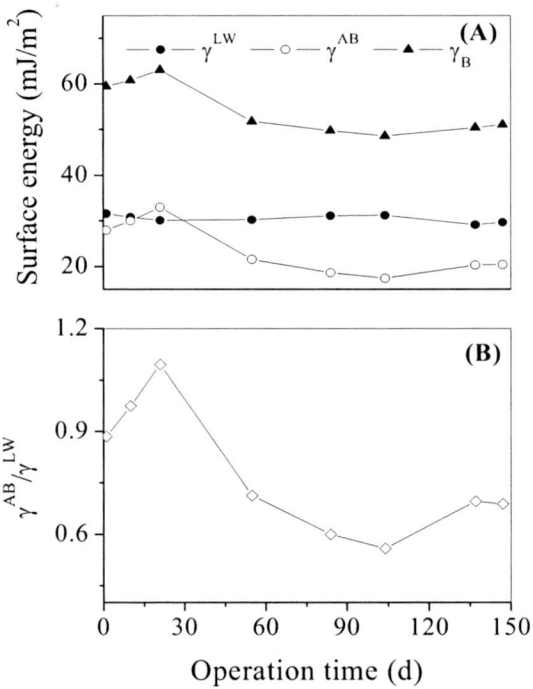

Figure 9 Changes of (A) surface energy; and (B) the ratio of polar to non-polar component of sludg

As shown in Figure 10, the zeta potential, the EPS content and the ratio of extracellular carbohydrates/proteins all had a positive effect on the surface energy of bacterium and the polar component of bacterial surface energy. This was partially consistent with our previous study about *Rhodopseudomonas acidophila* cultivation (Sheng and Yu, 2006). Yet, the correlation affinities in Figure 10 were not high. This indicates that, apart from the above parameters, the surface energy of sludge was also influenced by many other factors, such as density of primary agglomerates, cohesiveness, size, and the operating condition, etc (Artelt *et al.*, 2005).

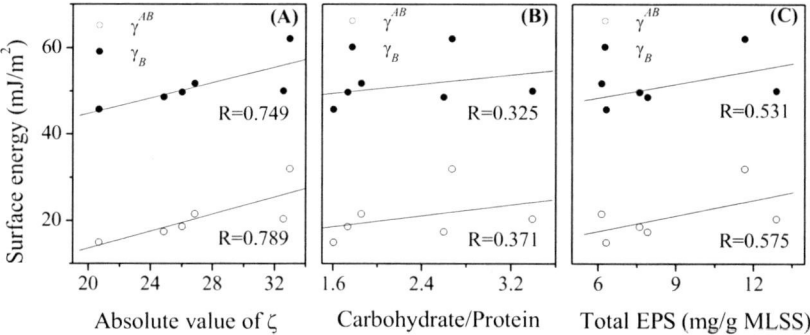

Figure 10 Correlation between surface energy of sludge and (A) zeta potential; and (B) the ratio of polysaccharides/proteins; and (C) total EPS content

Correlation between Sludge Storage and Granulation

From the metabolism point of view, the consumption of PHB involves many enzyme catalysis processes of different depolymerases. Glycosylation has been found in the dPHB depolymerases of *P. lemoignei* and those of fungi (Briese *et al.*, 1994).The carbohydrate moiety might become noncovalently attached to depolymerase during secretion of the enzyme across the cell wall. It is possible that glycosylation enhances the resistance of the exoenzyme to elevated temperature and/or to hydrolytic cleavage by proteases of competing microorganisms (Briese *et al.*, 1994; Jendrosske, *et al.*, 1995). It has been demonstrated in previous research that intracellular polymers like PHB have some relationships with extracellular polymers at the enzymatic level.

The granular sludge accumulated large amount of PHB and their surfaces were covered with some viscous substances (Figure 2D). Moreover, the increase in PHB content was observed accompanied with the significant increase in EPS (Figs. 1 and 5). The enhanced production of EPS in granules could have been induced by some so-called stressful culture conditions (Nichols *et al.*, 2004). So far, a number of operating parameters, including reactor type, substrate composition, substrate loading rate, hydraulic retention time, hydrodynamic shear force, settling time and feast-famine regime in SBR, culture temperature, and so on, have been recognized to stimulate bacteria to secrete more EPS (Liu *et al.*, 2006). The composition of EPS is also related to the characteristics of the feeding wastewater. Nitrogen limiting condition favors the production of EPS, which in turn accelerates anaerobic granulation (Punal *et al.*, 2000). In this study, PHB-storage microorganisms were selected under nitrogen-limited condition. When no external substrates is available, PHB could be consumed for the maintenance of microorganisms (e.g. the production of other cellular components, like extracellular carbohydrates) (Serafim *et al.*, 2004). Therefore, apart from the above factors, the storage-stressful culture condition might have also contributed to the production of EPS in this study.

It has been demonstrated that stressful operating conditions, in terms of high hydrodynamic shear force, short settling time/hydraulic retention time and periodical feast-famine period, can remarkably stimulate bacteria to produce more extracellular polysaccharides than proteins in SBR (Liu *et al.*, 2006; Su and Yu, 2005). This study proved that the metabolism of intracellular storages induces microbe to produce EPS, which favors the formation of aerobic granules. However, the physiological reason for the energy-consuming cycle of degradation and synthesis of PHB remains unknown. The microorganisms with stored material have a strong selective advantage over those with poor storage ability. As intrinsic processes that play a key role in selector-like systems (Martins *et al.*, 2004), storage and regeneration (depletion) should be considered in description of the metabolic processes during sludge granulation.

CONCLUSIONS

A two-step strategy was adopted to cultivate PHB-rich aerobic granular sludge in an SBR. Nitrogen-deficient synthetic wastewater was used for the cultivation. In the first step, both oxygen and ammonia were initially limited in order to enhance the PHB-storage ability of sludge. In the second step, granular sludge was cultivated to get a high PHB volumetric productivity. The experimental results indicate that limiting the ammonia availability in the medium favored the selection of organisms with high PHB storage capability. The PHB content of sludge increased to 43.1% in the first step and remained at 40% during the subsequent granulation process. During granulation, the settling ability of the PHB-rich sludge continuously improved. Meanwhile, the surface charge, EPS content and the sludge hydrophobicity all increased in this process. It is demonstrated in this work that the metabolism of intracellular storages induces microbe to produce EPS, which favor the formation of aerobic granules.

ACKNOWLEDGMENTS

The authors wish to thank the NSFC (50625825 and 50738006), the Key Special Program on the S&T for the Pollution Control and Treatment of Water Bodies (2008ZX07316-002 and 2008ZX07010-005), and the Anhui R&D Key Project (07010301022 and 08010302109) for the partial support of this study.

REFERENCES

APHA (1995). *Standard Methods for the Examination of Water and Wastewater*, 19th edn, American Public Health Association.

Artelt C., Schmid H. J. and Peukert W. (2005). On the impact of accessible surface and surface energy on particle formation and growth from the vapour phase. *Journal of Aerosol Science*, **36**(2), 147–172.

Beccari M., Majone M., Massanisso P. and Ramadori R. (1998). A bulking sludge with high storage response selected under intermittent feeding. *Water Research*, **32**(11), 3403–3413.

Ben-Jacob E., Cohen I. and Levine H. (2000). Cooperative self-organization of microorganisms. *Advances in Physics*, **49**, 395–554.

Beun J. J., Dircks K., van Loosdrecht M. C. M. and Heijnen J. J. (2002). Poly-*b*-hydroxybutyrate metabolism in dynamically fed mixed microbial cultures. *Water Research*, **36**, 1167–1180.

Beun J. J., Paletta F., Van Loosdrecht M. C. M. and Heijnen J. J. (2000). Stoichiometry and kinetics of poly-beta-hydroxybutyrate metabolism in aerobic, slow growing activated sludge cultures. *Biotechnology and Bioengineering*, **67**(4), 379–389.

Bos R., van der Mei H. C. and Busscher H. J. (1999). Physico-chemistry of initial microbial adhesive interactions – its mechanisms and methods for study. *FEMS Microbiology Review*, **23**, 179–230.

Briese B. H., Schmidt B. and Jendrossek D. (1994). *Pseudomonas lemoignei* has five poly (hydroxyalkanoic acid) (PHA) depolymerase genes: a comparative study of bacterial and eukaryotic depolymerases. *Journal of Environmental Polymer Degradation*, **2**, 75–87.

Chau H. and Yu P. H. F. (1999). Production of biodegradable plastics from chemical wastewater: a treatment. *Water Science and Technology*, **39**, 273–280.

Choi J. I. and Lee S. Y. (1997). Process analysis and economic evaluation for poly-3-hydroxybutyrate production by fermentation. *Bioprocess Engineering*, **17**, 335–342.

Dias J. M. L., Serafim L. S., Lemos P. C., Reis M. A. M. and Oliveira R. (2005). Mathematical modelling of a mixed culture cultivation process for the production of polyhydroxybutyrate. *Biotechnology and Bioengineering* **92**(2), 209–222.

Dionisi D., Majone M., Vallini G., Di Gregorio S. and Beccari M. (2006). Effect of the applied organic load rate on biodegradable polymer production by mixed microbial cultures in a sequencing batch reactor. *Biotechnology and Bioengineering*, **93**(1), 76–88.

Duncan-Hewitt W. C., Policova Z., Cheng P., Vargha-Butler E. I. and Neumann A. W. (1989). Semi-automatic measurement of contact angles on cell layers by a modified axis symmetric drop shape analysis. *Colloids and Surfaces A*, **42**(3–4), 391–403.

Frolund B., Palmgren R., Keiding K. and Nielsen P. H. (1996). Extraction of extracellular polymers from activated sludge using a cation exchange resin. *Water Research*, **30**(8), 1749–1758.

Jendrosske D., Frisse A., Behrends A., Andermann M., Kratzin H. D., Stanislawski T. and Schlegel H. G. (1995). Biochemical and molecular characterization of the *Pseudomonas lemoignei* depolymerase system. *Journal of Bacteriology*, **177**(3), 596–607.

Kessler B., Weusthuis R., Witholt B. and Eggink G. (2001). Production of microbial polyesters: fermentation and downstream processes. *Advances in Biochemical Engineering/Biotechnology*, **71**, 159–182.

Liu Y. Q., Tay J. H. and Moy Y. P. B. (2006). Characteristics of aerobic granular sludge in a sequencing batch reactor with variable aeration. *Applied Microbiology and Biotechnology*, **71**(5), 761–766.

Maier R. M., Pepper L. L. and Gerba C. P. (2000). *Environmental Microbiology*. Hardbound, Academic Press.

Majone M., Massanisso P., Carucci A., Lindrea K. and Tandoi V. (1996). Influence of storage on kinetic selection to control aerobic filamentous bulking. *Water Science and Technology*, **34**(5–6), 223–232.

Martins A. M. P., Pagilla K., Heijnen J. J. and van Loosdrecht M. C. M. (2004). Filamentous bulking sludge-a critical review. *Water Research*, **38**(4), 793–817.

Mikkelsen L. H. and Keiding K. (2001). Physico-chemical characteristics of full scale sewage sludges with implication to dewatering. *Water Research*, **36**, 2451–2462.

Nichols C. A. M., Garon S., Bowman J. P., Raguenes G. and Guezennec J. (2004). Production of exopolysaccharides by Antarctic marine bacterial isolates. *Journal of Applied Microbiology*, **96**, 1057–1066.

Punal A., Trevisan M., Rozzi A. and Lema J. M. (2000). Influence of C:N ratio on the start-up of up-flow anaerobic filter reactors. *Water Research*, **34**, 2614–2619.

Russ J. C. (1990). Computer assisted microscopy: the measurement and analysis of images. Plenum Press, New York.

Salehizadeh H. and Van Loosdrecht M. C. M. (2004). Production of polyhydroxyalkanoates by mixed culture: recent trends and biotechnological importance. *Biotechnology Advances*, **22**, 261–279.

Satoh H., Iwamoto Y., Mino T. and Matsuo T. (1998). Activated sludge as a possible source of biodegradable plastic. *Water Science and Technology*, **38**, 103–109.

Serafim L. S., Lemos P. C., Oliveira R. and Reis M. A. M. (2004). Optimization of polyhydroxybutyrate production by mixed cultures submitted to aerobic dynamic feeding conditions. *Biotechnology and Bioengineering*, **87**(2), 145–160.

Sheng G. P. and Yu H. Q. (2006). Relationship between the extracellular polymeric substances and surface characteristics of *Rhodopseudomonas acidophila*. *Applied Microbiology and Biotechnology*, **72**(1), 126–131.

Su K. Z. and Yu H. Q. (2005). Formation and characterization of aerobic granules in sequencing batch reactor treating soybean processing wastewater. *Environmental Science and Technology*, **39**, 2818–2828.

Third K. A., Newland M. and Cord-Ruwisch R. (2003). The effect of dissolved oxygen on PHB accumulation in activated sludge cultures. *Biotechnology and Bioengineering*, **82**(2), 238–250.

Tsuneda S., Aikawa H., Hayashi H., Yuasa A. and Hirata A. (2003). Extracellular polymeric substances responsible for bacterial adhesion onto solid surface. *FEMS Microbiology Letters*, **223**, 287–292.

Van Loosdrecht M. C. M., Pot M. A. and Heijnen J. J. (1997). Importance of bacterial storage polymers in bioprocess. *Water Science and Technology*, **35**, 41–47.

Wang J., Fang F. and Yu H. Q. (2007). Substrate consumption and biomass growth of *Ralstonia eutropha* at various S0/X0 levels in batch cultures. *Bioresource Technology*, **98**, 2599–2604.

Wang J. and Yu H. Q. (2007). Biosynthesis of polyhydroxybutyrate (PHB) and extracellular polymeric substances (EPS) by *Ralstonia eutropha* ATCC 17699 in batch cultures. *Applied Microbiology and Biotechnology*, **75**, 871–878.

Wendlandt K. D., Jechorek M., Helm J. and Stottmeister U. (2001). Producing poly-3-hydroxybutyrate with a high molecular mass from methane. *Journal of Biotechnology*, **86**(2), 127–133.

Wilen B. M., Jin B. and Lant P. (2003). The influence of key chemical constituents in activated sludge on surface and flocculating properties. *Water Research*, **37**, 2127–2139.

Zheng Y. M., Yu H. Q. and Sheng G. P. (2005). Physical and chemical characteristics of granular activated sludge from a sequencing batch airlift reactor. *Process Biochemistry*, **40**, 645–650.

Hybrid Process of Advanced Oxidation and Membrane Filtration Facing the Challenge of Future Urban Water Quality

J. Ma[a,b], Z.H. Wang[a], J. Jiang[a,b] and P.P. Wang[a]

[a]State Key Laboratory of Urban Water Resource and Environment, Harbin Institute of Technology, Harbin 150090, China
[b]National Engineering Research Centre of Urban Water Resources, Harbin 150090, China.
E-mail: majunhit@gmail.com

Abstract With the rapid increase of urbanization, highly urbanized areas will be found more commonly in most countries. As a result, urban water systems are under a huge strain. Providing quality water is becoming increasingly difficult due to high levels of pollution. Pollutants, both organic and inorganic, such as micro-organisms and treatment by-products, are all problematic. For cities, the more complex constituents of raw water will cause increasing challenges to drinking water quality in the future. Conventional water treatment processes have their limitations as they require longer and more complex advanced processes to achieve satisfactory water quality. In order to solve water quality problems for cities of the future, novel water treatment processes, with shorter treatment processes, lower costs and higher efficiency levels, must be applied. It has been proposed that the combination of advanced oxidation and membrane filtration might be the solution for solving water quality problems for the future. In this study, the recent developments in advanced oxidation and membrane processes are reviewed. The study examines them, in particular, in terms of their ability to meet demands for water quality in the future.

Keywords Future city, water quality, advanced oxidation, membrane filtration, combined processes

INTRODUCTION

As a highly efficient technology in removing (or degrading) the organic pollutants in water, the advanced oxidation process (AOP), is becoming more and more important in water quality control. Applied to membrane filtration, its unique properties in separating the impurities from water also make it a powerful tool in water purification.

At present, the composition of the impurities (or pollutants) in the influent of urban water treatment plants is getting more complex. This is due to the various anthropogenic discharges. Conventional water treatment technologies are gradually becoming less competent at removing newly appeared pollutants and meeting new rigid water quality standards. Advanced oxidation and membrane filtration technologies, however, seem to overcome all of these problems. In addition, the two types of technologies are mutually complementary. It is reasonable, therefore, to think that a hybrid process of advanced oxidation and membrane filtration might provide a new suitable water treatment technology to overcome the future challenges of urban water quality. The aim of this paper is to give a brief overview of this discussion.

CHALLENGES OF WATER QUALITY IN FUTURE CITIES
Water Quality in Urban Water Resources

Urban water has the following characteristics: high water quality demand; high load of water consumption; high safety requirements; high possibility of pollution; high cost of long distance water transportation; and a low ability of bio-restoration. The quality of urban water resources is affected by human activities in urban areas. Water is drawn from resources such as rivers, lakes, and groundwater. After purification, the clean water is distributed into urban areas for domestic and industrial uses. After usage, various wastewaters containing different pollutants are generated. In general, this wastewater would be treated in wastewater treatment plants. The effluent would then be discharged into rivers or lakes. In this process, the residual pollutants from the effluent of the wastewater treatment plants and the pollutants from directly discharged wastewaters will be introduced into the receiving water body.

Rapid urbanization is showing its impacts in the following ways: Urban discharge causes an increase of nutrients such as nitrogen, phosphorus and carbon in the water environment; Industrial discharge causes an increase of endocrine disrupting substances such as persistent organics, phametheticals and personal care products, as well as mutagenic substances etc; Agricultural runoff causes increased concentrations of pesticides, herbicides, fertilizers, humic acids, fluvic acids in the water environment; The global warming effect causes algae glooming, taste and odour, and toxin problems in the water environment; High tech R&D development might introduce nano-particles, bio-intrusion etc. problems to the water environment.

For one single city, or a less urbanized area, the impact of human activity on water resources is relatively easy to manage. However, for highly urbanized areas, more water quality problems emerge. Clearly, the water resources of the downstream cities may be polluted by the discharges from upstream cities to different extents. The self-cleaning capability of the water bodies in highly urbanized areas becomes weaker as the urbanization rate increases. When confronted by large amounts of wastewater discharge, the water quality will deteriorate. As a result, cities in highly urbanized areas, particularly those located downstream, are facing greater challenges with water quality. Such water quality challenges will only become more serious as the rate of urbanization increases.

Limitations of Current Water Treatment Processes

The conventional water treatment process generally includes four units: Coagulation, Sedimentation, Filtration, and Disinfection). It mainly focuses on removing particles, such as colloids, and bacteria rather than removing organic pollutants in the water. In order to deal with the newly arising problems, advanced treatment technologies, such as ozonation and activated carbon adsorption processes have been developed. These are based on the same conventional processes of coagulation, sedimentation, filtration and disinfection. However, there are several drawbacks to this. It requires long treatment processes and high costs. The difficulties in managing these technologies offset their value and limit their application in practice. Therefore, both researchers and engineers are seeking more effective and practically feasible water treatment technologies.

RECENT DEVELOPMENT IN AOP PROCESSES
Advanced Oxidation by Shortly Lived Mn and Fe Species

It has been found that the short lived manganese and iron species could initiate effective advanced oxidation. The intermediate mangenese and iron species, generated *in situ* have high oxidation power and have multiple functions for organic removal, enhanced coagulation and control of disinfection by products (Ma, 1990; 1994; 1996; 1997). As for the manganese species, they enhance the Mn(VII) oxidation of organic pollutants such as endocrine disrupting substances like pharmaceuticals and personal care products etc. However, all the effective species must be formed *in situ*. Previous research shows that the aged (or commercial) intermediates are not effective.

It was found that the presence of ligands significantly promoted the oxidation of phenolics by permanganate (Jiang 2009). This indicates the importance of manganese intermediates (Mn(INT)) formed *in situ* upon the reduction of permanganate. Without stabilizing agents, most of the Mn(INT) would decompose and become disproportionate. However, unlike phenolics and aromatic amines, the organics which prefer a two-electron oxygen atom transfer oxidation pathway, showed similar susceptibility to oxidation by permanganate in the absence of ligands. Small amounts of strong reducing additives can result in the fast generation of highly active Mn(INT) in the presence of ligands. This can therefore strengthen permanganate oxidation further.

In addition, the effects of humic acids (HA) and their different nominal molecular weight (NMW) fractions were studied (He 2009). It was found that phenol oxidation by permanganate was enhanced by the presence of humic acids under pH 4-8 range. The effects of humic acids on phenol oxidation by permanganate were dependent on humic acid concentration and permanganate/ phenol molar ratios. The high NMW fractions of humic acids enhanced phenol oxidation by permanganate at pH 7 more significantly than the low fractions of humic acid. It is expected that π-electrons of humic acids strongly influenced oxidative reactivity of permanganate. The humic fractions rich in aliphatic character, polysaccharide-like substances and the amount of carboxylate groups had less effect on phenol oxidation by permanganate.

Up until now, many Chinese water treatment plants and facilities have employed permanganate or composite chemicals containing permanganate for the removal of organics. The combined processes of permanganate oxidation and powdered activated carbon or granular activated carbon are also used in China for the treatment of polluted waters at a relatively low cost.

Iron species have performs in a similar way to manganese. Iron exists in various forms such as Fe(VI), Fe(V), Fe(IV), Fe(III), Fe(II), Fe(0) etc. The oxidation by Fe(VI) has been receiving increasing attention due to its green oxidant nature and the abundance of iron in the earth. In addition, it has no negative impacts on the environment and there are not by-product problems in water treatment. The oxidation of organic or inorganic pollutants by Fe(VI) may undergo various stages such as Fe(V), Fe(IV) and finally to Fe(III). The intermediate species formed during the oxidation might enhance both oxidation and the removal of algae (Ma 2002a) as well as coagulation (Ma 2002b).

Advanced Oxidations by Oxygenated Radicals

There are several types of radical related advanced oxidation. These mainly include: Ozone related advanced oxidation (O_3/H_2O_2, O_3/UV, O_3/TiO_2, O_3/MnO_2, $O_3/FeOOH$, $O_3/NiOOH$, O_3/CuO etc.); UV Related AOPs (UV/TiO_2, UV/O_3, UV/H_2O_2); Electronic beams; Ultrasonic waves radiation; an Plasmas. The point of these advanced oxidation technologies, is to use the generated •OH. It is a powerful non selective oxidant species, used to degrade the organic pollutants.

Although many AOPs have been developed, only a limited number of AOP processes have been used in full scale water treatment plants. In our previous research, it was found that the presence of a small amount of Mn (II) accelerated the degradation of atrazine by ozone. It was indicative that the hydrous manganese dioxide, formed *in situ,* had a similar effect as Mn(II) for the catalytic degradation of atrazine under pH 7.0 (Ma 1997; 1999; 2000). However, the commercially available manganese dioxide did not have any catalytic effect on the oxidation of atrazine by ozone. It is difficult to separate the hydrous manganese dioxide from the aqueous solution due to the small size of the catalyst. Granular activated carbon (GAC) was, therefore, used to support the amorphous manganese catalyst (MnO_x). This was prepared via the wet impregnation of GAC with permanganate in an aqueous solution. The catalyst prepared showed higher effectiveness on nitrobenzene removal compared with that of ozonation alone (Ma et al., 2004; Ma et al., 2005). To improve the stability of the active phase on the catalyst, MnO_2/ceramic honeycomb was prepared by a wet impregnation and calcination procedure. This also kept a good catalytic activity on nitrobenzene degradation (Zhao et al., 2008a). A catalyst with a highly hydroxylated surface (FeOOH) was developed further. It was found that the surface hydroxyl groups had a close relationship with the activity of the FeOOH in the catalyzed ozonation of nitrobenzene (Zhang 2008a). Furthermore, according to the comparative research on synthetic oxo-hydroxides such as α-FeOOH, β-FeOOH, γ-FeOOH and γ-AlOOH, it was proposed that not all surface hydroxyl groups of the oxo-hydroxides had the same high catalytic activity. Only the weak surface MeO-H bonds were favourable sites for promoting •OH generation from aqueous ozone (Zhang et al., 2008b). A nano-TiO_2 catalyst was prepared which improved the oxidation of nitrobenzene by ozone (Yang et al., 2007). Considering its potential risk to the environment, the direct usage of nano-size materials as catalysts or catalyst supporters is impossible. Therefore, three supported TiO_2 catalysts (TiO_2/silica-gel, TiO_2/haydite and TiO_2/zeolite) were prepared. All of them showed high activity on the mineralization of nitrobenzene, particularly for TiO_2/silica-gel (Yang et al., 2006). The presence of ceramic honeycomb also significantly improved the degradation of nitrobenzene by ozonation (Zhao et al., 2008b). A multiple elements metal oxides ceramic honeycomb catalyst was developed. This showed a more pronounced improvement on the benzophenone degradation, TOC removal and ozone utilization (Hou et al., 2006). It was found that the catalyst which could generate the hydroxyl radical was not an efficient catalyst for the degradation of intermediate oxidation products. However, CeO_2 was a better catalyst for the further oxidation of organic intermediate oxidation products. It was also found that CeO_2 has the ability to control the formation of bromate (Zhang et al., 2008c).

A multi-stage catalyzed ozonation was proposed, in which two or more kinds of catalysts would be used to achieve different purposes at different stages. The first part of the multi-stage catalyzed ozonation is for the removal of stable organic micropollutants, hence a catalyst with high R_{ct} (activated carbon, FeOOH, ceramic honeycomb and modified ceramic honeycomb, etc.) is used to obtain high degradation efficiency on stable organic micropollutants (Zhao, 2009a;

2009b; 2009c). The reaction time of conventional catalyzed ozonation is usually 10–20 min. Nevertheless, it has been reported that the majority of the removal of nitrobenzene, with a low concentration (< 60 μg L^{-1}), is usually completed in no more than 5 min during ozonation in the presence of several catalysts (Ma et al., 2004; Yang et al., 2007; Zhang 2008a; Zhao et al., 2008a). According to our previous results, FeOOH has a high R$_{ct}$ and a high activity on the ozonation of nitrobenzene. However, CeO$_2$ has almost no activity on the ozonation of nitrobenzene. That said, opposite results were observed on the degradation of oxalic acid, which is an important intermediate compound during the ozonation of nitrobenzene (Zhang 2008a; Zhao et al., 2008b). Thus a catalyst that could form complexes with a small molecular weight acid would be beneficial to the decrease of DOC. It has been proved that CeO$_2$ has a good performance on the catalytic degradation of intermediate organics and the minimization of bromate formation. It was assumed, therefore, that the combination of the above catalysts with CeO$_2$, through a physical or chemical method, could not only decrease the intermediate organics but also control the formation of bromate (Zhang et al., 2008c) if the treated raw water contained a certain concentrations of bromide.

Catalytic ozonation has been used in the oxidation of organic pollutants in over ten water and wastewater treatment plants in China. One typical water treatment plant, using catalytic ozonation, is Shijiuyang water works of Jiaxing in Zhejiang province of China. Here, the drinking water source is heavily polluted. A combination of catalytic ozonation and biological activated carbon was employed based on the conventional treatment processes. The operation results indicated that the biodegradability of treated water through catalyzed ozonation was increased more obviously than that of ozonation alone. So, the biological activated carbon that followed the catalyzed ozonation process has shown better performances on both COD$_{Mn}$ and DOC removal than that of ozonation alone. It is found that in catalyzed ozonation, the bromate formation could be minimized significantly compared with the use of ozonation alone (Zhang, et al., 2008d).

In addition to catalytic ozonation, the ultrasonic assisted oxidation also showed a better peromance in degrading the pollutants (Zhao 2009d). It was observed that there was a synergetic effect of ultrasound with dual fields for the degradation of nitrobenzene in aqueous solution.

ADVANCES IN MEMBRANE FILTRATION

Membrane filtration technology is attracting more and more attention. This is due to its many beneficial properties such as its short processes, high separation efficiency, small footprint, low chemical addition, easy operation, and strong adaptability etc. Thus, it is being widely accepted as a promising technology for future urban water treatment. However, problems such as the high pressure it requires, its high costs, high fouling rate, and short life time etc. limits its widespread application. Reducing the membrane pressure, increasing the ability of antifouling, and extending the membrane life are, therefore, important aspects for the application of membrane technology in water treatment. Some solutions to these problems are illustrated in Fig 1.

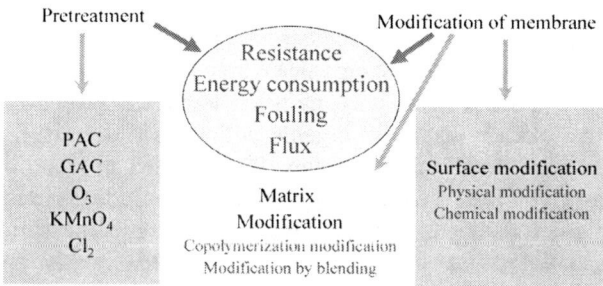

Figure 1 Proposed approaches for solving the problems of membrane filtration

Membrane fouling is the main problem encountered in the application of membrane filtration in water treatment. The principal foulants are considered to be those organics which can form a gel-like layer on the surface of the membrane. Pre-treatment by activated carbon (granular or powdered forms) is an accepted method in controlling the fouling of membranes. The addition of powdered activated carbon is an easy, low cost way to enhance the removal of microorganic pollutants, and to encourage the formation of biological films for enhancing the degradation of organics and the oxidation of ammonia in water. Granular activated carbon filtration is an effective pre-treatment means way to reduce the fouling of membranes and to extend the life of the membrane filtration. Pre-oxidation by ozone, permanganate or chlorine has also been used to remove organics or enhance the degradation of refractory organics by the following activated carbon process.

The modification of the membrane is the most important way to improve the membrane property. There are two types of membrane modification methods: surface modification and matrix modification. Surface modification uses surface treatment methods such as plasma, ultrawave-radiation, adsorption of high molecular weight substances or surface active reagent. This reduces the pressure, increases antifouling ability, and increases the flux of the membrane.

OVERVIEW OF HYBRID AOP AND MEMBRANE PROCESSES FOR WATER TREATMENT IN FUTURE CITIES

Membrane filtration is capable of removing the suspended solids and a large part of the colloids in water. These impurities mainly include: inorganic particles, algae, bacterial, virus, particulate organic carbon, some particulate complexes, and some high molecular weight molecules etc. However, advanced oxidation can efficiently decompose or convert the organic pollutants into harmless species. These impurities mainly include humic acid, fulvic acid, persistant organic pollutants, endocrine disrupting substances, pharmaceuticals, personal care chemical products and even some organics with a low molecular weight etc.

Advanced oxidation can be used as a pre-treatment means for membrane filtration. It can oxidize persistent organic pollutants and make them easier to be degraded before membrane filtration. Without this process, these organic pollutants may go through the membrane directly. Thus, advanced oxidation may be developed as a reliable pre-treatment method for membrane filtration.

It could be expected that a hybrid process of advanced oxidation and membrane filtration could effectively remove nearly all the pollutants in water resources. If a catalytic membrane is developed, both the degradation of organics and the control of membrane fouling may be achieved. A schematic example of a hybrid technology is shown in Fig 2. In this process, certain kinds of efficient catalysts would be fixed on the membrane. The membrane would thus have a catalytic property. During its separation of the relatively larger impurities in water, through sieving effect, some oxidant agents (take ozone for example) would be introduced into the water. The oxidation processes would then be catalyzed. This would lead to the generation of hydroxyl radicals (•OH), which is also a powerful oxidant species (ORP: 2.80 V). The generated •OH is a powerful species that can oxidize those stable and persistent organic pollutants non-selectively. This would also benefit the control of membrane fouling.

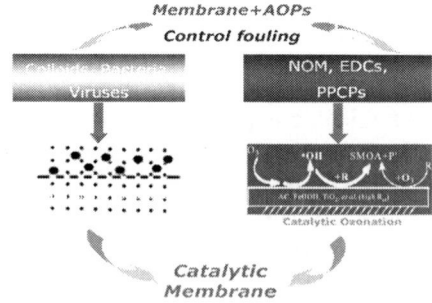

Figure 2 A hybrid process of advanced oxidation and membrane filtration

In recent years, nano-particle modification has received a great deal of attention. This is due to its combined properties of organic polymers and inorganic nanoparticles. A TiO_2 modified membrane, was prepared by a phase inversion method (Cao 2006). The surface of the TiO_2 modified membrane was denser than that of the neat PVDF membrane. It was found that TiO_2 nanoparticles embedded in the membrane intensified the crystal structure of PVDF molecules.

Based on this research, a novel membrane was developed by introducing FeOOH nanoparticles to the PVDF membrane which formed a catalytic membrane. The results indicated that some nanoparticles were formed on the surface of membrane which not only improved the membrane property but also made the membrane have a catalytic property.

A recent study also demonstrated that the modification of the membrane by some catalysts also made membrane have an increased reduction property (Wang 2008a; Wang 2008b). By using a catalytic membrane, some chlorinated organics were dechlorinated to a greater extent.

In fact, there is great flexibility in choosing the combination modes of these two effective technologies. This provides great scope for improving the capacity of this hybrid process in practical use.

CONCLUSIONS

Based on the analysis and discussions of water quality requirements and the progress of water treatment technologies, the following conclusions could be made: urbanization will cause a number of water quality and quantity problems; The hybrid process of advanced oxidation processes and membrane filtration will potentially solve the water quality problems; and the combination of the different technologies will depend on the specific requirements of each situation.

ACKNOWLEGEMENT

The support from the Natural Science Foundation of China (No. 50778049, 50821002) and the Science and Technology Ministry of China (2008ZX07421-002-03) is greatly appreciated.

REFERENCES

Cao X. C., Ma J. and Shi X. H. (2006). Effect of TiO2 Nan particle size on the performance of PVDF membrane. *Applied Surface Science*, **253**, 2003–2010.

Graham N. J. D., Jiang C. C., Li X. Z., Jiang J. Q. and Ma J. (2004). The influence of pH on the degradation of phenol and chlorophenols by potassium ferrate. *Chemosphere*, **56**(10), 949–956.

He D., Guan X. H., Ma J. and Yu M. (2009). Influence of different nominal molecular weight fractions of humic acids on phenol oxidation by permanganate. *Environmental Science and Technology*, **43**, 8332–8337.

Jiang J., Pang S. Y. and Ma J. (2009). Oxidation of triclosan by permanganate (Mn(VII)): Importance of ligands and in situ formed manganese oxides. *Environmental Science and Technology*, **43**, 8326–8331.

Hou Y. J., Ma J., Sun Z. Z., Yu Y. H. and Zhao L. (2006). Degradation of benzophenone in aqueous solution by Mn-Fe-K modified ceramic honeycomb-catalyzed ozonation. *Journal of Environmental Sciences-China*, **18**(6), 1065–1072.

Ma J. (1990). *Removal and Control of Organic Micropollutants from Water by Permanganate*. Ph.D. thesis, Dept of Municipal and Envrionmental Engineering, Harbin Architectural and Civil Engineering Institute, Harbin 150090, China.

Ma J., Li G. B. and Graham N. (1994). Efficiency and Mechanism of acrylamide removal by permanganate oxidation. *Journal of Water Supply Research and Technology-Aqua*, **43**(6), 278–295.

Ma J. and Graham N. J. D. (1996). Controlling the formation of chloroform by permanganate preoxidation-destruction of precursors. *Journal of Water Supply Research and Technology-Aqua*, **45**(6), 308–315.

Ma J., Graham N. and Li G. (1997a). Effect of permanganate preoxidation in enhancing the coagulation of surface waters-laboratory case studies. *Journal of Water Supply Research and Technology-Aqua*, **46**(1), 1–10.

Ma J. and Graham N. J. D. (1997b). Preliminary investigation of manganese-catalyzed ozonation for the destruction of atrazine. *Ozone-Science & Engineering*, **19**(3), 227–240.

Ma J. and Graham N. J. D. (1999). Degradation of atrazine by manganese-catalysed ozonation: Influence of humic substances. *Water Research*, **33**(3), 785–793.

Ma J. and Graham N. J. D. (2000). Degradation of atrazine by manganese-catalysed ozonation: Influence of radical scavengers. *Water Research*, **34**(15), 3822–3828.

Ma J. and Li G. B., Chen Z. L., Xu G. R. and Cai G. Q. (2001). Enhanced coagulation of surface waters with high organic content by permanganate preoxidation. *Water Science and Technology: Water Supply*, **1**(1), 51–61.

Ma J. and Liu W. (2002a). Effectiveness and mechanism of potassium ferrate(VI) preoxidation for algae removal by coagulation. *Water Research*, **36**, 871–878.

Ma J. and Liu W. (2002b). Effectiveness of ferrate (VI) preoxidation in enhancing the coagulation of surface waters. *Water Research*, **36**(20), 4959–4962.

Ma J., Sui M. H., Chen Z. L. and Wang L. N. (2004). Degradation of refractory organic pollutants by catalytic ozonation—Activated carbon and Mn-loaded activated carbon as catalysts. *Ozone-Science & Engineering*, **26**(1), 3–10.

Ma J., Sui M. H., Zhang T. and Guan C. Y. (2005). Effect of pH on MnOx/GAC catalyzed ozonation for degradation of nitrobenzene. *Water Research*, **39**(5), 779–786.

Ma J., Wang Z. H., Pan M. B. and Guo Y. F. (2009). A study on the multifunction of ferrous chloride in the formation of poly(vinylidene fluoride) ultrafiltration membranes. *Journal of Membrane Science*, **341**, 214.

Sun H. J., Liu H. L., Ma J., Wang X. Y., Wang B. and Han L. (2008). Preparation and characterization of sulfur-doped TiO2/Ti photoelectrodes and their photoelectrocatalytic performance. *J. Hazardous Materials*, **156**, 552–559.

Wang X. Y., Chen C., Liu H. L. and Ma J. (2008a). Preparation and characterization of PAA/PVDF membrane-immobilized Pd/Fe nanoparticles for dechlorination of trichloroacetic acid. *Water Research*, **42**, 4656–4664.

Wang X. Y., Chen C., Liu H. L. and Ma J. (2008b). Characterization and evaluation of catalytic dechlorination activity of Pd/Fe bimetallic nanoparticles. *Industrial & Engineering Chemistry Research*, **47**(22), 8645–8651.

Yang Y. X., Ma J., Zhang J., Wang S. J. and Qin Q. D. (2006). Degradation of trace nitrobenzene by nano-TiO2/silica-gel catalyzed ozonation in aqueous solution. *Acta Scientarium (in Chinese)*, **26**(8), 1258–1264.

Yang Y. X., Ma J., Qin Q. D. and Zhai X. D. (2007). Degradation of nitrobenzene by nano-TiO2 catalyzed ozonation. *Journal of Molecular Catalysis A-Chemical*, **267**(1–2), 41–48.

Zhang T. and Ma J. (2008a). Catalytic ozonation of trace nitrobenzene in water with syntheticgoethite. *Journal of Molecular Catalysis A-Chemical*, **279**(1), 82–89.

Zhang T., Li C. J., Ma J., Tian H. and Qiang Z. M. (2008b). Surface hydroxyl groups of synthetic α-FeOOH in promoting •OH generation from aqueous ozone: Property and activity relationship. *Applied Catalysis B-Environmental*, **82**(1–2), 131–137.

Zhang T., Chen W. P., Ma J. and Qiang Z. M. (2008c). Minimizing bromate formation with cerium dioxide during ozonation of bromide-containing water. *Water Research*, **42**(14), 3651–3658.

Zhang T., Lu J. F., Ma J. and Qiang Z. M. (2008d). Comparative study of ozonation and synthetic goethite-catalyzed ozonation of individual NOM fractions isolated and fractionated from a filtered river water. *Water Research*, **42**(6–7), 1563–1570.

Zhao L., Ma J., Sun Z. Z. and Zhai X. D. (2008a). Catalytic ozonation for the degradation of nitrobenzene in aqueous solution by ceramic honeycomb supported manganese. *Applied Catalysis B-Environmental*, **83**(3–4), 256–264.

Zhao L., Ma J. and Sun Z. Z. (2008b). Oxidation products and pathway of ceramic honeycomb-catalyzed ozonation for the degradation of nitrobenzene in aqueous solution. *Applied Catalysis B-Environmental*, **79**(3), 244–253.

Zhao L., Ma J., Sun Z. Z. and Zhai X. D. (2008c). Mechanism of influence of initial pH on the degradation of nitrobenzene in aqueous solution by ceramic honeycomb catalytic ozonation. *Environmental Science and Technology*, **42**(11), 4002–4007.

Zhao L., Ma J., Sun Z. Z. and Zhai X. D. (2009a). Mechanism of heterogeneous catalytic ozonation of nitrobenzene in aqueous solution with modified ceramic honeycomb. *Applied Catalysis B-Environmental*, **89**, 326–334.

Zhao L., Ma J., Sun Z. Z. and Liu H. L. (2009b). Enhancement mechanism of heterogeneous catalytic ozonation by cordierite supported copper for the degradation of nitrobenzene in aqueous solution. *Environmental Science and Technology*, **43**(6), 2047–2053.

Zhao L., Sun Z. Z., Ma J., Liu H. L. and Zhai X. D. (2009c). Novel relationship between hydroxyl radical initiation and surface group of ceramic honeycomb supported metals for the catalytic ozonation of nitrobenzene in aqueous solution. *Environmental Science and Technology*, **43**(11), 4157–4163.

Zhao L., Ma J. and Zhai X. D. (2009d). Synergetic effect of ultrasound with dual fields for the degradation of nitrobenzene in aqueous solution. *Environmental Science and Technology*, **43**(13), 5094–5099.

Shortcut Nitrification During the Start-Up in Biological Aerated Filter under Environmental Condition

LIU Yong-zheng[a], QIU Li-ping[a]* and ZHANG Shou-bin[a,b]

[a]School of Civil Engineering & Architecture, University of Jinan, Jiwei Road 106, Jinan 250022, China
E-mail: lipingqiu@163.com
[b]School of Municipal & Environmental Engineering, Harbin Institute of Technology, Harbin 150090, China

Abstract Shortcut nitrification was studied during the start-up in biological aerated filter (BAF) under environmental condition. Two different start-up modes are adopted in the test. The first BAF (No.1) started with condition that water temperature was 15~25 °C, ammonia nitrogen concentration was 25~35 mg/L, COD concentration was 20~30 mg/L, and filtration rate was 1 m/h, while the second BAF (No. 2) started under the same condition but 900 mL activated sludge was added as inoculated seed. The No. 2 BAF start-up with activated sludge had a high biofilm culturing speed. Removal rate of ammonia nitrogen was over 90% only after three days and stable in the following days, while No.1 BAF had a very low start-up speed and there was obvious removal efficiency after the eighth day. The No. 2 BAF also had a higher speed of ammonia removal rate in effluent than that in No.1 BAF under natural condition; Obvious nitrite accumulation had appeared in No.1 BAF which was not added activated sludge into and the highest nitrite accumulation rate was over 80%, but the state did not last stably for a long time and fell to a stable rate of 3% quickly after reaching the highest concentration.

Keywords Biological nitrogen removal, Biological aerated filter, Shortcut nitrification, Nitrite accumulation

INTRODUCTION

Biological nitrogen removal is considered to be economical and effective (Wang *et al.*, 2005; Liang *et al.*, 2004), a lot of deficiencies in conventional water treatment process are remedied since the theory of shortcut nitrification was proposed (Feng *et al.*, 2001). Ammonia is oxidized forming nitrite but not nitrate controlled in the process, and then reduced to N_2 directly in denitrification process. Shortcut nitrification has a perfect removal efficiency for high ammonia and low C/N ratio wastewater in water treatment (Yuan *et al.*, 2000). Nearly all the nitrite accumulation is achieved by controlling the condition which is beneficial to nitrosomonas in the laboratory and using the high ammonia wastewater in recent studies. However, it is not resolved that how to realize the obvious nitrite accumulation fast and easily and which can maintain a long time at high content for low ammonia municipal wastewater under environmental condition. BAF has a lot of advantages in wastewater treatment. However, it is unclear whether shortcut nitrification can generate in BAF recently. So the study is based on BAF, and the purpose is discussing the possibility of shortcut nitrification in BAF. We also try to resolve how to achieve stable nitrite accumulation for a long time at high concentration and how to get fast and easily for low ammonia municipal wastewater under environmental condition. Microbial characteristics in BAF during all the process were also studied.

MATERIALS AND METHODS

Reactor Set-Up

Two same up-flow BAF reactors were used in the test, both made up of acrylic cylinder with effective height of 200 cm and internal diameter of 10 cm, A 15 cm gravel supporting layer was placed at bottom and 100 cm clay pellets (particle size 3–5mm) were filled in BAF. Total 9 sample points are located at intervals of 150mm above supporting layer. Compressed air flow into BAF reactor via diffuser from the air inlet (100mm above the bottom). Schematic diagram of the experimental setup followed as Figure 1:

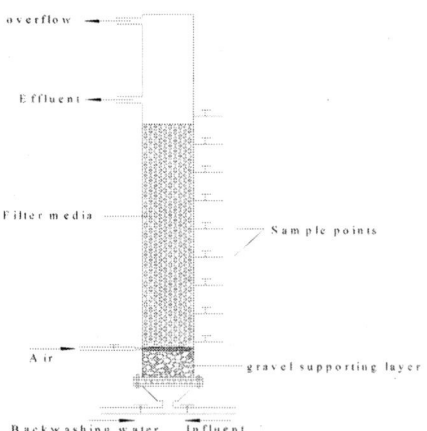

Figure 1 Schematic diagram of the experimental setup

Sludge and Organic Wastewater

The BAF was inoculated with sludge obtained from secondary sedimentation tank of Jinan Wastewater Treatment Plant. The characteristics of fed wastewater which consisted of 10% sewage and 90% synthetical domestic water were followed in Table 1.

Table 1 Wastewater characteristics

NH_4^+–N (mg/L)	COD (mg/L)	NO_2^-–N (mg/L)	NO_3^-–N (mg/L)	Temperature (°C)
25~35	20~30	0.2~3.4	0~1	15~25

Analysis

Test items of this study contain chemical oxygen demand (COD), ammonia nitrogen, nitrite nitrogen, nitrate nitrogen, dissolved oxygen (DO), pH and temperature. COD and DO was measured using National Standard Methods (potassium dichromate method and iodimetry respectively). Ammonia nitrogen, nitrite nitrogen and nitrate nitrogen was measured using spectrophotometer (Spectrumlab 752-S). pH and temperature was measured using PHS-3B pH Meter and thermometer respectively.

RESULTS AND DISCUSSION

Effect of Start-Up Mode on Biofilm Culturing Speed

Two different start-up modes are adopted in the test. The first BAF (No. 1) started with condition that water temperature was 15~25 °C, ammonia nitrogen concentration was 25~35 mg/L, COD concentration was 20~30 mg/L, and filtration rate was 1 m/h, while the second BAF (No. 2) started under the same condition but 900 mL activated sludge was added as inoculated seed. The two BAFs started at the same time. Firstly, the filters only aerated without wastewater discharge for 3 days, then flowed wastewater feeding and discharging continuously. This operation stage kept on for nearly 1 month without intermittence until the concentration of effluent ammonia nitrogen was very low and stable. The result of the experiment (Figure 2) shows the removal rate of ammonia nitrogen over 90% averagely, and mature biofilm had formed.

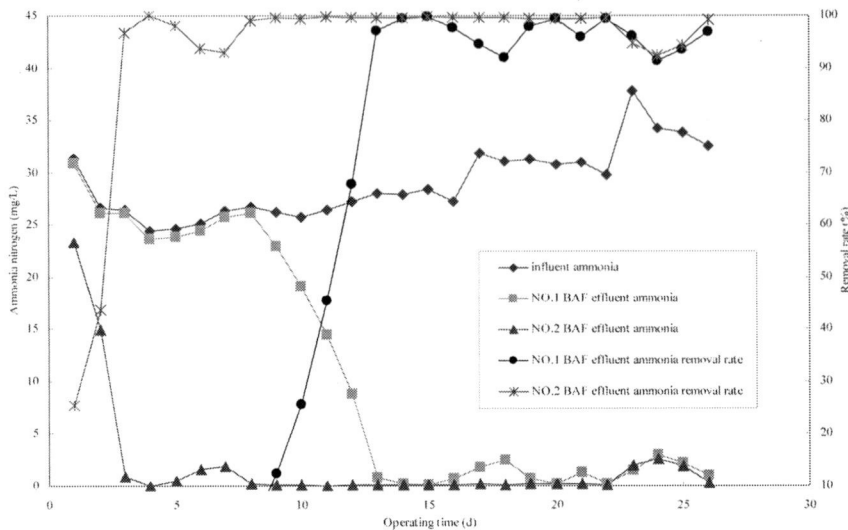

Figure 2 Removal rate of ammonia nitrogen in influent and effluent of BAF in biofilm culturing phase

The result of the experiment (Figure 2) shows that the No. 2 BAF start-up with activated sludge had a high biofilm culturing speed. Removal efficiency existed at the beginning, and had a high removal rate speed, removal rate of ammonia nitrogen was over 90% only after 3 days and stable in the following days, while No. 1 BAF had a very low start-up speed, there was obvious removal efficiency after the eighth day and the speed was very low, until the thirteenth day the removal efficiency of ammonia nitrogen became high and stable. The reasons for this are that there are a large number of microorganisms including ammonifiers and nitrobacteria in the activated sludge, and can recover activity quickly when the condition is fit for, growth and reproduction also occurred at the same time, so the quantity of ammonifiers and nitrobacteria became much larger than other microorganisms, and there was a high and stable removal rate of ammonia nitrogen only 3 days later. Whereas there were few microorganisms in No. 1 BAF at the beginning, so removal rate of ammonia nitrogen was very low until ammonifiers and nitrobacteria appeared and increased in the conformable condition on the eighth day. From then on, the concentration of effluent ammonia nitrogen gradually decreased accordingly.

The result of the experiment (Figure 2) also shows that No. 2 BAF start-up with activated sludge had a higher speed of ammonia removal rate in effluent than that in No. 1 BAF under natural condition. Ammonia removal rate was more than 90% on the third day in No. 2 BAF. Compared with No. 2 BAF, It costs 5 days to achieve a high and stable removal rate of ammonia nitrogen in No.1 BAF with a low removal speed. This is for a large number of unactivated ammonifiers and nitrobacteria were in No. 2 BAF, and they can recover activity quickly when the condition is fit for, and as a result of a large quantity they reproduced fast. Contrarily, there were few microorganisms in No. 1 BAF, so they can not propagate quickly even in the suitable environment.

Effect of Start-Up Mode on Nitrite Accumulation

Concentration of nitrite nitrogen and nitrate nitrogen were tested during the start-up process in BAF under environmental condition (pH 7.5~8.0, water temperature 15~25 °C and ample dissolved oxygen). The result of the test (Figure 3) shows that obvious nitrite accumulation had appeared in No.1 BAF which was not added activated sludge into and the highest nitrite accumulation rate was over 80%, but the state did not last stably for a long time and fell to a stable rate of 3% quickly after reaching the highest concentration. There was not obvious nitrite accumulation in No. 2 BAF from the beginning.

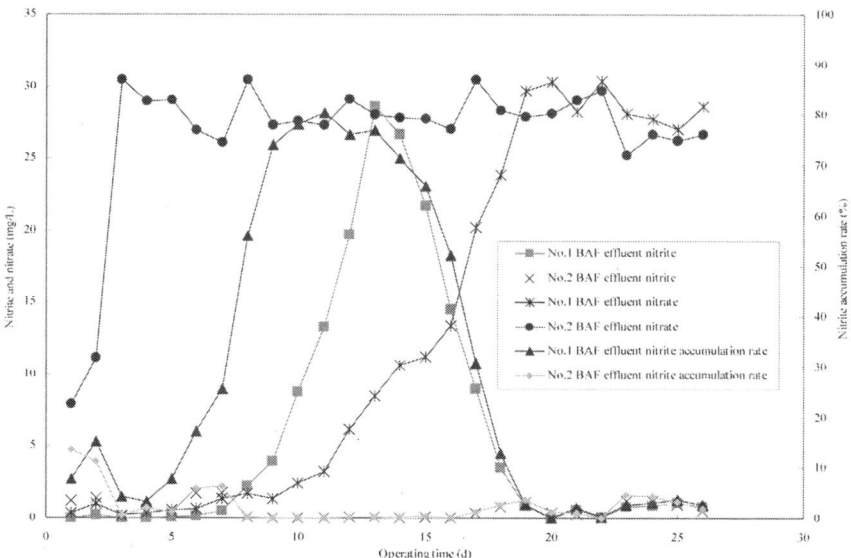

Figure 3 Nitrite accumulation rate of BAF in biofilm culturing phase

The probable reason may be that there were few microorganisms in No. 1 BAF at the beginning, so there was little ammonia nitrogen transformed into nitrite nitrogen and nitrate nitrogen by nitrosomonas and nitromonas. When nitrosomonas and nitromonas appeared and increased fast in the conformable condition from the eighth day, the concentration of nitrite nitrogen and nitrate nitrogen increased accordingly. At the beginning, nitrosomonas and nitromonas didn't affect each other for enough food, so nitrite nitrogen and nitrate nitrogen were both increasing, but as a result of their mass rearing, growth inhibition occurred for insufficient food a few days later. Nitromonas have a better adaptation than nitrosomonas for the environment, so nitrosomonas was inhibited and decreased gradually, which led to addition of nitrate nitrogen and reduction of nitrite nitrogen. Concentration of nitrate nitrogen was high and stable while that of nitrite nitrogen was very low from the nineteenth day, which shows that nitromonas had become dominant bacteria and nitrosomonas had been nearly washed out. In No. 2 BAF, there were a large number of unactivated ammonifiers and nitrobacteria, and can recover activity quickly when the condition was fit for, but the environment condition was more suitable for nitromonas than nitrosomonas, so nitrosomonas was inhibited at the beginning. It spent 3 days only to increase to the maximum quantity restricted by food for nitromonas since added into BAF, nitrate nitrogen of effluent reached a large stable quantity.

CONCLUSIONS

1. Start-up with inoculums has a higher biofilm culturing speed and is easier than natural start-up in BAF, and has a stable treatment efficiency.
2. Obvious nitrite accumulation has appeared in No.1 BAF which was not added activated sludge into, but the state did not last stably for a long time for the environment condition is more suitable for nitromonas.
3. It should be possible to achieve stable nitrite accumulation of a high content for a long time fast and easily for low ammonia municipal wastewater treatment with natural start-up in BAF by controlling the influencing factors of ammonifiers and nitrobacteria and operation conditions of the reactor reasonably under environmental condition.

ACKNOWLEDGEMENT

This work was partly funded by National Science Foundation of China (50978121), Natural Science Foundation of Shandong Province (Y2006B39), Project of Shandong Province Higher Education Science and Technology Program (J06I01).

REFERENCES

Feng Y. C., Wang J. L. and Qian Y. (2001). Developments of novel technology for biological nitrogen removal from wastewater. *Microbiology*, **28**(4), 88–91.

Liang Y. G., Zhang Z. Y. and Zhou Y. X. (2004). Shortened nitrification process for biological nitrogen removal. *Journal of Hefei University of Technology (Natural Science)*, **27**(10), 1292–1296.

Wang S. P., Peng Y. Z. and Yu D. S. (2005). Biological nitrogen removal by endogenous denitrification via nitrite at normal temperature. *Journal of Beijing University of Technology*, **31**(3), 298–302.

Yuan L. J., Peng D. C. and Wang Z. Y. (2000). Short-cut nitrification and denitrification. *China Water & Wastewater*, **16**(2), 29–31.

Alum Sludge-Based Constructed Wetland: Novelty, Benefits and Constraints

Y.Q. Zhao*, A.O. Babatunde, X.H. Zhao, J.L.G. Kumar and Y.S. Hu

Centre for Water Resources Research, School of Architecture, Landscape and Civil Engineering, University College Dublin, Newstead Building, Belfield, Dublin 4, Ireland
E-mail: yaqian.zhao@ucd.ie

Abstract A novel constructed wetland system (CWs), which employs dewatered alum sludge as the main substrate of the CWs, has been recently developed in University College Dublin, Ireland for the purpose of promoting the beneficial reuse of the water industrial "waste". Alum sludge refers to the by-product derived from the drinking water treatment process, which uses aluminium sulphate as major coagulant. The so called "alum sludge-based CWs" makes it possible to treat P-rich wastewater, thus converting the alum sludge from a "waste" into a value-added raw material for water pollution control. This paper outlines a short history of the development of the new and novel CW and then discusses the novelty, benefits and its constraints.

Keywords Alum sludge, constructed wetland, wastewater treatment

Introduction

In view of the wastewater treatment in rural areas and isolated industrial estates, constructed wetland (CW) system may be one of the most promising technologies pertaining to cost-effective wastewater treatment and sustainable development. In particular, increasing demands for decentralized systems for sewage treatment in rural areas and industrial settings offer great opportunity for wide application of CW. At the CW research group of University College Dublin (UCD), Ireland, the central theme of our work has been the development of a new generation of CW system which is simple and sustainable in construction but more efficient than conventional CW systems. The CW research group is the pioneer of the novel alum sludge-based CW system for wastewater treatment which has the capacity for enhanced removal of phosphorus (P) and organic matter, particularly from high strength wastewaters. A special novelty of the newly developed system is the use of a hitherto landfill designated by-product 'dewatered alum sludge' as the substrate in the CW as opposed to the traditional media of gravel, sand and local soils. In the last five years, the group has made considerable efforts to study the reuse of the dewatered alum sludge as a raw material in the civil and environmental engineering. The dewatered alum sludge is an inevitable by-product from potable water treatment facilities when aluminium sulphate is used as a coagulant for source water purification. Several projects have been conducted and experimented at different scales to explore the different approaches towards the innovative reuse of the alum sludge. A novel CW system of so called "alum sludge-based CW" to treat P-rich animal farm wastewater, has been developed with the new concept of "tidal flow" operation strategy, thus leading to a new generation of the CW. It is worth noting that the alum sludge has been converted from a "waste" into a value-added raw material for water pollution control (Zhao et al., 2008, 2009a, b).

This paper outlines a short history of the development of the new and novel CW and then discusses the novelty, benefits and its constraints.

SHORT HISTORY OF DEVELOPMENT

Prior to carrying out extensive work in the laboratory, a global review on waterworks sludges, including alum sludge, was carried out to ascertain the state of art as regards their beneficial reuse. It was immediately realised that although the current trend towards reusing the waterworks sludges with eleven different options is trialled at various stages, there is no attempt so far to reuse waterworks sludge, in particular alum sludge, as a main substrate to develop a CW system (Babatunde and Zhao, 2007). The review further showed that Al in the alum sludge accounts for ca. $29.7\pm13.3\%$ dry weight based on available data. Therefore, the development of the alum sludge-based CW was premised on the principle and fact that the alum sludge is mainly composed of Al which possesses strong affinity with wastewater pollutants, especially P. Consequently, the utilization of the alum sludge as a substrate for CW can potentially lead to increased treatment efficiency through the enhancement of adsorption and chemical precipitation processes by the Al ions in the alum sludge. This transforms the alum sludge into a valuable material for wastewater treatment. The development of the alum sludge-based CW was carried out in seven phases detailed briefly below:

Phase I: The P-adsorption characteristics and capacity of the alum sludge was examined in detail using batch adsorption tests. A wide range of P concentration simulating P levels in different types of wastewater was used and the adsorption behaviour for ortho-P, poly-P and organic-P was examined (Zhao et al., 2007; Yang et al., 2008). The maximum P adsorption capacity of the alum sludge was established to be three orders of magnitude higher than the conventional gravel used in CW and it also has a comparable capacity to other materials being tested as CW substrates.

Phase II & III: In order to ensure an efficient design of the system and enable a thorough understanding of the pollutant removal processes in the novel CW system, particularly adsorption and precipitation reactions on the alum sludge, two investigative phases were carried out. Firstly, a detailed investigation of the mechanism and characteristics of P adsorption onto the alum sludge along with the accompanying reactions were carried out. P adsorption onto the alum sludge was described as an inner sphere complexation during which phosphate ions are adsorbed onto the alum sludge by ligand exchange (Yang *et al.*, 2006). Secondly, an extensive characterization of the alum sludge was carried out to determine if it has physico-chemical characteristics consistent with recommended guidelines required for CW substrates. The results indicate the specific surface area of the alum sludge ranged from 28.0 to 41.4 m^2/g while X-ray Diffraction, Fourier transform infrared and energy-dispersive X-ray spectroscopies all indicate that the alum sludge is mainly composed of amorphous aluminium which influences its P adsorption capacity. The pH and electrical conductivity ranged from 5.9 to 6.0 and 0.104 to 0.140 dS/m respectively, and both showed that it should suitably support plant growth while the alum sludge also had a uniformity coefficient of 3.6 (Babatunde *et al.*, 2009).

Phase IV & V: Following on the batch and characterization tests, the development was focused on determining the optimal configuration for the novel alum sludge-based CW. In the first part of the configuration optimization, comparative studies were carried out using single stages for both horizontal and vertical flow systems. Consequently the development was to determine the most suitable substrate placement and configuration for the vertical flow alum sludge-based CW. Six individual model CW system were examined in the laboratory. The systems were operated in the 'tidal flow' mode which involves rhythmical filling and draining of the CW matrices with wastewater. The six individual model CWs had pea gravel at their infiltrative surface resulting in a proportion of gravel in the substrate ranging from 0% to 60%. Influent P loading into the systems ranged from 18 to 346 mg-P/l while a range of hydraulic loading rate from 1.23 to 1.86 m^3/m^2.d was used. Results show that the performance of the model CW systems decreased with increasing proportion of gravel in the systems and there was no distinctive beneficial advantage to delaying clogging irrespective of the infiltrative surface layer and configuration. The model systems with only 0–20% proportion of substrate by gravel achieved the best performance with P removal efficiency ranging from 82.1 to 92.7%.

Phase VI: The sixth phase of the development involved laboratory trials using a multi-stage format for the CW and employing 100% alum sludge as the substrate based on the outcome of the fifth phase of the investigations. Under a hydraulic loading rate of 1.3 m^3/m^2.d and a range of organic loading rate of 279–775 g-BOD$_5$/m^2.d and 361–1,029 g-COD/m^2.d, average removal efficiencies (± SD) of 90.6 ± 7.5% (BOD$_5$) and 71.8 ± 10.2% (COD) were achieved. P removal was exceptional with average removal efficiency of 97.6 ± 1.9% achieved for soluble reactive P at a mean influent concentration of 21.0 ± 2.9 mg/l. Clearly, the beneficial advantage of the CW system was seen, particularly in P removal which was high from the start of the trials and sustained throughout.

Phase VII: Finally, a field demonstration study of the newly developed CW was initiated. The demonstration study involved setting up a pilot scale of the newly developed system to treat wastewater from an animal farm located at the Lyons Estate, Newcastle, Co. Dublin, Ireland (Fig. 1). Alum sludge was used as the main substrate in the field system and the systems were planted with common reeds 'Phragmites australis'. The CWs consists of four identical cells in series (referred as stages) designed with a total treatment surface area of 3.42 m^2. Loadings up to 0.29 m^3/m^2.d (hydraulic) and 151 g-BOD$_5$/m^2.d (organic) have been applied across the entire system. Results show that the system was highly efficient for organics removal with removal efficiency up to 90% achieved for BOD$_5$ removal. More significantly, P removal was particularly fascinating with results indicating P removal efficiencies up to 99.8% (Fig. 2). The field demonstration study has further shown that the system demonstrates huge promise as a low-cost system of choice for treating P-rich wastewater.

Figure 1 The pilot-scale alum sludge-based CW system during set-up (left) and 7 weeks after set up with good growth of the common reed established (right)

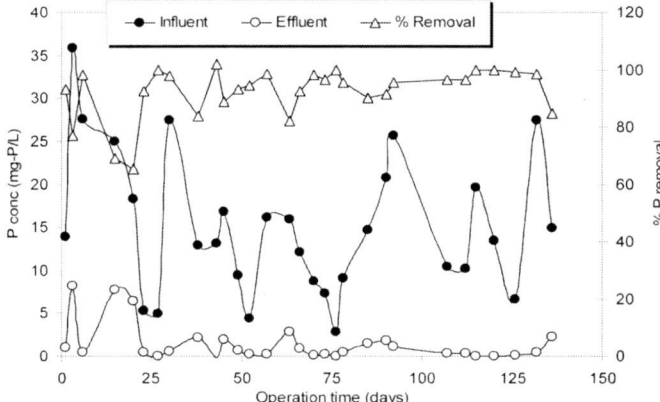

Figure 2 Influent and effluent P values and P removal efficiency in the pilot alum sludge-based CW system

NOVELTY

It is well known that the substrate in CW plays a key role for wastewater purification. Typically, local soils, sand and gravel have been used as a substrate in conventional CW systems. However, such materials are good for carriers of biofilm development for COD, BOD_5 removal, but feeble for P immobilization. Consequently, the discovery/invention of alternative substrate with considerable capacity for P adsorption in CW is an effective research topic in CW development. Several materials have been tested as possible substrate for CW particularly for enhancing P removal. However, the scale and economics of their use have not been justified, while attempts to scale up such preliminary investigation to field scale have been lacking. However, the use of alum sludge as a substrate is justifiable both economically and environmentally. Alum sludge is an inevitable by-product largely produced in water treatment plants worldwide and often designated for landfill disposal. It is locally, easily available and free of charge. The novelty of using the alum sludge as a P adsorbent and biofilm carrier to develop a novel CW that is capable of enhanced wastewater treatment stems from the win-win tactics of using "waste" for wastewater treatment. In particular, such reuse of the alum sludge offers a sustainable alternative to its landfill disposal, while at the same time, it serves to enhance wastewater treatment in CW system.

BENEFITS

There are a number of benefits associated with the use of alum sludge to develop the alum sludge-based CW. They include:

(1) Abundant available supply of dewatered alum sludge. Alum sludge is widely available in large quantities in all cities and metropolis worldwide that uses aluminium as a coagulant. Annual generation of the sludge is reported to be 182,000; 125,000; 34,000; 20,000 and 18,000 tonnes of the dry solids in the UK, Germany, Netherlands, Portugal and Ireland respectively (Babatunde and Zhao, 2007). In addition, giving the almost inevitable continual generation of the alum sludge as a result of the prevalent drinking water treatment processes used worldwide, sufficient quantities of the alum sludge for developing the alum sludge-based CW is guaranteed.

(2) Reduced capital investment of construction. Even though CW systems can offer comparative and/or higher treatment performance at low-cost when compared to other similar systems, the cost of the media used in the CW still accounts for a large chunk of the construction cost. According to US-EPA (1993), the cost of media for a constructed wetland system can account for up to 53% of the overall construction cost and yet, the media usually have very limited P adsorption capacity. This further implies that there is the likelihood of a recurring cost of media replacement when the media is saturated with P. Therefore, by using alum sludge, the overall capital investment of construction is reduced significantly, by possibly up to 53%, and at the same, the frequency and cost of media replacement will be drastically reduced

(3) Good P adsorption capacity (14.3–14.9 mg-P/g sludge at pH 7.0 (Yang *et al.*, 2008)) and excellent P removal efficiency (87±7% (Zhao *et al.*, 2009b)). The alum sludge has demonstrated excellent P adsorption capacity which is at least 1-3 orders of magnitude higher than typical substrates used in CW. It therefore has a distinct advantage in that in terms of P removal performance, it will conveniently outlast most other CW substrates

(4) Good carrier/material for biofilm attachment and thus excellent removal of COD (73±16%), BOD$_5$ (83±12%) (Zhao *et al.*, 2009b). In addition to being an excellent adsorbent for P, the alum sludge also possesses the advantage of being able to support biofilm growth and development which is very crucial for pollutant removal.

(5) Possibility of treating high strength wastewater. Treatment of high strength wastewater is often a technical challenge in conventional CW systems due to the high nutrient concentration (especially P) and oxygen demand of these wastewaters. However, by using the alum sludge-based CW system operated in tidal flow mode, it becomes feasible to treat such high strength wastewaters

as the alum sludge serves to greatly enhance P removal while the tidal flow operation strategy enhances greater oxygen transfer which leads to greater microbial degradation.

(6) Suitable growth medium for wetland plant (say common reed) (see Fig. 1). CW plants contribute to landscape fit of CW, while it is also commonly argued that they contribute to treatment processes within the CW through rhizofiltration and oxygen transfer through their stems. It is therefore very important that any CW substrate should be suitable for plant growth. The alum sludge was characterized as having suitable qualities to support plant growth and this has also been evidently demonstrated in the field (see Fig. 1).

(7) Considerable saving in landfill cost and area. The cost of landfilling alum sludge will almost certainly keep increasing with increasing population and demand for potable water, and also with increasingly stringent treatment regulations. Yet, this cost is significant. For instance, the total cost of disposing waterworks sludge in Netherlands stands at a staggering £30–£40 per annum (Babatunde and Zhao, 2007). It can thus be seen that considerable savings in landfill cost and area can be achieved through the alum sludge-based CW and other beneficial reuse options.

(8) Double transformation of alum sludge from a "waste" into a value-added raw material. The alum sludge-based CW effectively transforms the alum sludge firstly from a waste into a useful material for enhancing wastewater treatment in CW through its use as a substrate and then secondly upon its saturation in the CW, as a value-added raw material for use either a slow P release fertiliser (Babatunde and Zhao, 2009) or for P recovery (Zhao and Zhao, 2009).

CONSTRAINTS

Even though the alum sludge-based CW system has demonstrated obvious advantages over the conventional CW, there are constraints that should be carefully considered for the large scale applications. These include:

(1) Logistics of the use of the alum sludge at field scale viz a viz the application mode of the alum sludge and the typical P sorption capacity of the alum sludge in the field. Being an unconventional substrate material for CW systems, the application mode of the alum sludge (i.e. either fresh dewatered alum sludge; aged dewatered alum sludge; dried alum sludge; granulated alum sludge) in the CW should be carefully considered as this will influence the transport and handling logistics. Furthermore, adequate care should be paid to the field P loading capacity of the alum sludge used which could differ from computations based on laboratory work.

(2) Lifespan of the alum sludge regarding the saturation with P (although it has been preliminarily estimated as 4–17 years (Zhao *et al.*, 2009b)). This can be used as a guide, but it is yet to be fully verified on a field scale.

(3) Although alum sludge is derived from residual of treatment of raw water which contains mainly turbidity, colour, suspended clays and humic substances and therefore, it is relatively free and highly unlikely to contain a substantial quantity of toxic substances, there is still a degree of apprehension as regards possible release of substances/metals, particularly aluminium during its reuse. This is very crucial before any large scale application can be accepted. It is however very likely that the system will require continuous aluminium monitoring as the effects of aluminium level beyond the recommended safe limit could be lethal. This may imply additional cost for the CW operation. However, from our current field data, only 21% of samples analysed were above the recommended limit of 0.2 mg/l for Al discharge, meaning that although results from the field system suggest aluminium release, the level of Al in the effluents are generally well below the 0.2 mg/l limit in most cases.

(4) Clogging is an inevitable long-term operational drawback of CW systems, especially those operated in the vertical flow mode. While field experience of operating the CW system has so far shown no clogging, it is important to be aware of this as a possible operational challenge of the system in the long-term operation.

(5) Eventual disposal of the P saturated alum sludge. Although the alum sludge has a considerable high capacity for P adsorption, its capacity is still finite. Therefore, at the end of its service life, the P saturated sludge has to be disposed and this has to be planned for. However, as earlier mentioned, there is possibility for the use of the saturated sludge as a soil amendment (Babatunde and Zhao, 2009) or for P recovery (Zhao and Zhao, 2009).

ACKNOWLEDGMENTS

The authors would like to thank the following organizations for the financial support of the research. The Irish Environmental Protection Agency (for projects grants no. 2005-ET-S-7-M3 and 2005-ET-MS-38-M3); Enterprise Ireland (Proof of Concept Scheme project no. PC 20070308); The Irish Department of Agriculture and Food (Research Stimulus Fund project no. RSF 07 529). Ballymore Eustace Water Treatment Plant is thanked for kind assistance. Mr. P. Kearney, section head technician of the Water and Effluents Laboratory, UCD, and Dr Edward Jordan who is the Lyons Estate farm manager are also thanked for their invaluable technical assistance and numerous supports towards the field set-up during the research.

REFERENCES

Babatunde A. O. and Zhao Y. Q. (2007). Constructive approaches towards water treatment works sludge management: An international review of beneficial re-uses. *Critical Reviews in Environmental Science and Technology*, **37**(2), 129–164.

Babatunde A. O. and Zhao Y. Q. (2009). Forms, patterns and extractability of phosphorus retained in alum sludge used as substrate in laboratory-scale constructed wetland systems. *Chemical Engineering Journal*, **152**(1), 8–13.

Babatunde A. O., Zhao Y. Q., Burke A. M., Morris M. A. and Hanrahan J. P. (2009) Characterization of aluminium-based water treatment residual for potential phosphorus removal in engineered wetlands. *Environmental Pollution*, **157**(10), 2830–2836.

US-EPA (1993). *Subsurface Flow Constructed Wetlands for Wastewater Treatment: A Technology Assessment*. Publication of the United States Environmental Protection Agency.

Yang Y., Zhao Y. Q., Babatunde A. O., Wang L., Ren Y. X. and Han Y. (2006). Characteristics and mechanisms of phosphate adsorption on dewatered alum sludge. *Separation and Purification Technology*, **51**(2), 193–200.

Yang Y., Zhao Y. Q. and Kearney P. (2008). Influence of ageing on the structure and phosphate adsorption capacity of dewatered alum sludge. *Chemical Engineering Journal*, **145**(2), 276–284.

Zhao Y. Q., Babatunde A. O., Razali M. and Harty F. (2008). Use of dewatered alum sludge as a substrate in reed bed treatment systems for wastewater treatment. *Journal of Environmental Science and Health, Part A*, **43**(1), 105–110.

Zhao Y. Q., Babatunde A. O., Zhao X. H. and Li W. C. (2009a). Development of alum sludgebased constructed wetland: An innovative and cost-effective system for wastewater treatment. *Journal of Environmental Science and Health, Part A*, **44**(8), 827–832.

Zhao Y. Q., Razali M., Babatunde A. O., Yang Y. and Bruen M. (2007). Reuse of Aluminium-based water treatment sludge to immobilize a wide range of phosphorus contamination: Equilibrium study with different isotherm models. *Separation Science and Technology*, **42**(12), 2705–2721.

Zhao X. H. and Zhao Y. Q. (2009). Investigation of phosphorus desorption from P-saturated alum sludge used as a substrate in constructed wetland. *Separation and Purification Technology*, **66**(1), 71–75.

Zhao Y. Q., Zhao X. H. and Babatunde A. O. (2009b). Use of dewatered alum sludge as main substrate in treatment reed bed receiving agricultural wastewater: Long-term trial. *Bioresource Technology*, **100**(2), 644–648.

Using Sub-lethal UV-C Irradiation to Prevent *Microcystis aeruginosa* Blooming for Urban Stream

Yi Tao[a, b], Xihui Zhang[a]*, Xianzhong Mao[a] and Doris W. T. Au[b]

[a]Research Center for Environmental Engineering and Management, Graduate School at Shenzhen, Tsinghua University, Shenzhen 518055, China
[b]Department of Biology and Chemistry, City University of Hong Kong, Tat Chee Avenue, Kowloon, Hong Kong
Tel: +86 755 2603 6707; fax: +86 755 2603 6707
E-mail: zhangxh@sz.tsinghua.edu.cn

Abstract *Microcystis aeruginosa* is a typical harmful bloom-forming cyanobacterium, especially in China waters. This study aims to investigate the effects of sub-lethal UV-C irradiation on photosynthetic activity of *M. aeruginosa*. *M. aeruginosa* of 10^6 cells mL^{-1} were exposed to UV-C irradiation at 0~200 mJ cm^{-2} and subsequently incubated for 30 days under normal culture conditions. We assessed the cell-specific growth rates and pulse-amplitude-modulated parameters (rapid light response curves, its initial slope, and relative maximum electron transport rate). The results suggested that sub-lethal UV-C irradiation at 50–75 mJ cm^{-2} can suppress *M. aeruginosa* growth for 5–9 days. Sub-lethal UV-C irradiation induces suppression effect on photosynthetic activity. Photosynthetic activity is more sensitive to UV-C stress than growth activity. It is possible to adopt variation of photosynthetic activity to prewarning growth recovery.

Keywords *Microcystis aeruginosa*, sub-lethal UV-C irradiation, growth suppression, rapid light response curves, photosynthetic activity

INTRODUCTION

Cyanobacterial blooms in China waters have been increasingly reported (Jin *et al.*, 2005; Guo, 2007). Cyanobacterial blooms induce deoxygenation of the water leading to fish kills and degradation of water recreational value (Oliver and Ganf, 2000). Severe cyanobacterial blooms may influence water supply due to filter blockage, erosion of water supply pipelines and frequent backwash (Oliver and Ganf, 2000). Accumulated cyanobacterial cells produce large amount of toxin compounds (Falconer *et al.*, 1999) and odour compounds (Oliver and Ganf, 2000) that induce risks for human health. Current methods for cyanobacterial bloom treatment, such as adding copper sulfate (McKnight *et al.*, 1983) and chlorination (Daly *et al.*, 2007), are used for emergent situation after severe cyanobacterial bloom occurrence. In addition, most methods are immobile, and therefore of low effectiveness when applied in large scale waters such as lakes and reservoirs.

Ultraviolet irradiation at 254 nm (UV-C) is an alternative to prevent cyanobacteria blooms in lakes and reservoirs. It has been reported that growth of cyanobacterium species including *M. aeruginosa* and *Anabaena flosaquae* can be suppressed within several days by means of 30 s to 10 min UV-C irradiation (Sakai *et al.*, 2007). UV-C irradiation at 20 and 50 mJ cm^{-2} is sub-lethal to *M. aeruginosa* cells, as over 80% of the exposed cells remain intact. However, UV-C irradiation at 100 and 200 mJ cm^{-2} induced severe cell disintegration in more than 70% of the irradiated cells (Tao *et al.*, 2010). Usually, the UV-C equipment is simple to construct and easy to operate and conduct maintenance on, and so it is possible for it to be shipborne. Boats equipped with UV-C lamps were developed and applied to inactivate freshwater bloom of the dinoflagellate *Peridinium bipes* and red tide of the *Chattonella spp.* Moreover, UV-C technology will not cause secondary pollution (Haas, 1999).

There are only a few published studies regarding the suppression effects of UV-C irradiation on cyanobacterial growth. The objective of this study was to investigate variations of photosynthetic activity under sub-lethal UV-C irradiation during growth suppression.

METHODS AND MATERIALS

Microorganisms

Axenic culture of cyanobacterium *M. aeruginosa* (FACHB 905) was obtained from the Culture Collection of Freshwater Algae of the Institute Hydrobiology (FACHB-Collection; Wuhan, China). The uni-algal inoculants were cultured in autoclaved BG11 medium including NaNO$_3$ 1500 mg L^{-1}, K$_2$HPO$_4$·3H$_2$O 40 mg L^{-1}, MgSO$_4$·7H$_2$O 75 mg L^{-1}, CaCl$_2$·2H$_2$O 36 mg L^{-1}, C$_6$H$_8$O$_7$ (citric acid) 6 mg L^{-1}, Fe(NH$_4$)$_3$(C$_6$H$_5$O$_7$)$_2$ (ferric ammonium citrate) 6 mg L^{-1}, Na$_2$EDTA 1 mg L^{-1}, Na$_2$CO$_3$ 20 mg L^{-1}, H$_3$BO$_3$ 2.86 mg L^{-1}, MnCl$_2$·H$_2$O 1.81 mg L^{-1}, ZnSO$_4$·7H$_2$O 0.22 mg L^{-1}, CuSO$_4$·5H$_2$O 0.079 mg L^{-1}, CO(NO$_3$)$_2$·6H$_2$O 0.049 mg L^{-1}, Na$_2$MoO$_4$·2H$_2$O 0.39 mg L^{-1} at pH around 8.0. Cultures were incubated at 25 °C in an incubation chamber under controlled lighting. Fluorescent lamps (Philips) were used as a light source with an automated light/dark cycle of 12 h/12 h. Light intensity during the lighting phase was 2000 Lx.

UV-C irradiation and subsequent incubation

Within exponential growth phase, initial cell density of *M. aeruginosa* for UV-C treatment was set at approximately 1×10^6 cells mL^{-1}. A collimated beam apparatus equipped with a low-pressure UV-C lamp (40 W), as shown in Fig. 1, was adopted to accomplish UV-C exposure following recommended procedure of USEPA (USEPA, 2006). Briefly, 40 mL of each algal suspension was irradiated in glass petri dishes 90 mm in diameter. The absorbance of BG-11 medium at 254 nm was 0.09. The intensity of UV-C irradiation at the surface level of samples was determined by a UV sensor (RM12, Dr. Gröbel Elektronik Gmbh, Germany). The intensity value was stable at 0.41 mW cm^{-2} within the entire experiment. Therefore UV-C doses for this study were adjusted to 20, 50, 75, 100, and 200 mJ cm^{-2} by varying the exposure time at 49 s, 122 s, 183 s, 244 s and 488 s, respectively. The sample without UV-C irradiation was set as a control. Such UV-C exposure for each sample was repeated two times, and then the replication solutions were combined into 250 mL Erlenmeyer flasks and incubated under the same conditions described above. Samples were taken and measured immediately before and after UV-C irradiation (within 2 h, i.e. 0.1 d), as well as 1, 3, 5, 7, 9, 12, 15, 20, 25, and 30 d after UV-C irradiation.

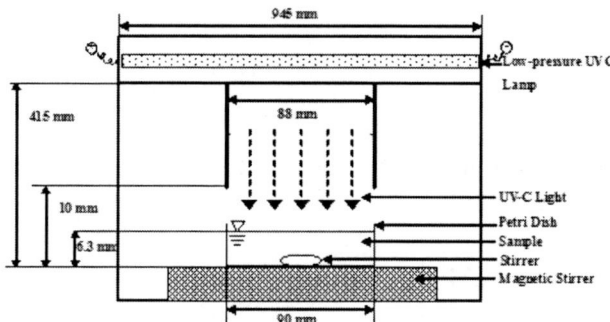

Figure 1 The Collimated beam apparatus for UV-C exposure

Algal growth analysis

Cell density of samples was measured with a fluorescent microscope (BX51, Olympus, Japan) using a haemocytometer. The effects of UV-C irradiation on algal growth of the species were estimated by the percent inhibition of average specific growth rate (I_r) which is defined by the equation 1 (OECD, 2006):

$$I_r(\%) = \frac{\mu_C - \mu_T}{\mu_C} \times 100 \qquad (2.1)$$

where, I_r is the percentage of inhibition of the average specific growth rate (μ); μ_C is the mean value for the average specific growth rate in the control group (d^{-1}); μ_T is the average specific growth rate for the UV-C treatment replicate (d^{-1}).

In vivo fluorescence measurements

Photosystem II fluorescence was assessed using a Pulse Amplitude Modulated (PAM) fluorimeter (Water-PAM, Walz, Effeltrich, Germany). Measuring, actinic and saturating light were provided by a red LED-lamp with peaks at 450 nm and half-bandwidth of 20 nm. The cells were dark adapted for 30 min before measurement. The sample was exposed to nine PAR irradiation steps increasing from 11 to 1281 μmol photons m^{-2} s^{-1}. A single saturating light pulse of 4000 μmol photons m^{-2} s^{-1} was applied after each step of every 30 s. The signals were collected using the WinControl Software (Walz). The effective quantum yield of photosystem II (Φ_{PSII}) was calculated according to Schreiber (2004), as follows:

$$\Phi_{PSII} = (F'_m - F_t)/F'_m \tag{2.2}$$

where F_t is the chlorophyll a fluorescence yield immediately prior to the saturating irradiation; F_m' is the maximal chlorophyll a fluorescence yield in illuminated state. The relative electron transport rate (rETR) was calculated as follows (Schreiber, 2004):

$$rETR = \Phi_{PSII} \times PAR \times 0.5 \times 0.84 \tag{2.3}$$

where PAR is the irradiance in μmol photons m^{-2} s^{-1}. Rapid light response curves (RLCs) were drawn using rETR versus PAR and characterized by fitting the model of Platt et al. (1980):

$$rETR = rETR_s\left[1 - \exp\left(-\frac{\alpha E}{rETR_s}\right)\right]\exp\left(-\frac{\beta E}{rETR_s}\right) \tag{2.4}$$

Based on RLCs, parameters including the initial slope in light response curve (α) and photoinhibition parameter (β) were determined, and then the maximum relative electron transport rate, rETR$_m$, was calculated as follows (Schreiber, 2004):

$$rETR_m = rETR_s\left(\frac{\alpha}{\alpha + \beta}\right)\left(\frac{\beta}{\alpha + \beta}\right)^{\beta/\alpha} \tag{2.5}$$

Statistical analysis

All samples were carried out in 3 replications. Data was presented in means$\pm$ standard deviation (SD). Figures were plotted using Microcal Origin (version7.5, Microcal Software Inc.). ANOVA from SigmaStat (version 3.5, Systat Software Inc.) was adopted to determine significant differences among control and treated samples. Difference was considered to be significant at $p < 0.05$.

RESULTS AND DISCUSSION

Effects of UV-C irradiation on growth characteristics of *M. aeruginosa*

The effect of UV-C irradiation on the growth of *M. aeruginosa* was investigated. Figure 2 presents the inhibition curve of *M. aeruginosa* during the incubation period after UV-C irradiation. For samples exposed to 20 mJ cm^{-2} UV-C, I_r values were around zero, indicating that there was no marked suppression effect on algal growth. Interestingly, for samples exposed to 50 mJ cm^{-2} UV-C, I_r values increased to maximum at day 5 and then gradually decreased to zero at day 20. This indicates that algal growth was significantly suppressed within 5 d and then restarted and recovered to control level from day 5 to day 30. For samples exposed to 75, 100, and 200 mJ cm^{-2} UV-C, I_r values reached maximum on day 9 and then declined toward control level.

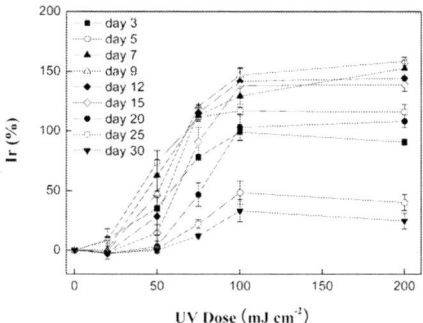

Figure 2 Effects of UV-C irradiation on growth characteristics of *M. aeruginosa*

For samples exposed to 20~100 mJ cm^{-2} UV-C, the increase of I_r values was UV-C dose-dependent. This suggests that the suppression effect can effectively be reinforced through increasing UV-C dose. However, there were no significant differences for I_r values between 100 mJ cm^{-2} and 200 mJ cm^{-2}. It is possible that the suppression effect reaches stabilization period when the UV-C dose increases to 100 mJ cm^{-2} and therefore stronger suppression effect is difficult to be achieved by further enhancing UV-C dose.

Bloom formation requires high cell density of cyanobacterial species. Usually, 2×10^4 cells mL^{-1}, 10^5 cells mL^{-1}, and 10^7 cells mL^{-1} cyanobacterial cells in water are regarded as low, moderate, and high risk levels for bloom formation, respectively (WHO, 2003). In this study it is supposed that the high risk level can be avoided by adopting sub-lethal UV-C irradiation to suppress algal growth. In this study, 50~75 mJ cm^{-2} UV-C irradiation can suppress *M. aeruginosa* growth within 5~9 d. Based on our former report, a UV dose of above 100 mJ cm^{-2} is a lethal dose and not suitable for a suppression objective (Tao *et al.*, 2010). Therefore, if the expected suppression time is 5 d, UV-C irradiation of around 50 mJ cm^{-2} should be the optimal choice. If the UV dose increases to 75 mJ cm^{-2}, the suppression effect will last for 9 d. In addition, the results based on 30 d confirmed that UV-induced suppression effect on *M. aeruginosa* growth is time-limited; it may require repeated UV-C treatment at the end of suppression time.

Effects of UV-C irradiation on photosynthetic characteristics of *M. aeruginosa*

As a photoautotrophic microorganism, *M. aeruginosa's* growth is tightly correlated with its photosynthetic activity. Effects of UV-C irradiation on photosystem II were investigated regarding the response of RLCs to ambient irradiance, as shown in Fig. 3. Variations of α and rETR$_m$ within the incubation period are presented in Fig. 4. Just after UV-C irradiation, 0.1 d, the initial slope α of samples exposed to 20, 50, 75, 100, and 200 UV-C decreased to 98%, 88%, 82%, 54%, and 36% of control samples, respectively. Similarly, the maximum relative electron transport rate, rETR$_m$, of UV-treated samples decreased to 96%, 72%, 70%, 38%, and 35% of control samples, respectively. This indicated that 20~200 mJ cm^{-2} UV-C treatments immediately induced damage to photosystem II, and that the damage was UV-C dose dependent. During the incubation period, RLCs of control samples kept in initial pattern. With a UV-C dose at 20 mJ cm^{-2}, both α and rETR$_m$ were not significantly different from that of the control since day 3. However, with a UV-C dose at 50 mJ cm^{-2}, both α and rETR$_m$ gradually decreased from day 1 to day 3, then increased again from day 5, and recovered to the same pattern as that of control from day 12. For samples exposed to UV-C doses at 75, 100, and 200 mJ cm^{-2}, both α and rETR$_m$ decreased to zero on day 3, kept at zero from day 3 to days 7~9, and then increased again from days 9~12.

RLCs are often adopted to assess photosynthetic activity. The slope at low PAR, α, reflects the maximal photosynthetic quantum yield and rETR$_m$ presents the capacity of the electron transport chain (Schreiber, 2004). In this study, the parallel decrease of α and rETR$_m$ indicates UV-induced disturbance and down-regulation of photosystem II. It is possible that excitation energy captured by the antennae is diverted from photochemistry pathway into fluorescence or heat dissipation pathways. Researchers reported similar observations of UV-B radiation on algal photosynthetic activity (Six *et al.*, 2007). However, it has yet been reported in UV-C studies. In this study, UV-C irradiation at 20~50 mJ cm^{-2} induced reversible suppression effect on photosynthetic activity. However, UV-C irradiation at

75~200 mJ cm⁻² caused total inhibition of photosynthetic activity. UV-C induced photosystem damage affects light absorption and transportation and depletes energy to recover. Therefore, a UV-C induced suppression effect on photosynthetic activity contributes to the growth suppression effect.

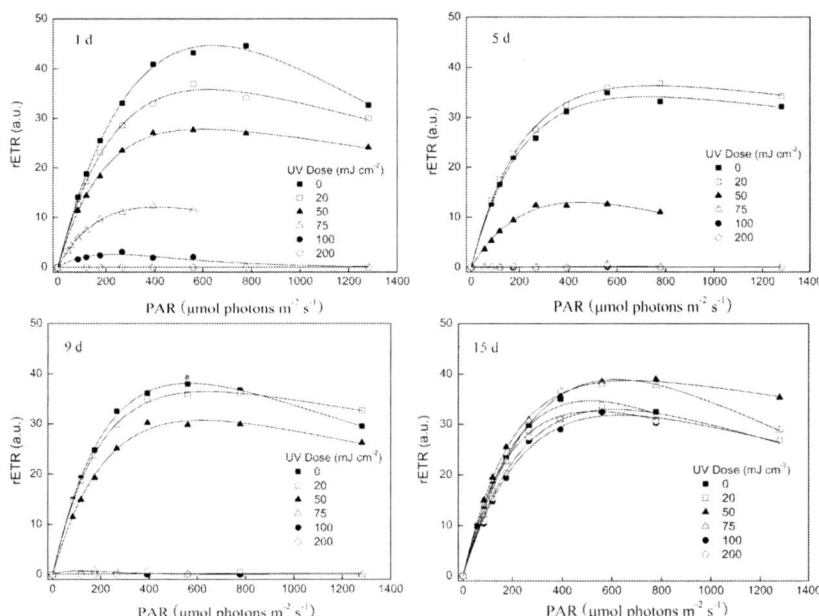

Figure 3   Rapid light response curves of *M. aeruginosa*

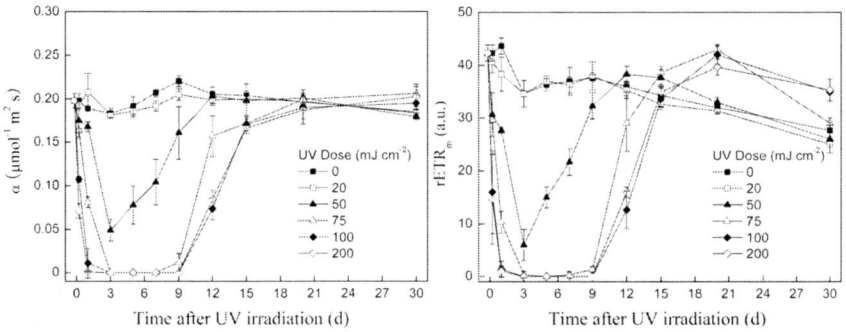

Figure 4 Variations of initial slope and relative maximum electron transport rate of *M. aeruginosa*

Adopting photosynthetic activity for prewarning of growth activity

As discussed above, sub-lethal UV-C irradiation of 50~75 mJ cm^{-2} can suppress *M. aeruginosa* growth for 5~9 d. It is possible to prevent bloom formation through repeated sub-lethal UV-C treatments. The authors attempted to adopt a variation of photosynthetic activity for prewarning of growth activity. Both growth curves of cell density and photosynthetic curves of rETR$_m$ were presented in Fig. 5. Just after UV-C irradiation, there was no significant difference among cell density results of control, 50 mJ cm^{-2}, and 75 mJ cm^{-2} samples. On the other hand, rETR$_m$ of 50 mJ cm^{-2} and 75 mJ cm^{-2} samples were marked lower than that of control. This suggests that photosynthetic activity is more sensitive to UV-C stress than growth activity. During the incubation period, the growth curve of samples exposed to UV-C doses at 50 mJ cm^{-2} increased from day 5, while the reflection point of photosynthetic curve appeared on day 3, 2 days before growth curve. Similarly, the reflection point of photosynthetic curves of samples exposed to UV-C doses at 75 mJ cm^{-2} occurred on day 7 and 2 days earlier than that of the growth curve on day 9. Therefore, it is possible to adopt variation of photosynthetic activity for prewarning of growth activity.

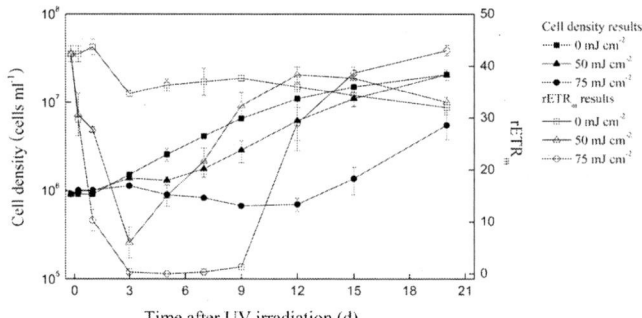

Figure 5 Comparisons between growth curve and photosynthetic activity curve of *M. aeruginosa*

CONCLUSION

The results suggest that sub-lethal UV-C irradiation at 50~75 mJ cm^{-2} can suppress *M. aeruginosa* growth for 5~9 d. If the expected suppression time is 5 d, UV-C irradiation of around 50 mJ cm^{-2} should be the optimal choice. The suppression time can be increased to 9 d when a UV-C dose at 75 mJ cm^{-2} is adopted. A UV dose above 100 mJ cm^{-2} is not suitable for suppression objectives. Sub-lethal

UV-C irradiation induces a suppression effect on photosynthetic activity that contributes to the growth suppression effect. Photosynthetic activity is more sensitive to UV-C stress than growth activity. It is possible to adopt a variation of photosynthetic activity for prewarning of growth activity.

ACKNOWLEDGEMENT

This work was supported by grants from National Natural Science Foundation of China (No. U0773002), National 863 High-tech Program (No. 2008AA06A413), National Water Grant (No. 2008ZX07423).

REFERENCES

Daly R. I., Ho L. and Brookes J. D. (2007). Effect of Chlorination on *Microcystis aeruginosa* cell integrity and subsequent microcystin release and degradation. *Environmental Science and Technology*, **41**(12), 4447–4453.

Falconer I., Bartram J., Chorus I., Goodman T. K., Utkilen H., Burch M. and Codd G. A. (1999). Safe levels and safe practices. In: Toxic Cyanobacteria in water: A Guide to their Public Health Consequences, Monitoring, and Management, I. Chorus and J. Bartram (eds.), WHO, London, pp. 161–182.

Guo L. (2007). Ecology-doing battle with the green monster of Taihu Lake. Science, **317**(5842), 1166–1166.

Haas C. N. (1999). Disinfection. In: *Water Quality and Treatment-A Handbook of Community Water Supplies*, R. D. Letterman (eds.), McGraw-Hill, INC, New York, pp. 944–1003.

Jin X. C., Xu Q. J. and Huang C. Z. (2005). Currents status and future tendency of lake eutrophication in China. *Science China Series* C, **48**, 948–954.

McKnight D. M., Chisholm S. W. and Harleman D. R. F. (1983). CuSO4 treatment of nuisance algal blooms in drinking water reservoirs. *Environmental Management*, 7, 311–320.

OECD (2006). Sixteenth Addendum, to the OECD Guidelines for the Testing of Chemicals, in Guideline 201: Alga, Growth Inhibition Test. OECD, Paris.

Oliver R. L. and Ganf G. G. (2000). Freshwater blooms. In: *The Ecology of Cyanobacteria*, B. A. Whitton and M. Potts (eds.), Kluwer Academic Publishers, the Netherlands, pp. 149–194.

Paltt T., Gallegos C. L. and Harrison W. G. (1980). Photoinhibition of photosynthesis in natural assemblages of marine phytoplankton. *Journal of Marine Research*, **38**, 687–701.

Sakai H., Oguma K., Katayama H. and Ohgaki S. (2007). Effects of low- or mediumpressure ultraviolet lamp irradiation on *Microcystis aeruginosa* and *Anabaena variabilis*. *Water Research*, **41**(1), 11–18.

Schreiber U. (2004). Pulse-Amplitude-Modulation (PAM) fluorometry and saturation pulse method: an overview. In: *Chlorophyll a Fluorescence*, Springer, the Netherlands, G. C. Papageorgiou and Govindjee (eds.), pp. 279–319.

Six C., Judovic J., Partensky F., Holtzendorff J. and Garczarek L. (2007). UV-induced phycobilisome dismantling in the marine picocyanobacterium *Synechococcus* sp. WH8102. *Photosynthesis Research*, **92**, 75–86.
Tao Y., Zhang X. H., Au W. T. D., Mao X. Z. and Yuan K. (2010). The effects of sub-lethal UV-C irradiation on growth and cell integrity of cyanobacteria and green algae. *Chemosphere*, **78**(5), 541–547.
USEPA (2006). Ultraviolet Disinfection Guidance Manual for the Final Long Term 2 Enhanced Surface Water Treatment Rule. Appendix C, pp. 1–12. http://www.epa.gov/safewater/disinfection/lt2/compliance.html
WHO (2003). Guidelines for safe recreational water environments, volume 1: Coastal and fresh waters, WHO, pp. 136–139.

The Application of Wasted Architecture Walling Materials Used as a Constructed Wetland Media

Lu Zhou[a], Zhaojun Huang[a,b] and Tao Li[a]

[a]Department of Environmental Science and Engineering, Tsinghua University Beijing 100084,
[b]School of Civil and Environmental Engineering, Beijing University of Science and Technology Beijing 100083
Tel: 86-10-62773079 Fax: 86-10-62785687; E-mail: zhoulu@tsinghua.edu.cn

Abstract In this study, three kinds of wasted architecture walling materials were selected and tested for their suitability as media for developing constructed wetlands. These materials included broken concretes, clay bricks, and foamed insulation bricks, all of which account for nearly 70% of architecture walling waste. Several characteristics of these materials were analyzed such as, external aperture distribution, elemental constitution and their potential for pollutant adsorption. Results showed that all of the materials could meet the application requirements of constructed wetlands. This was particularly evident in the case of the foamed insulation bricks which had the biggest capacity for pollutant adsorption. Using foamed insulation bricks as a media for a small subsurface flow constructed wetland, the removal of nitrogen and phosphorus from the secondary effluent was studied. The removal efficiency of total nitrogen could reach 30% at its best when it's average inlet concentration was 19.7 mg/L. For phosphorus, the removal efficiency of total phosphorous was 75% with the average inlet concentration of 1.0 mg/L. Offering good removal performances of nitrogen and phosphorus, waste walling materials could provide suitable media for constructed wetlands.

Keywords Architecture walling waste, constructed wetlands, media, wastewater

INTRODUCTION

Rapid urbanization has led to a lot of old architecture being dismantled. As a result, there is a large amount of architecture walling waste. Usually, most architecture walling waste is sent to landfill sites for disposal. Not only does this neglect the residual use-value of the walling waste, but it also occupies a large amount of land. If parts of the architecture walling waste could be used as different media for constructed wetlands, it could solve several problems: It would save soil from landfill sites; it would reduce environment pollution from architecture walling waste; it could protect natural resources and save the energy effectively.

In this paper three kinds of architecture walling waste were selected. The chaacteristics of each one have been studied and tested as a potential media for developing constructed wetlands.

MATERIALS AND METHODS

Selection of Architecture Walling Waste for Experiment

Broken-concretes, clay bricks and foamed insulation bricks accounts for more than 70% of the total amount of architecture walling waste (Sun, 2003). They were, therefore, all selected for the test. Amongst all of the wasted walling, broken concrete accounts for most of the debris. This is produced by the mixing of cement, sand and water proportionately. The amount of clay bricks is also considerable. However, whilst those that are still intact are reused, the broken ones are generally sent to landfill sites as architecture waste.

The foam-flyash-concrete block, also called foamed insulation brick, is a new kind of porous lightweight building material. It is produced by mixing fly-ash, magnesium oxychloride cement, and some effective modifier. With proper technology control, and normal atmospheric conditions, it solidifies at room temperature. Here, the fly-ash is the main raw material and the magnesium oxychloride cement acts as a cementing agent. Foam-flyash-concrete blocks are used as a common building material, effective for heat preservation. It is often used to build filler walls or non- partition bearing walls. The three kinds of walling waste tested in this study were taken from a dismantled hotel in the campus of Tsinghua University.

Pre-Treatment of Architecture Walling Waste Samples for Physico-Chemical Analysis

After the samples were crushed, they were screened and collected respectively, keeping the sample granules at a diameter between 1–2 millimetres. The sample granules were then washed by distilled water. Finally, the granules were dried at 105 °C in an oven, before the dry granules were moved into a dessicator for the experiment.

Analysis Methods

The porosity and aperture of the three kinds of walling waste were measured by an Auto Pore IV 9500 mercury injection apparatus (Micromeritics, USA), and the elementary analysis was completed by a XRF-1700 X-ray fluorescence spectrometer (Shimadzu, Japan). The water quality analysis referred to correlated methods and standards (Ministry of Environmental Protection of China, 2002).

Description of the Constructed Wetland in the Study

One small subsurface flow constructed wetland was constructed to study its removal performance of nitrogen and phosphorus pollutants. The data is listed in Table 1. The length of the inlet zone or outlet zone was 0.5 meter. The inlet water is the secondary effluent from one wastewater treatment plant in Beijing.

Table 1 The parameters of the constructed wetland in the study

Item	Bottom slope/ %	Length/ m	Width/ m	Media depth /cm			Main plants
				Sand	Foamed insulation bricks (particle size: 30~60mm)	Clay bricks (particle size: above 60mm)	
Parameters/ description	1	5.5	1.5	20	60	20	*Iris pseuda-corus*

ANALYSIS OF BASIC PROPERTIES OF ARCHITECTURE WALLING WASTE SAMPLES

Analysis of the Aperture Architecture Walling Waste Samples

The performance of nutrient adsorption and micro-organism adhesion growth of different media usually depends on the surface features of the media. Features such as the specific surface area, pore diameter and porosity should all be accounted for. The specific surface area of a medium is the sum of the surface pore area of one unit weight of the medium (the common unit is m^2/g). Generally, the bigger the specific surface area is, the bigger the adsorption capacity is. The pore diameter, in a medium, is the diameter of the pores in the surface of the medium. It can be expressed by media pore diameter of volume or area. Proper pore diameter is good for microorganism growth. Porosity is the ratio of pore volume to total media volume. This could be used to reflect the compactability of the media. The bigger the porosity is, the more room there is for the micro-organisms in the media. The test results of the porosity and pore diameter of the three selected walling waste materials are listed in Table 2.

Table 2 Apertures features of the tested architecture walling waste materials

Material types	Median pore diameter of pore volume/Å	Median pore diameter of surface area/Å	Porosity/%
Broken-concretes	3082	70	71.72
Clay bricks	78556	47	47.92
Foamed insulation bricks	293669	3127131	35.39

The data in Table 2 indicates that the broken-concrete and clay bricks had a bigger porosity but a smaller median pore diameter. It was smaller than 10 μm when calculated by pore volume and smaller than 8 nm when calculated by surface area. These conditions would inhibit the formation and growth of micro-organisms in the pores. In the case of foamed insulation bricks, however, whilst it has smaller porosity, its pore diameter is greater. Under these conditions, the potential for good micro-organism growth is greater.

Table 3 Composition of elements in the selected walling waste samples (expressed in the form of oxide except fluorine)

Items	Broken-concrete	Clay bricks	Foamed insulation bricks
SiO_2	40.54%	57.03%	50.13%
CO_2	18.60%	—	—
Al_2O_3	8.41%	13.73%	5.52%
Fe_2O_3	3.03%	6.78%	1.60%
CaO	21.11%	12.74%	38.04%
K_2O	1.66%	2.58%	1.56%
Na_2O	1.15%	4.01%	0.56%
MgO	3.59%	—	1.70%
TiO_2	0.54%	0.85%	0.28%
F	—	—	0.08%
P_2O_5	0.13%	0.16%	0.06%
SrO	0.07%	0.11%	0.07%
SO_3	0.97%	0.22%	0.35%
MnO	0.07%	0.15%	0.05%
Cr_2O_3	0.08%	0.08%	—
ZnO	—	—	—
ZrO_2	—	0.05%	—
Rb_2O	—	0.02%	—
ZnO_2	0.03%	—	—
Total	100%	100%	100%

Element Analysis in Architecture Walling Waste Samples

The elemental composition of walling waste is an important physical and chemical property to consider, when used as media for constructed wetlands. The analysis of elements is helpful in deducing whether or not the selected waste material has a potential to send new pollutants into the water. It is also useful to find out whether the material has any beneficial components which might strengthen the chemical absorption performance. The element analysis results of the three different materials, completed by an X-ray fluorescence spectrometer (Shimadzu, Japan), are recorded in Table 3.

The results showed that all of the elements of the three selected waste materials belong to the elements of common stone. The waste materials, therefore, would not send new pollutants into the water. Yuan (Yuan *et al.*, 2005) suggested that the phosphorus adsorption ability of a material in its basic condition depended, mainly, on the content of calcium in the material. Higher calcium content leads to greater phosphorus adsorption capacity and better phosphorus removal efficiency from wastewater. The results indicated that each kind of architecture walling waste material had a high calcium content. In foamed bricks, there was 38.04%, in the form of CaO. Furthermore, the selected waste materials all contain various amounts of aluminium and iron which could improve the removal efficiency of pollutants from water.

Adsorption Ability Analysis of Architecture Walling Waste Materials

When choosing a suitable media type for constructed wetlands, it is important to evaluate the phosphorous adsorption ability of the media. In this study, phosphorus adsorption isotherm tests were conducted on the three different materials: broken-concrete, clay bricks and foamed insulation bricks.

For adsorption happening on solid surface at a constant temperature, *Freundlich* and *Langmuir* equations are used to describe the relationship between the adsorption capacity of the solid surface and the equilibrium concentration of solute in solvent. When the initial solute concentration is lower, the adsorption law will match the *Freundlich* equation better (Drizo, 1999). In the test, the initial phosphorus concentration was set as 0.5 mg/L, then 1.0 mg/L and then 2.0 mg/L. The phosphorus adsorption of the three selected walling waste materials could be accounted for by the *Freundlich* equation. The results are listed in Table 4 (In the table: C_0 is the initial phosphorus concentration in the test water, mg/L; k is *Freundlich* capacity coefficient; 1/n is the *Freundlich* strength factor; R is the correlation coefficient.)

Table 4 Phosphorus adsorption isotherm equation and related parameters for different walling waste samples

Sample type	C_0/TP mg/L	k	1/n	R^2
Clay bricks	0.5	7.48×10^3	1.72	0.99
	1.0	9.93×10^4	1.31	0.94
	2.0	5.85×10^6	0.86	0.96
Broken-concrete	0.5	8.80×10^2	1.38	0.95
	1.0	9.74×10^2	0.87	0.90
	2.0	8.80×10^3	0.66	0.91
foamed insulation bricks	0.5	1.48	0.81	0.99
	1.0	3.33×10^3	0.63	0.94
	2.0	1.44×10^6	0.51	0.95

Table 4 indicates that the foamed insulation bricks have the greatest phosphorus adsorption capacity, while the clay bricks have the smallest.

PERFORMANCE OF NUTRIENT REMOVAL IN CONSTRUCTED WETLANDS WITH ARCHITECTURE WALLING WASTE MATERIALS AS MEDIA

In this study, the subsurface flow of the constructed wetland was running continuously with the hydraulic load as 0.07 m/d. Three sample points were set uniformly along the direction of the water flow in the wetland. The nutrient concentrations in the water was analyzed at both the inlet point and the sample points every specific time interval. The mean values of the removal effectiveness of total phosphorus (TP) and total nitrogen (TN) were recorded along the length of the wetland. The results are shown in Figure 1 and Figure 2. The average removal rates of TP and TN were 45.7% and 31.0% respectively. When the average COD_{Cr} concentration in the inlet water was 32.0 mg/L, the constructed wetland had almost no obvious organics removal. This is because organics were always non-biodegradable in the inlet water, previously treated in the treatment plant.

According to the EPA statistic of nutrient removal efficiency in existing constructed wetlands, the removal efficiency of nitrogen and phosphorus gradually rises in the first five years of the wetland being in operation. Wetland plants usually take several years to develop roots and reproduce. Because of this they are unable to display the effects in an established 'wetland' immediately. The depth and density of roots in a wetland could directly impact the growth and amount of microorganism (USEPA, 2000).

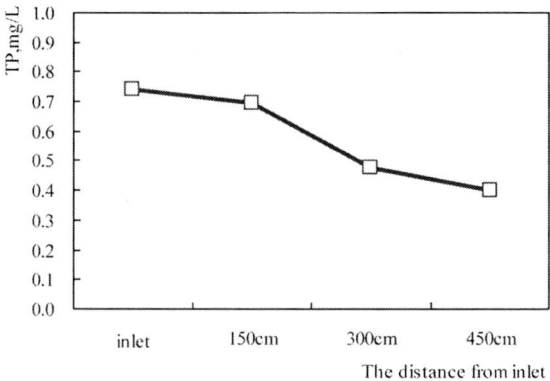

Figure 1 TP removal efficiency in the constructed wetland

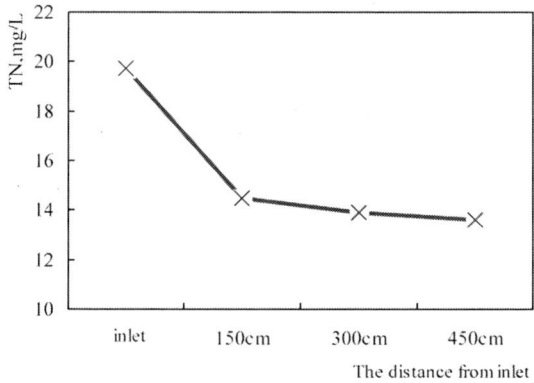

Figure 2 TN removal efficiency in the constructed wetland

The hydraulic load rate could severely affect the pollution removal efficiency. In order to evaluate the relationship between phosphorus removal and hydraulic load, three hydraulic load rates were set during the test. The inlet concentration of phosphorus was kept at 1.0 mg/L. Table 5 displays the related parameters.

Table 5 Different hydraulic load rates and handling capacities

Hydraulic load rate (HLR), m/d	0.22	0.07	0.03
Capacity per cubic meter, m³/d	1.81	0.6	0.26

Figure 3 shows the phosphorus removal efficiency at different hydraulic loads. When the hydraulic load was 0.22 m/d, the removal rate of TP was about 30% and the outlet TP concentration was about 0.7 mg/L. However, when the hydraulic load rate was 0.07 m/d, the removal rate of TP rose to 58% and the effluent TP concentration was less than 0.5 mg/L. With a rate higher than 75%, the TP removal rate was at its greatest when the hydraulic load rate reduced to 0.03 m/d, and the effluent concentration was below 0.2 mg/L. This is because a smaller hydraulic load rate could supply a longer hydraulic residence time, allowing both the plants and the media to absorb the pollutants.

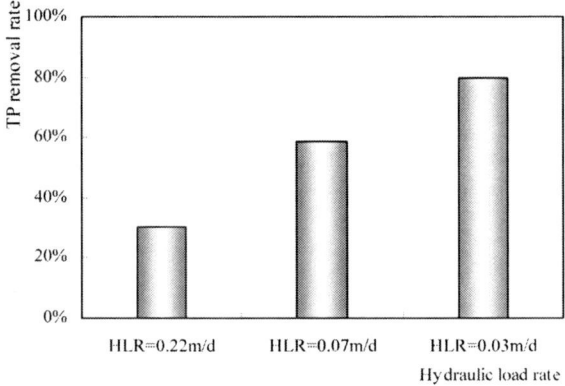

Figure 3 TP removal rate at different hydraulic loads

CONCLUSION

Three kinds of architecture walling waste such as waste broken-concrete, clay bricks and foamed insulation bricks were shown to have an effective pollutant removal performance when used as media for a constructed wetland. When foamed insulation bricks were used as the main media, the nitrogen and phosphorus removal efficiency was tested with the secondary effluent from a municipal sewage treatment plant as inlet water. The results indicated that the constructed wetland could significantly decrease the nutrient content of the secondary effluent. The TP removal rate could reach 70% when the inlet TP concentration was 1.0 mg/L, and the TN removal rate could reach 30% when the TN concentration was 19.7 mg/L in the influent. In summary, the waste walling materials that were selected could be used in the construction of constructed wetlands.

ACKNOWLEDGMENT

The authors would like thank Centre for Asian Studies in Tsinghua University for financial support in the study.

REFERENCES

Drizo A., Frost C. A., Grace J. and Smith K. A. (1999). Physico-chemical screening of phosphate-removing substrates for use in constructed wetland systems. *Water Research*, **33**(17), 3595–3602.

Ministry of Environmental Protection of China. (2002). *Water and Wastewater Monitoring and Analysis Methods*. China Environmental Science Press, Beijing.

Sun Y. (2003). Use and handling of construction waste. *Building Technique Development*, **8**(1), 71–73. (in Chinese)

USEPA Office of Research and Development. (2000). *Manual of Constructed Wetlands Treatment of Municipal Wastewaters*, Cincinnati.

Yuan D., Jing L. J., Gao S. X., Yin D. Q. and Wang L. S. (2005). Analysis on the removal efficiency of phosphorus in some substrates used in constructed wetland systems. *Environmental Science*, **26**(1), 51–55.

Mass Balance and Energy Consumption Calculation in Partial Nitrification Process

Cao Jiashun[a,b], Cai Jianming[a], Fang Fang[b], Zhou Bibo[c] and Jiang Leina[b]

[a]National Engineering Research Center of Water Resources Efficient Utilization and Engineering Safety, Hohai University, Nanjing 210098
[b]Key Laboratory for Integrated Regulation and Resources Exploitation on Shallow Lakes, Ministry of Education, Hohai University, Nanjing 210098
[c]Designing institute of Hohai University, Nanjing 210098
E-mail: caojiashun@163.com

Abstract Simultaneous nitrification and denitrification (an average efficiency at 77.83%) have been achieved by a modified integrative A_n/O bioreactor in treating municipal wastewater. In this study, the existence of partial nitrification was verified in aeration area by water quality monitoring and mass balance calculations of carbon in stable phase as well as nitrogen. The value of nitrogen removal through partial nitrification was 75.78%. Through the calculation of different denitrification forms, all kinds of denitrification processes have been quantified. Compared to the traditional activated sludge process, the process in the study could save oxygen demand and carbon sources corresponding to 81.76% and 47.36%, respectively, by an energy consumption analysis.

Keywords A_n/O bioreactor, partial nitrification, SND, mass balance, energy consumption

INTRODUCTION

Nitrite accumulation during nitrification was discovered by Voets (Voets, *et al.*, 1975) during research of high ammonia nitrogen wastewater from 1975, and the conception of shortcut nitrification and denitrification was thus first brought forward. Partial nitrification was a control step of reaction velocity. Nitrification rested on the nitrite reaction stage and straight into denitrification without the conversion of nitrite into nitrate. Restraining of NOB (nitrite-oxidizing bacteria) was the key point of shortcutting nitrification and denitrification in order to lead to nitrite accumulation steadily during nitrification. Because of the synergistic multiplication effects of AOB (ammonia-oxidizing bacteria) and NOB, NOB could not be excluded solely. Nitrite accumulation could only be achieved through quantity or activity imbalance of AOB and NOB in hybrid systems. Research on partial nitrification mainly focuses on SBR all over the world (Blackburne, *et al.*, 2008; Sánchez, *et al.*, 2008), but less on continuous flow reactors.

© 2010 IWA Publishing. *Water Infrastructure for Sustainable Communities: China and the World*. Edited by Xiaodi Hao, Vladimir Novotny and Valerie Nelson. ISBN: 9781843393283. Published by IWA Publishing, London, UK.

Simultaneous nitrification and denitrification (SND), that nitrification and denitrification reactions were concurrent in the same reactor and same operation conditions have been reported all over the world (Masuda *et al.*, 1991; Hyungseok *et al.*, 1999; Lu, 2002; Yang *et al.*, 2002). Research on SND has always concentrated on SND via nitrate, but much less on SND via nitrite, and many rested on bench scales. SND via nitrite was a new biological nitrogen removal process with the collective advantage of the two processes mentioned above (Ruiz *et al.*, 2006; Guo *et al.*, 2009). However, limited reports were available on comparisons of SND via nitrite in continuous flow, of which most used synthetic wastewater as the object of study. It was still doubtful whether SND via nitrite would achieve a satisfactory performance in a large-scale plant using municipal wastewater. In particular, the amount of partial nitrification could not be quantified by traditional mass balances, with the result of causing the biological nitrogen removal process to be ambiguous for researchers. Moreover, DO concentration was one of the most essential control parameters for economical and practical considerations. Previous studies had shown that low dissolved oxygen (DO) concentration was benefit for partial nitrification as well as SND (Garrido *et al.*, 1997; Kuai *et al.*, 1998; Chuang *et al.*, 2007). However, the amount of energy being saved could not be quantified, also due to reasons mentioned above. So an improvement of mass balance was necessary in order to answer these questions. Thus, denitrification modes of analysis and energy consumption calculation could be carried out.

In our study, we applied a modified integrative A_n/O process in order to explore the application of the process in WWTP around Taihu Basin and quantify the contribution of partial nitrification to the efficiency of nitrogen removal. The process was a new intensive and effective wastewater treatment process combined with nitrogen, phosphorus and organic matter removal. It was a transmutation of the A^2O process combined with the A_n/O (oxygen-limited) bioreactor and sedimentation tank. An air lift was used as a substitute for a conventional submerged stirrer, and moreover, soft tubular micropore aeration instead of traditional discal or tubular micropore aeration was laid on the bottom of the aeration zone. Primary sedimentation tank effluent of Lucun WWTP in Wuxi was studied. The existence of shortcut simultaneous nitrogen removal was verified through the process of carbon and nitrogen mass flow analysis in this paper. At the same time, the effect of the energy-saving technology was revealed through the process of energy consumption analysis.

MATERIALS AND METHODS
Pilot Plant

In this study, the reactor integrated inclined lamellar sedimentation and A_n/O bioreactor was used as equipment, which was altered by a container with a size of $12.03 \times 2.35 \times 2.39$m $(L \times B \times H)$. Available volume and water depth of the bioreactor was 58m³ and 2.19m, respectively (Figure 1). A submersible pump was used for influent, and the airlift was used for mixing liquid and sludge recycles. An online dissolved oxygen monitor was settled at the end of the aeration zone. Valve opening of the blower could be adjusted by the enactment of DO range. The sludge of the pilot was operated at low DO (0.05–0.45 mg·L⁻¹). The average mixed liquor suspended solids (MLSS) concentration was within a range of 5–8 g/L during the experimental period. The influent quantity and hydraulic retention time (HRT) was controlled at 5m³/h and around 10h, respectively. The sludge of the pilot plant was supported by the aeration tank of Lucun WWTP.

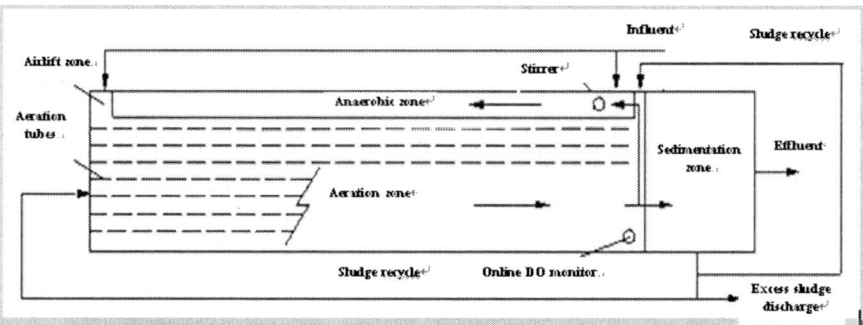

Figure 1 Schematic diagram of pilot plant

Wastewater Quality

The influent of the experiment was the primary sedimentation tank effluent of Lucun WWTP, including about 40% industrial wastewater and 60% municipal wastewater. The wastewater characteristics are described in Table 1.

Table 1 Wastewater characteristics

	Range (mg/L)	Average value		Range (mg/L)	Average value
pH	6.92~7.83	7.49	TN	31.3~67.0	47.7
COD	254~948	536	NH_4^+–N	22.4~45.8	33.1
BOD_5	85.6~276	164	NO_3^-–N	0.011~0.037	0.017
SS	100~608	351	NO_2^-–N	0.014~0.060	0.029

Analytical Methods

The Chemical oxygen demand (COD), Biological oxygen demand (BOD_5), Total nitrogen (TN), NH_4^+-N, NO_3^--N, NO_2^--N, Suspended solids (SS), MLSS and Mixed liquor volatile suspended solids (MLVSS) were measured according to Standard Methods (APHA, 1998). Water temperature and DO were obtained by use of a mercury thermometer and online DO monitor (ZULLIG), respectively.

MASS BALANCE

Mass Flow

Mass balance (Nowak *et al.*, 1999) was an effective approach to evaluating the operation conditions which responded to the operation and process efficiency of the entire WWTP. The existence of nitrogen removal via partial nitrification during stable phase could be illustrated by carbon and nitrogen mass balance. Mass flow in the study is shown in Figure 2.

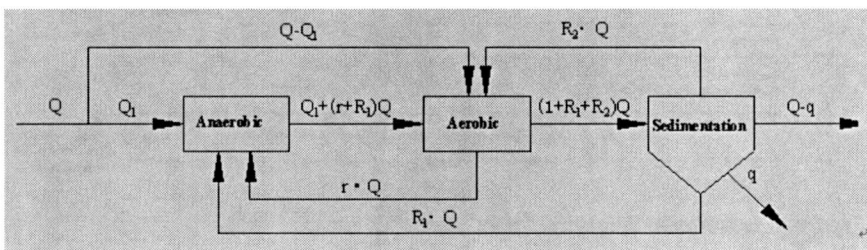

Figure 2 Mass flow of the pilot plant during stable phase: $Q_1=0.3Q$; $r=400\%$; $R_1=15\%$; $R_2=35\%$.

Formula of OU_C

The formula of OU_C (total mass of COD oxidized in the system) could be expressed as follows when nitrification was fully implemented via nitrate:

$$\left\{ \begin{array}{c} OU_C = OU_{C,O2} + OU_{C,ND} \\ OU_{C,O2} = O_{OU} \times V_{aer} \times 24/1000 - 4.57 \times O_{Nox} \\ OU_{C,ND} = 2.86 \times M_{dent,T} \end{array} \right. \qquad \text{Eq. (1)}$$

Where 4.57 represents nitrification (via nitrate) oxygen demand (NOD); and 2.86 is oxygen equivalence factor for nitrate denitrification completely.

Calculation of Nitrite Accumulation Ratio

The nitrite accumulation ratio (NAR) was calculated according to the following equation:

$$NAR(\%) = \frac{[NO_2^-]}{[NO_2^-]+[NO_3^-]} \times 100\% \qquad \text{Eq. (2)}$$

Calculation of SND Efficiency

The efficiency of SND was calculated as follows (Katie $et\ al.$, 2003):

$$SND(\%) = \left(1 - \frac{[NO_{x\ produced}^-]}{[NH_{4\ removal}^+]}\right) \times 100\% \qquad \text{Eq. (3)}$$

Oxygen Transfer Efficiency

The oxygen transfer efficiency could be calculated using Eq. 4 (Feng $et\ al.$, 2007):

$$\frac{E_{p2}}{E_{p1}} = \beta \cdot \left(\frac{h_2}{h_1}\right)^{\frac{2}{3}} \cdot \frac{N_1}{N_2} = 0.9 \cdot \left(\frac{h_2}{h_1}\right)^{\frac{2}{3}} \cdot \frac{1}{1} = 0.9 \cdot \left(\frac{h_2}{h_1}\right)^{\frac{2}{3}} \qquad \text{Eq. (4)}$$

Where E_p represents Oxygen transfer efficiency; β is a coefficient, $\beta=0.9$; h is depth; N means theoretical power.

RESULTS AND DISCUSSION

Nitrogen Removal Performance

Low DO was a critical factor in achieving SND (Pochana and Keller, 1999; Zhao $et\ al.$, 1999), as well as partial nitrification (Abeling $et\ al.$, 1992; Ma $et\ al.$, 2009). It was also beneficial for higher oxygen transfer and became one of the most cost-effective and sustainable biological nitrogen removal processes. Nitrite accumulation ratio, SND efficiency and TN removal efficiency were discussed in Figure 3. Each experimental point was obtained in steady-phase conditions after stabilizing DO concentration. The 1st day in the cycles was the 71st day of the research. A high ammonia accumulation was achieved in the study and NAR varied from 55.50% to 98.15%. Generally, nitrification mainly occurred in the aerobic zone, while denitrification was mostly achieved in the anoxic zone. However, the modified integrative A_n/O bioreactor without an anoxic zone in this study showed TN removal efficiency in the range of 58.09–87.06%. In fact, a large amount of denitrification happened in the aerobic zone by chemical detection. The uniform distributed aeration tubes under the aeration zone made dissolved oxygen evenly distributed, creating a relatively stable environment for microbial growth. It was in an anoxic state for the activated sludge aeration zone in which a large environment is concerned. At the same time, it was full of oxygen in the micro-aerobic environment. In this condition, the aerobic zone was conducive to nitrification bacteria and denitrification bacteria coexisting simultaneously. Simultaneous nitrification and denitrification occurred and displayed a wonderful efficiency, which showed an average efficiency of 77.83%. Denitrification and partial nitrification efficiency could not be fully explained, though good removal efficiency had been achieved through chemical monitoring. Thus, mass balance was needed for further illumination.

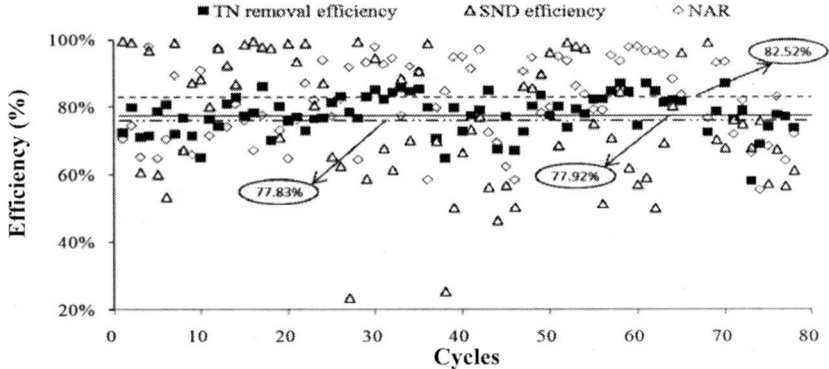

Figure 3 Nitrogen removal performance during stable phase

Mass Balance

An important aspect about steady-state data was that the system should be operated for an extended period necessary to obtain a "steady" operating condition. Even in this condition there will be fluctuations of monitored parameter values from day to day. Therefore, it was necessary to use average data obtained over an extended period after attaining steady state for mass balance calculation. This will account for small fluctuations in response. Also, averaging over an extended period avoided the necessity to include accumulation terms in mass balance calculations.

Nitrogen mass balance (Barker *et al.*, 1994) mainly includes nitrogen entering the system (influent nitrogen) and leaving the system in three parts as effluent, denitrification and excess sludge discharge. Analysis of nitrogen mass balance was shown in Figure 4. Taking into account the system in the aeration zone of nitrification (ammonia nitrogen having been removed) as well as denitrification. Therefore, the traditional model of the anoxic-aerobic system was not suitable for calculating the gross amount of TN removal via denitrification to N_2.

Two components (effluent nitrogen and sludge containing nitrogen content were known by chemical experiments) could be utilized for speculating the amount of N_2 according to mass balance. Nitrogen removal efficiency shown in Figure 3 was 77.92%. Effective denitrification was demonstrated by denitrification efficiency at 58.47%. In accordance with the McKinney theory (Zhang and Lin, 2000) on relations among three kinds of microbial metabolic activities, one-third of the degradable organic compounds in influent (expressed in BOD) were oxidized to inorganic substances and energy, and the others were synthesized to new cells. About 20% of indecomposable residues was discharged with excess sludge through endogenous metabolic reactions. Thus, BOD for the life of the cells were only 2/3×80%=53.33% to total. Average BOD was 164 mg/L according to the measurement during stable phase. The quantities of BOD for cells behavior were 53.33%×164=87.46mg/L. It was useful to consider the assumptions made in performing a COD balance on a denitrification system. The implicit assumption with an oxygen equivalence factor of 2.86 for nitrate denitrified was that denitrification was achieved completely via nitrate. If denitrification was carried out via nitrite, fewer electrons would be transferred per unit nitrate denitrified, and the equivalence factor would be 1.71.

$$NO_3^- + 5H \rightarrow 1/2N_2 + H_2O + OH^- \qquad NO_2^- + 3H \rightarrow 1/2N_2 + H_2O + OH^-$$

If the nitrogen was deoxidized in the form of nitrate, the effluent BOD requirements for deoxidizing per liter of nitrogen required $47.7 \times 58.47\% \times 2.86 = 79.77$mg/L (where 47.7 mg/L represents the average concentration of influent TN). Similarly, if the nitrogen was deoxidized in the form of nitrite, the effluent BOD requirements for deoxidizing per liter of nitrogen required $47.7 \times 58.47\% \times 1.71 = 47.69$mg/L. Although the DO concentration at the end of the aeration zone was lower than 0.45mg/L, biodegradation of organic matter was very active due to 77.4% of the TN removal efficiency. If all the nitrogen was removed via nitrate, the calculation result showed that only 7.69 mg/L of organic matter was used to meet the needs of phosphorus removal (the study for TP average removal efficiency was 87.98% which was temporarily absent from this discussion). Such a high nitrogen and phosphorus removal efficiency obtained by the activated sludge system was obviously contrary to the necessary amount of organic matter. Therefore, a considerable portion of nitrogen was converted to nitrogen gas through denitrification from nitrite directly but not nitrate in the SND process. However, the specific rates of SND via nitrite here were still unable to be determined (these will be discussed later).

COD mass balance was generally used to describe carbon mass balance. Carbon balance could help to have an accurate grasp of the sludge load, the required amounts of oxygen and other important process control parameters. COD entering the system mainly consisted of the influent COD and COD generated by autotrophic nitrification bacteria growth. COD leaving the system consisted of three parts: excess sludge discharge, effluent and CO_2 emissions transformed by organic substrates. Analysis of the carbon mass balance was shown in Figure 5.

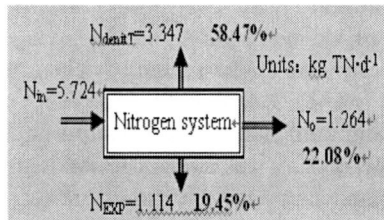

Figure 4 Quantitative graphs of nitrogen mass balance

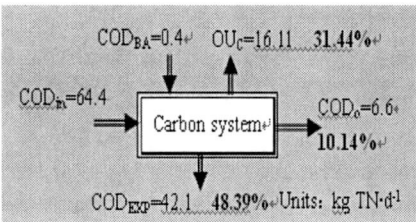

Figure 5 Quantitative graphs of carbon mass balance

From the carbon mass balance analysis, COD left the system by effluent, excess sludge and organic substrate converted to CO_2 accounting for 10.14%, 48.39% and 24.89%, respectively. The excess sludge accounted for a large part of COD, which might be due to the large amount of SS in influent. This part of the SS was finally disposed of by sedimentation. Because the nitrification and denitrification processes were both carried out in the low aerobic zone, the conventional method of mass balance made it difficult to measure COD oxidization in concrete terms, whether the electron acceptor was nitrate or nitrite (i.e. the electrons which were transferred from the organic material to the electron acceptor). Additionally, a partial nitrification phenomenon was verified to occur in the system. So it is hard to identify the part of the COD using oxygen as an electron acceptor that converted from organic substrate to CO_2. As COD departing from the system by effluent and excess sludge discharge was known, COD oxidized from organic substrate converting to CO_2 (we called it OU_C) could be obtained by mass balance. The following calculation of the nitrogen removal efficiency via shortcut nitrification and denitrification could be carried out by the backstepping method of OU_C. The formula of OU_C had been shown in Section 2.4.2, when nitrification was achieved via nitrate completely.

The present study involved partial nitrification, but whether nitrification occurred via nitrite entirely could not be ensured. So setting up an unknown percentage of θ (θ ranges between 0 and 1) in order to indicate the amount of nitrification via nitrite, then $(1-\theta)$ could be expressed as the amount of nitrification via nitrate.

The formula for OU_C changed to:

$$\left\{ \begin{array}{c} OU_C = OU_{C,O2} + OU_{C,ND} \\ OU_{C,O2} = O_{OU} \times V_{aer} \times 24/1000 - [1.71 \times \theta + 4.57 \times (1-\theta)] \times O_{Nox} \quad \text{Eq. (5)} \\ OU_{C,ND} = [1.71 \times \theta + 2.86 \times (1-\theta)] \times M_{dent,T} \end{array} \right.$$

where $OU_C = 16.11$kg $COD \cdot d^{-1}$; $O_{OU} = 17.1$ O_2 mg/(L·h); $V_{aer} = 43.2$m^3;

$$O_{NOx} = 3.444 \text{kg N/d}; \quad M_{dent,T} = 3.347 \text{kg N/d}.$$

Substituted into the above-mentioned values were: $\theta = 75.78\%$. Nitrogen removal via the shortcut way accounted for 75.78%, and the other 24.22% were removed via the complete way.

It can be seen from the study that the average BOD/TP$=164/14.2=11.5<17$ showed an inadequate carbon source for phosphorus removal, the average BOD/TKN$=164/(47.7-0.2)=3.45<4$ made it also difficult to satisfy the demands of nitrogen removal. The calculation results revealed that a deficiency of organic carbon would affect the system's capabilities, but the actual running results proved to be a satisfactory nitrogen and phosphorus removal efficiency which had excellent relationships with the above-mentioned partial denitrification.

Analysis of Denitrification Modes

Nitrification modes could be divided into partial nitrification (nitrification via nitrite) and complete nitrification (nitrification via nitrate) according to whether nitrification was complete or not. Similarly, according to whether nitrification and denitrification happened simultaneously, denitrification could be divided into simultaneous nitrification and denitrification (SND) and traditional nitrification and denitrification (TND). Combinations of these two classifications could be described in four different forms of denitrification.

SND efficiency accounted for 77.83% which has been calculated in section 3.1. In accordance with section 3.2 above, nitrogen removed through denitrification accounted for 58.47% and the average rate that partial nitrification contributed to TN removal was 75.78%.

Energy Consumption Analysis

Through the mass balance, energy-saving was mainly achieved on TN removal in a short-cut way. Studies have shown that in the present experimental conditions compared with other activated sludge processes, the process had certain energy-saving advantages.

The process of nitrification was generally carried out in two steps. First step: The oxidation of ammonia into nitrite, $NH_4^+ + 3/2O_2 \rightarrow NO_2^- + 2H^+ + H_2O$. Second step: The oxidation of nitrite into nitrate, $NO_2^- + 1/2O_2 \rightarrow NO_3^-$. A possible way of optimization is to carry out partial nitrification, consisting in stopping the oxidation of ammonia at the stage of the nitrite NO_2^-, and then to treat nitrites by denitrification. In this way, it is possible to save 25% of the oxygen uptakes. This study maintained low DO conditions (the maximum DO was under 0.45 mg/L) compared to DO in the aeration zone of traditional processes of mostly higher than 2.0 mg/L. The results showed a significant energy-saving effect. In addition, compared to conventional activated sludge processes, the plant at least saved the volume of O_2: $1 - 0.45/2 \times (1 - 75.78\% \times 25\%) = 81.76\%$. The aeration tank was the largest energy consuming structure in WWTP, and the Blower system accounted for 40~50% (calculated as 40%) in general of the total power consumption. The process saved power consumption: $40\% \times 81.76\% = 32.70\%$.

Denitrification started directly from the nitrite in shortcut nitrogen removal process could save $2.86/4.57 = 62.5\%$ of the carbon source and increase the efficiency of denitrification. In this study, compared to conventional activated sludge processes, the carbon source for nitrogen removal could save $75.78\% \times 62.5\% = 47.36\%$, corresponding to the denitrification rates. It could also save plant volume as well as area and construction cost.

In the stable phase of the study, oxygen transfer efficiency was at 29.53% on average (Figure 6). Considering the depth of the pilot plant was only 2.19m, calculation as the application works according to the actual water depth of 6m, the efficiency of oxygen transfer could achieve at: $E_{p2}=0.9$ $(h_2/h_1)^{2/3}$ $E_{p1}=0.9 \times (6/2.19)^{2/3} \times 29.53 = 52.04\%$.

As a result of SND via nitrite, alkalinity consumption during nitrification and production during denitrification mostly occurred in the aeration zone. Neutralization of the two processes was able to effectively maintain pH stability, which could also reduce the concentration of nitrate nitrogen in order to reduce the secondary settling tank sludge floating and sludge recycle on the effects of anaerobic phosphorus release.

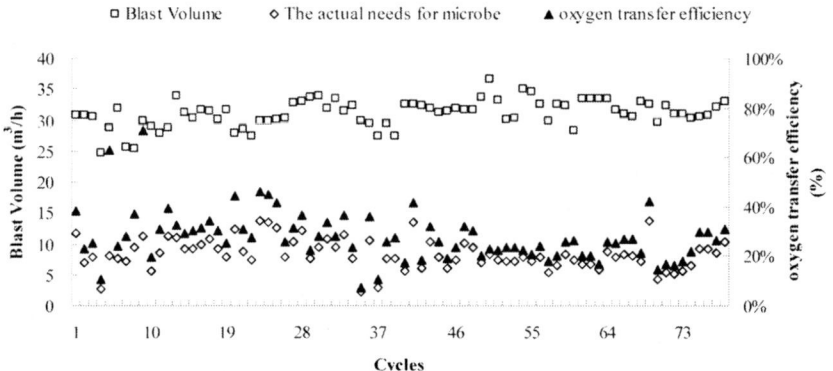

Figure 6 Analysis of oxygen transfer efficiency

CONCLUSIONS

In the pilot plant research, SND efficiency, mass flow, denitrification modes and energy consumption were investigated to analyze shortcut simultaneous nitrogen removal. Denitrification accounted for 58.47% of TN removal efficiency by component analysis of nitrogen, which predicated denitrification efficiency. Through the McKinney theory, shortcut nitrogen removal might happen in the system. Emissions from the excess sludge accounted for 48.39% of COD by component analysis of carbon due to high SS in influent. Moreover, 75.78% as an average rate of denitrification efficiency was contributed to shortcut nitrogen removal. It was concluded from the calculation of SND efficiency that 77.83% of nitrogen removal was achieved by SND. Among these, four kinds of denitrification, including traditional/simultaneous nitrification and denitrification via nitrate or nitrite, have been quantified. 34.49% of TN was removed by SND via nitrite. Compared to the traditional activated sludge process, a modified integrative A_n/O bioreactor could save carbon sources for denitrification, oxygen demand and power consumption at an average rate of 47.36%, 81.76% and 36.79%, respectively. Because of depth restrictions in the pilot plant, oxygen transfer efficiency did not produce a satisfactory result.

ACKNOWLEDGEMENTS

This research was financially supported by "Eleventh Five-Year" national scientific and technological project "Research of saving energy and reducing consumption on key technologies of Municipal Wastewater Treatment Plant A2O Process" (2006BAC19B01) and Jiangsu Province Natural Science Foundation of China (BK2007181). We also wish to thank the staff of Lucun WWTP, for their support.

REFERENCES

Abeling U. and Seyfried C. F. (1992). Anaerobic-aerobic treatment of high-strength ammonium wastewater—nitrogen removal via nitrite. *Water Science and Technology*, **26**(5–6), 1007–1015.

APHA (1998). *Standard Methods for the Examination of Water and Wastewater*, 20th edn, American Public Health Association, American Water Works Association and Water Environment Federation, Washington DC, USA.

Barker P. S. and Dold P. L. (1994). COD and nitrogen mass balances in activated sludge systems. *Water Research*, **29**(2), 633–643.

Blackburne R., Yuan Z. and Keller J. (2008). Demonstration of nitrogen removal via nitrite in a sequencing batch reactor treating domestic wastewater. *Water Research*, **42**(8–9), 2166–2176.

Chuang H. P., Ohashi A., Imachi H., Tandukar M. and Harada H. (2007). Effective partial nitrification to nitrite by down-flow hanging sponge reactor under limited oxygen condition. *Water Research*, **41**(2), 295–302.

Feng J. S. and Wan Y. S. (2007). A relation between blasting aeration performance and submersion depth of aerator. *Environmental Engineering*, **25**(1), 19–21. (in Chinese)

Garrido J. M., van Benthum W. A., van Loosdrecht M. C. and Heijnen J. J. (1997). Influence of dissolved oxygen concentration on nitrite accumulation in a biofilm airlift suspension reactor. *Biotechnology and Bioengineering*, **53**(2), 168–178.

Guo J. H., Peng Y. Z., Wang S. Y., Zheng Y. N., Huang H. J. and Wang Z. W. (2009). Longterm effect of dissolved oxygen on partial nitrification performance and microbial community structure. *Bioresource Technology*, **100**(11), 2796–2802.

Kuai L. and Verstraete W. (1998). Ammonium removal by the oxygen limited autotrophic nitrification-denitrification system. *Applied and Environmental Microbiology*, **64**(11), 4500–4506.

Lu X. W. (2002). Theory and practice of simultaneous nitrification and denitrification. *Environmental Chemistry*, **21**(6), 564–570. (in Chinese)

Ma Y., Peng Y. Z., Wang S. Y., Yuan Z. G. and Wang X. L. (2009). Achieving nitrogen removal via nitrite in a pilot-scale continuous pre-denitrification plant. *Water Research*, **43**(3), 563–572.

Masuda S., Watanabe Y. and Ishiguro M. (1991). Biofilm properties and simultaneous nitrification and denitrification in aerobic rotating biological contactors. *Water Science and Technology*, **23**(7–9), 1355–1363.

Nowak O., Franz A., Svardal K., Muller V. and Kuhn V. (1999). Parameter estimation for activated sludge models with the help of mass balances. *Water Science and Technology*, **39**(4), 113–120.

Pochana K. and Keller J. (1999). Study of factors affecting simultaneous nitrification and denitrification (SND). *Water Science and Technology*, **39**(6), 61–68.

Ruiz G., Jeison D., Rubilar O. and Chamy R. (2006). Nitrification denitrification via nitrite accumulation for nitrogen removal from wastewaters. *Bioresource Technology*, **97**(2), 330–335.

Sánchez O., Bernet N. and Delgenès, J.-P. (2007). Effect of dissolved oxygen concentration on nitrite accumulation in nitrifying sequencing batch reactor. *Water Environment Research*, **79**(8), 845–850.

Third K. A., Burnett N. and Cord-Ruwisch R. (2003). Simultaneous nitrification and denitrification using stored substrate (PHB) as the electron donor in an SBR. *Biotechnology and Bioengeenering*, **83**(6), 706–720.

Voets J. P., Vanstaen H. and Verstraete W. (1975). Removal of nitrogen from highly nitrogenous wastewaters. *Water Pollution Control*, **47**(2), 394–398.

Yang Q., Liu S. Q. and Gan S. Y. (2002). Study on simultaneous nitrification and denitrification (SND) in carrousel oxidation ditch. *Chinese Journal of Environmental Sciences*, **23**, 40–43. (in Chinese)

Yoo H., Ahn K. H., Lee H. Y., Lee K. H., Kwak Y. J. and Song K. G. (1999). Nitrogen removal from synthetic wastewater by simultaneous nitrification and denitrification (SND) via nitrite in an intermittently-aerated reactor. *Water Research*, **33**(1), 145–154.

Zhang Z. J. and Lin R. Z. (2000). *Wastewater Engineering*. China Architecture & Building Press, Beijing, China. (in Chinese)

Zhao H. W., Mavinic D. S., Oldham W. K. and Koch F. A. (1999). Controlling factors for simultaneous nitrification and denitrification in a two-stage intermittent aeration process treating domestic sewage *Water Research*, **33**(4), 961–970.

A Study on a Coupling Bioreactor for the Treatment of Domestic Wastewater and Mechanisms of Sludge Reduction

Li Jun [a], Wang Chunrong [b], Zhou Ting [a], Zhang Xuesong [a] and Li Zebing [a]

[a]Key Laboratory of Beijing for Water Quality Science & Water Environment Recovery Engineering, Beijing University of Technology, Beijing, 100022, China
[b]China University of Mining and Technology, Beijing,100083, China
E-mail: jglijun@bjut.edu.cn

Abstract Based on the separate flow theory and the multi-phase reaction principle, this paper examines the use of a coupling bioreactor for wastewater treatment and sludge reduction. Results showed that the coupling bioreactor sustained a COD removal efficiency of at least 85%. The COD loading rate increased from 0.49kgCOD/ (m³·d) to 1.93kg COD/ (m³·d). By changing operation parameters, it was found that an excellent effluent quality was obtained. When under Micro-aeration, Anoxic and Aeration conditions, a total nitrogen (TN) removal rate of 86.7% was obtained. With HRT 8h and 43% volume ratio of anoxic section, it has an effluent TN concentration lower than 15mg. With the coupling bioreactor, a lower sludge production rate also took place. There was only a mean sludge production rate of 0.068MLSS/kgCOD. In the liquid phase of the dynamic cyclic closed reactor, TN, TP and TC all increased. This was accompanied by the emission of CH_4 gas, from which it can be inferred that the sludge trapped by porous carriers must have biodegraded under the conditions of the inner anaerobic environment. Thus, less or even no sludge was produced.

Keywords Coupling bioreactor, Sludge reduction, Flow separate, Porous carrier

INTRODUCTION

Biological wastewater treatment, particularly the activated sludge system, has been applied to wastewater plants all over the world. However, problems arise due to its higher yielding of sludge. The sludge produced, needs separating, condensing, digesting, dehydrating, and further disposal. This results in higher capital and operating costs. It accounts for about 40% (drying) and 65% (incineration) of the total cost of the wastewater treatment plant [Low, 1999; Campbell, 2000]. As a result of these expensive costs, very few wastewater plants in China have put into operation adequate facilities for the removal sludge. In addition, the domestic wastewater treatment systems in China also have poor facilities for sludge treatment and disposal. This is due to the scattered housing and the need for small scale systems in residential areas. So, the need to find cost-effective methods for the removal of sludge has become one of the most urgent problems in China.

The concept of sludge reduction was put forward in the 1990s. The idea was to discharge the minimum biomass in the whole wastewater treatment system by using physical, chemical, and biological methods. In order to reduce the biomass, in essence, a low production rate of microorganisms was to be used (endogenous respiration, oxidation, anaerobic digestion, and so on) [Lee, 1996; Ghyoot, 2000]. A coupling bioreactor for wastewater treatment was, therefore, developed. Based on the flow separation and multi-phase reaction principles, it mainly investigated the effluent quality and the mechanisms of sludge reduction.

The Theory of Flow-Separation

If there is a flow velocity difference between the two sides of an object and the vectors are the same, the object moves only in one direction. However, an object submerged in water, as a kind of viscous matter, will move forward with high flow velocity, and on the contrary, it will turn into the direction with a slower velocity. So the result of flow is that the object will move and accumulate in the position of the slow velocity during this process.

The phenomena of the flow-separation theory can be commonly observed in daily life: dust that gets attached to the bottom of the tables; silt accumulating on the bottom of a river where there are microbial communities; and objects thrown into a river moving towards the river banks (see Fig. 1).

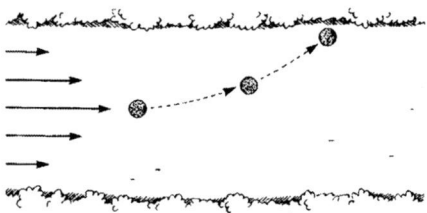

Figure 1 Move of object in the river

MATERIALS AND METHODS

Test Equipment and Materials

The porous carriers

Based on the theory of flow separation, a novel porous carrier was adopted in order to provide a flow-separation field to capture suspended matter. It also provides a survival site for the microbial community. The porous carrier is a stony spherical structure which has a diameter of 10cm. Its porosity is as high as 0.6 (see Fig. 2). Thus, a greater biomass can be easily accumulated in it. It also provides a fluid field for capturing particles and an environment for all kinds of living microorganisms. When wastewater is passed through the porous carrier, due to resistance, the velocity of the water in the inner carrier is slowed. However, the velocity between the porous carriers remains fast. Particles in the wastewater, therefore, move from the place with the highest velocity to that of the lowest velocity. The particles are then accumulated into the pores. Through the numerous movement and accumulation processes, solid particles are separated from the water.

Figure 2 Test media

Test equipment

A plug-flow bioreactor and a dynamic cyclic closed reactor will be set up for wastewater treatment and sludge reduction (see Fig. 3).

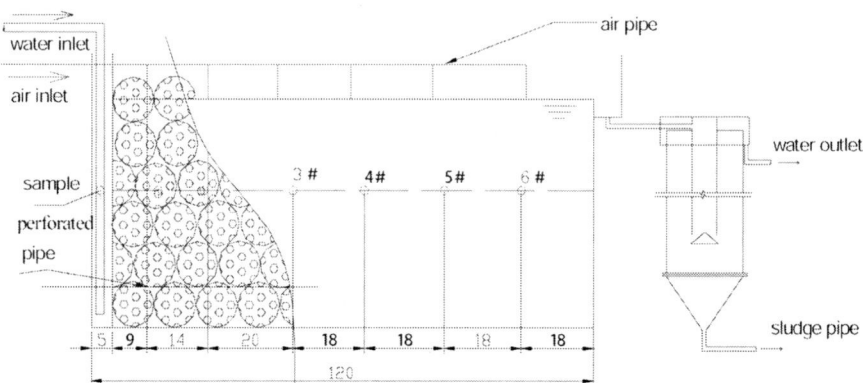

Figure 3 Schematic diagram of experiment

The plug-flow bioreactor that was used in this experiment had dimensions of: length, 120cm, width 25cm, height, 60cm, work height, 50cm and total work volume, 180L (see Fig. 3). 6 perforated pipes were laid out along the length, and 7 sampling points, each with a diameter of 1cm, were set along the length: #0, #1, #2, #3, #4, #5, #6, and #7. This was 30cm high from the bottom of reactor.

The dynamic cyclic closed reactor had a diameter of 18cm, a height of 50cm and a funnel height of 20cm. It had a total volume of 17L and a work volume of 13.8L. There were 4 sampling ports along the lateral wall of the reactor, each with a diameter of 1cm. Meanwhile, the reactor was closed without oxygen, that is to say, in anaerobic conditions. 8 porous carriers, covered with the anaerobic biofilm were selected from the push-flow reactor. The wastewater then flowed circularly by using a pump. In addition, a gas sampling port was used with a water seal on top of the reactor for investigating the emission of gas.

Wastewater and Test Methods

Wastewater

The wastewater used was domestic wastewater, taken from Beijing University of Technology. The quality of the wastewater is shown in table 1. It was noted that one characteristic of the wastewater was its low C/N ratio of 3. It has a high NH_4^+–N concentration of 70mg and TN concentration of 100mg/L.

Table 1 Characteristics of wastewater

Item	COD	BOD5	NH_4^+–N	SS	TN	pH	Alkalinity
Range	200–400	95–215	50–90	60–150	90–110	7.2–8.0	320–500

Test methods

The experiment was carried out in the municipal engineering laboratory of Beijing University of Technology. The test methods for all the items were carried out based on the "Monitoring Analysis Method of Water and Wastewater" [China NEPA, 1989].

Scheme for the experiment

During the experiment, different operating modes were studied, including:

Operating mode 1: Complete-aeration (with different air-flow of 0.8m³/h, 0.6m³/h and 0.4m³/h);

Operating mode 2: Micro-aeration (with DO of 1~2mg/L) + anoxic + aeration (with the mean DO of 3.5mg/L);

Operating mode 3: Return process without adding carbon source;

Operating mode 4: Return process with carbon source addition;

At each process, different parameters were adjusted, such as the volume percentage of the anoxic section, the HRT, the influent organic loading, the DO, the C/N and so on. This was done in order to obtain the optimal conditions for removing nitrogen as well as lowering sludge yields. The results show that an excellent effluent can bé obtained in the operating mode of micro-aeration, anoxic and aeration. These conditions are particularly effective in terms of nitrogen removal efficiency. The treatment efficiency was, therefore, mainly studied in operating mode 2 in this paper.

RESULTS AND DISCUSSION

The Efficiency at Different HRT

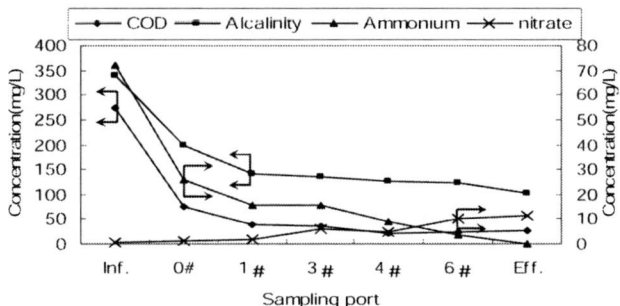

Figure 4 The variation of COD, NH$_4^+$–N and NO$_3^-$–N along the sampling port at HRT=13h

During the experiment, under the conditions of operating mode 2, micro-aeration was controlled from sampling port #2. Here, the DO 1~2mg/L, can help nitrification and lower consumption of carbon sources. The anoxic zone of 28% was obtained with #3 aerator shut off; the air was then passed through from #4 to #6 sampling port with 3.5mg/L DO. Next, the test was carried out at different HRT(13h, 8h and 5.3h). The results are shown as follows:

Fig. 4 shows the variation of COD, NH$_4^+$–N and NO$_3^-$–N along the sampling port at HRT=13h (the corresponding COD volumetric loading was 0.49Kg/m^3·d). It had a mean aeration intensity of 0.3 m^3/h, from which it was found that the inlet COD was 274.2 mg/L, with the outlet COD of 26.1mg/L and the COD removal efficiency of 90.5%. About 70% alkalinity of raw water was consumed during the experiment with the inlet alkalinity of 340.3 mg/L and the outlet alkalinity of

102.2 mg/L. Meanwhile the outlet NH_4^+–N was 0 mg/L with 100% of NH_4^+–N removal efficiency, while the inlet NH_4^+–N was 72.4 mg/L, and the outlet NO_3^-–N was 11.2 mg/L.

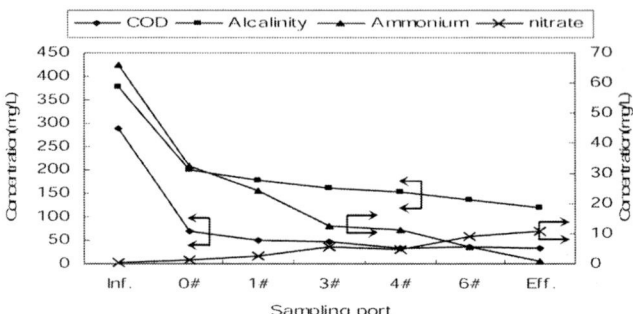

Figure 5 The variation of COD, NH_4^+–N and NO_3^-–N along the sampling port at HRT=8h

Fig. 5 shows the variation of COD, NH_4^+–N and NO_3^-–N along the sampling port at HRT=8h (the corresponding COD volumetric loading was 0.78Kg/m³·d) with the mean aeration intensity of 0.3 m³/h. It was found that the inlet COD was 288.6 mg/L with the outlet COD of 33.2 mg/L and COD removal efficiency of 88.48%. The outlet NH_4^+–N was 0.7mg/L (the inlet NH_4^+–N of 66.3mg/L) with NH_4^+–N removal efficiency of 99%, while the outlet NO_3^-–N was 10.82 mg/L).

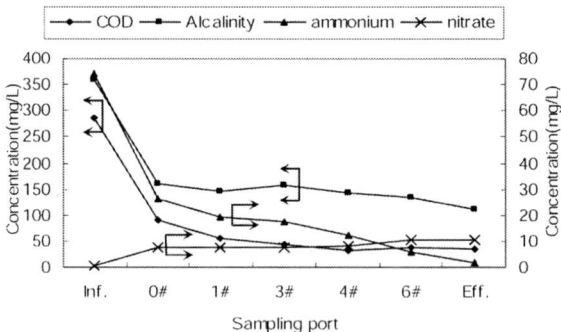

Figure 6 The variation of COD, NH_4^+–N and NO_3^-–N along the sampling port at HRT=5.3h efficiency of 88.48%. The outlet NH_4^+–N was 0.7mg/L (the inlet NH_4^+–N of 66.3mg/L) with NH_4^+–N removal efficiency of 99%, while the outlet NO_3^-–N was 10.82 mg/L).

Fig. 6 is the variation of COD, NH_4^+–N and NO_3^-–N along the sampling port at HRT=5.3h (the corresponding COD volumetric loading was 1.29Kg/m³·d) with the mean aeration intensity of 0.3 m³/h. The results were shown to be as follows: the inlet COD was 285.1 mg/L, with the outlet COD of 34.31 mg/L and COD removal efficiency of 87.97%. The inlet NH_4^+–N was 74.32 mg/L with the outlet NH_4^+–N of 1.8 mg/L and NH_4^+–N removal efficiency of 97.58%. The outlet NO_3^-–N was 10.78 mg/L. During this period, the inlet and outlet alkalinity was 430.486 mg/L and 110.235 mg/L respectively. About 74.39% alkalinity of raw water was consumed.

Compared with the above figures, it was found that the reactor has an excellent COD removal with the HRT of 13h, 8h and 5.3h. The outlet NH_4^+N, obtained at the HRT of 13h and 8h, was less than 1mg/L. However, it was as high as 2.1mg/L at the HRT of 5.3h. Meanwhile the results also showed that an outlet NO_3^-–N of less than 15mg/L was obtained at each HRT. In addition, it was found that, under conditions of micro-aeration, alkalinity was also well indicated. The coupling bioreactor presented an excellent effluent quality (effluent NH_4^+–N< 5mg/L and TN removal efficiency> 75%) with outlet alkalinity in the range of 80mg/L to 150mg/L and about 60% consumption of raw alkalinity. However, the effluent alkalinity increased remarkably, as high as 200mg/L, when a poor effluent quality was obtained.

The Efficiency in Different Volume Percentages of Anoxic Section at HRT=8h

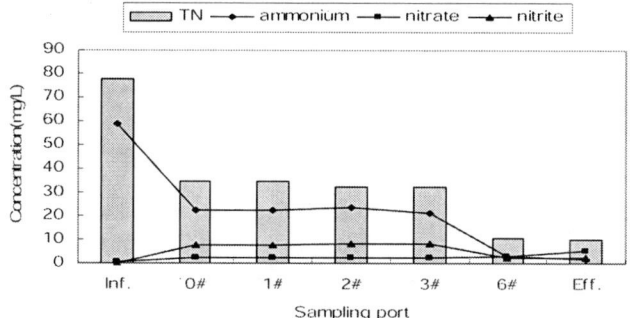

Figure 7 Variation of nitrogen compounds along the sampling ports with the anoxic ozone volume enhanced from 28% to 43%

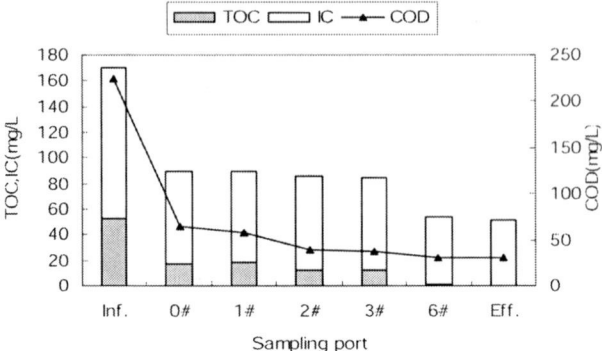

Figure 8 Variation of TOC, IC, COD along the sampling ports with the anoxic ozone volume enhanced from 28% to 43%

When the volume percentage of anoxic section was kept at 28% and HRT=8h, the outlet COD was 33.2 mg/L with 88.5% of COD removal efficiency (inlet COD 288.6mg/L). The outlet NH_4^+–N and NO_3^-–N was 9.7 mg/L and 10.8 mg/L respectively, with the inlet NH_4^+–N of 66.3 mg/L. The TN decreased from 94.2 mg/L to 16.7 mg/L with 82.3% of the TN removal efficiency.

After almost one year in operation, the anoxic section volume increased from 28% to 43%. This is because the aerator failed between the #4 and #5 sampling ports. During this period, the corresponding COD volumetric loading was 0.78Kg/m^3·d , HRT=8h. However, an excellent effluent quality was obtained with 85% and 81.11% of TOC and TN removal efficiency respectively. Moreover, the NO_2^-–N variation was also investigated at steady-state (Fig. 7). It was found that the NO_2^-–N concentration increased gradually from #0 (7.4mg/L)~#1 (7.9mg/L)~#3 (8.1mg/L)~ #6 (8. 3 mg/L). The TN removal efficiency accounted for 28.36% due to no aeration at #4 and #5 sampling ports. It must be noted that no NO_3^-–N accumulation took place during the course of the operation, with NO_3^-–N concentration of figures of: #0 (2.46mg/L); #1 (2.42mg/L); #3 (2.6mg/L) and #6 (2.93 mg/L). However, when aeration was carried out from the #6 sampling port, the effluent NO_3^-–N reached 5.12mg/L. This is because a part of NH_4^+–N was oxidized into NO_3^-–N.

Fig. 8 shows the variation of TOC, IC and COD along the sampling ports with the anoxic ozone volume enhanced from 28% to 43%. It was found that there was no COD removal from #3 to #6 sampling ports, which indicated that there was no conventional heterotrophic de-nitrification in the coupling bioreactor; the biodegradable COD had been exhausted during the course of the former aeration, so the carbon source was not enough for traditional de-nitrification. Compared with Fig. 7, it was found that the nitrogen loss mainly took place between #3 and #6 sampling ports. However, there was gradual variation between #0 and

#3 sampling ports where free ammonia was higher. The NH_4–N concentration reached 21.23mg/L at #3 sampling port, then increased to 3.21mg/L at #6 sampling port. However, the NO_2^-–N concentration was kept at 8mg/L from #1 to #3 sampling ports. It then decreased to 2.16mg/L at #6 sampling port. It was deduced, therefore, that the NO_2^-–N + NH_4^+–N→N_2 took place.

To summarise, excellent de-nitrification and high TN removal efficiency occurred; the volume percent of anoxic section increased from 28 to 43%.

Analysis of Sludge Reduction

The coupling bioreactor ran for 49 days under the conditions of micro-aeration, anoxic and aeration. In this time, a total sludge production of 312.278g was recorded, with a sludge yielding rate of 0.068gMLSS/g COD. The COD loading rate was 0.94Kg/m³·d. Moreover, a lower effluent turbidity was obtained during this period, in the range from 7 to 15NTU. Results recorded an effluent SS of less than 20mg/L.

It was confirmed that porous carriers and the formation of anaerobic environments were the key factors in sludge reduction. The dynamic cyclic closed reactor was, therefore, set up to investigate the mechanisms of sludge reduction.

Closed operation

Fig. 9 demonstrates the variation of the parameters in the dynamic cyclic closed reactor during the course of the operation. The value of each parameter, except IC, decreased within 6 days after the reactor was setup. This is attributed to the biofilm being attached to the porous carrier. Biodegraded organic and inorganic matter was, therefore, increased during the course of the process. Thus the yield of sludge accumulated on the porous carrier was lower, and its degradation was slight without gas emission from water seals.

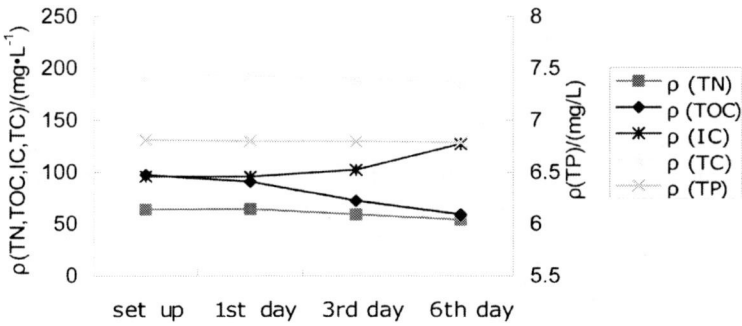

Figure 9 The variation of parameters in dynamic cyclic bioreactor

The Sludge Trap

After one week of operating, sludge was added every 12 days, into the top of the reactor. The phenomenon was investigated in terms of porous carriers (see Fig. 10). The same volume of wastewater was discharged from the reactor before the sludge was added.

Figure 10 The state of sludge trapped by the porous carrier

Fig. 10 shows the sludge state in the reactor, immediately after the sludge was added. The wastewater in it was turbid. However, after 3 days, the sludge was trapped by the porous carriers efficiently and the wastewater clarified (see Fig.10 right).

The variation of matter in liquid phase

After sludge had been regularly added to the reactor, the level of sludge degradation was investigated. 4L of mixed sludge was added every time with an MLSS of 3.13g/L. After 3 operation periods, the parameters in the liquid phase were tested (see Fig. 11).

Fig. 11 shows that TN and TP increased slightly. Its also demonstrates that TOC decreased when IC and TC increased. It can be inferred that, after the sludge degraded, the C in the solid phase was transferred into the gaseous and liquid phase. TP can only exist in solid and liquid phases. The increase of TP concentration in the liquid phase confirmed that the sludge trapped by the porous carrier had degraded under anaerobic conditions. The P in the sludge was partially transferred into the liquid phase.

There are 3 fates for nitrogen in wastewater treatment. One is that nitrogen is absorbed by micro-organisms. Another is that nitrogen still exists in a liquid phase, and the third is that it is emitted in terms of NOx, (even if in very small amounts). The slight increase of TN in Fig. 11, therefore, shows that the sludge trapped by the porous carrier took place through anaerobic decomposition.

In order to confirm the anaerobic decomposition of sludge further, the variation of matter in the gas phase was tested.

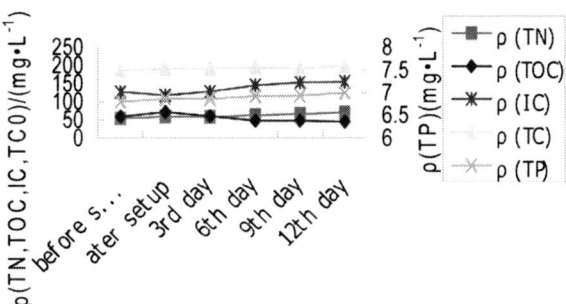

Figure 11 the variations of parameters in liquid phase

CONCLUSION

Based on the separate flow theory and the coupling function in porous carriers, the coupling bioreactor, demonstrates efficient treatment on low C/N domestic wastewater. An excellent effluent quality and a good level of sludge reduction was obtained under the conditions of micro-aeration (DO=1~2mg/L), anoxic and aeration (DO=3.5mg/L).

1. The COD removal efficiency was higher than 82%. With HTR decreased from 13h to 5.3h, and the outlet NO_3^-–N decreased from 30mg/L to 10mg/L, a mean NH_4^+–N effluent of 0.98mg/L can be obtained.
2. A stable COD removal as high as 85% was obtained at HRT=8h with the volume percentage of the anoxic section increased from 28% to 43%. However, the removal efficiency of NH_4^+–N decreased from 99% to 96.8% with the effluent NO_3^-–N decreased from 10.8mg/L to 5.1mg/L. Its TN removal efficiency increased from 82.3% to 86.7%. In summary, as the volume percentage of the anoxic section increased, excellent de-nitrification and high TN removal efficiency took place.
3. The coupling bioreactor showed a total sludge production of 312.278g with a sludge yielding rate of only 0.068gMLSS/gCOD. It demonstrated an effluent turbidity, somewhere in the range of 7 to 15 NTU and an effluent SS of less than 20mg/L was obtained.

4. In the dynamic cyclic reactor, TN and TP increased slightly, and TOC dropped as IC and TC increased. It can be inferred that in the porous carrier, the anaerobic degradation of sludge took place.

ACKNOWLEDGEMENTS

This paper was financially supported by the National Science Foundation of China (50678008), the Science Foundation of Beijing (8092007), and the State Water Pollution Control and Management of Major Special Science and Technology (2008ZX07314-008), (2008ZX07314-009).

REFERENCES

Campbell H. W. (2000). Sludge management-future issues and trends. *Water Science and Technology*, **41**(8), 1–8.
Ghyoot W. and Verstraete W. (2000). Reduced. sludge production in a two-stage membrane-assisted bioreactor. *Water Research*, **34**(1), 205–215.
Lee N. M. and Welander T. (1996). Reducing sludge production in aerobic wastewater treatment through manipulation of the ecosystem. *Water Research*, **30**(8), 1781–1790.
Low E. W. and Chase H. A. (1999). Reducing production of excess biomass during wastewater treatment. *Water Research*, **33**(5), 1119–1132.

Membrane Combination Technique on Treatment and Remediation of Heavy Metals Polluted Water Body

Zhang Lin-nan[a,b]*, Wu Yan-jun[b] and Li Zhenshan[b]

[a]School of Science, Shenyang University of Technology, 110021, Shenyang, China
[b]Department of Environmental Sciences and Engineering, Peking University, Beijing 100871, China
E-mail: zhanglinnan@iee.pku.edu.cn

Abstract Heavy metals have been released in large quantities into the environment due to rapid industrialization and have created a major global concern. The process of urbanization and industrialization is accompanied by increased automobile and industrial emissions of heavy metals to surrounding water bodies. Traditional treatment processes are either incapable of reducing metal concentration to the levels regulated by law or prohibitively expensive and difficult to operate. In this paper, a hybrid process which integrates electrolysis and LPRO was proposed to treat the synthetic wastewater containing Heavy metals ions. Mechnism of treatment and remediation of heavy metals polluted water body was investigated. The influnce of electrolysis voltage, pH, and electrolysis time on the metal recovery efficiencies were studied. Relationship between trans-membrane pressure drop (ΔP), additions ratio, initial Heavy metals concentration on operating efficency ,stability of membrane and the possibility of water reuse were studied.

Keywords Heavy metal, water body, membranes, electro-winning, remediation

INTRODUCTION

Heavy metals have been released in large quantities into the environment due to rapid industrialization and have created a major global concern. The process of urbanization and industrialization is accompanied by increased automobile and industrial emissions of heavy metals to surrounding water bodies (Castelblanque and Salimbeni, 2004, Chang *et al.*, 2007). Copper, cadmium, zinc, nickel, lead, mercury and chromium are often detected in industrial wastewaters, which originate from metal plating, mining activities, smelting, pigment manufacture, printing and photographic industries, etc.(Kadirvelu *et al.*, 2001; Williams *et al.*, 1998; Reeve, 2007). Unlike organic wastes, heavy metals are non-biodegradable and they are also subject to bio-magnification and can be accumulated in living tissues, causing various diseases and disorders (Kamitani and Kaneko, 2007). As a result, there has been a tightening of the regulations for heavy metals in point source wastewater discharges and treatment and remediation of heavy metal wastewater to harmless concentrations is an important goal.

Traditional treatment processes are either incapable of reducing metal concentration to the levels regulated by law (such as precipitation processes) or prohibitively expensive and difficult to operate (such as ion exchange, activated carbon adsorption, liquid membrane extraction , biosorption) (Namasivayam and Senthilkumar, 2002; Wang *et al.*, 2006). Electrochemical method is the most promising processes for heavy metal recovery, however, low current efficiency; high energy consumption were the main problems especially with low concentration heavy metal waste water treatment (Chen G. H, 2004). Membrane separation processes is a powerful process developed to remove various contaminants and gain widespread acceptance in the treatment of industrial and municipal waste-water, as well as in groundwater remediation (Ujang and Anderson, 1998; Juang and Shiau ,2000 ; Ozaki *et al.*, 2002; Petrov and Nenov, 2004)). Membrane process can also increase the concentration of heavy metal content in the retentate water making it easy for electro-deposition recovery of heavy metals. Combination of these two techniques can treat heavy metals polluted waters body efficiently, at the same time, recovery of heavy metals and implementation of reuse of the polluted water can be achieved. Fig1 shows a typical flow sheet of electro-membrane combination treatment process.

In this paper, a hybrid process which integrates electrolysis and LPRO was proposed to treat the synthetic wastewater containing Heavy metals ions(Cu^{2+} as sample ion). Mechanism of treatment and remediation of C^{2+} polluted water body was investigated.

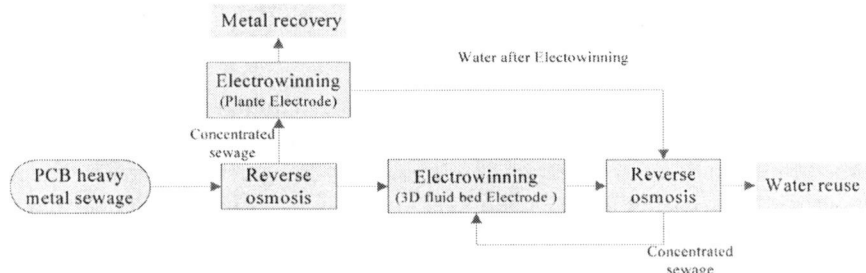

Figure 1 Simplified flow sheet of typical Electro-membrane combination process

Materials and Methods

Water samples were obtained from water body near a print circuit board (PCB) factory in Shenzhen, South China. The water quality was: pH 7–8, COD 100 mg/L, concentration of Cu^{2+} 10–50 mg/L. Synthetic water samples were prepared by adding different amounts of copper sulfates (AR, Merck,German) and other additions(such as EDTA, PPG, and sodium dodecyl sulfate) into distilled water to simulate the real polluted water. The composition of synthetic raw polluted water is shown in Table 1.

Table 1 Composition of synthetic water

Parameter	Range	Mean
pH	7–8	–**
Cu^{2+} (mg/L)	10–100	20
COD* (mg/L)	50–120	100
EDTA (mg/L)	5–20	10

* Caused by EDTA, PPG, and sodium dodecyl sulfate, ** usually around 7.0.

The filtration equipment was manufactured by SAIBO Industry (Wuxi, China). This laboratory scale membrane filtration unit was used to carry out the experiments. Solution samples were filtered with disk membranes (Φ76) manufactured by GE-Osmonics (USA), in which a polyamide selective layer is supported on polysulfone layers. To improve the removal efficiency of metal ions, the effects of additions anionic surfactant sodium dodecyl sulfate (SDS) and EDTA (AR, Merck, German) were also tested.

The electro-winning cell reactor was made of acrylics. Cell was packed with a pair of electrodes (graphite and stainless steel plates employed as cathode and anode, respectively). To enhance the mass transfer of Cu^{2+} onto the cathode, special active carbon particles were packed between the electrodes.

Metallic ions were determined using atomic absorption spectrophotometer (AAS, Vario 6, and Germany). Conductivity and salinity of water were measured using conduct-meter (EC215, Hanna Instrument, Germany). Morphology of membranes and electro rods were observed using Quanta 200 environmental scanning electron microscopy (ESEM)(FEI company, Holland), elements composition were ascertained by energy dispersive X-ray spectroscopy (EDX)(EDAX, USA). The surface of the membrane were characterized by AFM(SPA400, Seiko instruments Inc., Japan).

RESULTS AND DISCUSSION

Influence Factors on Membrane Separating Processes

The purpose of evaluating the effects of various parameters were to explore the suitable conditions for the maximum retention of metal with the highest possible permeate flux. Thus the retention of metal and the permeate flux were the main criteria for evaluation of performance of membrane processes. After metal ion concentrations of permeate and feed solutions were measured, retention values (R) were calculated according to following Esq. (1) and (2).

$$R = \frac{C_f - C_p}{C_f} \times 100\% \tag{1}$$

Where, C_p and C_f is the concentration of metal ion in the permeate in the feed solution, respectively.

$$SF = \frac{J}{\Delta p} = \frac{1}{\mu R_t} \tag{2}$$

Where, SF is membrane specific flux; J is membrane flux; Δp is pressure difference between high and low pressure sides of the membrane; μ is viscosity of permeate solution; Rt is total resistance in membrane filtration process. Table 2 shows the permeability of low pressure RO and ultra-filtration membranes. The resistance in membrane filtration process Rt in LPRO process was quite stable in the whole Δp range studied, while in UF process Rt increased with the increasing pressure.

Cu^{2+} concentration of retentive flux increased as the feed raw water Cu^{2+} concentration increase. When feed Cu^{2+} concentration is less than 50 mg/L, LPRO Cu^{2+} concentration on permeated water always had concentrations below 5 mg/L. As feed Cu^{2+} concentration increased, Cu^{2+} concentration of the permeated water increased slowly, but not more than 15 mg/L. Considering the shorter membrane filtering time and distance, the result of the removal efficiency results were quite good.

Table 2 Permeability of low presure reverse osmosis and ultra-filtration membranes

ΔP (MPa)		0.1	0.2	0.3	0.4	0.5	0.6
SF(L/(m²·h·MPa)	UF	60	40	33	30	32	30
	RO	20	20	20	20	20	22
Rt (10^{14} m⁻¹)	UF	0.5	0.75	0.96	1.0	0.98	0.98
	RO	1.5	1.5	1.5	1.5	1.5	1.5

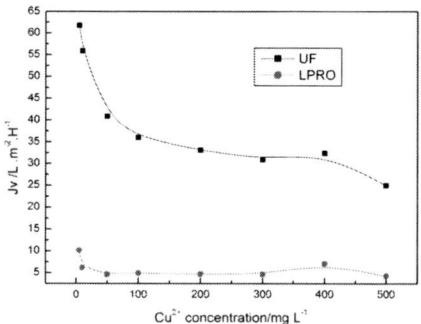

Figure 2 Effect of feed Cu^{2+}concentration on permeate flow and membrane resistance

As shown in Fig. 3b, Cu^{2+} was removed fairly well from the wastewater by LPRO. The concentration of Cu^{2+} in the product water (permeate) from low pressure RO process was reduced to an average value of 1 mg/L in two level series LPRO process for an initial feed concentration range 10–100 mg/L. Each step has a Cu^{2+} removal efficiency of 85%. On the other hand, the removal efficiency of Cu^{2+} by UF ranged from 65% to 80%. In contrast, as the feed Cu^{2+} concentration increased, Cu^{2+} concentration of retentive flux increased gradually. There was little difference between RO and UF retentive flux Cu^{2+} concentration, but the Cu^{2+} concentration in RO and UF permeate flux was quite different. Surfactant was added and its concentration is adjusted so that micelles are formed. The permeate water will contain very little, if any, of the feed surfactant components in UF process and even less residual surfactant in RO process, while the retentate will contain most of the surfactant and the solutes. The selective rejection of metal ions having the same electrical charge as the surfactant ion cannot be expected, since the main binding force is due to an electrostatic attraction between metal and surfactant.

Table 3 Effect of SDS and EDTA on Cu^{2+} removal and recovery

Addition (mg/L)	ΔP (Mpa)	R (%)		J (L/m^2 ·h)		PWC*(µS/cm)	
		UF	RO	UF	RO	UF	RO
Blank 0	0.6	68	85	17	14	60	20
EDTA, 20	0.6	75	82	18	12	120	100
EDTA, 60	0.6	76	83	20	12	200	100
SDS, 60	0.6	86	92	16	10	180	80
SDS, 600	0.6	99	99.5	15	8	360	106

*: Rejection rate; **: Permeated water conductivity

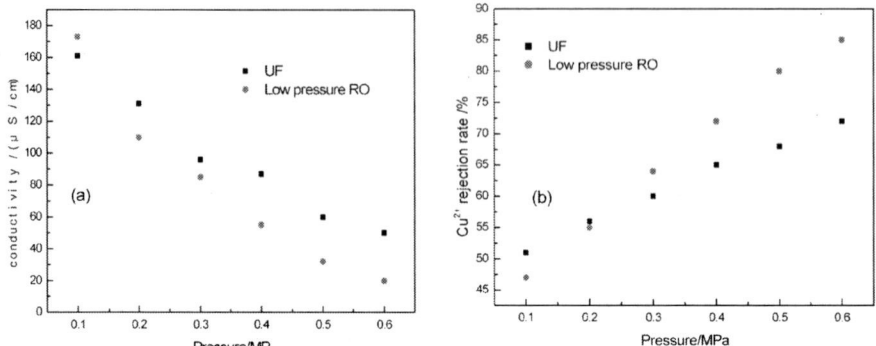

Figure 3 Relation between permeated water conductivity (a) rejection ratio (b) with ΔP

Mechanism of the enhancement in the LPRO process

For reverse osmosis membranes, the solution-diffusion mechanism has been widely used to characterize the transport of solvent and permeating solutes through the active asymmetric membrane. In the absence of any coupling, salt rejection(R) are given as, $R = 100 - \dfrac{100K_s}{C_f \cdot K_w} \dfrac{\Delta c}{\Delta p - \Delta \pi}$, where K_w, K_s, ΔP, $\Delta \pi$, Δc, and C_f represent water permeability coefficient, salt permeability constant, water pressure across the membrane, osmotic pressure difference of the two solutions across the membrane, salt concentration difference across the membrane, salt concentration of feed solution. In confected water experiment, zinc and copper sulfate was the only salt in the aquatic solution. As a result, R in Equation represent rejection coefficient of metal ions. In the equation above, the Ks, C_f and K_w are constant in a specific experiment. Subsequently, the metal-free surfactants were able to retain more metal ions entering the reactor afterward.

When the concentration polarization occurred on the surface of the membrane, the concentration of heavy metal ions near the surface of RO membrane increased, while the organic layer separate the zinc ions from the membrane, causes the osmotic pressure of the zinc to membrane increased little. Thus, zinc ions permeate through the RO membrane varied little and the rejection coefficient of zinc ions increased sharply.

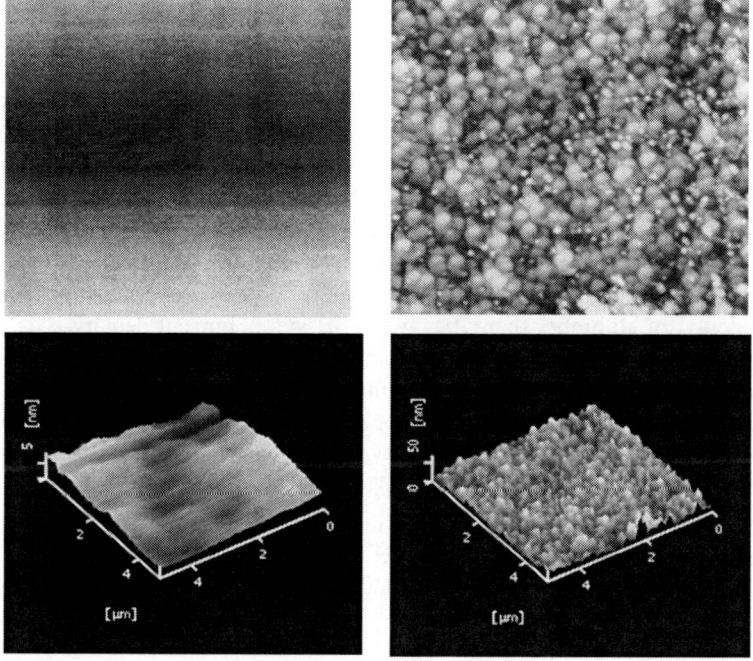

Figure 4 AFM of membrane. Left: virgin RO membrane. Right: dimension micrograph of (a).

Table 4 Energy dispersive X-ray spectroscopy (EDX) analysis result from Fig. 4

Element (At %)	Blank	48 h
CK	81.08	89.37
OK	12.82	07.70
CuK	–	00.84

Metal removal and recovery in electro-winning processes

The electrolysis process liberates metal ions from the solution metal complexes by electrodepositing metals onto the cathode. In order to maximize the removal ratio of metal ions from the membrane concentrate and the possibility for copper recovery, the solutions were treated by electro-winning process where graphite was used as anode and stainless as cathode. To enhance the mass transfer of copper ions onto the cathode, special activated carbon particles were packed between the electrodes, which formed a 3D packet electro-winning cell. When concentration of heavy metals in solution declined, current efficiency in electro-winning decreased too.

In order to get a higher current efficiency in heavy metals recovery, it is necessary to determine the appropriate electro-deposition time and the minimum final concentration of heavy metals. During electro-deposition process, concentration of electrolyte solution decrease steadily, but the decline was more pronounced in the first 30 min, and then decreased slowly. Therefore, 60 min electrolysis is suitable. It is apparent from Fig. 5 that the reaction rate changes after the second-hour of the experimental run at 10 A/m². Figure 5 shows that the copper removal efficiencies are all higher than 90% when pH is 6.0, current density is 10 A/m², and the system operated at the hydraulic residence time (HRT) of 60 min. Increasing permeate flux enlarges the amount of fouling materials transported to the membrane surface, resulting in more sever irreversible membrane fouling.

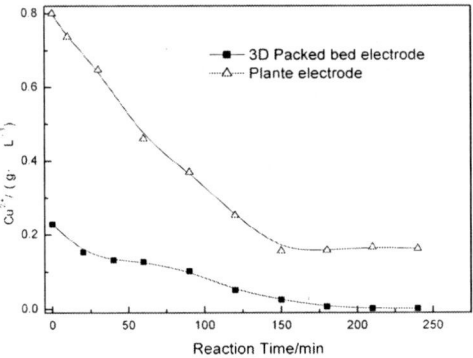

Figure 5 Copper ion removal efficiency in different electrode cells (10 A/m²), pH=6

CONCLUSIONS

1. The average Cu^{2+} removal efficiency in LPRO process is about 85%. On the other hand, the removal efficiency of Cu^{2+} by UF ranged from 50% to 70%. The addition of SDS can improve the Cu^{2+} removal efficiency which can reach as high as 90% to 99.5%. Addition of EDTA had little side influence on the membrane filter performance especially in LPRO process.
2. Conductivity of permeate from LPRO can be less than 100 µS/cm, with low heavy metals concentrations. Thus heavy metals in wastewater can be treated effectively, by this hybrid process with recovery of heavy metals and at the same time implementation of water reuse.

3. Use of electro-reduction technology, using 3D electrode cell average copper removal efficiency are ranging from 70 to 95% and use plante electrode cell Cu^{2+} concentration can be reduced to 5 mg/L, Cu^{2+} concentration can be less than 0.5 mg/L.

ACKNOWLEDGMENTS

This work was supported by the National Special Program on Water (No.2008ZX07212-01) and the National Natural Science Foundation of China (No.20877001) and the China Postdoctoral Foundation (No. 20070420255), The authors are grateful to Professor Shankha K.Banerji from University of Missouri-Columbia for his kind help.

REFERENCES

Castelblanque J. and Salimbeni (2004). NF and RO membranes for the recovery and reuse of water and concentrated metallic salts from waste water produced in the electroplating process. *Desalination*, **167**, 65–73.

Chang F. C., Lo S. L. and Ko C. H. (2007). Recovery of copper and chelating agents from sludge extracting solutions. *Separation and Purification Technology*, **53**, 49–56.

Chen G. H. (2004). Electrochemical technologies in wastewater treatment. *Separation and Purification Technology*, **38**, 11–41.

Juang R. S., Xu Y. Y. and Chen C. L. (2003). Separation and removal of metals ions from dilute solutions using micellar enhanced ultra-filtration. *Journal of Membrane Science*, **218**, 257–267.

Kadirvelu K., Thamaraiselvi K. and Namasivayam C. (2001). Adsorption of nickel(II) from aqueous solution onto activated carbon prepared from coirpith. *Separation and Purification Technology*, **24**(3), 497–505.

Kamitani T. and Kaneko N. (2007). Species-specific heavy metal accumulation patterns of earthworms on a floodplain in Japan. *Ecotoxicology and Environmental Safety*, **66**, 82–91.

Namasivayam C. and Senthilkumar S. (1999). Adsorption of copper (II) by "waste" Fe (III)/Cr(III) hydroxide from aqueous solution and radiator manufacturing industry wastewater. *Separation and Purification Technology*, **34**(2), 201–217.

Ozaki H., Sharma K. and Saktaywin W. (2002). Performance of an ultra-low pressure reverse osmosis membrane (ULPROM) for separating heavy metal: effects of interference parameters. *Desalination*, **144**(1–3), 287–294.

Petrov S. and Nenov V. (2004). Removal and recovery of copper from wastewater by a complex ultra-filtration process. *Desalination*, **162**(1–3), 201–209.

Reeve D. J. (2007). Environmental improvements in the metal finishing industry in Australia. *Journal Cleaner Production*, **15**(8–9), 756–763.

Ujang Z. U. and Anderson G. K. (1998). Performances of low pressure reverse osmosis membrane (LPROM) for separating mono and divalent ions. *Water Science and Technology*, **38**(4–5), 521–528.

Wang X. J., Chen L., Xia S. Q., Zhao J., Chovelon J.-M. and Renault N. J. (2006). Biosorption of Cu(II) and Pb(II) from aqueous solutions by dried activated sludge. *Mineral Engineering*, **19**(9), 968–971.

A Pilot-Scale Solar Photocatalysis Reactor with Immobilized Catalyst

S.H. Lin and Z.H. Jing

College of Civil Engineering, Nanjing Forestry University, Nanjing 210037, China
E-mail: franklinsh@126.com

Abstract A novel solar photocatalytic reactor was developed, where a low concentrating compound parabolic concentrator was applied as a solar collecting component. The catalyst TiO_2 supported on glass fiber mesh by sol-gel method was used as catalyst, and cold cathode low-pressure mercury lamps were used as assisting artificial UV sources. The photonic collection efficiency factor was estimated to be about 0.759 by potassium ferrioxalate actinometries. Phenol in tap water was treated with high efficiency, and bacteria could be destroyed by solar photocatalysis as well. The solar photocatalysis technique has a promising prospective in water treatment for green building programme.

Keywords Solar photocatalysis, immobilized catalyst, potassium ferrioxalate actinometer, drinking water

INTRODUCTION

TiO_2 photocatalysis, as a well-known advanced oxidation process (AOP), has been studied as an effective process for aqueous organic contaminants and water disinfection (Hoffmann *et al.*, 1995; Fujishima *et al.*, 2000; Wist *et al.*, 2002; Watts *et al.*, 1995; Rincon and Pulgarin, 2003). The artificial generation of UV radiation contributes to a large portion of the operating, capital and maintenance costs of a photocatalytic reaction system because of the utility consumption and periodic replacement of the UV lamps (Alex *et al.*, 2003). It is estimated that in the solar spectrum, approximately $50 W/m^2$ radiation from the sun that can be used for the excitation of TiO_2 reaches the surface of the earth under AM 1.5 conditions (Dillert *et al.*, 1999), which makes solar TiO_2 photocatalysis an increasingly attractive technique for water treatment.

However, the suspension of TiO_2 needs separation steps such as filtration and decantation, which makes the technology unsuitable for real practice, especially for treatment of drinking water.

The aim of this work is to evaluate the efficiency of a new pilot-scale solar photocatalytic reactor with immobilized catalyst, which was developed for *green building*.

MATERIALS AND METHODS

Photoreactor

All the experiments were carried out using a compound parabolic concentrator reactor (CPCR) with immobilized catalyst (Figure 1). The reactor consists of modules mounted on a fixed platform tilted at an angle of 31° (local latitude) and connected in series so that the water flows directly from one to another and finally to a tank. A centrifugal pump then returns the water to the collectors at a rate of 16L/min. Due to the special design, almost all the UV radiation arriving at the CPC aperture area (each collector surface is 0.16 m², and the total collector surface is 0.96 m²) can be collected, and the reflected UV light is distributed around the back of the photoreactor, so that most of its circumference is illuminated. The outer reactor tube is borosilicate glass with an internal diameter of 48 mm and a thickness of 2 mm, and the quartz glass inner tube has an outer diameter of 23 mm and a thickness of 1.5mm. The two layers of TiO_2 photocatalyst supported on glass fiber mesh with sol-gel method are fixed into the photoreactor, except for potassium ferrioxalate actinometeries. 6 cold cathode low-pressure mercury lamps (210W) were installed in inner tubes.

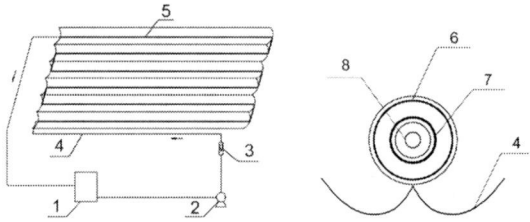

(a) flow chart of the new reactor   (b) schematic diagram of transect

(c) photograph of the new reactor

Figure 1 Schematic diagram of solar photocatalytic reactor with immobilized catalyst 1. water tank 2. centrifugal pump 3. flowmeter 4. compound parabolic collector 5. photoreactor 6.outer catalyst layer 7. inner catalyst layer 8. cold cathode low-pressure mercury lamp

All the solar photocatalysis experiments were carried out near noon, when the solar radiation was relatively stable. Solar radiation was determined during the experiments by means of a SUV-2 radiometer (Photoelectric Instrument Factory of Beijing Normal University), also mounted on a 31° fixed-angle platform (the same angle as the CPC). Some combination of the data from several experiments performed at different times and their comparison was obtained by the application of Eq. (1).

$$Q_{UV,n} = Q_{UV,n-1} + \Delta t_n \overline{UV}_{G,n} \frac{A}{V} \tag{1}$$

Where $\Delta t_n = t_n - t_{n-1}$, t_n is the experimental time for each sample, $UV_{G,n}$ is the average incident radiant flux UV_G during t_n, A represents the collector surface, V the total reactor volume and Q_{UV}, n is the accumulated energy (kJ/L) incident on the photoreactor per unit of volume of aqueous sample taken at t_n (Malato et al., 2000).

Reagents and Analytical Determinations

$K_3Fe(C_2O_4)_3$ was prepared according to the literature (Rabek, 1982; Steven, 1973). Fe^{2+} and phenol was measured by a Shimadzu UV2550 uv-vis spectrophotometer (1, 10-phenanthroline spectrophotometric method, 4-aminoantipyrene spectro-photometric method, respectively). All reagents were analytical grade.

Before all drinking water experiments, the tap water was filtered by a Manganese sand bed to remove Mn^{2+}, Fe^{2+}, which could poison TiO_2. Bacteria were inoculated by tap water flowing through saturated PAC bed for disinfection experiments. To determine the concentration of bacteria in samples, serial dilutions of the samples in autoclaved and distilled water were carried out when necessary, with every sample then being plated 3 times on Petri dishes containing agar medium. The inoculated samples were incubated at 37°C for 20–24 h and colony-forming units (CFUs) were visually identified and reported as average CFU/mL.

RESULTS AND DISCUSSIONS

Evaluation of optical performance

Potassium ferrioxalate actinometer was chosen to characterize the optical performance of the new reactor. The device was filled with actinometric solution (0.006M $K_3Fe(C_2O_4)_3$ and 0.1N sulphuric acid). The solution was homogeneized in the dark (the modules were covered). Following this, the covers were taken out and sunlight illuminated the modules, taking this time as zero and being C_0 the Fe^{2+} concentration. Samples were taken during the time, and the concentration of Fe^{2+} ($C(t)$) was measured by 1, 10-phenanthroline spectrophotometric method. Simultaneously, the radiation arriving at the CPCs was measured by a radiometer.

The radiation arriving at the CPCs (F_a, eins/s) can be expressed as:

$$F_a = \sum_{\lambda_{min}}^{\lambda_{max}} F_\lambda(t) \tag{2}$$

$F_\lambda(t)$ is the photonic flow (eins/s) reaching the external surface of CPCs modules for each wavelength. λ_{min} is the minimum wavelength (300 nm) in the spectrum of the radiation source and λ_{max} is the maximum wavelength (577 nm) that can be absorbed by the solution into the photoreactors. However, only part of the radiation arriving the CPCs modules enters the photoreactor because of the characteristics of the system (ϕ_{ef}, efficiency factor, related to reactor geometry and position with respect to the sun) and the transmittance of the outer borosilicate reactor tube ($\phi_{T\lambda}$, transmittance factor, transmittance in the active actimometric range provided by the manufacturer). The radiation entering the CPCs modules (F_e, eins/s) can be expressed as:

$$F_e = \phi_{ef} \sum_{\lambda_{min}}^{\lambda_{max}} F_\lambda(t) \cdot \phi_{T\lambda} \tag{3}$$

Planck's formula should be introduced into eqn (3) to transform the power units of solar radiation ($I(t)$, W/m^2) measured by radiometer into einstein units. Thus considering the collectors area (S, 0.16 m^2), eqn (3) can be transformed in:

$$F_e = \frac{\phi_{ef}S}{N_a h c_v} \sum_{\lambda_{min}}^{\lambda_{max}} \lambda f_\lambda I(t) \phi_{T\lambda} \tag{4}$$

N_a being the Avogadro number (6.023×10^{23}), h the Planck's constant (6.63×10^{-34} Js), c_v the light rate (3×10^8 m/s), f_λ spectral distribution, i.e. the fraction power of global radiation between 300–577nm.

According to the Beer-Lambert law, the radiation absorbed by the actinometer can be experssed as:

$$F_{abs,\lambda} = F_{e,\lambda}(1-10^{-\mu_\lambda D}) \tag{5}$$

$F_{e,\lambda}$ the radiation entering the CPCs modules at wavelength λ (eins/s), and $F_{abs,\lambda}$ the radiation absorbed by the actinometer at wavelength λ (eins/s). D being the optical pathway, assumed to be equal to the outer tube diameter (cm), μ_λ absorption coefficient of actinometic solution at wavelength λ (per cm), acquired by scanning the actinometric solution using Shimadzu UV2550 uv-vis spectrophotometer and displayed in Table 1. Thus the total absorbed radiation can be expressed as:

$$F_{abs} = \frac{\phi_{ef}S}{N_a hc_v} \sum_{\lambda_{min}}^{\lambda_{max}} \lambda f_\lambda I(t)(1-10^{-\mu_\lambda D})$$ (6)

When the actinometric solution concentration is relatively low, Fe^{2+} formation rate is zero order with respect to the radiation absorbed by the actinometer. Thus taking into account the quantum yield (ϕ_λ, mol/eins) at each wavelength for this actinometer, Fe^{2+} formation rate can be expressed as:

$$r_a = \frac{dc}{dt} = \frac{\phi_{ef}S}{V_s N_a hc_v} \sum_{\lambda=300}^{577} \lambda f_\lambda I(t)\phi_{T\lambda}(1-10^{-\mu_\lambda D})\phi_\lambda$$ (7)

V_s being the total volume of the system (L). The integration of eqn (7) is:

$$c(t) = c_0 + \frac{\phi_{ef}S}{V_s N_a hc_v} W \sum_{\lambda=300}^{577} \lambda f_\lambda \phi_{T\lambda}(1-10^{-\mu_\lambda D})\phi_\lambda$$ (8)

The parameter W has been calculated as:

$$W = \sum_{t=0}^{t} I(t)\Delta t$$ (9)

Δt being the interval of time between radiation measurement (s). The data of chemical actinometries would fit eqn (8) well, and the only parameter ϕ_{ef} can be obtain from the slope of the fitted straight line K_L.

$$K_L = \frac{\phi_{ef}S}{V_s N_a hc_v} \sum_{\lambda=300}^{577} \lambda f_\lambda \phi_{T\lambda}(1-10^{-\mu_\lambda D})\phi_\lambda$$ (10)

Variations of different parameters with the wavelength and potassium ferrioxalate actinometries summary are shown in Table 1 and Table 2, respectively. Linear fitting of Fe^{2+} concentration to accumulated radiation as shown in Figure 2, three estimated value for ϕ_{ef} can be obtained, these being 0.735, 0.747, 0.796. The average photonic collection efficiency factor ϕ_{ef} was about 0.759. Results indicated that the optical performance of the new photocatalysis reactor with immobilized catalyst was similar to that of CPCR in PSA (Curco et al., 1996).

Table 1 Variation of different parameters with the wavelength

λ/nm	f_λ	μ_λ/(1/cm)	ϕ_λ/(mol/eins)	$\phi_{T\lambda}$
300	0.0002081	≥5①	1.24	0.45
313	0.0068965	≥5①	1.24	0.67
334	0.031476	≥5①	1.23	0.86
366	0.069397	4.516	1.21	0.91
405	0.12405	1.095	1.14	0.92
436	0.13691	0.268	1.11	0.92
468	0.11223	0.050	0.93	0.92
480	0.081901	0.026	0.94	0.92
509	0.18626	0.006	0.86	0.92
546	0.16679	0.002	0.15	0.92
577	0.08389	0.003	0.013	0.92

① Actinometric solution (0.006mmol/L) was not diluted to acquire the absolute value between 300-334nm because the incident light would be completely absorbed when light path was more than 1cm.

Table 2 Chemical actinometries summary

Actinometry	V_s/ L	S/m²	Experimental period	Conversion/ %	K_L/((mM · m²)/ kJ)	R^2
1	10	0.16	11:17/11:19	7	0.0180	0.9909
2	10	0.16	11:28/11:32	15	0.0183	0.9965
3	10	0.16	11:21/11:24	12	0.0195	0.9935

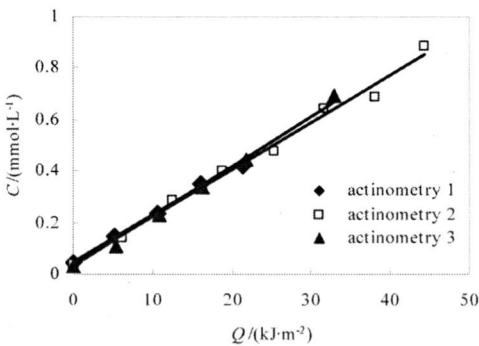

Figure 2 Linear fitting of Fe^{2+} concentration to accumulated radiation

Degradation of phenol in drinking water

Removal of phenol, a typical environmental priority pollutant in tap water, was studied. Some physicochemical characteristics of polluted tap water are presented in Table 3. The results, as shown in Figure 3, were that phenol was degraded effectively under different UV irradiation. When photocatalysis was excited by solar UV irradiation, 80% of the phenol was removed within 100 min under sunny conditions (22.7 W/m²), while that was within 120 min under 18.0 W/m². Even under overcast conditions (14.1 W/m²), it took no more than 180 min to degrade the same amount of phenol. With assistance from UV irradiation, the time span was less than 80 min to break down 80% of the phenol molecules. However, it only took 60 min to achieve the same effect under combined irradiation (assisting UV irradiation and 5.4 W/m² solar UV irradiation). The results were satisfying, although the competition of NOM (TOC value was 4.0 ~ 4.2 mg/L) for •OH slow the degradation rate of phenol. Therefore, the new solar reactor can make all-weather stable operation with good compatibility to different weathers.

Table 3 Some physicochemical characteristics of polluted tap water

Parameter	Unit	Value
TOC	mg/L	4.0 ~ 4.2
UV254	–	0.092 ~ 0.096
pH	–	7.1 ~ 7.5
Temperature	–	20 ~ 24

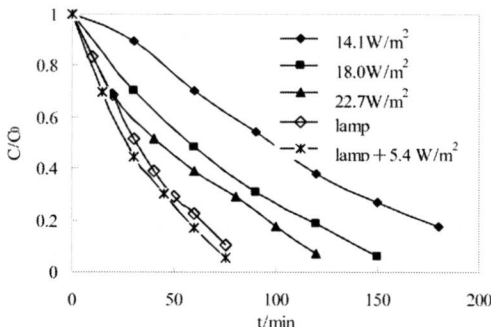

Figure 3 Removal of phenol in tap water under all-weather operation (C_0=2mg/L)

Photocatalytic disinfection of drinking water

Experiments in drinking water disinfection were also carried out to evaluate the efficiency of the solar photocatalytic reactor when applied to real water treatment. Results are shown in Figure 4 (a): when the initial bacterial load was about 4.0 ~ 6.0 × 10^2CFU/mL, under sunny (Exp. 2, 17.1W/m^2) and mostly sunny conditions (Exp. 3, 12.4 W/m^2), bacteria could be deactivated to less than 100 CFU/mL after 100 ~ 120 min treatment, which meets the requirement of *drinking water quality standards of China*. When solar UV radiation was 6.6 W/m^2 (Exp. 1), the bacteria could be deactivated to meet this standard after 180 min treatment.

A typical method of comparing the inactivation results is to report the values of UV intensity and exposure (or residence) time together as UV dose. In our case, the dose can then be calculated from the average solar UV intensity and the residence time in the irradiated part of the reactor (Figure 4(b)). And no matter what weather it was, the solar UV dose necessary to meet the drinking water quality standards was about 3.5kJ/L.

The disinfection rate with immobilized catalyst was much slower than that with slurry catalyst (Rincon and Pulgarin, 2004). However, considering that more rapid and better disinfection performance could be ensured by artificial UV irradiation (254nm), the solar reactor with immobilized catalyst is promising in drinking water disinfection.

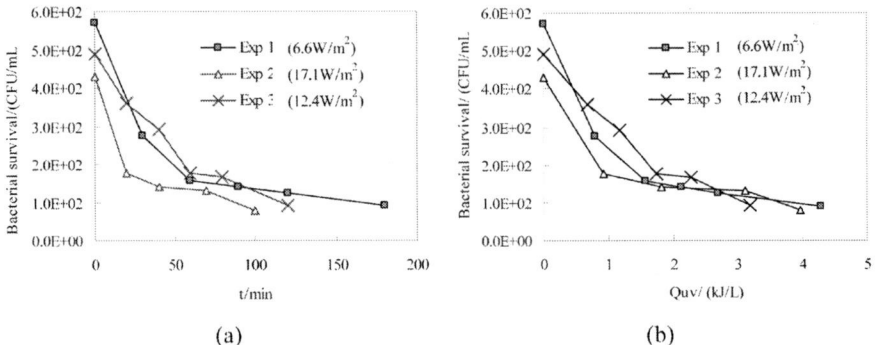

(a)   (b)

Figure 4 Photocatalytic disinfection of tap water. The graph (a) shows the bacteria survival vs. time and the graph (b) the bacterial survival vs. Q_{UV}.

CONCLUSIONS

The new developed solar photocatalytic reactor with immobilized catalyst was evaluated. The optical performance was assessed by potassium ferrioxalate actinometries, and the photonic collection efficiency factor ϕ_{ef} was about 0.759. Phenol and bacteria in tap water were destroyed efficiently by solar photocatalysis. The all-weather reactor has a promising prospective in *green building* programmes.

REFERENCES

Chan A. H.C. Chan C. K., Barford J. P. and Porter J. F. (2003). Solar photocatalytic thin film cascade reactor for treatment of benzoic acid containing wastewater. *Water Research*, **37**, 1125–1135.

Curco D., Malato S., Blanco J., Gimenez J. and Marco P. (1996). Photocatalytic degradation of phenol: comparison between pilot-lant-scale and laboratory results. *Solar Energy*, **56**(5), 387–400.

Dillert R., Cassano E. A., Goslich R. and Bahnemann D. (1999). Large scale studies in solar catalytic wastewater treatment. *Catalysis Today*, **54**, 267–282.

Fujishima A., Rao T. N. and Tryk D. A. (2000). Titanium dioxide photocatalysis. *Journal of Photochemistry and Photobiology C: Photochemistry Reviews*, **1**, 1–21.

Hoffmann M. R., Martin S. T., Choi W. Y. and Bahnemann D. W. (1995). Environmental applications of semiconductor photocatalysis. *Chemical Reviews*, **95** (1), 69–96.

Malato S., Blanco J., Richter C. and Maldonado, M. I. (2000). Optimization of preindustrial solar photocatalytic mineralization of commercial pesticides: Application to pesticide container recycling. *Applied Catalysis B: Environmental*, **25**, 31–38.

Murov, S. L. (1973). *Handbook of photochemistry*, Marcel Dekker Inc. New York.

Rabek J. F. (1982). *Experimental Methods in Photochemistry and Photophysics (Part 2)*, John Wiley & Sons, Inc., New York.

Rincon A. G. and Pulgarin C. (2003). Photocatalytical inactivation of E. coli: effect of (continuous–intermittent) light intensity and of (suspended–fixed) TiO2 concentration. *Applied Catalysis B: Environmental*, **44**, 263–284.

Rincon A. G. and Pulgarin C. (2004). Field solar E. coli inactivation in the absence and presence of TiO2: is UV solar dose an appropriate parameter for standardization of water solar disinfection? *Solar Energy*, **77**, 635–648.

Watts R. J., Kong S. H., Orr M. P., Miller G. C. and Henry B. E. (1995). Photocatalytic inactivation of coliform bacteria and viruses in secondary wastewater effluent. *Water Research*, **29**, 95–100.

Wist J., Sanabria J., Dierolf C., Torres W. and Pulgarin C. (2002). Evaluation of photocatalytic disinfection of crude water for drinking-water production. *Journal of Photochemistry and Photobiology A: Chemistry*, **147**, 241–246.

Concentration of Endocrine Disruptors in the Surface Water of Agricultural Fields and Irrigation Systems at Two Representative Study Sites of the Lower Mekong Delta, Vietnam – Preliminary Results

N.T. Hoa, L. T. A. Hong and J. Clemens

Author's address: Plant Nutrition, Institute of Crop Science and Resource Conservation – University of Bonn, Karlrobert-Kreiten-Strasse, 53115 Bonn, Germany
E-mail: thaihoa.nguyen@gmail.com

Abstract In this study, the concentration of Endocrine Disruptors (EDs) in terms of 17β-estradiol (E2) equivalent were monitored in the surface water of three different categories of water bodies: irrigation canals; agricultural fields; and fishponds. The study was conducted in Cantho city and in the Dongthap province of the Lower Mekong Delta, Vietnam. Its duration was from Aug 2008 to Mar 2009. E2 concentrations were determined by Yeast Estrogen Screen (YES) assay analysis after solid phase extraction. The median E2 equivalent concentrations ranged from lower than the Detection Limit (DL) to 1.23 and 0.7 ng/L in Cantho and Dongthap respectively. The median concentration in Cantho ranged from 0.24 to 1.04 ng/L for the irrigation canals, DL to 1.23 ng/L for the agricultural fields (fruit, vegetable and rice fields), and DL to 0.88 ng/L for fishponds. In Dongthap, it was from DL to 0.7 ng/L for the irrigation canals, DL to 0.09 ng/L for the rice fields, and 0.05 to 0.45 ng/L for the fishponds. The concentration obtained in the more urbanization city (Cantho) was higher than in the agricultural province (Dongthap). Agricultural runoff, in this study, did not seem to be a source of EDs. The higher concentration in irrigation canals compared to other surface water sampling categories in this study indicated that there were other sources of EDs that influenced its concentration in the canals. The most likely source was sewage discharge.

Keywords Endocrine Disruptors, YES assay, Mekong Delta, Vietnam

INTRODUCTION

Recently, endocrine disruption has become a public, political, and scientific issue. There are concerns that chemicals in the environment might exert profound and deleterious effects on wildlife populations, and that human health is inextricably linked to the health of the environment. The health effects attributed to endocrine disrupting compounds include problems such as: reduced fertility; male and female reproductive tract abnormalities; skewed male/female sex ratios; loss of fetus; menstrual problems; changes in hormone levels; early puberty; brain and behavior problems; impaired immune functions; and various cancers.

Endocrine Disruptors (EDs) have a variety of chemical compositions. They can be both naturally occurring and man-made. The sources of EDs to the environment, therefore, are also diverse. However, as reported in a number of papers, the main sources of EDs can be found in agricultural runoff (Shore *et al.*, 2004; Campbell *et al.*, 2006) and industrial and municipal effluents (Desbrow *et al.*, 1998; Routledge *et al.*, 1998; Ying *et al.*, 2002; Shore *et al.*, 2004; Fernandez et al., 2007).

This report focuses on the EDs from agricultural fields and irrigation canals in the Mekong Delta, Vietnam. The ED concentration of the surface waters of agricultural fields, fishponds and irrigation canals in Catho City and in the Dongthap Province were monitored for six months from August 2008 to March 2009.

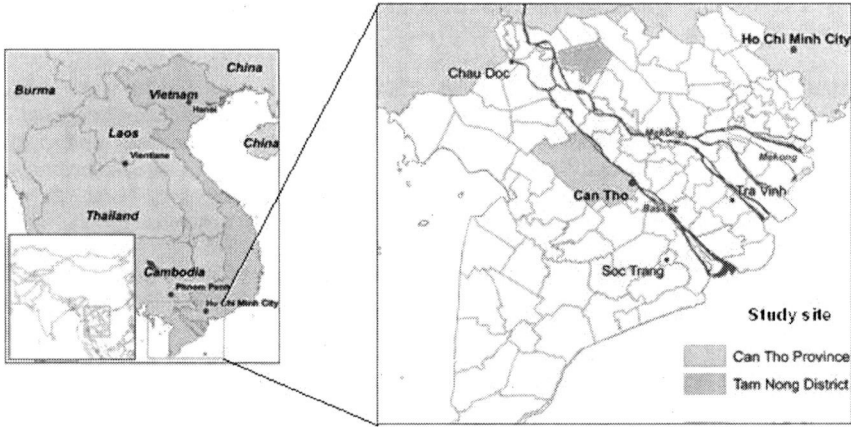

Figure 1 Map of the study sites

MATERIALS AND METHODS

The agricultural fields, fishponds and irrigation canals, from the highly intensive agricultural province, Dongthap, and a sub-urban area of the most urbanization and industrialization city in the Lower Mekong Delta in Vietnam, Cantho city, were selected to monitor EDs in terms of the 17β-estradiol (E2) concentrations (EEC) in surface waters. The study ran from Aug 2008 to Mar 2009. The sampling site in the Dongthap province was located in the upper stream of the Mekong

Delta in Vietnam, whilst the sampling site in Cantho city was located in the centre of the Delta (Figure 1). The sampling was repeated every four weeks, for Cantho city, and six weeks, for the Dongthap province.

A Yeast Estrogen Screen (YES) assay was used to analyse EDs, and the results obtained in terms of E2 equivalent. In principle, in the YES assay, the DNA sequence of the human estrogen receptor (hER) is integrated into the yeast genome. This also contains expression plasmids, carrying estrogen-responsive sequences (ERE) which control the expression of the reporter gene *lac-Z* (encoding the enzyme β-galactosidase). Thus, in the presence of estrogens, β-galactosidase is synthesized and secreted into the medium, where it causes a colour change from yellow to red (Routledge *et al.*, 1996). This analysis method was adopted from Routledge and Sumpter in 1996 where the estrogenicity of water samples caused by different compounds is converted into E2 equivalent concentrations.

The analytical procedure of E2 equivalent is essentially composed of sample filtration (Glass fiber with core sizes are 0.2–0.6 µm and ≈ 8 µm respectively), solid phase extraction (strata - X 33µm Polymeric Reserved Phase (200 mg/ 6 mL)), and finally the YES assay.

For the extraction, 5 ml of methanol was added to a 1 little water sample which was shaken well before filtration. 500 ml of the filtrated water sample was used. The extraction step was carried out by using a 12 unit manifold vacuum. The solid phase extraction column (SPE) was activated with 8 ml of methanol and 8 ml of water/methanol (0.5%) consecutively. The water sample was loaded onto the SPE with the flow of 5 to 6 ml per min. The SPE column was finally washed with 10 ml of methanol/water (1:1), followed by acetone/water (1:2) before it was dried by nitrogen gas (99.995%). The elution was done by being passing through 10 ml of methanol. The extract was then dried under a vacuum at 40°C before being re-dissolved in 500 µl of methanol and stored in 1.5 ml, vial at 4°C, before YES analysis.

RESULTS AND DISCUSSION

In Cantho city, the EEC obtained from the Cantho and Balang rivers ranged from lower than the detection limit, approximately 0.05 ng/L (DL) to 0.77 ng/L (median from DL to 0.37 ng/L) (Figure 2). The EEC from canals (directly supplied water for the fields) that connect those rivers to agricultural fields was slightly higher than in the rivers and ranged from DL to 2.57 ng/L (median from 0.24 to 1.04 ng/L) (Figure 3). The concentration's variations are due to the multiple sampling points. The lower concentrations and narrower variations in the Cantho and Balang rivers, compared to those of the canals, (Figure2 and 3) might be caused by the larger discharges in these rivers.

The EEC from the agricultural fields varied depending on the type of agricultural field. The highest median EEC was found in the fruit fields (DL to 1.25 ng/L, median DL to 1.23 ng/L) followed by vegetable fields (DL to 1.79 ng/L, median 0.10 to 0.81 ng/L) and rice fields (DL to 1.50 ng/L, median DL to 0.34 ng/L), (figure 4 showed only median values). The concentrations from the agricultural fields, however, where the water is directly supplied by canals, are somewhat lower. The target compounds might have been removed from the surface water in the fields by sedimentation processes, which would have consequently lowered the EEC.

The EEC in fishponds ranged from DL to 2.66 ng/L (median from DL to 0.88 ng/L) (figure 4 showed the median values only). In the river (sampling point at the water intake gate of the fishponds) that supplied water for the fishponds, the EEC ranged from DL to 0.48 ng/L (median from DL to 0.24 ng/L). It seemed that the EEC in the fishpond was greater than the reference sampling point in Cantho.

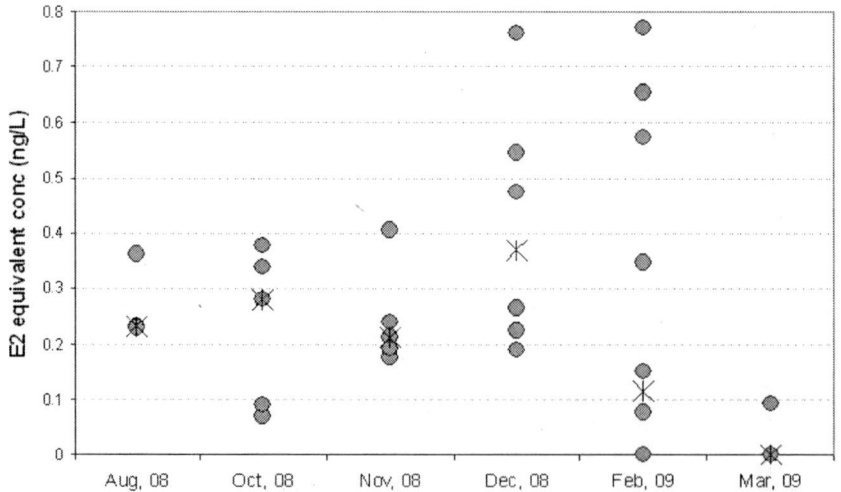

Figure 2 E2 equivalent concentration in Cantho and Balang rivers in Cantho city (dots and crosses are concentrations and median concentrations respectively)

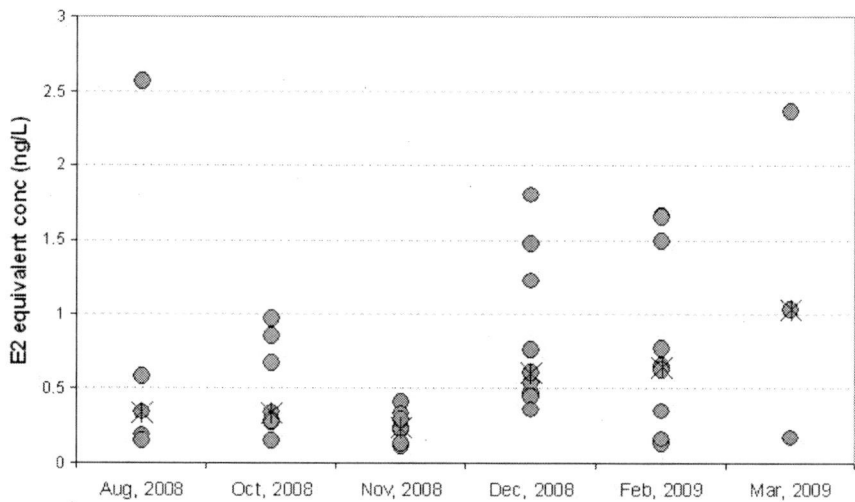

Figure 3 E2 equivalent concentration in canals directly supplied water for agricultural fields in Cantho city (dots and crosses are concentrations and median concentrations respectively)

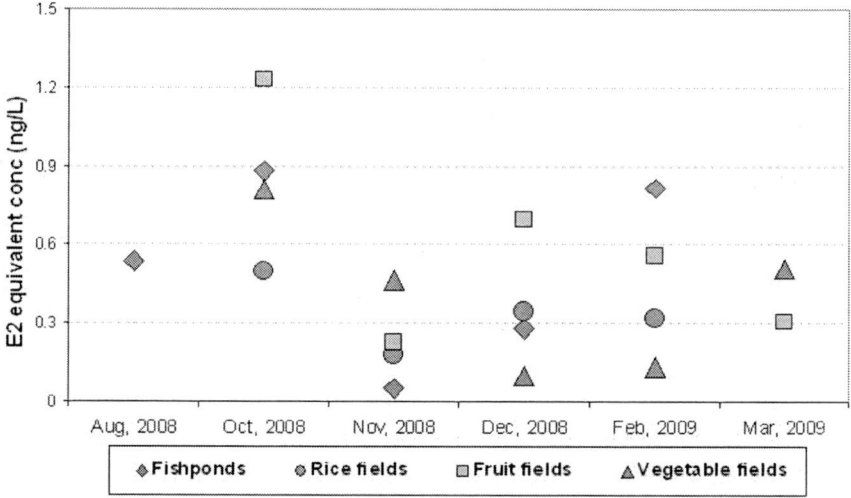

Figure 4 Median E2 equivalent concentration in fishponds and agricultural fields in Cantho city

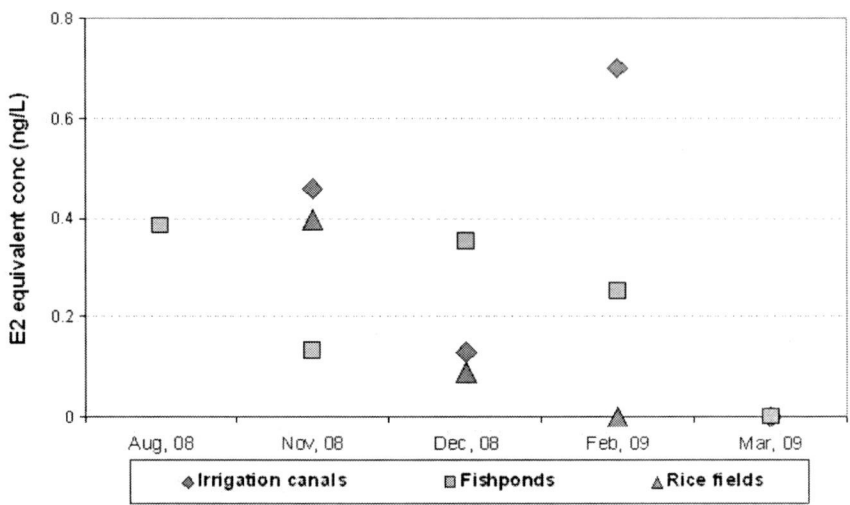

Figure 5 Median E2 equivalent concentration in fishponds, irrigation canals and rice fields in Dongthap province

In Dongthap, the EEC from rice fields ranged from DL to 0.52 ng/L (median from DL to 0.09 ng/L) from Nov 2008 to Mar 2009. The EEC in the irrigation canals that supplied water to the rice fields was higher than those in the rice fields and ranged from DL to 1.79 ng/L (median from DL to 0.70 ng/L), (Figure 5 showed only the median values). The rice fields seem to have the ability to remove EEC in their respective surface water, perhaps through an absorption process in the soil.

The EEC in the fishponds of the Dongthap province ranged from 0.05 to 0.45 ng/L (median from 0.13 to 0.38 ng/L), (Figure 5 showed only the median values) during the monitoring period from Aug 2008 to Feb 2009. In Mar 2009, the fish was harvested and the EEC in the fishpond was DL. The EEC taken from the reference sampling point for the fishponds was lower than those of the actual fishponds and ranged from DL to 0.08 ng/L (median from DL to 0.07 ng/L).

The EEC in irrigation canals was the highest in relation to the fishponds and the rice fields. A similar result was obtained in for the case study in Cantho city. These results imply that there might be other sources of EDs influencing the EEC in irrigation canals. Due to the poor sanitation in these sites, the most likely suspect for this is domestic wastewater which often discharges directly into surface waters.

The preliminary results have shown that the EEC, monitored from a sub-urban area in Cantho (DL to 2.66 ng/L), was higher than in the agricultural area in

Dongthap (DL to 1.79 ng/L). Here, the EEC is comparable to the EEC in aquatic environments as reported in some texts (Danish EPA, 2005). In addition, more samples in Cantho exceeded 1 ng/L than in Dongthap. It is hard to judge the potential effects of EEC on respective water bodies in this study. However, it is reported by Routledge *et al.*, 1998 that in the case of rainbow trout, concentrations of E2 as low as 1 ng/L elicited a vitellogenic response.

CONCLUSION

The higher EEC obtained in Cantho city compared to the Dongthap province might be resultant of the increased urbanization in Cantho. Moreover, the current amount of discharged domestic wastewater, and water from the fishponds contributing to water bodies seems to be the reason why the EEC in irrigation canals is higher than in agricultural fields.

Agricultural runoff in the Mekong Delta seems not to be a source of EDs. EDs in the surface water of the agricultural fields, particularly rice fields, seems to be removed, perhaps by adsorption into the soil.

REFERENCES

Campbell C. G., Borglin S. E., Green F. B., Grayson A., Wozei E. and Stringfellow W. T. (2006). Biologically directed environmental monitoring, fate, and transport of estrogenic endocrine disrupting compounds in water: A review. *Chemosphere*, **65**, 1265–1280.

Desbrow C., Routledge E. J., Brighty G. C., Sumpter J. P. and Waldock M. (1998). Identification of estrogenic chemicals in STW effluent. 1. Chemical fractionation and in vitro biological screening. *Environmental Science and Technology*, **32**(11), 1549–1558.

Environmental Protection Agency (EPA) -Danish Ministry of the Environment. (2005). Survey of estrogenic activity in the Danish aquatic environment. Environmental Project Nr. 977–2005.

Fernandez M. P., Ikonomou M. G. and Buchanan I. (2007). An assessment of estrogenic organic contaminants in Canadian wastewaters. *Science of the Total Environment*, **373**, 250–269.

Routledge E. J., Sheahan D., Desbrow C., Brighty G. C., Waldock M. and Sumpter J. P. (1998). Identification of estrogenic chemicals in STW effluent. 2. In vivo responses in trout and roach. *Environmental Science and Technology*, **32**, 1559–1565.

Routledge E. J. and Sumpter J. P. (1996). Estrogenic activity of surfactants and some of their degradation products assessed using a recombinant yeast screen. *Environmental Toxicology and Chemistry* **15**(3), 241–248.

Shore L. S., Reichmann O., Mordechai S., Wenzel A. and Litaor M. I. (2004). Washout of accumulated testosterone in a watershed. *Science of the Total Environment*, **332**(1–3), 193–202.

Ying G.-G., Kookana R. S. and Ru Y.-J. (2002). Occurrence and fate of hormone steroids in the environment. *Environment International*, **28**, 545–551.

Investigation of Effects of Specific Bacteria on the Bioremediation of Fu-tian River

W.B. Jin[a], Y.S. Chen[b] and Y.N. Yang[b]

[a]Department of Environmental Engineering, Harbin Institute of Technology Shenzhen Graduate School, Shenzhen, P. R. China. 518055
[b]Department of Environmental Engineering, Beihang University, 37th Xueyuan Rd, Haidian District, P. O. Box 106, Beijing, P. R. China. 100083
E-mail: 13828830095@139.com

Abstract Due to the direct deposition of sewage from industry and the daily lives of human beings, the water in Fu-tian River, Shenzhen city, has been severely polluted, and so sewage treatments of this river are imminent. In this paper, the effects of specific bacteria were investigated on the bioremediation of Fu-tian River. After studying for almost twenty months, the following results were obtained: 1) Two different specific bacteria were independently developed by the members of our research group, referred to as bacteria A and bacteria B. The specific bacteria cannot only meet the application needs, but at low cost. 2) Experimental results revealed that adding these specific bacteria can form a better ecological structure and significantly improve the quality of the water in Fu-tian River. Under normal conditions the COD_{Cr} removal effect is obvious and the removal percentage is 45–98%. On average the COD_{Cr} in the effluent is 33mg/L. But in the rainy season the removal effect of COD_{Cr} is not obvious and the removal percentage was 10–81%. The removal percentage of SS in Fu-tian River is 6-81%. This biofilm treatment system did not obviously remove the TN and TP, but the NH_4^+–N removal percentage was the highest, at 90%. 3) According to the DGGE results of biofilms in Fu-tian River, the biological diversity and total biomass of biofilms became abundant and increased from the upstream to the downstream, and from the surface to the bottom of Fu-tian River.

Keywords Sewage treatment, bioremediation, specific bacteria, DGGE

INTRODUCTION

The water in Fu-tian River, Shenzhen city, has been severely polluted due to the direct deposition of untreated water from industry and the daily lives of the population, as well as some effluents that do not meet standards of direct deposition even after treated. Therefore, treatment of polluted water in this river is imminent. At present, domestic and overseas engineering technologies of sewage treatment mainly include sewage flushing (Williams, *et al.*, 1995, William, *et al.*, 2002), flow augmentation (Whalen, *et al.*, 2002), aeration technology (Lombardia *at el.*, 2001), adding bacteria (Davis *et al.*, 2003), biofilm technology (Allision *et al.*, 1992, Akiyoshi *et al.*, 1996, Arcangelis *et al.*, 1992), and so on.

Until now, almost none of the specific bacteria have been applied in practice to the treatment of polluted rivers in the cities of China, since the river treatments are restricted by the peripheral conditions of the rivers. Therefore, a comprehensive application of multiple treatment technologies is needed for the river pollution treatment. Among them, an important application of technology is the addition of specific bacteria.

Flocculants have been widely used in wastewater treatments and other industries due to their high efficiency and low cost. However, most synthetic polymer flocculants have been proved to cause health and environmental problems (Cairns, *et al.*, 1995, Gong, *et al.*, 1999, Plonka, *et al.*, 1999). Specific bacteria, as an alternative choice, have gained more attention recently since they are environmentally friendly, biodegradable, and nontoxic. Specific bacteria have the potential to be applied in a wide range of industrial processes, such as wastewater treatments (Davis, *et al.*, 2003, Bruce, *et al.*, 2002), industrial processes (Jianlong, *et al.*, 2002), removal of heavy metals (Lombardia *et al.*, 2001), and other such areas.

The goal of this study is to investigate the effects of specific bacteria on the water purification of Fu-tian River in Shenzhen city. According to the conditions of the Fu-tian River, finding specific bacteria with low cost and high efficiency is the key to successful treatment. Specific bacteria in the biofilm have been optimally chosen based on lab-scale experiments, field pilot tests, and in situ tests which are used as guidance for practical applications.

Denaturing Gradient Gel Electrophoresis (DGGE) analysis is a promising, simple, and rapid tool to detect the biological diversity and total biomass of biofilms. Changes occurred in DGGE profiles, including the variation in band intensity as well as gain or loss of bands (Crocetti, *et al.*, 2000). The Biological diversity and total biomass of biofilms were analyzed by DGGE in this study. This paper also investigated the effects of specific bacteria and the distribution of the biotic community in the biofilm after adding bacteria to reveal the effects of specific bacteria.

MATERIALS AND PROCEDURES

Schematic of Processing of Water Treating

Figure 1 represents the planar graph of Fu-tian River; the schematic of the processing water treatment for Fu-tian River is shown in Fig. 2.

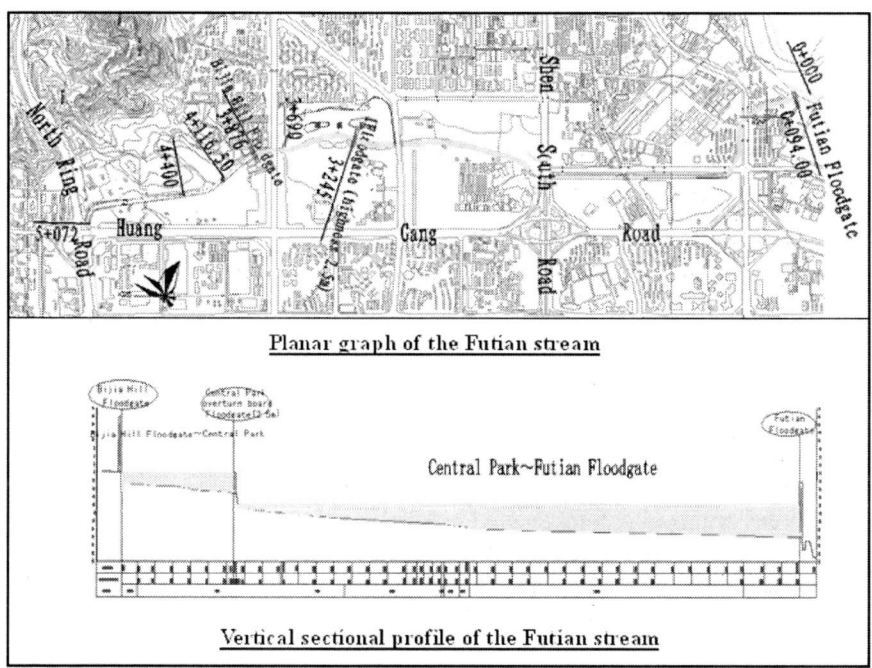

Figure 1 Planar graph of the Fu-tian River

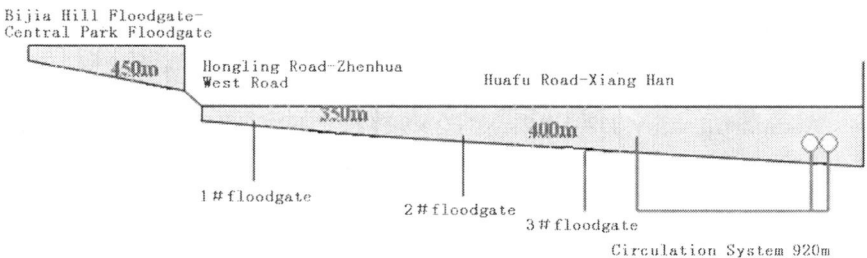

Figure 2 Schematic of the processing water treatment for Fu-tian River

Design of Water Gates of Fu-tian River

A newly constructed water gate in Fu-tian River for water holding and controlling of sewage water processing, is located in front of the hydraulic drop pump of Central Park. The first section is 656 m long. The volume of water in the river between the Bijia Hill floodgate and first floodgate is 12,000 m³. The second section, between the first floodgate and Fu-tian floodgate, is 3,165 m long, 15–17 m wide, and has a water depth of 0.5–4.0 m.

Water Quality Improvement Based on the Process

The design includes installing fibrous braids filler between the first floodgate and the Bijia Hill floodgate (First processing section), adding the specific bacteria, and oxygen aeration. As a result of this, there are three sewage discharge outlets (#1, #2, #3) at the downstream of Fu-tian River. The volume of water is about 40,000 m³ between #1 and #3 sewage discharge outlets. The lengths are 350m and 400m, respectively. Finally, the water reflux circulating equipment and fibrous braids fillers are installed, specific bacteria are added, and oxygen is aerated.

The main design parameters are as follows:

- Daily sewage flow: 15,000 m³/d;
- Maximum processing volume: 40,000 m³;
- Fibrous braids filler total length L = 1,207 m, spacing 2.5 m;
- Average width of Fu-tian River W = 16 m;
- Average effective water depth H = 2.1 m (0.5–4.0 m);
- Hydraulic retention time: HRT = 64 h;
- Packing rate of filler = 0.9%;
- Biodegradation of organic compounds oxygen demand: $[O_2]_1 = 31.25$ kg O_2/h;
- Nitrification of ammonia oxygen demand: $[O_2]_2 = 29.38$ kg O_2/h;
- Sediment oxygen demand: $[O_2]_3 = 31.25$ kg O_2/h;
- Aerator oxygenate quantity: $[O_2]$aerator = 88.7 kg O_2/h;

Addition of Specific Bacteria

- Addition quantity: D2 = (LF2 × TF2 × STTF2)/SF2 = 0.103 (T/d);
- Active agents: H2 = D2×0.25 = 0.026 (T/d), SF2 = 1/(Q2×0.00000264);
- Addition quantity in one week: D20 = (H2+D2) ×7 = 0.129×7 = 0.9 T.

Activation of specific bacteria: According to specified proportions of the addition of bacteria A and B to drinking water, aerate and control the temperature to remain between 15.5–32°C (the ideal temperature being 24°C). Bacterium A requires a continual aeration for 48 hours, and bacterium B requires a continual aeration for 7 days. The goal of activation is to increase the quantity of bacteria and improve water quality, as well as to produce enzymes for Bacterium A to decompose macromolecule organisms, and for Bacterium B to increase the quantity of nitrifying and denitrifying bacteria.

At the beginning, these specific bacteria are added to Fu-tian River with the packing filler by the sequencing batch method. After one month, specific bacteria are added to the river by the continuous method. During the rainy season, the situation is different, and the system impact is also different. Adding the specific bacteria quickly restores the system. If the system is under continuous impact, the system will function improperly. At this time the specific bacteria must be added at 2.5 – 5 times the normal dosage in order to restore the system.

Design of Water Reflux Circulation

The reflux circulating pump is downstream from the Fu-tian River. After the water treatment, it transports the treated water to the upper stream, which is the terminal of the third processing section. This causes the circulatory flow moving from the upper stream downstream, enhancing the oxygen effect, keeping the water clean and stabilized. In order to guarantee purification of the river, the amount of dissolved oxygen in every section must be higher than 2mg/L. The two diving pumps (one working, one on standby) enhance the denitrification as well as improving the rate of phosphorus removal, downstream in front of the Fu-tian River floodgate. The diving pump type is: 150QW200–22A, $Q = 117–238 \, m^3/h$, H = 15–21 m, N = 22kw.

Analysis of the Chemical Properties of the Wastewater

Table 1 Analysis of the physical and chemical index of water

Item	Examination method	Instrument name and type	Remark
TN	Alkaline potassium persulfate digestion-UV spectrophotometric	Thermo Helions Gamma	GB 11894–89
TP	Ammonium molybdate spectrophotometric	Thermo Helions Gamma	GB 11893–89
NH_3-N	Nessler's reagent colorimetric	Thermo Helions Gamma	GB 7479–87
Turbidity	Boiler and cooling system	DR/890	GB/T 12151
BOD_5	TWT rapid determination	BOD Tanker	DIN 38409T51
COD_{cr}	HACH Quick sealing cataytic digestion-spectrophotometry	COD Quick determination meter (DR/890)	HJ/T 399–2007
SS, VSS	Gravimetric Method	Drying Oven	GB 11901–89

DGGE Profiling

Polymerase Chain Reaction Amplification: This method of genomic DNA extraction was used according to Roling, *et al.*, (2002). The parameters used for amplification of 16S rDNA of Bacteria are listed in Table 2. The PRBA341F to 518R DNA region contains one variable loop of rRNA. Amplification mixtures with bacteria primers (10μmol/L P_1 2μL, 10μmol/L P_2 2μL) had a final volume of 50 μL and contained 2×Taq Master Mix test kits (containing 0.1U Taq Polymerase/μL, 500μM dNTP each, 20mM Tris-HCl [pH8.3], 100mM KCl, 3mM $MgCl_2$) 25μL, DNA template 2μL, ddH_2O 19μL. The reaction began with an initial 94°C denaturation for 5 min, followed by 30 cycles of 94°C for 45 s, 50°C for 45 s, 72°C for 90 s, and a final extension at 72°C for 10 min. It was then held at 4°C. The presence of PCR products was confirmed by electrophoresis on 1.5% agarose gels stained with ethidium bromide. The marker was 100bp DNA ladder. To confirm the reproducibility of the method, all amplifications were repeated at least three times in the Beihang University Laboratory.

Table 2 Parameter for the amplification of 16S rRNA genes of Bacteria or Archaea.

Primer	16S rDNA target (base number)	Primer sequence5'–3'
P1	V3(341F/534R)	5'-CGCCCGCCGCGCGGCGGGCG-GGGGGGGCCCAC GGGGGGCC TACGGGAGGAGCAG-3'
P2		5'-ATTACCGCGGATGCTGG-3'

A DGGE analysis was carried out in a D-Gene apparatus (Bio-Rad Dcode ™). Briefly, PCR products were resolved on 8% (w/v) polyacrylamide gels in 1×TAE (20 mM tris-Cl, 10 mM acetate, 0.5 mM Na_2EDTA, pH 7.4) using denaturing gradients ranging from 30% to 55%. Gradients varied according to the parameters used for amplification, the nucleotide composition of the resultant products, and the apparatus used for DGGE. Electrophoresis was carried out at a low voltage (25V) for 20 min, then 120V for 3 to 4.5 hours. An electrophoresis buffer (1×TAE) was maintained at 60°C. Gels were then stained with 0.1% $AgNO_3$ on a UV transilluminator and photographed (Bio-Rad Amp Gene).

Statistical Analysis of DGGE Banding Patterns

DGGE gels were processed and analyzed with Quantity One 4.6 software (Bio-Rad, USA). Banding patterns were normalized with the reference pattern included in all the gels. Their reproducibility was evaluated using Sorenson's similarity index: $Cs = 2N_{AB}/(N_A + N_B)$, where Cs is the Sorenson's similarity, N_A the total number of peaks (bands) in profile A, N_B the total number of peaks (bands) in profile B, and N_{AB} is the number of matching peaks in profiles A and B. To assess similarities between whole profiles, the binary matrix was analyzed by clustering, using unweighted pair groups with mathematical averages (UPGMA) and a Dice coefficient of similarity. The Shannon-Wiener index of diversity (H) was used to determine the diversity of bacteria present biofilm sampled from carriers in the Fu-tian River. This index was calculated by the following equation:

$$H = -\sum_{i=1}^{s} Pi \times \log_e Pi$$

(where s is the number of species in the sample and Pi is the proportion of species i in the sample).

RESULTS AND DISCUSSION

Since the treatment project began in August 2006, the water quality in Fu-tian River has continually improved. In a short period of time, the quality of the water became so high that the water in some sections of this river has achieved the surface water V-class standard, as shown in Figure 3(b).

(a)   (b)

Figure 3 Fu-tian River before treatment (a) and after treatment (b)

Monitoring data is shown in Table 3. At the beginning, a bacterium was added to the river with the packing filler by the sequencing batch method. During the steady running period the bacterium was added to the river upstream by the continuous method. In the rainy season, if the system was impacted continuously, the system might have collapsed. To prevent this, more specific bacteria (2.5–5 times the regular amount) had to be added to restore the system. The added specific bacteria formed the biofilm covers on the surface of fibrous braids filler which improved the removal efficiency. The addition of the specific bacteria was the key factor in changing the water quality of Fu-tian River.

In Fu-tian River, the DO levels, particularly in stations with untreated water, reached the anoxic on January 16th, January 21st, and April 12th, 2007. Ammonium, TN, TP and COD decrease from upstream to downstream, whereas DO increase, showing a large improvement (Table 3), due to the aerator effect in the Fu-tian River. Dissolved oxygen was positively correlated with TN and negatively with NH_4^+–N and COD.

As shown in Table 3, COD_{Cr} varied at every floodgate of the Fu-tian River Center Park. In the rainy season the sewage and non-point source pollutants increased. As a result, the removal effect of COD_{Cr} was not obvious. But during the period from January 23rd to March 20th, 2007, the COD_{Cr} removal effect was marked, with a removal percentage of 45–98%. The lowest COD_{Cr} concentration was 6mg/L. After rainfall, a large amount of non-point source sewage flowed into the Fu-tian River, causing the volume of the contaminated water to increase by several times – or even several dozen times – the regular level. The volume of contaminated water and organic loading have changed within a large range, and therefore an accurate estimation is difficult to make. This water treatment project experienced a number of serious impacts from August 2006 to June 2007. During the period of January 9th – February 8th, 2007, sewage water hugely increased and the amount of pollutants in the water, such as chicken and duck feathers, was high. COD tended to decrease from upstream to downstream, but COD in the raw water was not so high, and the COD_{Cr} removal effect was not obvious. The removal percentage was 10–81%, and the average content of COD in effluent was 33mg/L.

The turbidity of Fu-tian River has also changed at every floodgate (as shown in Table 3). From January 16 to April 14, 2007, the SS tended to decrease from upstream to downstream, and the removal effect was obvious. The removal percentage was 6-81%, and the lowest content of SS was 4mg/L.

In this river, nitrogen and phosphorus are the principal elements creating water eutrophication. Eutrophication is one of the main problems that most rivers encounter. Concentrations of N and P exceeding the allowed values will cause the eutrophication, which brings a series of problems of the contamination of the water in rivers. In Fu-tian River, discharged sewage had very high levels of N and P, and also contained some pollutants from the long rainy season. Therefore, the water in the river must be controlled from the source of the river by the sewage treatment plant to effectively eliminate N and P. For in situ remediation, N and P that has already entered the water cannot be forecasted. In this paper, it is shown

Table 3 Data monitored from January to April 2007. Units in mg L^{-1}, except for temperature (°C)

Date	Site	DO	Temperature	NH$_4^+$–N	TN	TP	COD	SS
January 9th	Raw water	3.75	16.7	5.14	12.28	0.77	42	14
	Upstream	3.63	19.3	6.43	10.89	0.70	27	15
	Middlestream	3.77	19.1	6.43	11.08	0.91	25	15
	Downstream	6.02	19.0	6.17	12.56	0.56	32	13
January 12th	Raw water	0.36	18.5	7.33	12.10	1.02	147	25
	Upstream	2.19	21.9	5.79	13.49	0.79	31	11
	Middlestream	2.33	21.7	7.07	14.51	0.33	37	15
	Downstream	4.77	21.6	6.04	9.50	0.52	47	13
January 23rd	Raw water	0.31	18.4	19.38	20.36	0.51	208	69
	Upstream	1.89	21.9	7.71	14.33	0.13	50	20
	Middlestream	2.81	21.3	7.71	14.23	0.20	42	19
	Downstream	3.34	20.7	9.63	16.00	0.20	65	20
March 13th	Raw water	0.20	20.4	19	31.10	1.15	280	21
	Upstream	4.28	21.7	7.9	16.60	1.04	6	9
	Middlestream	4.90	21.6	7.6	19.20	0.82	32	10
	Downstream	7.33	21.1	7.8	16.00	0.44	16	13
March 20th	Raw water	1.34	22.6	21.71	22.31	2.91	68	53
	Upstream	2.30	21.8	3.64	10.80	1.28	32	16
	Middlestream	4.61	21.5	5.58	13.03	1.61	41	18
	Downstream	6.28	21.3	3.64	13.77	0.97	14	16
April 12th	Raw water	0.15	27.0	11.12	18.29	0.86	82	39
	Upstream	1.77	27.1	8.91	15.61	0.79	56	25
	Middlestream	6.65	28.4	5.10	6.66	0.73	34	15
	Downstream	6.81	29.0	4.48	6.80	0.44	31	15
April 14th	Raw water	0.79	21.4	6.36	8.49	0.91	133	16
	Upstream	1.47	20.1	5.98	8.09	0.82	113	13
	Middlestream	7.25	22.0	3.83	5.15	0.75	35	15
	Downstream	6.85	23.0	3.57	5.22	0.43	35	7
April 15th	Raw water	0.41	20.6	4.78	6.80	0.91	112	19
	Upper reaches	1.16	20.7	4.17	6.39	0.90	100	18
	Middle reaches	7.17	20.7	3.10	3.56	0.35	36	13
	Lower reaches	8.07	20.9	3.64	5.41	0.39	21	4

(Continued on next page)

Table 3 Data monitored from January to April 2007. Units in mg L⁻¹, except for temperature (°C) (*Continued*)

Date	Site	DO	Temperature	NH_4^+–N	TN	TP	COD	SS
April 18th	Raw water	0.65	25.0	8.19	9.69	0.59	86	15
	Upstream	1.01	24.3	7.99	8.66	0.57	81	19
	Middlestream	7.97	24.4	5.58	7.42	0.36	32	10
	Downstream	8.55	24.1	4.78	7.37	0.27	37	14
April 21st	Raw water	1.17	25.7	6.33	7.42	0.36	77	18
	Upstream	2.65	26.1	6.25	6.70	0.40	53	16
	Middlestream	6.67	26.0	5.80	6.08	0.33	45	11
	Downstream	7.00	26.4	5.31	5.36	0.24	31	15

How this problem was solved by using the specific bacteria. However, as shown in Table 3, this system could not effectively remove the TN, and the range of variation was quite large. In the rainy season, the removal rate of TN dropped. The reason was the huge amount of water in the rainy season, causing the decrease of the concentration of TN. Generally speaking, this treatment method is not good for the effective elimination of TN.

The variation of NH_4^+–N in each section of Fu-tian River at every sample point is shown in Table 3. It can be seen that in March, the average dissolved oxygen was 5.2mg/L, so biodegradation was prevailed by nitrification; in other words, the denitrification was weak. During the treatment period, the removal effect became better and better. Relating with the analysis to TN, if COD_{Cr} was low in the water of Fu-tian River, then the denitrification cannot work well due to the lack of sufficient carbon sources. Therefore it can be seen from the monitoring data shown in Table 3 that from January 12th to April 21st, the removal percentage of NH_4^+–N reached 90%, but the removal effect of TN was not so high.

As Table 3 shows, the change in the concentration of TP is a result of the climate warming. Massive detergents, including P surfactants along with the sewage discharge to Fu-tian River in the river bottom, leave little sediment. Therefore this system also did not significantly remove the TP. From January to February the trend was one of escalation, but after the March rainy season, the removal rate of TN dropped. The reason was that the capacity of the rainy season is huge, causing the concentration of TP to be diluted.

Denaturing Gradient Gel Electrophoresis Analysis

The amplified regions of the 16S rDNA, from bases 338 (PRBA338F) to 518 (PRUN518R), are shown in Fig. 4(a). The DGGE profiles indicates that there are several dozen different bacterial 16S rDNA nucleotide sequences, which are amplified from Fu-tian River water DNA, as shown in Fig. 3(b). The profiles from upstream showed many differences with different situations. The DGGE fingerprints of Fu-tian River from different sample points because of differences in band intensities. For example, in Fig. 4(b), Lanes 1, 2, and 3 are different from each Gradients varied with the primers used for amplification, the other, which illustrates spatial heterogeneity. When profiles are composed of distinct bands, comparisons can be made between gels by including molecular markers as standards. Based on the presence and absence of bands in each sample, similarity coefficients were determined for the DGGE profiles generated using the primers (Table 1).

In general, the bacterial structure of the Fu-tian River differs at upper reaches Fig. 3(b) (Lanes 1–3). The community structure of the middle reaches or lower reaches were different from each other Fig. 3(b) (Lanes 4–6) or (Lanes 7–9).

Products show distinct differences in different sample points in the microbial community composition because of the number and the varying distances that the DGGE products migrated in the gradient gel. Ecological studies were conducted to distinguish communities from different ecosystems and to determine numerically dominant phylotypes. The biodiversity analysis was conducted with DGGE correlation software, and the results of the analysis are illustrated in Fig. 5. It can be seen from the figure that in the upstream, the biological diversity of biofilms consists of 18 species in the surface layer. There are 26 species in the middle layer and 29 species in the bottom layer. In the middlestream, the biological diversity of biofilms includes 31 species in the surface layer, 33 species in the middle layer, and 31 species in the bottom layer. In the downstream, the biological diversity and total biomass of biofilms includes 35 species in the surface layer, 40 species in the middle layer, and 42 species in the bottom layer.

The number of bands that comprise the DGGE patterns indicated that there was a high diversity of Bacteria PCR amplification products in Fu-tian River. The profile is distinct, showing some strong bands that indicate the presence of numerically dominant bacteria populations. We are still able to qualitatively distinguish between the complex DGGE profiles of the upstream and middlestream and also between the downstream. According to the DGGE results of biofilms in Fu-tian River, the biological diversity of biofilms in the Fu-tian River become abundant and increased from upstream to downstream, and from the surface layer to the bottom layer.

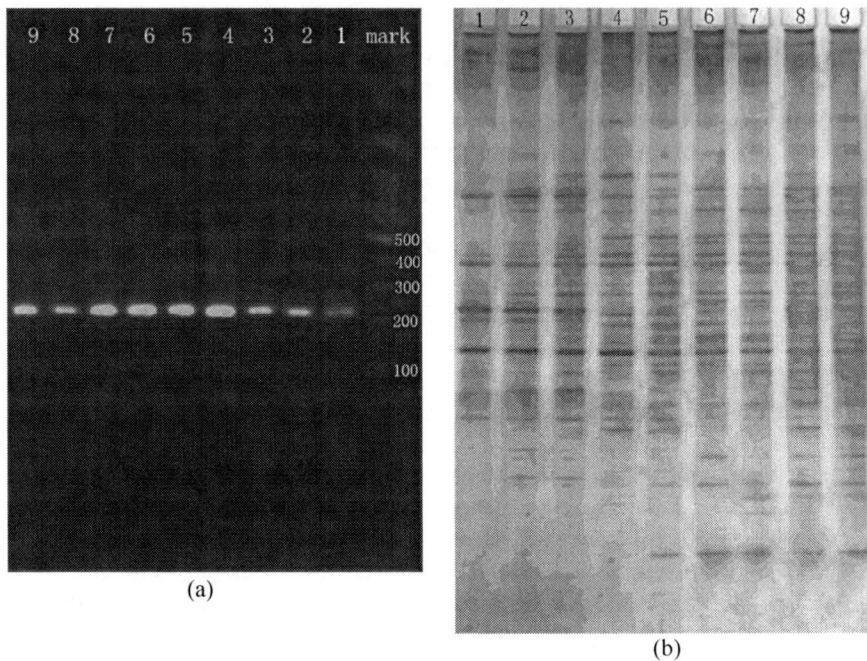

(a)

(b)

Figure 4 (a) PCR profiles (Bases 341F to 534R relative to the E. coli rRNA sequence), (b) Denaturing gradient gel electrophoresis analysis of Bacteria 16S rDNA polymerase chain reaction products.(1,2,3 the surface ,middle and bottom layers of biofilm from upstream; 4,5,6 the surface, middle and bottom layers of biofilm from middle stream; 7,8,9 the surface, middle and bottom layers of biofilm from downstream)

These combined observations show that the presence of a few dominant species will produce DGGE profiles that can distinguish different sample points with different relative proportions of populations in the two communities. It is possible that the communities are more readily differentiated because only a subset of the 16S rDNA from the bacterial community is being amplified. This can be beneficial if the communities are too complex and one merely wishes to see whether two sample points can be differentiated. A statistical analysis was possible with bacteria communities in our study because dominant populations were readily observed. Therefore, the technique is able to differentiate among ecosystems. In the upstream of Fu-tian River, the bacteria diversity was greatly reduced. The limited bacterial diversity in all the upstream is indicated by the few products observed by DGGE, as shown in Fig. 4(b) (Lanes 1–3). There are few dominant species in the upstream, which will not produce a stable ecosystem. The middlestream is a transitional zone, having an obvious peak value and certain dominant species, but other species also have very strong competition in this position. It is not a stable stage within the biological community. In the downstream, as indicated by the many other products observed by DGGE in Fig. 4(b) (Lanes 7–9), there are more dominant species. This time the species, through a competitive balance, enter the concordance evolutionary and stable community stage, which is also indicated by the clearer water downstream.

Table 4 Shannon-Wiener index of bacteria population in biofilms from carriers of the Fu-tian River

sample	1	2	3	4	5	6	7	8	9
Shannon-Wiener index	2.47	2.99	3.17	3.08	3.25	3.17	3.25	3.33	3.29

The results revealed that biological diversity varied in different locations of Fu-tian River and depth of carriers. In the first instance, the biological diversity index increased from upstream to downstream. The mean value of biological diversity index upstream is 2.88; in the middle stream and downstream it increased to 3.17 and 3.29, respectively. Secondly, the biological diversity varied in the changes of depth, for the upstream, the Shannon-Wiener index of surface layers was 2.47, the index of middle and bottom layers increased to 2.99 and 3.17, respectively. For the middle stream and downstream, the index of the middle layer was higher than the surface and bottom layers.

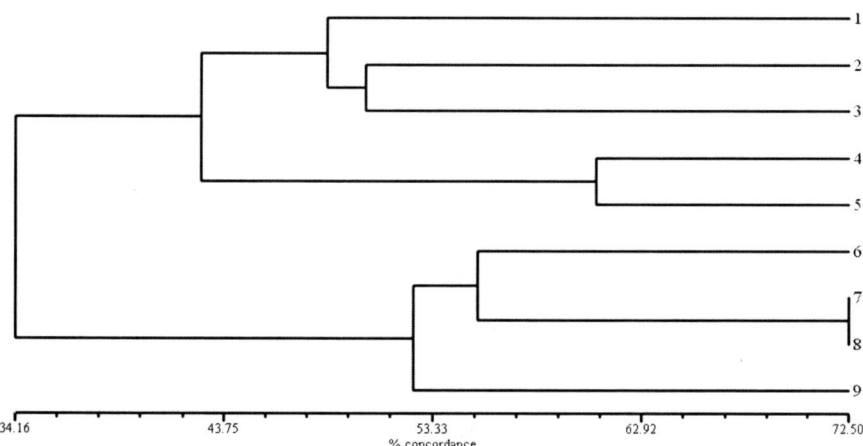

Figure 5 Cluster analysis of DGGE profile, biofilm from individual carrier from upstream to downstream

There were intense differences as well as the appearance of new bands in different profiles. The overall similarity between the 9 biofilm profiles on the cluster analysis was 34.16%, with clustering below 72.5%. The 9 biofilm samples were divided into three clusters. Bioflim samples from upstream classified cluster 1, bioflim samples from downstream and bottom layer of middle stream classified cluster 2, and bioflim samples from middle stream (expect bottom layer) classified cluster 3. The difference in overvall similarity values between each cluster in Fig 5, was 8.6% for cluster 1, 18.6% for cluster 2 and 26.6% for cluster 3. Thus the bacteria communities structure of biofilm are closely related to sampling sites of River.

ACKNOWLEDGEMENT

The work was supported by Shenzhen government, from the Fu-tian River Water Treatment Research Program of Shenzhen City

REFERENCES

Allision D. G. and Gilvert P. (1992). Bacterial biofilms. *Science Progress Oxford*, **76**, 305–321.

Akiyoshi O. and Hideki H. A. (1996). Novel concept for evaluation of biofilm adhesion strength by applying tensile force and shear force. *Water Science and Technolgy*, **34**, 201–211.

Arcangelis J. P. and Arvin E. (1992). Toiuene biodegradation and biofilm growth in an aerobic fixed-film reactor. *Applied Microbiology and Biotechnology*, **37**, 510–517.

Bruce E. R. and Perry L. M. (2002). *Environmental Biotechnology: Principles and Application*, Qinghua University Publishers, Beijing.

Cairns J. and Pratt J. R. (1995). Ecological restoration through behavioral change. *Restoration Ecology*, **3**(1): 51–53.

Magdaleno A., Puig L., de Cabo L., Salinas C., Arreghini S., Korol S. Bevilacqua S., López L. and Moretton J. (2001). Water pollution in an urban Argentine river. *Bulletin of Environmental Contamination and Toxicology*, **67**, 408–415.

Crocetti G. R., Hugenholtz P., Bond P. L., Schuler A., Keller J., Jenkins D. and Blackall L. L. (2000). Identification of polyphosphate accumulating organism and design of 16S rRNA-directed probes for their detection and quantitation. *Applied and Environmental Microbiology*, **66**(3), 1175–1182.

Davis N. M., Weaver V., Parks K and Lydy M. J. (2003). An assessment of water quality, physical habitat, and biological integrity of an Urban stream in Wichita, Kansas, prior to restoration improvements (phase I). *Archives of Environmental Contamination and Toxicology*, **44**, 351–359.

GB 11894–89, National Standards of P. R. China, State Environmental Protection of China (2005). Total nitrogen (TN), Alkaline potassium persulfate digestion-UV spectrophotometric method [S] (in Chinese).

GB 11893–89, National Standards of P. R. China, State Environmental Protection of China, (2005). Total phosphorus (TP), Ammonium molybdate spectrophotometric method [S] (in Chinese).

GB 7479–87, National Standards of P. R. China, State Environmental Protection of China (2005). Ammonium (NH3-N), Nessler's reagent colorimetric method [S] (in Chinese).

GB/T 12151 National Standards of P. R. China, State Environmental Protection of China (2005). Turbidity, Analysis of water used in boiler and cooling system[S] (in Chinese).

GB 11901–89, National Standards of P. R. China, State Environmental Protection of China, (2005). Suspended substance (SS), Gravimetric method [S] (in Chinese).

Jianlong W., Xiangchun Q., Liping H., Yi Q. and Hegemann W. (2002). Microbial degradation of quinoline by immobilized cells of Burkholderia pickettii. *Water Research*, **36**(9), 2288–2296.

Lombardi A. T., Garcia J. R. and Mozeto A. A. (2001). Bioleaching of metals from anaerobic sewage sludge: effects of total solids, leaching microorganisms, and energy source. *Journal of Environmental Science and Health. Part A,. Toxic/Hazardous Substances & Environmental Engineering*, **36** (5), 793–806.

Plonka A. and Bogus W. (1999). Radiation Physics and Chemistry, **56**, 623.

Roling W. F. , Milner, M. G., Jones, D. M., Lee K., Daniel F., Swannell R. J. and Head I. M. (2002). Robust hydrocarbon degradation and dynamics of bacterial communities during nutrient enhanced oil spill bioremediation. *Applied and Environmental Microbiology*, **68**(11), 5537–5548.

Rutherfurd I. D., Jerie K. and Marsh N. (2000). A rehabilitation manual for Australian Rivers, Land and Water Resources. Research and Development Corporation, Cooperative Research Centre for Catchment Hydrology 1–192.

Whalen P. J., Toth L. A., Koebel J. W. and Strayer P. K. (2002) Kissimmee river restoration: A case study. *Water Science and Technology*, **45**(11), 55–62.

William J. M., Jean C., et al., (2002). Ecological engineering applied to river and wetland restoration. *Ecological Engineer*, **18**, 529–541.

Williams T. C. and Dee P. E. (1995). A citizen's approach to integrated river basin management. *Water Science and Technology*, **32**(5–6), 169–174.

PART SEVEN

Future Outlook

Network Infrastructure – Cities of the Future

Valerie I. Nelson

Coalition for Alternative Wastewater Treatment, PO Box 7041, Gloucester, MA 01930, USA
E-mail: valerie.i.nelson@gmail.com

Abstract The genius of science and design in the 21st Century is the discovery of "smart, clean, and green" ways to capture the value of resources. "Smart" because they unlock the complex designs of nature and use information and signaling to achieve efficiencies. "Clean" because they capture and use resources and methods that don't involve significant externalities in extraction or disposal. And, "green" because they rely to a much higher degree on vegetation, and in the process begin to restore the natural ecosystem and its wide and deep benefits.

Keywords Decentralized, cities, biomimicry, integrated resource management

TRADITIONAL WATER MANAGEMENT

Traditional water management has relied on a low-tech, industrial-scale engineering and economic model mostly developed in the 1800's. With a goal of public health protection, big pipe systems were built to transport clean water into and wastewater out of urban neighborhoods. This model which produced important health and ecological gains for our communities has also shown a down side.

In recent years, a concern has been growing that this "paradigm" of big-pipe water management is not sustainable, both from a natural resource and an economic perspective. The appropriation of huge volumes of water from the ecosystem and its release as partially-treated effluent into rivers, lakes, and oceans has been increasingly disruptive to those ecosystems. Population growth, climate change, agricultural practices, energy and other practices will challenge this approach further.

Signs of stress are seen in falling groundwater levels and decreasing dry-weather stream flows (and unnatural peak flows during wet weather), destructive eutrophication of lakes and estuaries, disappearance of wetlands, increasing dead zones in coastal areas, and other catastrophic changes in hydrological functions. Climate change is expected to exacerbate patterns of droughts and heavy rainfalls, putting both water supplies and flood control measures at risk. Reductions in evapotranspiration from vegetation destruction are being studied as potentially significant contributors to global warming.

Drinking water systems lose huge amounts of water (a US average of 20%) from their leaky distribution pipes, existing treatment technologies were not designed to eliminate emerging biological and chemical contaminants that are increasingly found in sourcewaters, and treating all water to new and more stringent standards is both increasingly difficult and expensive. Except for the small amount of water needed for potable uses, the delivery and treatment of entire, ever increasing, supplies is extremely wasteful of energy, chemicals and money. Most cities and towns have been unwilling to charge ratepayers the full cost of repairing and replacing the existing, often inadequate infrastructure, so collapsing pipes and breakdowns in delivery systems and treatment plants have become more frequent, while innovation is minimally on the radar screen.

THE BALTIMORE CHARTER FOR SUSTAINABLE WATER SYSTEMS

The 2007 Baltimore Charter for Sustainable Water Systems asserts an alternative approach to water management that "mimics and works with nature." Natural systems create an abundance of value and diversity, where species cooperate and one species' waste is another species' resource. These naturally-balancing ecosystems have been steadily deteriorating under a century-long highly-disruptive human extraction and use of resources in the industrial era.

An emerging paradigm relies instead on design principles found in nature: in particular, integrated systems, efficiency and reuse, and adaptation to local context. Many of the new high-performance treatment technologies, such as membranes, "mimic" biological and chemical designs that scientists are discovering in nature (biomimicry). Just as recently found in the energy arena, there are alternative approaches that can restore natural resource patterns and functions found across a landscape. These new design approaches create a wealth of services and benefits at the local level and can help restore the ecological and societal well-being of the global Commons as well.

Opportunities also exist in integrated design, rather than in narrower specialized thinking and practice. To paraphrase, the "sum of the conventional parts" in the traditional approach has been much less than the "whole" in infrastructure services. Integrated design can increase productivity of the larger system, while also serving the separate functional needs of the parts.

Another resource to be tapped from nature is the efficiency and high-performance of its organisms and systems. Biologists and chemists are looking more and more to nature for models to re-engineer products and processes. Membranes in nature, for example, are inherently more efficient than those used in water and wastewater treatment, because of active rather than passive transport mechanisms inherent in biological versions.

Finally, as Ian McHarg wrote in the late 1960's, by locating activities in the most appropriate places in a watershed, natural resource "streams of value" can be tapped with less cost and disruption. McHarg laid out guidelines for locating farms, ports, forests, wildlife corridors, cities, etc. There are lessons to be learned, as well, from "networks" of "nodes" and "links" in nature that assure resilience and adaptability to external shocks to the system.

NETWORK INFRASTRUCTURE

A birds-eye view of the new infrastructure would reveal "networks" of decentralized, repurposed, and at times hybridized systems. Some of the innovative treatment and resource recovery technologies would be "embedded" in subdivisions, apartment complexes, or individual homes and offices. Other functions would be taken over by vegetative "green infrastructure", such as green roofs and walls, trees, and swales along roads and restored streams, riparian areas, and wetlands. Water and sewer lines might be slip-lined and repurposed for potable or reclaimed water, water storage and distribution, and heat recovery. Monitoring and control technologies would be key elements in managing these systems and in protecting public health and the environment.

These engineered and green networks mimic the natural systems of nodes and links in nature, where water both recycles and supports life at a local scale, but also is a linkage and transport mechanism across a landscape and into the atmosphere. Adopting these systems in cities and towns can cost less to provide water and sanitation services than current approaches and can also add significant benefits in terms of air quality, energy savings and production, recreation, beauty and aesthetics, increased property values, and jobs. Innovative pricing, incentives, and new performance-based regulatory mechanisms will be required to ensure that these sustainable practices are adopted and that the remaining watershed and global "externalities" are also addressed by developers, homeowners, industries, and municipalities.

Some leading-edge infrastructure experts are now suggesting that these "networks" of engineered and green energy and water systems need to be integrated and also be co-engineered with transportation, solid waste, buildings, and other urban infrastructure management. The lessons of nature are that such integration can lead to significant synergies of design, cost-savings, and an abundance of positive benefits for society.

For example, an "eco-block" incorporating architectural innovations, wind and solar power, green roof and wall cooling, rainwater harvesting, water reuse and energy recovery, and nutrient recycling into community gardens, can be nearly "off-the-grid" in both energy and water, and can be located at transportation "hubs". These new designs of infrastructure may cost less in dollars and will both improve the quality of life in urban communities and begin to protect and restore the ecological Commons.

Paralleling the shift in technologies will be a shift in the institutions and markets for resource management. Municipal utilities evolved for each single-service "monopoly" in the form of separate centralized systems for water supply, stormwater transmission, and wastewater discharge (and in some cases energy generation/distribution). But embedded and green infrastructure "nodes" in homes, subdivisions, and commercial establishments engage a wide range of private firms, non-profit groups, and other city agencies (such as parks and recreation, housing, job training, etc), and the developer and property-owner will have many more choices for technologies and design and ongoing maintenance services. Municipalities and other local governments can anticipate more complex and highly-productive new roles in coordinating municipal utilities and agencies internally and in overseeing the new private and non-profit sector externally through ordinances, incentives, education, and inspections.

A new policy framework for cities and towns of the future will be necessary to maximize the strengths of new markets, but also to direct those markets toward protection and restoration of the Water Commons, rather than to "commodify" water. Current policies protect public health in important ways, but also impede the discovery of efficiencies and adoption of innovative technologies and designs. Market forces do need to be unleashed, but only if goals, incentives, and safeguards are in place to advance the public interest, including the health and functioning of ecosystems and communities.

Finally, the solutions to water management in the 21st Century will require a high level of interdisciplinary collaboration and broad public engagement. Here also, nature serves as a model for the benefits of collaboration and cooperation in society, as opposed to the specialization and hyper-individualism of the 20th Century. Networks of conversations and pilot projects will serve as the foundation for creative invention and enhancement of the "Common Wealth."

The International Water Association and Cities of the Future
A Work Program on Behalf of a New Paradigm

Steve Moddemeyer and Lee Roberts

Seattle Public Utilities and Department of Planning and Development,
CollinsWoerman, Seattle, USA
E-mail: smoddemeyer@collinswoerman.com

Abstract Cities of the Future is an initiative of the International Water Association (IWA). Global networks of working groups have been established, addressing fundamental issues in future urban design. The programme includes engineering elements such as integrated treatment technologies and smart networks. It also includes planning and institutional elements, including new partnerships in spatial planning, land use and developer interaction, institutional reform and a new functional aesthetic of urban design land and waterscapes. The workgroups will rely upon IWA's robust assemblage of specialist groups as well as new partnerships both within and outside the water sector. These partnerships will require recruiting beyond IWA's traditional areas of expertise and an increased awareness of the issues and leaders who are driving similar change in other sectors. The energy sector, the social sector, and transportation sectors will each play a particular role as collaborators to achieve the kinds of changes that are required to meet the challenge. A number of global conferences and symposia will be employed to share the ideas of the Cities of the Future Programme. Leaders in other disciplines will also be collaborators and partners in this emerging programme of change.

Keywords Cities of the future, Integrated treatment technologies, Energy, Water infrastructure

INTRODUCTION

Cities are facing rapid urbanization, fuelled further by the rising population and global income growth. This time of change is made even more uncertain by the emerging effects of climate change. Too many cities keep themselves busy with the day to day routine of working life and fail to notice the gathering clouds in the distance. A few cities, however, are now stopping to think and make new choices that will affect their ability to survive and thrive in the 21st Century. These are the Cities of the Future. These cities make informed choices in aspects such as: capital investment and infrastructure; regulations and incentives; and land use and open space. The choices they make and their ability to be ready for the future will affect many sectors that provide resources to urban areas. To be successful in the future, cities will need to concentrate their efforts on the services they provide and think about how to manage risk in a time of accelerating change.

As an international leader in the water sector, the International Water Association (IWA) can help cities, utilities and research communities to work together to create robust and resilient responses to these imminent changes. However, the responses that appear to be most appropriate will require new kinds of partnerships, new relationships, and a new sense of the interconnectivity between the sectors, the people, and the ecosystems that support them.

Towards that end, vanguard programs are required for stimulating advances in the leading edge of urban water efficiency, resource recovery and ecological sustainability. The emerging Cities of the Future program will have a broad focus that addresses these challenges across the spectrum of lower, middle and high income countries.

What is the Problem?

By 2050, the world's population will grow from the current 6.8 billion to around 9 billion. While this population growth alone is enough to inspire a reconsideration of our use of water and resources, it is important to note that most of this growth will occur in developing regions. As the population of developing regions undergoes continuous rapid growth, we must ask, what model of infrastructure is most appropriate? In a world of constrained resources and water security, what model should this new growth follow?

Developed regions have demonstrated huge gains in human health through modern infrastructure systems. Although these systems were first developed during the age of Rome, they do not offer a sustainable model for the rest of the world to follow. The average resource use footprint of one person, living in a developed region, is currently three times larger than the per capita resources available on Earth. This presents a major problem: if the environment and lifestyle of every person on the planet were to instantaneously upgrade to the standards of Europe or the US, we would need the resources of three Earths in order to serve us all; we have just one.

As developing regions rightfully push for improved health, welfare, and access to resources, it seems that our development pattern on this planet cannot be sustained. All of us, in both developed and developing regions, must seek new efficiencies and new ways of living in order to create and maintain a high standard of living. Health, economic opportunity, and social wellbeing must be provided in such way that allows the world's environments to regain balance and vitality.

We are in a period of intense change. We are moving from a linear, predictable world into a new world where the pace and direction of change are uncertain. We are moving from technologies that emphasize separation and specialization to those in which connection and collaboration are essential. We are moving from a mindset in which discrete problems could be solved individually into one where we must solve interlinked problems with integrated solutions. We are beginning to see the world as a finite, closed-loop system in which nothing is isolated, nothing is externalized, and nothing is waste.

We are in the Discovery Phase

The water community is a critical piece in the jigsaw of systems and regulations that drive the city forward. A holistic solution to cities involves parallel advancements in land use patterns, development and financing, design standards, and government regulations and incentives. The water community, in isolation, cannot solve these complex problems – nor can any other sector, without involving water. IWA, therefore, can play a pivotal role in identifying, understanding, and addressing the problems of our time.

How can IWA Help?

The Cities of the Future program was designed to seek, disseminate, and implement solutions to these complex problems. Cities of the Future brings together leaders of the water sector with the following beliefs:

1. that we can manage water (drinking water, stormwater and sanitation services) meet tomorrow's challenges;
2. that the technologies, tools, and practices available to us today can enable transformational approaches to urban water management;
3. that we would have a different set of design objectives for urban water management if we began with a blank slate;
4. that each city is unique and requires a context-specific water management solution, but that a common set of practices can be applied to arrive at the best solution for each city;
5. that the water management solutions for cities of the future cannot be developed by the water sector alone, but must be developed with a much broader range of partners and stakeholders to produce solutions that are flexible, adaptable and robust.

The aim of Cities of the Future is to equip water professionals with the skills and tools needed to make water a central part of the urban design process. The long-term vision is that water managers, if equipped with these tools, can routinely collaborate with other professionals and the local community to redesign water management integral with other city services to enhance life both within and beyond the urban environment.

To this end, IWA and Cities of the Future, sponsors periodic conferences and symposia that bring together technical experts, decision-makers, thought leaders and advocates. Through these conferences and symposia, as well as other events, Cities of the Future focuses on three broad categories of work: knowledge development; dissemination and communication; and influence.

To move forward within and across these categories, the Cities of the Future program has convened a number of Workgroups in various fields:

FRAMEWORK DOCUMENT

Task: To develop an overarching framework that will become the basis for extending the emerging Cities of the Future regime into practice.

The framework being developed is intended to improve the understanding of the main issues and emerging strategies of Cities of the Future. It involves the agreement of a common set of terminology and language used by groups and contextualizes the different elements of the program, thus setting the stage for an informed discussion. This framework will facilitate the uptake and interpretation of IWA's views by practitioners, regulators, policy makers, academics and others in the water sector and outside it. It will provide an overall reference point for professionals and organizations who wish to carry out work in this area and for them to have a joint overall framework complementing their own and other partners, tools and approaches.

FOOTPRINT DEVELOPMENT

Task: To develop an urban water/energy metabolic framework, providing agreed definitions, measures, and targets. This information will assist cities and utilities to identify potential systemic changes that could reduce the energy and water footprint of the provision of water in urban areas.

This Workgroup is responsible for developing an urban water/energy metabolic framework that will have definitions, measures, and targets for urban footprints. The metabolic framework will seek to characterize the processes, interactions, flows, and relationships of energy and water in an urban context. The aim is to provide a better understanding of how complex urban systems can both increase, and diminish, the impacts of urbanization on the global environment. The task group will review existing research on this topic and recommend definitions, measures, and targets for consideration. This information will support the development of whole systems and models and should be based upon the presentation and analysis of case study materials.

CASE STUDIES

To provide case studies of "Cites of the Future-Aspirant" (COFA) cities. The goal would be to format case study data using common formats, whilst drawing on the footprint project.

Cities who are preparing for the future will have a range of innovations, developments, or institutional arrangements. The purpose of this task is to gather the best-in-class examples and make them available to a broader audience. Some case studies will have occurred entirely outside of the water sector. However, they could still provide excellent examples of the kinds of practices, innovations, and governance models that the Cities of the Future programme represents. The working group will be asked to develop a prototype case study model that could be applied throughout IWA. Results will be collected into a case-study library and be made available through publication online or otherwise.

CITIES OF THE FUTURE ALLIANCES

Task: To create a select group of cooperating cities who will share experience and new innovations. The concept is to develop a formalized 'Cities of the Future Alliance' between the parties.

Some cities have a great deal of history and experience in innovating new programs, policies or infrastructure investments. Through collaboration and information sharing, the leaders of these cities might be able to accelerate the efforts of others. The early adopters of such new policies will be the most likely to participate in an alliance among and between cities. The intent of this work package is to facilitate the emergence of an alliance of Cities of the Future through inter-city agreements and memoranda. Potential participating cities will need to understand the criteria that are developed and the benefits of participation. This work product will build upon the footprint and framework tasks outlined above.

INTEGRATED TREATMENT TECHNOLOGIES

Task: To develop a family of integrated treatment approaches, capable of achieving high levels of efficiency (water, land, energy) and resource recovery. Approaches could range from centralized to decentralized solutions in both greenfield and brownfield circumstances.

This workgroup will work around identifying emerging integrated treatment approaches that are consistent with the intent and objectives of the Cities of the Future programme. The team will develop reports, white papers, and/or conference papers, highlighting appropriate emerging technologies. This should be considered an on-going process as new technologies and approaches are developed.

SMART NETWORKS

Task: To develop new and modified network designs complementary to different treatment approaches and capable of achieving high levels of efficiency (water, land, energy) and resource recovery.

SPATIAL PLANNING

Task: To develop ongoing dialog between city/regional planners and water resource/system engineers on behalf of an integrated planning processes.

In many cases, land use planning is viewed as a separate process from infrastructure planning. The typical assumption is that the infrastructure should be developed to meet the land use objectives of the land use authority. However, optimization between land use and infrastructure can be precluded if a collaborative planning process is not in place. The issue here is more than just a mere collaboration, as land use and water planners are not familiar with the approaches of the Cities of the Future initiative. This work package will build upon the products developed to create a higher level of understanding and fluency so that planners from all backgrounds have a good knowledge of resilient and robust land use/infrastructure systems. This can be accomplished through dialogue, publication, presentation, and the development of prototypes.

LAND USE AND DEVELOPER INTERACTION

Task: To bring water options/economics into the decision-making of the development community including developers and their designers, architects, financial institutions, and government regulators.

Too often in the development chain, one or more parties can prevent the adoption of an emerging appropriate technology. The developer may not be familiar with new techniques and might avoid implementing them to minimize risk of investment. The architects, designers, financial institutions and the regulators can stop a good idea from being implemented at any time. Building up the markets and capacity of the emerging Cities of the Future paradigm will require communication, education, exploration and implementation strategies throughout the development chain.

INSTITUTIONAL REFORM

Task: To develop and discover institutional options for the delivery of water services in line with evolving system options.

This is a parallel project/process to the engineering and planning elements above. Participants will search for case studies and models of governance change.

URBAN DESIGN LANDSCAPES AND WATERSCAPES

Task: To identify the best examples of cities whose urban water infrastructure is designed to become a city/neighborhood amenity and to document the processes and drivers behind the innovation.

Potzdamer Platz in Berlin, Dockside Green in Victoria, BC, and Cheonggyecheon, in Seoul, are three examples where there is integration between infrastructure requirements and public and private aesthetic needs. Potzdamer Platz was allowed zero rainwater runoff, so the developers and designers merged water management into the design of the buildings. The developer of Dockside Green in Victoria, BC disagreed with the Provincial government who said that it was acceptable to release primary treated sewage into the Straits of Juan de Fuca. Instead, a membrane bioreactor was installed into the development for onsite treatment and reuse. The entire system was paid for by the increase in sales value of condominium units which had a view of the beautifully landscaped water feature that was more than half septic tank effluent blended with harvested rainwater. Cheonggyecheon, in Seoul, was an urban renewal project combined sewer overflow project that together create a powerful urban amenity in a dense area of Seoul. In these three examples, the work group will be exploring the context and situation that created these powerful outcomes. Newer projects that are getting designed have an urban cooling strategy woven in as well.

Next Steps

Major upcoming conferences and events will provide venues for continuing the work of Cities of the Future Workgroups. Upcoming publications and public presentations will provide opportunities for disseminating these ideas to the larger community.

- Next Cities of the Future events:
  - **World Water Congress in Montreal**, September 19–24, 2010
  - **Cities of the Future Stockholm**: Sustainable Urban Planning and Water Management, 22–25 May, 2011
  - **Cities of the Future Xi'an**: Technologies for Integrated Urban Water Management, 15–18 September 2011
  - **Cities of the Future North America**, 2012 IWA/WEF collaboration
  - **Cities of the Future Turkey**: Making Cities of the Future a Reality (working title), 2013
- IWA Cities of the Future Book Series
- Cities of the Future national online presentation and discussion series

Index

© 2010 IWA Publishing. *Water Infrastructure for Sustainable Communities: China and the World*. Edited by Xiaodi Hao, Vladimir Novotny and Valerie Nelson. ISBN: 9781843393283. Published by IWA Publishing, London, UK.

Lightning Source UK Ltd.
Milton Keynes UK
UKOW04n0441100414

229696UK00001B/12/P